Good
Clinical Practice

Pharmaceutical, Biologics, and
Medical Device
Regulations and Guidance Documents

Concise Reference

Volume 2: Guidance

Good Clinical Practice

*Pharmaceutical, Biologics, and Medical Device
Regulations and Guidance Documents*

Concise Reference

Volume 2: Guidance

Good Clinical Practice: Pharmaceutical, Biologics, and Medical Device Regulations and Guidance Documents Concise Reference; Volume 2, Guidance

Copyright © 2010 by PharmaLogika, Inc

PharmaLogika

PharmaLogika, Inc.
PO Box 461
Willow Springs, NC 27592

www.pharmalogika.com

Author / Editor: Mindy J. Allport-Settle

Published by PharmaLogika, Inc.

Printed in the United States of America. First Printing.

ISBN 0-9821476-8-6
ISBN-13 978-0-9821476-8-9

Contents

Part II

Selected FDA GCP/Clinical Trial Guidance Documents: Institutional Review Boards (IRBs) and Informed Consent **203**

Part III

Selected FDA GCP/Clinical Trial Guidance Documents: Drugs and Biologics 235

Part IV

Selected FDA GCP/Clinical Trial Guidance Documents: Medical Devices........ 411

Part V

Selected FDA GCP/Clinical Trial Guidance Documents: Manufacturing Requirements for Investigational Products 543

Preface

About this Book

This book (*Volume 2 of a two volume set*) is designed to be a unified reference source for the U.S. Food and Drug Adminstration's Good Clinical Practice guidance documents (selected FDA GCP/Clinical Trial Guidance Documents[1]) for pharmaceutical, biologics and medical device products. Volume 1 of this set is a concise reference source for the associated regulations.

This book is designed to be a unified reference source for pharmaceutical, biologics and medical device products. Guidance documents are grouped by topic:

- General
- Institutional Review Boards (IRBs) and Informed Consent
- Drugs and Biologics
- Medical Devices
- Manufacturing Requirements for Investigational Products
- Electronic Data

The included *Overview and Orientation* (Chapter 2 of this book) is designed to provide a foundation for understanding the background of the FDA, its guidelines, its relationship with other countries and its relationship with researchers, clinicians, manufacturers and distributors.

This book is designed to be used both as a reference for experienced industry representatives and as a training resource for those new to the industry.

Included Documents and Features

Selected FDA GCP/Clinical Trial Guidance Documents Grouped by Topic:

- *FDA Overview and Orientation*
- *Introduction to GCP*
- *Part I: General*
- *Part II: Institutional Review Boards (IRBs) and Informed Consent*

---

[1] Available on the FDA website at: http://www.fda.gov/ScienceResearch/SpecialTopics/RunningClinicalTrials/GuidancesInformationSheetsandNotices/ucm219433.htm

- *Part III: Drugs and Biologics*
- *Part IV: Medical Devices*
- *Part V: Manufacturing Requirements for Investigational Products*
- *Part VI: Electronic Data*

Reference Tools

- *Part VII: Combined Glossary and Index for all Quality Guidance Documents*

About the Reference Tools

FDA Overview and Orientation and the Introduction to GCP

The FDA overview provides the reader with a brief history of the Food and Drug Administration (FDA) and explains not only what good clinical practice is, but why it exists and how it came to be. The introduction to GCP provides a guide to the FDA's approach to good clinical practice and clinical trial management.

Combined Glossary

The Combined Glossary includes all of the glossaries from each regulation and guidance listed alphabetically rather than by document.

When a word or term appears multiple times in the regulation and guidance documents, the word will appear multiple times in the Combined Glossary if there is a variation in the definition. Each duplicate entry is indented to highlight that it is a duplicate and the earliest reference to the entry is listed first. The source for each entry is bracketed (i.e., [21 CFR § 50]) for ease of reference. While the definitions are similar from one regulatory or guidance document to the next, they are not always identical.

Combined Index for all Regulations and Documents

The index is composed of a list of both words and terms specific to the Food and Drug Adminstration (FDA) good clinical practice regulations and guidance.

Pharmaceutical, biotechnology, and medical device companies use terminology that combines scientific and technical jargon with legal phrases and concepts.

The index provides keywords and terminology as a tool to easily locate specific references across all documents rather than having to rely on memory or paging through each document individually.

FDA Overview and Orientation

The Food and Drug Administration (FDA)

The United States Food and Drug Administration (FDA) is responsible for protecting and promoting the nation's public health.

The programs for safety regulation vary widely by the type of product, its potential risks, and the regulatory powers granted to the agency. For example, the FDA regulates almost every facet of prescription drugs, including testing, manufacturing, labeling, advertising, marketing, efficacy and safety. FDA regulation of cosmetics, however, is focused primarily on labeling and safety. The FDA regulates most products with a set of published standards enforced by a combination of facility inspections, voluntary company reporting standards, and public and consumer watchdog activity.

The FDA frequently works in conjunction with other Federal agencies including the Department of Agriculture, Drug Enforcement Administration, Customs and Border Protection, and Consumer Product Safety Commission.

Historical Origins of Federal Food and Drug Regulation

Prior to the 20th century, there were few federal laws regulating the contents and sale of domestically produced food and pharmaceuticals before the 20th century (with one exception being the short-lived Vaccine Act of 1813). Some state laws provided varying degrees of protection against unethical sales practices, such as misrepresenting the ingredients of food products or therapeutic substances.

The history of the FDA can be traced to the latter part of the 19th century and the U.S. Department of Agriculture's Division of Chemistry (later Bureau of Chemistry). Under Harvey Washington Wiley, appointed chief chemist in 1883, the Division began conducting research into the adulteration and misbranding of food and drugs on the American market. Although they had no regulatory powers, the Division published its findings from 1887 to 1902 in a ten-part series entitled Foods and Food Adulterants. Wiley used these findings, and alliances with diverse organizations (such as state regulators, the General Federation of Women's Clubs, and national associations of physicians and pharmacists) to lobby for a new federal law to set uniform standards for food and drugs to enter into interstate commerce.

Wiley's advocacy came at a time when the public had become alert to hazards in the marketplace by journalists and became part of a general trend for increased federal regulation in matters pertinent to public safety during the Progressive Era.[1] The 1902 Biologics Control

---

[1] A History of the FDA at www.FDA.gov.

Act was put in place after diphtheria antitoxin was collected from a horse named Jim who also had tetanus, resulting in several deaths.

The 1906 Food and Drug Act and creation of the FDA

In June 1906, President Theodore Roosevelt signed into law the Food and Drug Act, also known as the "Wiley Act" after its chief advocate.[2] The Act prohibited, under penalty of seizure of goods, the interstate transport of food which had been "adulterated," with that term referring to the addition of fillers of reduced "quality or strength," coloring to conceal "damage or inferiority," formulation with additives "injurious to health," or the use of "filthy, decomposed, or putrid" substances. The act applied similar penalties to the interstate marketing of "adulterated" drugs, in which the "standard of strength, quality, or purity" of the active ingredient was not either stated clearly on the label or listed in the United States Pharmacopoeia or the National Formulary. The act also banned "misbranding" of food and drugs.[3] The responsibility for examining food and drugs for such "adulteration" or "misbranding" was given to Wiley's USDA Bureau of Chemistry.[4] Strength, quality, identity, potency, and purity (SQuIPP) are currently the key product safety standards, with only two measures added since 1906 Act.

Wiley used these new regulatory powers to pursue an aggressive campaign against the manufacturers of foods with chemical additives, but the Chemistry Bureau's authority was soon checked by judicial decisions, as well as by the creation of the Board of Food and Drug Inspection and the Referee Board of Consulting Scientific Experts as separate organizations within the USDA in 1907 and 1908 respectively. A 1911 Supreme Court decision ruled that the 1906 act did not apply to false claims of therapeutic efficacy,[5] in response to which a 1912 amendment added "false and fraudulent" claims of "curative or therapeutic effect" to the Act's definition of "misbranded." However, these powers continued to be narrowly defined by the courts, which set high standards for proof of fraudulent intent.[6] In 1927, the Bureau of Chemistry's regulatory powers were reorganized under a new USDA body, the Food, Drug, and Insecticide organization. This name was shortened to the Food and Drug Administration (FDA) three years later.[7]

The 1938 Food, Drug, and Cosmetic Act

By the 1930s, muckraking journalists, consumer protection organizations, and federal regulators began mounting a campaign for stronger regulatory authority by publicizing a list of injurious products which had been ruled permissible under the 1906 law, including radioactive beverages, cosmetics which caused blindness, and worthless "cures" for diabetes and tuberculosis. The resulting proposed law was unable to get through the Congress of the United States for five years, but was rapidly enacted into law following the public outcry over the 1937

---

[2] A History of the FDA at www.FDA.gov.

[3] Text in quotation marks is the original text of the 1906 Food and Drugs Act and Amendments.

[4] A History of the FDA at www.FDA.gov.

[5] United States v. Johnson (31 S. Ct. 627 May 29, 1911, decided).

[6] A History of the FDA at www.FDA.gov.

[7] Milestones in U.S. Food and Drug Law History at www.FDA.gov.

Elixir Sulfanilamide tragedy, in which over 100 people died after using a drug formulated with a toxic, untested solvent. The only way that the FDA could even seize the product was due to a misbranding problem: an "Elixir" was defined as a medication dissolved in ethanol, not the diethylene glycol used in the Elixir Sulfanilamide.

President Franklin Delano Roosevelt signed the new Food, Drug, and Cosmetic Act (FD&C Act) into law on June 24, 1938. The new law significantly increased federal regulatory authority over drugs by mandating a pre-market review of the safety of all new drugs, as well as banning false therapeutic claims in drug labeling without requiring that the FDA prove fraudulent intent. The law also authorized factory inspections and expanded enforcement powers, set new regulatory standards for foods, and brought cosmetics and therapeutic devices under federal regulatory authority. This law, though extensively amended in subsequent years, remains the central foundation of FDA regulatory authority to the present day.[8]

Early FD&C Act amendments: 1938-1958

Soon after passage of the 1938 Act, the FDA began to designate certain drugs as safe for use only under the supervision of a medical professional, and the category of 'prescription-only' drugs was securely codified into law by the 1951 Durham-Humphrey Amendment.[9] While pre-market testing of drug efficacy was not authorized under the 1938 FD&C Act, subsequent amendments such as the Insulin Amendment and Penicillin Amendment did mandate potency testing for formulations of specific lifesaving pharmaceuticals.[10] The FDA began enforcing its new powers against drug manufacturers who could not substantiate the efficacy claims made for their drugs, and the United States Court of Appeals for the Ninth Circuit ruling in Alberty Food Products Co. v. United States (1950) found that drug manufacturers could not evade the "false therapeutic claims" provision of the 1938 act by simply omitting the intended use of a drug from the drug's label. These developments confirmed extensive powers for the FDA to enforce post-marketing recalls of ineffective drugs.[11] Much of the FDA's regulatory attentions in this era were directed towards abuse of amphetamines and barbiturates, but the agency also reviewed some 13,000 new drug applications between 1938 and 1962. While the science of toxicology was in its infancy at the start of this era, rapid advances in experimental assays for food additive and drug safety testing were made during this period by FDA regulators and others.[12]

---

[8] A History of the FDA at www.FDA.gov.
[9] A History of the FDA at www.FDA.gov.
[10] Milestones in U.S. Food and Drug Law History at www.FDA.gov
[11] A History of the FDA at www.FDA.gov.
[12] A History of the FDA at www.FDA.gov.

Good Manufacturing Practices vs. Current Good Manufacturing Practices

The terms "Good Manufacturing Practices (GMPs)" and "Current Good Manufacturing Practices (CGMPs or cGMPs[13])" are often used interchangeably both in industry and by FDA inspectors.

"Good Manufacturing Practices" generally refers to the legal mandates detailed in Title 21 of the Code of Federal Regulations (21CFR). "Current Good Manufacturing Practices" refers not only to the legal requirements, but to the guidance provided by the FDA and the standards practiced in industry that are not memorialized as law.

Organizational Structure

The FDA is an agency within the United States Department of Health and Human Services responsible for protecting and promoting the nation's public health. It is organized into the following major subdivisions, each focused on a major area of regulatory responsibility:

- The Office of the Commissioner (OC)
- The Center for Drug Evaluation and Research (CDER)
- The Center for Biologics Evaluation and Research (CBER)
- The Center for Food Safety and Applied Nutrition (CFSAN)
- The Center for Devices and Radiological Health (CDRH)
- The Center for Veterinary Medicine (CVM)
- The National Center for Toxicological Research (NCTR)
- The Office of Regulatory Affairs (ORA)
- The Office of Criminal Investigations (OCI)

How does the FDA communicate with Industry?

Code of Federal Regulations[14]

The Code of Federal Regulations (CFR) is the codification of the general and permanent rules and regulations (sometimes called administrative law) published in the Federal Register by the executive departments and agencies of the Federal Government of the United States. The CFR is published by the Office of the Federal Register, an agency of the National Archives and Records Administration (NARA).

The CFR is divided into 50 titles that represent broad areas subject to Federal regulation. Title 21 is the portion of the Code of Federal Regulations that governs food and drugs within the

[13] The lower case "c" was coined in industry to differentiate between the law, emphasized with capital letters, and the current accepted industry practice not mandated by law.
[14] Available CFR Titles on GPO Access at http://www.access.gpo.gov/nara/cfr/cfr-table-search.html#page1

United States for the Food and Drug Administration (FDA), the Drug Enforcement Administration (DEA), and the Office of National Drug Control Policy (ONDCP).

It is divided into three chapters:

- Chapter I — Food and Drug Administration
- Chapter II — Drug Enforcement Administration
- Chapter III — Office of National Drug Control Policy

Guidance Documents

Guidance documents represent the Agency's current thinking on a particular subject. They do not create or confer any rights for or on any person and do not operate to bind FDA or the public. An alternative approach may be used if such approach satisfies the requirements of the applicable statute, regulations, or both. For information on a specific guidance document, please contact the originating office. Another method of obtaining guidance documents is through the Division of Drug Information.

Federal Register

The Federal Register (since March 14, 1936), abbreviated FR, or sometimes Fed. Reg.) is the official journal of the federal government of the United States that contains most routine publications and public notices of government agencies. It is a daily (except holidays) publication.

The Federal Register is compiled by the Office of the Federal Register (within the National Archives and Records Administration) and is printed by the Government Printing Office.

There are no copyright restrictions on the Federal Register as it is a work of the U.S. government. It is in the public domain.[15]

Citations from the Federal Register are [volume] FR [page number] ([date]), e.g., 65 FR 741 (2000-10-01).

Direct Communication and Letters

The FDA interacts with consumers, health professionals, and industry representatives through letters, meetings (requested by either the FDA or the industry representatives), and telephone calls.

While not all questions can be answered over the phone, the FDA prefers telephone interactions over physical meetings (when a teleconference can reasonably replace a face-to-face meeting).

---

[15] The Federal Register at the GPO, online in both text and PDF, from 1994 on.

www.FDA.gov

The FDA maintains a website at www.fda.gov that is focused on three key audiences:

- consumers
- health professionals
- industry representatives

Through collaboration with users in testing site-wide designs, FDA.gov provides online access to its guidance documents, communication with industry, consumers, and health professionals. Information is categorized by topic, with related subjects consolidated in sections on the site.

Additionally, FDA.gov provides a search engine for Title 21 of the CFR that makes finding keyword references throughout the title more accessible.

Conferences

The FDA routinely sends speakers to industry conferences where they are available to answer questions on their particular area of expertise.

False Statement to a Federal Agency

The U.S. Code of Federal Regulations (CFR) makes it a federal crime for anyone willfully and knowingly to make a false or fraudulent statement to a department or agency of the United States. The false statement must be related to a material matter, and the defendant must have acted willfully and with knowledge of the falsity. It is not necessary to show that the government agency was in fact deceived or misled. The issue of materiality is one of law for the courts. The maximum penalty is five years' imprisonment and a $10,000 fine.

A person may be guilty of a violation without proof that he or she had knowledge that the matter was within the jurisdiction of a federal agency. A businessperson may violate this law by making a false statement to another firm or person with knowledge that the information will be submitted to a government agency. Businesses must take care to avoid exaggerations in the context of any matter that may come within the jurisdiction of a federal agency.

CFR Title 21 - Food and Drugs: Parts 1 to 1499[16]

General

(1) General enforcement regulations

(2) General administrative rulings and decisions

(3) Product jurisdiction

---

[16] All of the 21CFR regulations can be searched online for no charge at http://www.accessdata.fda.gov/scripts/cdrh/cfdocs/cfcfr/cfrsearch.cfm

(5) Organization

(7) Enforcement policy

(10) Administrative practices and procedures

(11) Electronic records; electronic signatures

(12) Formal evidentiary public hearing

(13) Public hearing before a public board of inquiry

(14) Public hearing before a public advisory committee

(15) Public hearing before the commissioner

(16) Regulatory hearing before the Food and Drug Administration

(17) Civil money penalties hearings

(19) Standards of conduct and conflicts of interest

(20) Public information

(21) Protection of privacy

(25) Environmental impact considerations

(26) Mutual recognition of pharmaceutical good manufacturing practice

(50) Protection of human subjects

(54) Financial disclosure by clinical investigators

(56) Institutional review boards

(58) Good laboratory practice for nonclinical laboratory studies

(60) Patent term restoration

(70) Color additives

(71) Color additive petitions

(73) Listing of color additives exempt from certification

(74) Listing of color additives subject to certification

(80) Color additive certification

(81) General specifications and general restrictions for provisional color additives for use in foods, drugs, and cosmetics

(82) Listing of certified provisionally listed colors and specifications

(83-98) [reserved]

(99) Dissemination of information on unapproved/new uses for marketed drugs, biologics, and devices

Foods

(100) General

(101) Food labeling

(102) Common or usual name for nonstandardized foods

(104) Nutritional quality guidelines for foods

(105) Foods for special dietary use

(106) Infant formula quality control procedures

(107) Infant formula

(108) Emergency permit control

(109) Unavoidable contaminants in food for human consumption and food-packaging material

(110) Current good manufacturing practice in manufacturing, packing, or holding human food

(111) Current good manufacturing practice in manufacturing, packaging, labeling, or holding operations for dietary supplements

(113) Thermally processed low-acid foods packaged in hermetically sealed containers

(114) Acidified foods

(115) Shell eggs

(119) Dietary supplements that present a significant or unreasonable risk

(120) Hazard analysis and critical control point (HACCP) systems

(123) Fish and fishery products

(129) Processing and bottling of bottled drinking water

(130) Food standards: general

(131) Milk and cream

(133) Cheeses and related cheese products

(135) Frozen desserts

(136) Bakery products

(137) Cereal flours and related products

(139) Macaroni and noodle products

(145) Canned fruits

(146) Canned fruit juices

(150) Fruit butters, jellies, preserves, and related products

(152) Fruit pies

(155) Canned vegetables

(156) Vegetable juices

(158) Frozen vegetables

(160) Eggs and egg products

(161) Fish and shellfish

(163) Cacao products

(164) Tree nut and peanut products

(165) Beverages

(166) Margarine

(168) Sweeteners and table syrups

(169) Food dressings and flavorings

(170) Food additives

(171) Food additive petitions

(172) Food additives permitted for direct addition to food for human consumption

(173) Secondary direct food additives permitted in food for human consumption

(174) Indirect food additives: general

(175) Indirect food additives: adhesives and components of coatings

(176) Indirect food additives: paper and paperboard components

(177) Indirect food additives: polymers

(178) Indirect food additives: adjuvants, production aids, and sanitizers

(179) Irradiation in the production, processing and handling of food

(180) Food additives permitted in food or in contact with food on an interim basis pending additional study

(181) Prior-sanctioned food ingredients

(182) Substances generally recognized as safe

(184) Direct food substances affirmed as generally recognized as safe

(186) Indirect food substances affirmed as generally recognized as safe

(189) Substances prohibited from use in human food

(190) Dietary supplements

(191-199) [reserved]

Drugs

(200) General

(201) Labeling

(202) Prescription drug advertising

(203) Prescription drug marketing

(205) Guidelines for state licensing of wholesale prescription drug distributors

(206) Imprinting of solid oral dosage form drug products for human use

(207) Registration of producers of drugs and listing of drugs in commercial distribution

(208) Medication guides for prescription drug products

(209) Requirement for authorized dispensers and pharmacies to distribute a side effects statement

(210) Current good manufacturing practice in manufacturing, processing, packing, or holding of drugs; general

(211) Current good manufacturing practice for finished pharmaceuticals

(216) Pharmacy compounding

(225) Current good manufacturing practice for medicated feeds

(226) Current good manufacturing practice for type A medicated articles

(250) Special requirements for specific human drugs

(290) Controlled drugs

(299) Drugs; official names and established names

New Drugs and Over-the-Counter Drug Products

(300) General

(310) New drugs

(312) Investigational new drug application

(314) Applications for FDA approval to market a new drug

(315) Diagnostic radiopharmaceuticals

(316) Orphan drugs

(320) Bioavailability and bioequivalence requirements

(328) Over-the-counter drug products intended for oral ingestion that contain alcohol

(330) Over-the-counter (OTC) human drugs which are generally recognized as safe and effective and not misbranded

(331) Antacid products for over-the-counter (OTC) human use

(332) Antiflatulent products for over-the-counter human use

(333) Topical antimicrobial drug products for over-the-counter recognized as safe and effective and not misbranded

(335) Antidiarrheal drug products for over-the-counter human use

(336) Antiemetic drug products for over-the-counter human use

(338) Nighttime sleep-aid drug products for over-the-counter human use

(340) Stimulant drug products for over-the-counter human use

(341) Cold, cough, allergy, bronchodilator, and antiasthmatic drug products for over-the-counter human use

(343) Internal analgesic, antipyretic, and antirheumatic drug products for over-the-counter human use

(344) Topical OTIC drug products for over-the-counter human use

(346) Anorectal drug products for over-the-counter human use

(347) Skin protectant drug products for over-the-counter human use

(348) External analgesic drug products for over-the-counter human use

(349) Ophthalmic drug products for over-the-counter human use

(350) Antiperspirant drug products for over-the-counter human use

(352) Sunscreen drug products for over-the-counter human use [stayed indefinitely]

(355) Anticaries drug products for over-the-counter human use

(357) Miscellaneous internal drug products for over-the-counter human use

(358) Miscellaneous external drug products for over-the-counter human use

(361) Prescription drugs for human use generally recognized as safe and effective and not misbranded: drugs used in research

(369) Interpretative statements re warnings on drugs and devices for over-the-counter sale

(370-499) [reserved]

Veterinary Products

(500) General

(501) Animal food labeling

(502) Common or usual names for nonstandardized animal foods

(509) Unavoidable contaminants in animal food and food-packaging material

(510) New animal drugs

(511) New animal drugs for investigational use

(514) New animal drug applications

(515) Medicated feed mill license

(516) New animal drugs for minor use and minor species

(520) Oral dosage form new animal drugs

(522) Implantation or injectable dosage form new animal drugs

(524) Ophthalmic and topical dosage form new animal drugs

(526) Intramammary dosage forms

(529) Certain other dosage form new animal drugs

(530) Extralabel drug use in animals

(556) Tolerances for residues of new animal drugs in food

(558) New animal drugs for use in animal feeds

(564) [reserved]

(570) Food additives

(571) Food additive petitions

(573) Food additives permitted in feed and drinking water of animals

(579) Irradiation in the production, processing, and handling of animal feed and pet food

(582) Substances generally recognized as safe

(584) Food substances affirmed as generally recognized as safe in feed and drinking water of animals

(589) Substances prohibited from use in animal food or feed

(590-599) [reserved]

Biologics

(600) Biological products: general

(601) Licensing

(606) Current good manufacturing practice for blood and blood components

---

(607) Establishment registration and product listing for manufacturers of human blood and blood products

(610) General biological products standards

(630) General requirements for blood, blood components, and blood components, and blood derivatives

(640) Additional standards for human blood and blood products

(660) Additional standards for diagnostic substances for laboratory tests

(680) Additional standards for miscellaneous products

Cosmetics

(700) General

(701) Cosmetic labeling

(710) Voluntary registration of cosmetic product establishments

(720) Voluntary filing of cosmetic product ingredient composition statements

(740) Cosmetic product warning statements

(741-799) [reserved]

Medical Devices

(800) General

(801) Labeling

(803) Medical device reporting

(806) Medical devices; reports of corrections and removals

(807) Establishment registration and device listing for manufacturers and initial importers of devices

(808) Exemptions from federal preemption of state and local medical device requirements

(809) In vitro diagnostic products for human use

(810) Medical device recall authority

(812) Investigational device exemptions

(813) [reserved]

(814) Premarket approval of medical devices

(820) Quality system regulation

(821) Medical device tracking requirements

(822) Postmarket surveillance

(860) Medical device classification procedures

(861) Procedures for performance standards development

(862) Clinical chemistry and clinical toxicology devices

(864) Hematology and pathology devices

(866) Immunology and microbiology devices

(868) Anesthesiology devices

(870) Cardiovascular devices

(872) Dental devices

(874) Ear, nose, and throat devices

(876) Gastroenterology-urology devices

(878) General and plastic surgery devices

(880) General hospital and personal use devices

(882) Neurological devices

(884) Obstetrical and gynecological devices

(886) Ophthalmic devices

(888) Orthopedic devices

(890) Physical medicine devices

(892) Radiology devices

(895) Banned devices

(898) Performance standard for electrode lead wires and patient cables

Mammography

(900) Mammography

Radiological Health

(1000) General

(1002) Records and reports

(1003) Notification of defects or failure to comply

(1004) Repurchase, repairs, or replacement of electronic products

(1005) Importation of electronic products

(1010) Performance standards for electronic products: general

(1020) Performance standards for ionizing radiation emitting products

(1030) Performance standards for microwave and radio frequency emitting products

(1040) Performance standards for light-emitting products

(1050) Performance standards for sonic, infrasonic, and ultrasonic radiation-emitting products

Regulations under Certain Other Acts

(1210) Regulations under the federal import milk act

(1230) Regulations under the federal caustic poison act

(1240) Control of communicable diseases

(1250) Interstate conveyance sanitation

(1251-1269) [reserved]

(1270) Human tissue intended for transplantation

(1271) Human cells, tissues, and cellular and tissue-based products

(1272-1299) [reserved]

Controlled Substances

(1300) Definitions

(1301) Registration of manufacturers, distributors, and dispensers of controlled substances

(1302) Labeling and packaging requirements for controlled substances

(1303) Quotas

(1304) Records and reports of registrants

(1305) Orders for schedule I and II controlled substances

(1306) Prescriptions

(1307) Miscellaneous

(1308) Schedules of controlled substances

(1309) Registration of manufacturers, distributors, importers and exporters of List I chemicals

(1310) Records and reports of listed chemicals and certain machines

(1311) Digital certificates

(1312) Importation and exportation of controlled substances

(1313) Importation and exportation of list I and list II chemicals

(1314) Retail sale of scheduled listed chemical products

(1315) Importation and production quotas for ephedrine, pseudoephedrine, and phenylpropanolamine

(1316) Administrative functions, practices, and procedures

Office of National Drug Control Policy

(1400) [reserved]

(1401) Public availability of information

(1402) Mandatory declassification review

(1403) Uniform administrative requirements for grants and cooperative agreements to state and local governments

(1404) Governmentwide debarment and suspension (nonprocurement)

(1405) Governmentwide requirements for drug-free workplace (financial assistance)

(1406-1499) [reserved]

Introduction to GCP

Good Clinical Practice

FDA regulates scientific studies that are designed to develop evidence to support the safety and effectiveness of investigational drugs (human and animal), biological products, and medical devices.[17] Physicians and other qualified experts ("clinical investigators") who conduct these studies are required to comply with applicable statutes and regulations. These laws and regulations are intended to ensure the integrity of clinical data on which product approvals are based and to help protect the rights, safety, and welfare of human subjects.

The following resources are provided to help investigators, sponsors, and contract research organizations who conduct clinical studies on investigational new drugs comply with U.S. law and regulations covering good clinical practice (GCP).

General Information

- General Information for Clinical Investigators
- From test-tube to patient: New drug development in the United States
- Information and Activities Targeted Toward Specific Audiences and Populations

Institutional Review Boards (IRBs)

- Institutional Review Boards (IRBs) and Protection of Human Subjects in Clinical Trials
- Guidance for Institutional Review Boards and Clinical Investigators
- FDA Compliance Program 7348.809 - BIMO for Institutional Review Boards

FDA Regulations

- Institutional Review Boards (21 CFR Part 56)
- Human Subject Protection (Informed Consent) (21 CFR Part 50)
- Additional Safeguards for Children involved in Clinical Investigations of FDA-Regulated Products (Interim Rule) (21 CFR Part 50, subpart D)
- Investigational New Drug Application (IND Regulations) (21 CFR Part 312)

---

[17] Available on the FDA website at
http://www.fda.gov/AboutFDA/CentersOffices/CDER/ucm090259.htm

Sponsors, Monitors, and Contract Research Organizations

- Guidances and Enforcement Information
- Investigational New Drug Application (IND Regulations) (21 CFR Part 312)
- Regulations for Applications for FDA Approval to Market a New Drug (NDA Regulations) (21 CFR Part 314)

Office of Good Clinical Practice

The Office of Good Clinical Practice[18] is the focal point within FDA for Good Clinical Practice (GCP) and Human Subject Protection (HSP) issues arising in human research trials regulated by FDA. The Office of Good Clinical Practice:

- Advises and assists the Commissioner and other key officials on GCP and HSP issues arising in clinical trials that have an impact on policy, direction, and long-range goals.
- Leads, supports, and administers FDA's Human Subject Protection/Bioresearch Monitoring Council that manages and sets agency policy on bioresearch monitoring (BIMO), GCP, and HSP affecting both clinical and non-clinical trials regulated by FDA. The office also coordinates and provides oversight of working groups established by this Council.
- Coordinates FDA's Bioresearch Monitoring program with respect to clinical trials, working together with all FDA Centers as well as FDA's Office of Regulatory Affairs (ORA)
- Plans and conducts training and outreach programs, both internally and externally.
- Serves as a liaison with other Federal agencies (e.g., the HHS Office for Human Research Protection (OHRP) and the Veterans Administration), outside organizations, regulated industry, and public interest groups on BIMO, GCP, and HSP policies and regulatory matters.
- Contributes to international Good Clinical Practice harmonization activities.

Running Clinical Trials

Adherence to the principles of good clinical practices (GCPs), including adequate human subject protection (HSP) is universally recognized as a critical requirement to the conduct of research involving human subjects. [19] Many countries have adopted GCP principles as laws and/or regulations. The Food and Drug Administration's (FDA's) regulations for the conduct of clinical trials, which have been in effect since the 1970s, address both GCP and HSP. These FDA regulations and guidance documents are accessible from this site. International GCP

---

[18] Available on the FDA website at
http://www.fda.gov/AboutFDA/CentersOffices/OC/OfficeofScienceandHealthCoordination/GoodClinicalPracticesProgram/default.htm
[19] Available on the FDA website at
http://www.fda.gov/ScienceResearch/SpecialTopics/RunningClinicalTrials/default.htm

---

guidance documents on which FDA has collaborated and that have been adopted as official FDA guidance are also be found here. Finally, this site includes links to other sites relevant to the conduct of clinical trials, both nationally and internationally.

Bioresearch Monitoring

FDA's bioresearch monitoring (BIMO) program conducts on-site inspections of both clinical and nonclinical studies performed to support research and marketing applications/submissions to the agency. Links to the compliance programs for each inspection type and contact information for each Center's BIMO program are also accessible from this site.

Bioresearch Monitoring Program (BIMO)

FDA uses Compliance Program Guidance Manuals (CPGM) to direct its field personnel on the conduct of inspectional and investigational activities. [20] The CPGM's described below form the basis of FDA's Bioresearch Monitoring Program. The purpose of each program is to ensure the protection of research subjects and the integrity of data submitted to the agency in support of a marketing application.

- CPGM for Clinical Investigators

- CPGM for Sponsors, Monitors, and Contract Research Organizations

- CPGM for Good Laboratory Practice (Non-Clinical Laboratories)

- CPGM for In-Vivo Bioequivalence Compliance Program 7348.001

- CPGM for Institutional Review Boards

Regulations

FDA Regulations Relating to Good Clinical Practice and Clinical Trials

- FDA regulations governing the conduct of clinical trials describe good clinical practices (GCPs) for studies with both human and non-human animal subjects.[21]

- Electronic Records; Electronic Signatures (21 CFR Part 11)

- Protection of Human Subjects (Informed Consent) (21 CFR Part 50)

- Financial Disclosure by Clinical Investigators (21 CFR Part 54)

- Institutional Review Boards (21 CFR Part 56)

- FDA IRB Registration Rule (21 CFR 56.106)

---

[20] Available on the FDA website at
http://www.fda.gov/ScienceResearch/SpecialTopics/RunningClinicalTrials/ucm160670.htm
[21] Available on the FDA website at
http://www.fda.gov/ScienceResearch/SpecialTopics/RunningClinicalTrials/ucm155713.htm#FDARegulati
ons

- FDA IRB Registration Rule (21 CFR 56.106)

- Good Laboratory Practice for Nonclinical Laboratory Studies (21 CFR Part 58)

- Investigational New Drug Application (21 CFR Part 312)

- Foreign Clinical Trials not conducted under an IND (21 CFR 312.120)

- Expanded Access to Investigational Drugs for Treatment Use

- Charging for Investigational Drugs

- Form 1571 (Investigational New Drug Application)

- Form 1572 (Statement of Investigator)

- Applications for FDA Approval to Market a New Drug (21 CFR Part 314)

- Bioavailability and Bioequivalence Requirements (21 CFR Part 320)

- New Animal Drugs for Investigational Use (21 CFR Part 511)

- New Animal Drug Applications (21 CFR Part 514)

- Applications for FDA Approval of a Biologic License (21 CFR Part 601)

- Investigational Device Exemptions (21 CFR Part 812)

- Premarket Approval of Medical Devices (21 CFR Part 814)

Preambles to GCP Regulations

Each time Congress enacts a law affecting products regulated by the Food and Drug Administration, the FDA develops rules to implement the law. The FDA takes various steps to develop these rules, including publishing a variety of documents in the Federal Register announcing the FDA's interest in formulating, amending or repealing a rule, and offering the public the opportunity to comment on the agency's proposal. The Federal Register notice explains the legal issues and basis for the proposal, and provides information about how interested persons can submit written data, views, or arguments on the proposal. Any comments that are submitted are addressed in subsequent publications that are part of the agency's decision-making process.

The "preamble" to each of these publications includes all of the printed information immediately preceding the codified regulation. The preamble provides information about the regulation such as why the regulation is being proposed, the FDA's interpretation of the meaning and impact of the proposed regulation, and in those cases where the agency has solicited public comment, the agency's review and commentary on those comments. The preamble can also include an environmental impact assessment, an analysis of the cost impact, comments related to the Paperwork Reduction Act, and the effective date of the implementation or revocation (as the case may be) of the regulation.

The documents listed below include the various publications that contributed to the development of final rules related to FDA's regulations on good clinical practice and clinical trials.[22]

Parts 50 and 56

- Protection of Human Subjects; Informed Consent (January 27, 1981)

- Protection of Human Research Subjects; Standards for Institutional Review Boards for Clinical Investigations (January 27, 1981)

- Protection of Human Research Subjects; Clinical Investigations Which May Be Reviewed Through Expedited Review Procedure Set Forth in FDA Regulations (January 27, 1981)

- Protection of Human Subjects; Informed Consent; Standards for Institutional Review Boards for Clinical Investigations (November 1988)

- Federal Policy for the Protection of Human Subjects (June 18, 1991)

- FDA Policy for the Protection of Human Subjects (June 18, 1991)

- Protection of Human Subjects; Informed Consent; Proposed Rule (September 21, 1995)

- Protection of Human Subjects; Informed Consent, Part II (October 2, 1996)

- Protection of Human Subjects; Informed Consent and Waiver of Informed Consent Requirements in Certain Emergency Research; Final Rules (October 2, 1996)

- Protection of Human Subjects; Informed Consent (December 22, 1995)

- Protection of Human Subjects; Informed Consent Verification; Final Rule (November 5, 1996)

- Categories of Research That May Be Reviewed by the Institutional Review Board (IRB) Through an Expedited Review Procedure (November 9, 1998)

- Protection of Human Subjects; Informed Consent, Exception from general requirements (October 5, 1999)

- Additional protections for children (66 FR 20589-600, April 24, 2001)

- Exception from General Requirements for Informed Consent (71 FR 32827, June 7, 2006) [PDF]

Part 54

- Financial Disclosures by Clinical Investigators (63 FR 5233, February 2, 1998)

- Financial Disclosures by a Clinical Investigator(63 FR 72171-81, December 31, 1998)

---

[22] Available on the FDA website at
http://www.fda.gov/ScienceResearch/SpecialTopics/RunningClinicalTrials/ucm155713.htm#FDARegulati ons

---

Part 210

- Current Good Manufacturing Practice Regulations and Investigational New Drugs (January 17, 2006)

Parts 312 and 314

- Proposed New Drug, Antibiotic, and Biologic Drug Product Regulations (June 9, 1983)

- New Drug and Antibiotic Regulations (February 22, 1985) New Drug, Antibiotic, and Biologic Drug Product Regulations (March 19, 1987)

- Investigational New Drug Applications and New Drug Applications (September 8, 1995)

- Investigational New Drug Applications and New Drug Applications (February 11, 1998)

- Disqualification of a Clinical Investigator (February 16, 1996)

- Disqualification of a Clinical Investigator (September 5, 1997)

- Expedited Safety Reporting Requirements for Human Drug and Biological Products (October 7, 1997)

- Clinical Hold for products intended for life threatening conditions (65 FR 34963-71, June 1, 2000)

Part 320

- Retention of BE and BA testing samples (April 28, 1993)

Part 812

- Quality System Regulations (October 7, 1996)

- Treatment Use of Investigational Devices (September 18, 1997)

- Withdrawal of Intraocular Lenses Regulation (Part 813) (January 29, 1997)

- Disqualification of Clinical Investigators (March 14, 1997)

- Modifications to the List of Recognized Standards (July 12, 1999)

- Modifications to the Medical Device and/or Study Protocol (November 23, 1998)

Part 814

- Medical Devices; Humanitarian Use Devices Part V (June 26, 1996)

- 30-Day Notices and 135-Day PMA Supplement Review (October 8, 1998)

- Humanitarian Use of Devices (November 3, 1998)

Miscellaneous

- Part 11 Electronic Records; Electronic Signatures (March 20, 1997)

- Use of Bioresearch Monitoring Information System (BMIS) (63 FR 55873-6, October 19, 1998)

- Presiding officer (66 FR 45317-8, August 28, 2001)

Part I

Selected FDA GCP/Clinical Trial Guidance Documents: General

Process for Handling Referrals to FDA under 21 CFR 50.54: Additional Safeguards for Children in Clinical Investigations

Process for Handling Referrals to FDA under 21 CFR 50.54:[1]

Additional Safeguards for Children in Clinical Investigations:

Guidance for Clinical Investigators, Institutional Review Boards and Sponsors

U.S. Department of Health and Human Services
Food and Drug Administration

December 2006

Contains Nonbinding Recommendations

This guidance is intended to assist clinical investigators, Institutional Review Boards (IRBs), sponsors, and other interested parties in understanding the FDA's process for handling clinical investigations that include children as subjects and that have been referred to FDA for review under 21 CFR 50.54.

Note: *This guidance represents the Food and Drug Administration's (FDA's) current thinking on this topic. It does not create or confer any rights for or on any person and does not operate to bind FDA or the public. You can use an alternative approach if the approach satisfies the requirements of the applicable statutes and regulations. If you want to discuss an alternative approach, contact the appropriate FDA staff. If you cannot identify the appropriate FDA staff, call the appropriate number listed on the title page of this guidance.*

I. Introduction

This guidance is intended to assist clinical investigators, Institutional Review Boards (IRBs), sponsors, and other interested parties in understanding the Food and Drug Administration's (FDA or agency) process for handling clinical investigations that include children as subjects and that have been referred to FDA for review under 21 CFR 50.54. This guidance describes the procedures FDA generally intends to follow in handling such clinical investigations and in reaching final determinations pursuant to that regulation. It is based in part on FDA's experience to date with such referrals. The Department of Health and Human Services (HHS)

---

[1] Available on the FDA website at:
http://www.fda.gov/downloads/RegulatoryInformation/Guidances/UCM126577.pdf

has human subject protection regulations that also govern research involving children and supported or conducted by HHS. (See 45 CFR Part 46 Subpart D.) This guidance also addresses situations in which a clinical investigation is subject to both 21 CFR 50.54 and 45 CFR 46.407.[2]

FDA's guidance documents, including this guidance, do not establish legally enforceable responsibilities. Instead, guidances describe the FDA's current thinking on a topic and should be viewed only as recommendations, unless specific regulatory or statutory requirements are cited. The use of the word should in FDA's guidances means that something is suggested or recommended, but not required.

II. Background

FDA adopted 21 CFR Part 50 Subpart D, "Additional Safeguards for Children in Clinical Investigations", as an interim final rule in April 2001 (21 CFR Part 50, Subpart D) (Subpart D) (See 66 FR 20598, April 24, 2001). Under these regulations an IRB must review clinical investigations that involve children as subjects and are covered by Subpart D and must approve only those clinical investigations that satisfy the criteria described in 21 CFR 50.51, 50.52 or 50.53, and the conditions of all other applicable sections of Subpart D. If an IRB does not believe that a clinical investigation within the scope described in 21 CFR 50.1 and 56.101 and involving children as subjects meets the requirements of 21 CFR 50.51, 50.52 or 50.53, the clinical investigation may proceed only if:

- The IRB finds and documents that the clinical investigation presents a reasonable opportunity to further the understanding, prevention, or alleviation of a serious problem affecting the health or welfare of children; and

- The Commissioner of Food and Drugs (Commissioner), after consultation with a panel of experts in pertinent disciplines (e.g., science, medicine, education, ethics, law) and following opportunity for public review and comment, determines either:

  - The clinical investigation in fact satisfies 21 CFR 50.51, 50.52 or 50.53, or

  - The following three conditions described in 21 CFR 50.54 are met:

    1. The clinical investigation presents a reasonable opportunity to further the understanding, prevention, or alleviation of a serious problem affecting the health or welfare of children.

    2. The clinical investigation will be conducted in accordance with sound ethical principles; and

---

[2] The HHS Subpart D regulations are implemented and interpreted by HHS's Office for Human Research Protections (OHRP) and are nearly identical to FDA's regulations. OHRP has posted on its website information pertaining to its process for handling clinical protocols that are subject to 45 CFR 46.407, and to 45 CFR 46.407 as well as FDA's Subpart D regulations. See www.hhs.gov/ohrp/children/guidance_407process.html.

3. Adequate provisions are made for soliciting the assent of children and the permission of their parents or guardians as set forth in §50.55.

III. Review Process

In FDA-regulated clinical investigations involving children, the agency makes every effort to protect the rights, safety, and welfare of those children. In addition, the agency strives to achieve the basic goals of adherence to sound ethical principles, transparency through public and expert input, efficiency, timeliness, clarity, and consistency. These goals are also endorsed by the Subcommittee on Research Involving Children of the Secretary's Advisory Committee on Human Research Protections (SACHRP).[3] FDA believes that these goals are best served by having a clear, efficient, and comprehensive process for referrals by IRBs under 21 CFR 50.54.

A. Overview of Process

FDA will use its advisory committee process to evaluate clinical investigations referred for review under 21 CFR 50.54. This process will provide transparency and help ensure expert input as well as public participation. When FDA receives a referral from an IRB, the agency will determine if the clinical investigation described in the protocol is regulated by FDA as described in 21 CFR 50.1(a). If so, the agency will determine whether the referral is complete and can be the Pediatric Ethics Subcommittee (the Subcommittee) of the Pediatric Advisory Committee (PAC) and present the referral to the Subcommittee for its consideration. At the next open meeting of the PAC, the Subcommittee chair (or another member of the Subcommittee, if the chair is unavailable) will present to the PAC the Subcommittee's recommendation(s) regarding the referred clinical investigation. After discussion and deliberation, the PAC will make its recommendation(s) regarding the referral to the Commissioner of Food and Drugs. The Commissioner will then make the final determination as to whether the clinical investigation meets the requirements of FDA's Subpart D regulations. This process is discussed in greater detail below.

B. Public Participation

Public review and comment is required under 21 CFR 50.54. Furthermore, FDA believes that public participation is important to help ensure the comprehensiveness and integrity of the review of clinical investigations referred under that regulation and the protection of the children who may be enrolled in the investigations. In order to encourage public participation, FDA intends to establish an agency docket for each accepted referral and solicit public comment on the proposed investigation. The materials in the docket for each investigation will include the referral documents sent by the IRB, related agency correspondence, public comments on the referral, the transcripts of both the Subcommittee meeting and the PAC meeting, and the final determination of the Commissioner regarding the referral under 21 CFR 50.54. Materials submitted to the docket will be available through FDA's Division of Dockets Management or through FDA's Division of Freedom of Information.

---

[3] See Recommendations for OHRP/FDA Harmonization Subcommittee on Research Involving Children of SACHRP, January 29, 2004 minutes, page 9.

In addition, the Subcommittee and PAC meetings will be open to the public. At the Subcommittee meeting, the chair will also present a summary of the public comments received.

C. Scheduling the Subcommittee and Advisory Committee Meetings

When FDA receives an appropriate and complete referral[4] from an IRB under 21 CFR 50.54, the Office of Pediatric Therapeutics (OPT) will coordinate with the Executive Secretary[5] of the PAC regarding the scheduling of the Subcommittee and PAC meetings. FDA generally expects to schedule Subcommittee meetings directly preceding a PAC meeting. The PAC is currently scheduled to meet approximately three times per year, but FDA will consider calling additional meetings as necessary.

D. Submitting a Referral

The agency encourages IRBs considering referring a clinical investigation under § 50.54 first to discuss with the sponsor whether there are appropriate modifications to the protocol that would allow the clinical investigation to be approved under another provision of Subpart D.

IRBs should send referrals under 21 CFR 50.54 of clinical investigations regulated by FDA, as described in 21 CFR 50.1(a), to FDA's OPT at opt@fda.gov, or Office of Pediatric Therapeutics, Office of the Commissioner, FDA, 5600 Fishers Lane, RM 13B-45, HFG-2, Rockville, MD 20857. FDA may also receive copies of referrals from OHRP in situations where an IRB has referred a clinical investigation to OHRP but not to FDA. Similarly, if an IRB submits a clinical investigation to FDA but not to OHRP, FDA will usually send it to OHRP so that OHRP can determine whether the clinical investigation is also subject to HHS jurisdiction. If a clinical investigation is subject to both HHS and FDA jurisdiction, then, as discussed below in Section III.O., the referral process will be conducted jointly by FDA and OHRP under 21 CFR 50.54 and 45 CFR 46.407.

E. Documents to Include in the Referral

If an IRB decides to refer a clinical investigation for consideration under 21 CFR 50.54, the IRB should submit its finding (required under 21 CFR 50.54(a)) that the clinical investigation presents a reasonable opportunity to further the understanding, prevention, or alleviation of a serious problem affecting the health or welfare of children.

The referral should also include the following:

- An explanation as to why the IRB does not believe that the clinical investigation meets the requirements of 21 CFR 50.51, 50.52, or 50.53. FDA believes that in most cases this probably will be explained in the IRB's minutes from its discussions of the protocol at issue.

---

[4] See discussion of the joint review process with OHRP in section III.O
[5] The Executive Secretary is the Designated Federal Official (DFO) who coordinates the activities of the advisory committee and serves as the link between committee members, FDA, industry and the public.

- The research protocol and, if the clinical investigation is being conducted under an Investigational New Drug application (IND) or Investigational Device Exemption (IDE), the IND or IDE number assigned by FDA;

- All informed consent documents, including the parental permission form and, if being used, the assent forms and/or a description of the assent process; and

- Any other informative supporting documents, such as IRB minutes pertinent to the clinical investigation, correspondence between the IRB and the investigator, product labeling, and the investigator's brochure.

In addition, the IRB should ensure that the submitted documents consistently describe the clinical investigation. **We strongly encourage IRBs to submit electronic versions of all documents.**

These documents are needed to allow FDA to complete its initial assessment of the clinical investigation to determine if the study is subject to FDA jurisdiction, and to provide the Subcommittee and the PAC with complete information regarding the referral. The referring IRB should provide these documents in a form the IRB would find approvable, but for the issue(s) that prompted the referral under 21 CFR 50.54. Providing these documents in "IRB-approvable" condition will allow the Subcommittee and the PAC to focus on the issue(s) that prompted the referral, and not on other matters that the IRB is able to resolve itself.

When OPT receives a referral, it will contact the referring IRB to confirm receipt of the referral. This will begin the process of exchanging information about the referral process between OPT, the IRB, and the investigator. Either in that initial contact or subsequently, OPT will advise the IRB as to whether the clinical investigation is subject to FDA jurisdiction and, if it is, whether the referral is complete, and thus acceptable for review, or whether any information is missing from the submission. Depending on the circumstances of the research, OPT may ask for additional information, which may include, for example, information regarding the past use of the investigational article in children or adults, documents regarding continuing review, drug preparation protocol, or the certificate of analysis for the chemical being studied. During this exchange OPT will explain the procedures the agency intends to follow and will encourage the IRB to ask any questions it might have about the requirements of 21 CFR 50.54.

Following this exchange, the IRB may decide to withdraw the referral from consideration. The reasons for withdrawal could include, for example, a misunderstanding of the requirements of the Subpart D regulations and the applicability of 50.54 to the clinical investigation at issue, i.e., a determination by the IRB after further analysis that the clinical investigation can in fact proceed under another provision in the Subpart D regulations. The agency encourages the IRB to document the reasons for withdrawal of the referral in its request to do so. FDA suggests that an IRB in doubt about whether to refer a clinical investigation under 50.54 consult with OPT as soon as possible via email at opt@fda.gov.

F. Assessment of Referral

Following receipt of the referral, FDA will determine whether the protocol is FDA-regulated (generally within 2 weeks after receipt) and inform the referring IRB in writing of the determination. If the clinical investigation was referred to FDA by OHRP, FDA will also inform OHRP of that determination.

The considerations under FDA's IND regulations (see 21 CFR Part 312) and IDE regulations (see 21 CFR Part 812) that are used to determine whether an IND or IDE can go into effect are related to, but distinct from, the considerations under Part 50 regarding the protection of human subjects. Therefore, IRBs should note that, although some clinical investigations are exempt from the requirement that they need be conducted under an IND (see 21 CFR 312.2(b)), those that are exempt are still subject to 21 CFR Parts 50 and 56 and thus may trigger the Subpart D process under section 50.54. Similarly, a clinical investigation of a medical device may be exempt under 21 CFR 812.2(c) from the requirement that it be conducted under an IDE, but the clinical investigation may still be subject to the Subpart D process. Furthermore, as indicated in 21 CFR 50.1(a), Part 50 also applies to certain investigations that are not subject to sections 505(i) or 520(g) of the Federal Food, Drug, and Cosmetic Act (21 USC §§ 355(i), 360j(g)).

G. Acceptance of Referral

Following receipt of the referral, OPT will then ask the referring IRB for any other relevant documents and inform the IRB that FDA intends to post the documents in the docket created for the referral, and on the agency's website. If the IRB believes that any documents it has included in its referral may be considered confidential by the sponsor, the IRB will be responsible for obtaining any necessary permission from the sponsor to post the documents. If an IRB (or sponsor) objects to the public posting of the documents, the agency will be unable to proceed with the referral.

H. Multi-Site Clinical Investigations

In some circumstances clinical investigations referred for review under 21 CFR 50.54 may be conducted at multiple sites. In the event that an IRB at one or more of the clinical sites refers the investigation for review, it should notify the sponsor. In such a situation, FDA strongly encourages the sponsor to notify the IRB(s) and investigator(s) at all of the other sites of the referral. Once an IRB has made a referral under 50.54, pursuant to that regulation the investigation may not proceed (i.e., no subjects may be recruited or enrolled in the investigation) at any site for which that IRB has responsibility.

Once IRBs at other sites have been notified by the sponsor of the referral, they may choose to allow the clinical investigation to proceed at their sites, or they may choose to await the outcome of the 50.54 referral. In making this determination, FDA encourages IRBs to consider the implications of continuing the investigation during the pendency of the review and determination under 21 CFR 50.54. For example, at the conclusion of the review, it is possible that the Commissioner could determine that the investigation cannot proceed because one or more of the criteria in 21 CFR 50.54(b) are not met. Alternatively, the Commissioner might determine that the investigation can proceed but only with modifications of the protocol or the informed consent documents and procedures. At the sites where the clinical investigation goes forward, the IRBs of record retain all normal responsibilities for review and oversight.

I. Federal Register Notices

Following a referral, FDA generally will issue three Federal Register notices:

- A solicitation of public comments on the referral, along with background information for the referral, notice of the establishment of a docket for the referral, and directions for how to access the docket. Generally, FDA will provide approximately 4 weeks for submission of comments to the docket;

- The announcement of the Pediatric Ethics Subcommittee meeting; and

- The announcement of the Pediatric Advisory Committee meeting.

J. Composition of the Pediatric Ethics Subcommittee

FDA will select the members of the Subcommittee in accordance with 21 CFR 50.54(b) and other relevant federal laws, including the Federal Advisory Committee Act (FACA) (5 U.S.C. App. 2)(1972). The Subcommittee will include at least two members of the PAC. Whenever possible, the selected members of the PAC will have clinical expertise relevant to the subject matter under discussion. FDA will also invite individuals to serve on the Subcommittee who have expertise and/or experience relevant to the clinical investigation being discussed. As a general matter, OPT will make these selections in collaboration with the relevant FDA Center(s) and review division(s). Additional individuals will be invited according to the principles set forth in 21 CFR 50.54(b), such that the Subcommittee consists of a "panel of experts in pertinent disciplines (for example: science, medicine, education, ethics, law)" (21 CFR 50.54(b)). As a general matter, FDA also believes that usually it will be helpful to include a patient advocate/community representative and a statistician on the Subcommittee.

The agency notes that there will be no cap on the number of members of the Subcommittee, and the agency will include as many members as necessary to ensure that the relevant expertise is represented.

K. The Subcommittee Meeting

FDA intends to invite a representative of the referring IRB (to be selected by the IRB) and the investigator to attend the Subcommittee meeting and to make a presentation to the Subcommittee regarding the referred protocol. The agency encourages the IRB representative and the investigator to attend the meeting to help the Subcommittee understand the clinical investigation and basis for the referral. One or more representatives from the FDA review division responsible for reviewing the clinical investigation may also attend to answer questions from the Subcommittee. As appropriate, the agency may invite additional individuals to make presentations to the Subcommittee regarding either the referred protocol or issues raised in it (for example, the sponsor of the clinical investigation, or, in a situation involving a multi-site trial, a representative from an IRB that has approved the protocol). Although a referral may be made because of a particular aspect of the clinical investigation, the Subcommittee and the

PAC, as well as FDA (and OHRP if they are involved[6]), usually will consider the clinical investigation in its entirety.

FDA anticipates that a Subcommittee meeting usually will last for one day and a typical agenda will include:

- Call to order

- Meeting statement

- Description of Subpart D expert panel process

- Overview, charge to panel

- Background of clinical investigation and protocol-specific scientific/medical issues

- Investigator comments

- IRB comments, including identification of specific ethical issues which led to the referral

- Summary of public comments

- Open public hearing

- Presentation of questions

- Expert panel discussion

- Discussion, agreement, and vote on recommendation(s) (summary of deliberations) that will be presented to PAC.

The Subcommittee will operate by majority vote. Meeting transcripts and recommendations from the Subcommittee meeting will be made available to the public in the docket established for the referral and on the agency's website.

L. PAC Meeting

At the next meeting of the PAC (usually shortly after the Subcommittee meeting), the chair of the Subcommittee (or another member of the Subcommittee, if the chair is unavailable) will present the Subcommittee's recommendation(s) regarding whether the clinical investigation meets the requirements of Subpart D and other relevant requirements of Part 50. The agency also encourages the IRB representative and the investigator to attend the PAC meeting such that they are available to answer questions about the referred protocol. After discussion of the clinical investigation and the recommendation(s), the PAC will vote on its recommendation(s) to the Commissioner regarding whether the proposed clinical investigation may proceed under 21 CFR Part 50 Subpart D. The PAC's recommendations may include changes that the PAC believes are necessary for the clinical investigation to proceed as well as other suggested changes to improve the investigation. In cases where FDA and OHRP are working jointly to review a clinical investigation subject to both FDA's and HHS' regulations, the PAC

---

[6] See discussion of joint review process with OHRP in section III.O.

recommendation(s) also will be made to the Secretary. (See Section III.O., for more details on studies subject to joint FDA-HHS reviews.)

The PAC will operate by majority vote. Meeting transcripts and recommendation(s) from the PAC meeting will be made available to the public in the docket established for the referral and on the agency's website. After the PAC meeting, the chair will usually summarize the recommendation(s) of the PAC in a letter to OPT.

M. Transmittal Memorandum

OPT will draft and send a transmittal memorandum to the Commissioner outlining the PAC and Subcommittee recommendation(s), and including any recommendations/comments that FDA may have. The transmittal memorandum also will include any other necessary supporting documentation. The transmittal memorandum will request that the Commissioner make a final determination as to whether, and if so, under what provisions of Subpart D, the clinical investigation may proceed. After the Commissioner has made a final determination, OPT will forward that determination to the referring IRB and will post it in the docket created for the referral and on the agency's website.

N. Referral Process Timing

OPT projects that this referral process will take approximately six months to complete; complex submissions may take longer to process (e.g., protocols involving a joint referral with OHRP).

O. FDA-DHHS joint review under 21 CFR 50.54 and 45 CFR 46.407

In light of the need for both consistency and efficiency, the process for IRB referrals of proposed clinical investigations that are both HHS-conducted or supported and FDA-regulated will function in essentially the same manner as the process for FDA-only referrals. In such cases:

- HHS, through OHRP, will participate in the selection of members for the Subcommittee.
- After the Subcommittee makes a recommendation to the PAC, the PAC will make its recommendation(s) to both the Secretary of HHS and the Commissioner of Food and Drugs.
- OPT will forward the Commissioner's determination to OHRP.
- OHRP will send a transmittal memo with its recommendation, the Commissioner's determination, and all supporting documents, including the recommendation(s) of the PAC and the summary of the Subcommittee meeting, to the Secretary for a determination as to whether the clinical investigation may be conducted or supported by HHS.

Establishment and Operation of Clinical Trial Data Monitoring Committees

Establishment and Operation of Clinical Trial Data Monitoring Committees[1]

Guidance for Clinical Trial Sponsors

U.S. Department of Health and Human Services
Food and Drug Administration
Center for Biologics Evaluation and Research (CBER)
Center for Drug Evaluation and Research (CDER)
Center for Devices and Radiological Health (CDRH)

March 2006

Contains Nonbinding Recommendations

OMB Control No. 0910-0581

Expiration Date: 6/30/2012

See additional PRA statement in Section 8 of this guidance

Note: This guidance is intended to assist sponsors of clinical trials in determining when a data monitoring committee (DMC) is needed for study monitoring, and how such committees should operate.

1. Introduction and Background

This guidance discusses the roles, responsibilities and operating procedures of Data Monitoring Committees (DMCs) (also known as Data and Safety Monitoring Boards (DSMBs) or Data and Safety Monitoring Committees (DSMCs)) that may carry out important aspects of clinical trial monitoring. This guidance is intended to assist clinical trial sponsors in determining when a DMC may be useful for study monitoring, and how such committees should operate. We recognize that in many clinical trials the sponsor delegates some decision-making regarding the design and conduct of the trial to some other entity such as a steering committee (see Section 3.2) or contract research organization (CRO) (see 21 Code of Federal Regulations (CFR) 312.3(b)). This document, while pertaining primarily to the sponsor with regard to trial management and decision-making, may also be relevant to any individual or group to whom the sponsor has delegated applicable management responsibilities (see Section 3). This guidance finalizes the draft guidance entitled "Guidance for Clinical Trial Sponsors:

---

[1] Availableon the FDA website at: http://www.fda.gov/downloads/ RegulatoryInformation/Guidances/UCM127073.pdf

On the Establishment and Operation of Clinical Trial Data Monitoring Committees" dated November 2001.

Sponsors of studies evaluating new drugs, biologics, and devices are required to monitor these studies (see 21 CFR 312.50 and 312.56 for drugs and biologics, and 21 CFR 812.40 and 21 CFR 812.46 for devices). Various individuals and groups play different roles in clinical trial monitoring. One such group is a DMC, appointed by a sponsor to evaluate the accumulating outcome data in some trials.[2]

A clinical trial DMC is a group of individuals with pertinent expertise that reviews on a regular basis accumulating data from one or more ongoing clinical trials. The DMC advises the sponsor regarding the continuing safety of trial subjects and those yet to be recruited to the trial, as well as the continuing validity and scientific merit of the trial. When a single DMC is responsible for monitoring multiple trials, the considerations for establishment and operation of the DMC are generally similar to those for a DMC monitoring a single trial, but the logistics may be more complex. For example, multiple conflict of interest determinations may be needed for each DMC member.

Many different models have been proposed and used for the operation of DMCs. Although different models may be appropriate and acceptable in different situations, experience has shown that some approaches have particular advantages or disadvantages. In this document, we highlight these advantages and disadvantages, with particular attention to the setting in which investigational products are being evaluated for possible marketing approval in well-controlled

clinical trials. The intent of this guidance document is to ensure wide awareness of acceptable practices and of potential concerns regarding operation of DMCs that may arise in specific situations.

FDA's guidance documents, including this guidance, do not establish legally enforceable responsibilities. Instead, guidances describe FDA's current thinking on a topic and should be viewed only as recommendations, unless specific regulatory or statutory requirements are cited.

The use of the word should in FDA guidances means that something is suggested or recommended, but not required.

1.1. History of DMCs

DMCs have been a component of some clinical trials since at least the early 1960's. DMCs were initially used primarily in large randomized multicenter trials sponsored by federal agencies, such as the National Institutes of Health (NIH) and the Department of Veterans Affairs (VA) in the U.S. and similar bodies abroad, that targeted improved survival or reduced risk of major morbidity (e.g., acute myocardial infarction) as the primary objective. In 1967, an NIH external advisory group first introduced the concept of a formal committee charged with

---

[2] Some government agencies that sponsor clinical research have required the use of DMCs in certain clinical trials. Current FDA regulations, however, impose no requirements for the use of DMCs in trials except under 21 CFR 50.24(a)(7)(iv) for research studies in emergency settings in which the informed consent requirement is excepted.

reviewing the accumulating data as the trial progressed to monitor safety, effectiveness, and trial conduct issues in a set of recommendations to the then-National Heart Institute. (Heart Special Project Committee, 'Organization, Review and Administration of Cooperative Studies (Greenberg Report): A Report from the Heart Special Project Committee to the National Advisory Heart Council, May 1967;' Controlled Clinical Trials, vol. 9, 137-148, 1988.) The recommendation for the establishment of such committees was based on the recognition that interim monitoring of accumulating study data was essential to ensure the ongoing safety of trial participants, but that individuals closely involved with the design and conduct of a trial may not be able to be fully objective in reviewing the interim data for any emerging concerns. The involvement of expert advisors external to the trial organizers, sponsors, and investigators was intended to ensure that such problems would be addressed in an unbiased way by the trial leadership. The operational and functional aspects of these committees, based on experience over several decades, were discussed in a 1992 NIH workshop (Ellenberg, S., Geller, N., Simon, R. and Yusuf, S. (eds.): Practical Issues in Data Monitoring of Clinical Trials (workshop proceedings). Statistics in Medicine, 12:415-616, 1993.)

Few trials sponsored by the pharmaceutical/medical device industry incorporated DMC oversight until relatively recently. The increasing use of DMCs in industry-sponsored trials is the result of several factors, including:

- The growing number of industry-sponsored trials with mortality or major morbidity endpoints;

- The increasing collaboration between industry and government in sponsoring major clinical trials, resulting in industry trials performed under the policies of government funding agencies, which often require DMCs;

- Heightened awareness within the scientific community of problems in clinical trial conduct and analysis that might lead to inaccurate and/or biased results, especially when early termination for efficacy is a possibility, and need for approaches to protect against such problems;

- Concerns of IRBs regarding ongoing trial monitoring and patient safety in multicenter trials.

1.2. Current Status

DMCs are currently used in a variety of situations, and different models of operationhave been employed. Although no single model may be optimal for all settings, and there is not necessarily consensus about the optimal model in any given setting, there are advantages and disadvantages with respect to some of the different approaches that are in use.

As noted above, government agencies that sponsor clinical research, such as the NIH and the VA, have required the use of DMCs in certain trials. Current FDA regulations, however, impose no requirements for the use of DMCs in trials except under 21 CFR 50.24(a)(7)(iv) for research studies in emergency settings in which the informed consent requirement is excepted. FDA believes that the issues discussed in this document arise in trials with both private and public sponsorship. We recognize that the potential conflicts of interest faced by government sponsors can be different from those of industry sponsors, so that the implications for the approach to monitoring, particularly with regard to confidentiality and independence issues (see Section 4.2 and Section 6), may also differ to some extent. Nevertheless, we believe that the discussion of advantages and disadvantages of various approaches to DMC operation is relevant to all trials in which the use of a DMC is appropriate, regardless of the sponsor's funding (i.e., public or

private sector), the investigational setting of the trial (academic or other), trial size, or the phase of development. In general, DMC models used in federally funded trials that are established in accordance with policies of the funding agencies are acceptable to FDA.

2. Determining Need for a DMC

All clinical trials require safety monitoring, but not all trials require monitoring by a formal committee that may be external to the trial organizers, sponsors, and investigators. As noted earlier, DMCs have generally been established for large, randomized multisite studies that evaluate treatments intended to prolong life or reduce risk of a major adverse health outcome such as a cardiovascular event or recurrence of cancer. DMCs are generally recommended for any controlled trial of any size that will compare rates of mortality or major morbidity, but a DMC is not required or recommended for most clinical studies. DMCs are generally not needed, for example, for trials at early stages of product development. They are also generally not needed for trials addressing lesser outcomes, such as relief of symptoms, unless the trial population is at elevated risk of more severe outcomes (see Sections 4.4.1.5 and 4.4.2 for further discussion).

Although the value of a DMC is well accepted in settings such as those described above, it is important to recognize that DMCs add administrative complexity to a trial and require additional resources, so we recommend that sponsors limit the use of a DMC to the circumstances described in Section 2.1. There are several factors to consider when determining whether to establish a DMC for a particular trial. These factors, discussed below, relate primarily to safety, practicality, and scientific validity.

2.1. What is the Risk to Trial Participants?

A fundamental reason to establish a DMC is to enhance the safety of trial participants in situations in which safety concerns may be unusually high, in order that regular interim analyses of the accumulating data are performed. We recommend that sponsors consider using a DMC when:

- The study endpoint is such that a highly favorable or unfavorable result, or even a finding of futility, at an interim analysis might ethically require termination of the study before its planned completion;

- There are a priori reasons for a particular safety concern, as, for example, if the procedure for administering the treatment is particularly invasive;

- There is prior information suggesting the possibility of serious toxicity with the study treatment;

- The study is being performed in a potentially fragile population such as children, pregnant women or the very elderly, or other vulnerable populations, such as those who are terminally ill or of diminished mental capacity;

- The study is being performed in a population at elevated risk of death or other serious outcomes, even when the study objective addresses a lesser endpoint;

- The study is large, of long duration, and multi-center.

In studies with one or more of these characteristics, the additional oversight provided by a DMC can further protect study participants. In other studies, such as short-term studies for relief of symptoms as noted above, such committees are generally not warranted.

2.2. Is DMC Review Practical?

A second consideration is whether DMC review is practical. If the trial is likely to be completed quickly, the DMC may not have an opportunity to have a meaningful impact. In short-term trials with important safety concerns, however, a DMC may still be valuable. In such cases, in order for the DMC to be informed and convened quickly in the event of unexpected results that raise concerns, special mechanisms would have to be developed to permit DMC evaluation and input. Alternatively, the trial could build in "pauses" so that interim data could be reviewed by a DMC before an additional cohort of participants would be enrolled.

2.3. Will a DMC Help Assure the Scientific Validity of the Trial?

A third consideration in the decision of whether to have a DMC for a trial is whether a DMC can help assure scientific validity (and perception of such) of the trial. Trials of any appreciable duration can be affected by changes over time in the understanding of the disease, the affected population, and the standard treatment used outside the trial. These external changes may prompt an interest in modifying some aspects of the trial as it progresses. When a DMC is the only group reviewing unblinded interim data, trial organizers faced with compelling new information external to the trial may consider making changes in the ongoing trial without raising concerns that such changes might have been at least partly motivated by knowledge of the interim data and thereby endanger trial integrity. Sometimes accumulating data from within the trial (e.g., overall event rates) may suggest the need for modifications. Recommendations to change the inclusion criteria, the trial endpoints, or the size of the trial are best made by those without knowledge of the accumulating data (with the exception of changes the DMC might recommend on the basis of emerging safety concerns, as discussed in Section 4.4.1.2). When the trial organizers are the ones reviewing the interim data, their awareness of interim comparative results cannot help but affect their determination as to whether such changes should be made. Changes made in such a setting would inevitably impair the credibility of the study results. This problem will be addressed more fully in Section 6.3.

3. DMCs and Other Oversight Groups

Several different groups and individuals may assume or share responsibility for various aspects of clinical trial monitoring and oversight, and it is important to recognize the different roles they play. These groups are all components of a system that assists sponsors in conducting trials that are ethical and that produce valid and credible results. The sponsor of a clinical trial takes responsibility for and initiates the investigation (21 CFR 50.3(e); 21 CFR 312.3; 21 CFR 812.3(n)). Typically, the sponsor holds the Investigational New Drug Application or Investigational Device Exemption (IND/IDE) (21 CFR 312.40(a)(1); 21 CFR 812.40).[3]

The responsibilities delegated to steering committees or contract research organizations (CROs) by a manufacturer and/or funding agency can vary considerably. It is important that the responsibilities and authorities of the product manufacturer, the funding organization (if different) and any other entity be clearly defined and understood by all parties at the start of the endeavor. Potential conflicts of interest of each party, especially sponsors and clinical investigators (see 21 CFR Part 54) should be carefully considered when determining roles and responsibilities.

3.1. Institutional Review Boards

An institutional review board (IRB) is responsible for evaluating a trial to determine, among other things, whether "[r]isks to subjects are minimized" and "[r]isks to subjects are reasonable in relation to anticipated benefits" (21 CFR 56.111(a)). An IRB's evaluation entails review of the study protocol, relevant background information, the informed consent document, proposed plans for informing participants about the trial, and any other procedures associated with the trial. To determine whether risks to subjects are minimized by "using procedures which are consistent with sound research design" (21 CFR 56.111(a)(1)(i)), an IRB may appropriately request nformation about the approach to trial monitoring, including the statistical basis for early termination, when relevant, and what steps the sponsor is taking to minimize the risks to patients. As part of its oversight, therefore, an IRB may appropriately inquire as to whether a DMC has been established and, if so, seek information about its scope and composition.

For ongoing trials, the IRB is responsible for considering information arising from the trial that may bear on the continued acceptability of the trial at the study site(s) it oversees (see 21 CFR 56.103). A DMC, on the other hand, generally has access to much more data than the IRB during the trial, including interim efficacy and safety outcomes by treatment arm, and makes recommendations with regard to the entire trial. Given its obligation to minimize the risks to patients, an IRB may take action based on information from any appropriate source, including recommendations from a DMC to the sponsor. A trial may have multiple IRBs, each responsible for the patients at a single site, but only one DMC. Under 21 CFR 56.103, 21 CFR

---

[3] This guidance document may also be relevant to parties who participate in leadership roles in a clinical investigation other than sponsors, including funding organizations and/or others who share decision-making authority for a trial. The sponsor may be an individual, committee, company, university, or government agency, or some combination, that holds the IND or IDE and/or has responsibility for designing, initiating, funding, managing, coordinating, continuing and/or concluding the clinical trial. If a product manufacturer initiates a trial and delegates decision-making authority to a steering committee on which it has a representative, the manufacturer and the steering committee may also share certain responsibilities typically held by a sponsor. When the holder of the IND or IDE is also a study investigator, that individual is considered a sponsor-investigator (21 CFR 312.3(b); 21 CFR 812.3(o)).

312.66, 21 CFR 812.40, and 21 CFR 812.150(a), individual investigators (or the sponsor of investigational devices) are responsible for assuring that IRBs are made aware of significant new information that arises about a clinical trial. Such information may include DMC recommendations to the sponsor that are communicated to IRB(s), either directly or through individual investigators or sponsors. Additionally, it may be useful for sponsors to ensure that IRBs are informed when DMCs have met, even when no problems have been identified and the DMC has recommended continuation of the trial as designed.

3.2. Clinical Trial Steering Committees

In some clinical trials the sponsor may choose to appoint a steering committee; this committee may include investigators, other experts not otherwise involved in the trial, and, usually, representatives of the sponsor. A sponsor may delegate to a steering committee the primary responsibility for designing the study, maintaining the quality of study conduct, ongoing monitoring of individual toxicities and adverse events, and, in many cases, writing study publications. When there is a steering committee, the sponsor may elect to have the DMC communicate with this committee rather than directly with the sponsor. Interactions between the steering committee and the DMC consist primarily of discussions during "open sessions" (see Section 4.3) of DMC meetings and the communication of recommendations following each DMC review of the trial. More extensive interactions might occur when early termination is being considered, or when external forces (e.g., announcement of results of related studies) impact the ongoing trial.

3.3. Endpoint Assessment/Adjudication Committees

Sponsors may also choose to establish an endpoint assessment/adjudication committee (these may also be known as clinical events committees) in certain trials to review important endpoints reported by trial investigators to determine whether the endpoints meet protocol-specified criteria. Information reviewed on each presumptive endpoint may include laboratory, pathology and/or imaging data, autopsy reports, physical descriptions, and any other data deemed relevant. These committees are typically masked to the assigned study arm when performing their assessments regardless of whether the trial itself is conducted in a blinded manner. Such committees are particularly valuable when endpoints are subjective and/or require the application of a complex definition, and when the intervention is not delivered in a blinded fashion. Although such committees do not share responsibility with DMCs for evaluating interim comparisons, their assessments (if performed at frequent intervals throughout the trial with results incorporated into the database in a timely manner) help to ensure that the data reviewed by DMCs are as accurate and free of bias as possible.

3.4. Site/Clinical Monitoring

The sponsor or a group under contract to the sponsor generally performs site/clinical monitoring of a clinical trial to assure high quality trial conduct. They perform "on site" monitoring of individual case histories, assess adherence to the protocol, ensure the ongoing implementation of appropriate data entry and quality control procedures, and in general assess adherence to good clinical practices. In blinded studies, these monitors remain blinded to study arm assignment.

3.5. Others with Monitoring Responsibilities

In addition to those described above, other groups have important monitoring responsibilities. Study investigators, of course, have the front-line responsibility for identifying potential adverse effects experienced by study participants, adjusting the intervention accordingly and reporting the experience to the sponsor. The sponsor is responsible for monitoring and analyzing these investigator reports and relaying them as required to FDA, other regulatory authorities (as appropriate) and other investigators (21 CFR 312.32(c), 21 CFR 812.40). The sponsor and FDA, respectively, also review adverse experience reports from all trials of a given product (21 CFR 312.32(c); 21 CFR 812.150(b)). In addition, for medical device studies, sponsors are responsible for ensuring that FDA and any reviewing IRB(s) are promptly informed of significant new information about an investigation (21 CFR 812.40). For drug and biologic studies, sponsors must notify IRBs, as well as FDA and other investigators, if the sponsor withdraws the IND for a safety reason (21 CFR 312.38(c)).

4. DMC Establishment and Operation

4.1. Committee Composition

The selection of DMC members is extremely important, as DMC responsibilities relate to the safety of trial participants. A poorly constituted DMC may fail to note problems that should be addressed, or may make recommendations that are unwarranted or whose consequences are inadequately considered, thereby undermining the safety of participants as well as the value of the trial. The ability of DMCs to provide the anticipated additional assurance of patient safety and trial integrity therefore depends on appropriate selection of DMC members.

The sponsor and/or trial steering committee generally appoint members of a DMC. Factors to consider in the selection of individuals to serve on a DMC typically include relevant expertise, experience in clinical trials and in serving on other DMCs, and absence of serious conflicts of interest as discussed below. The objectives and design of the trial and the scope of the responsibilities given to the DMC determine the types of expertise needed for a particular DMC.

Most DMCs are composed of clinicians with expertise in relevant clinical specialties and at least one biostatistician knowledgeable about statistical methods for clinical trials and sequential analysis of trial data. For trials with unusually high risks or with broad public health implications, the DMC may include a medical ethicist knowledgeable about the design, conduct, and interpretation of clinical trials. Prior DMC experience is important when considering the committee as a whole; it is highly desirable that at least some members have prior DMC service. Prior DMC experience is particularly important for the statistical DMC member if there is only one statistician serving on the DMC.

Some trials may require participation of other types of scientists. Toxicologists, epidemiologists, and clinical pharmacologists, for example, could be included in particular cases when such expertise appears important for informed interpretation of interim results.

One or more individuals (often non-scientists) who may help bring to the DMC the perspectives of the population under study may be a useful addition in some settings. Generally, such a DMC member would not also be a participant in the trial, since awareness of the accumulating data could affect compliance or other aspects of trial participation. Rather, the member could be someone with the disease or condition under study or a close relative of such an individual, for example.

Appropriate representation of gender and ethnic groups may be of particular importance for some trials. DMCs for international trials will usually include representatives from at least a subset of participating countries or regions; however, it is often not feasible to have every participating country represented on the DMC. For the reasons discussed at the beginning of Section 4.1, we recommend that the primary criterion for selecting all appointees should be their respective expertise and experience. An important practical consideration would be their ability to commit to attending DMC meetings and to maintaining confidentiality of the interim results they have reviewed (see Section 4.2). A DMC may have as few as 3 members, but may need to be larger when representation of multiple scientific and other disciplines, or a wider range of perspectives generally, is desirable. For logistical reasons, sponsors typically wish to keep the DMC as small as possible, while still having representation of all needed skills and experience. Some redundancy may be desirable, however, in scientifically and/or ethically complex trials, trials of long duration in which DMC attrition might be anticipated, or in trials in which the DMC must meet fairly frequently so that not all members would likely be able to attend all meetings.

Conflicts of interest deserve special consideration in choosing individuals to serve on a DMC. The most obvious conflict is financial interest that could be substantially affected by the outcome of the trial. (See Section 6 for further discussion. See also Department of Health and Human Services, Financial Relationships and Interests in Research Involving Human Subjects: Guidance for Human Subject Protection, available at http://www.hhs.gov/ohrp/humansubjects/ finreltn/fguid.pdf.)

Investigators entering subjects into the trial have a different type of conflict of interest— their knowledge of interim results could influence their conduct of the trial. An investigator who is aware of early trends might change his or her pattern of recruitment, or modify his or her usual way of monitoring the status of participants. We therefore recommend that DMC members for a given trial not include investigators in that trial.

Individuals known to have strong views on the relative merits of the interventions under study may have an "intellectual" conflict of interest and might not be able to review the data in a fully objective manner; such individuals may therefore not be optimal DMC members. We recommend that sponsors avoid appointing to a DMC any individuals who have relationships with trial investigators or sponsor employees that could be considered reasonably likely to affect their objectivity.

We recommend that sponsors establish procedures to:

- Assess potential conflicts of interest of proposed DMC members;

- Ensure that those with serious conflicts of interest are not included on the DMC;

- Provide disclosure to all DMC members of any potential conflicts that are not thought to impede objectivity and thus would not preclude service on the DMC;

- Identify and disclose any concurrent service of any DMC member on other DMCs of the same, related or competing products.

The sponsor often appoints the DMC chair, but may seek advice from trial investigators or trial steering committee members. Prior DMC experience is more important for the chair than for other DMC members, as members will look to the chair for leadership on administrative as well as scientific issues. Sponsors will typically want to select a chair who is capable of facilitating discussion, integrating differing points of view, and moving toward consensus on recommendations to be provided to the sponsors. Sponsors may also want to be assured that a potential chair is willing to make a firm commitment to participate for the duration of the trial (or for the term of the appointment, for chairs of DMCs monitoring multiple trials).

4.2. Confidentiality of Interim Data and Analyses

As described in 21 CFR 314.126(b)(5) (drugs) and 21 CFR 860.7(f)(1) (devices), sponsors of well-controlled studies should take appropriate measures to minimize bias.[4]

Knowledge of unblinded interim comparisons from a clinical trial is generally not necessary for those conducting or sponsoring the trial; further, such knowledge can bias the outcome of the study by inappropriately influencing its continuing conduct or the plan of analyses. Unblinded interim data and the results of comparative interim analyses, therefore, should generally not be accessible by anyone other than DMC members or the statistician(s) performing these analyses and presenting them to the DMC (see id.).

Consistent with 21 CFR 314.126(b)(5) (drugs) and 21 CFR 860.7(f)(1) (devices), sponsors should establish written procedures, which may be included in the DMC charter, to ensure the minimization of bias, such as maintaining confidentiality of the interim data (see Section 4.3.1.4). Sponsors may, of course, also address such confidentiality issues in written agreements between the sponsor and members of the DMC as well as written agreements between the sponsor and investigators.

Even for trials not conducted in a double-blind fashion, where investigators and patients are aware of individual treatment assignment and outcome at their sites, the summary evaluations of comparative unblinded treatment results across all participating centers would usually not be available to anyone other than the DMC. Section 6 addresses the particular confidentiality issues for the statistician/statistical team performing the interim analyses.

---

[4] All discussions in this guidance relating to adoption of procedures for the minimization of bias refer to the minimization of bias in adequate and well-controlled clinical trials for drugs, as described in 21 CFR 314.126, and well-controlled clinical trials for devices, as described in 21 CFR 860.7(f).

4.2.1. Interim Data

Interim comparative data, whether treatment assignment is revealed or coded, will be most securely protected from inadvertent or inappropriate access by the sponsor or its project team if the data are prepared for analysis by a statistical group that is independent of the sponsor and investigators—that is, the group is not otherwise involved in the trial design or conduct and has no financial or other important connections to the sponsor or other trial organizers (see Section 6). The lead investigators, the study steering committee, and/or the sponsor generally develop the analytical plan (often collaboratively), but problems can arise when these same individuals are involved in the actual preparation of the interim results, for reasons discussed in Section 6.4. They may, however, work with the statistician who will be preparing and presenting the interim analyses prior to the first analysis of unblinded data to develop a template for the interim reports. Procedures should be established to safeguard confidential interim data from the project team, investigators, sponsor representatives, or anyone else outside the DMC and the statistician(s) performing the interim analyses (see 21 CFR 314.126(b)(5) (drugs) and 21 CFR 860.7(f)(1) (devices)).

Although assigning responsibility for interim analysis to individuals employed by the sponsor is generally discouraged, such assignment may be appropriate if sufficiently secure procedures are in place to credibly ensure that the results of such analyses are not revealed to other sponsor employees or to anyone other than DMC members. We recommend that a description of such procedures be included in the DMC charter (see Section 4.3).

4.2.2. Interim Reports to the DMC

We recommend that any part of the interim report to the DMC that includes comparative effectiveness and safety data presented by study group, whether coded or completely unblinded, be available only to DMC members during the course of the trial, including any follow-up period—that is, until the trial is completed and the blind is broken for the sponsor and investigators. If interim reports are shared with the sponsor, it may become impossible for the sponsor to make potentially warranted changes in the trial design or analysis plan in an unbiased manner (see Section 6.3). Even aggregate data on safety and efficacy may be informative; these data may be needed for some trial management functions (e.g., sample size adjustments, centralized endpoint assessment), but are best limited to those who cannot otherwise carry out their trial management responsibilities.

In some cases (for example, in open-label trials with special concerns about safety), there may be a rationale for the sponsor and/or investigators to have access to the ongoing comparative safety data to ensure continuous monitoring. Such access should be specified and justified in the study protocol and understood by the DMC (see 21 CFR 314.126(b)(5) (drugs) and 21 CFR 860.7(f)(1) (devices)).

In many cases, the DMC receives reports in two parts: an "open" section, which presents data only in aggregate and focuses on trial conduct issues such as accrual and dropout rates, timeliness of data submission, eligibility rates and reasons for ineligibility; and a "closed" section, in which the comparative outcome data are presented. The open section of these reports is usually provided to sponsors, who may convey any relevant information in these reports to investigators, IRBs, and other interested parties, as the data presented in the "open" section are not likely to bias the future conduct of the trial and are often important for improving trial management.

4.3. Establishing a Charter Describing Standard Operating Procedures

DMCs typically operate under a written charter that includes well-defined standard operating procedures. Such charters are important for the same reason that study protocols and analytical plans are important—they document that procedures were prespecified and thereby reduce concerns that operations inappropriately influenced by interim data could bias the trial results and interpretation. The sponsor may draft this charter and present it to the DMC for agreement, or the DMC may draft the charter with subsequent concurrence by the sponsor. Topics to be addressed would normally include a schedule and format for meetings, format for presentation of data, specification of who will have access to interim data and who may attend all or part of DMC meetings, procedures for assessing conflict of interest of potential DMC members, the method and timing of providing interim reports to the DMC, and other issues relevant to committee operations. FDA may request that the sponsor submit the charter to FDA well in advance of the performance of any interim analyses, ideally before the initiation of the trial (see 21 CFR 312.23(a)(6)(iii)(g); 21 CFR 312.41(a); 21 CFR 812.150(b)(10)). In such cases, FDA would usually consider the charter when FDA reviews the study protocol.

4.3.1. Considerations for Standard Operating Procedures

4.3.1.1. Meeting Schedule and Format

The initial frequency of DMC meetings will depend on the expected rate of accrual and event occurrence at the time the trial is designed as well as the perceived risk of the experimental and/or control interventions. Annual meetings may be adequate for some studies; other trials will require more frequent review. Occasionally, there may be a need for extra meetings, when, for example, there is concern about potentially emerging safety problems, or when important new information external to the trial arises. The study protocol will generally describe the schedule of interim analyses to be considered by the DMC, or the considerations that will determine the timing of meetings (e.g., a plan for interim analysis after a certain number of primary outcomes have been reported). The study protocol will also typically describe the statistical approach to the interim analysis of trial data. To minimize the potential for bias, these descriptions should be complete before the conduct of any unblinded interim analyses (see 21 CFR 314.126(b)(5) (drugs) and 21 CFR 860.7(f)(1) (devices)).

Face-to-face meetings are generally preferable, but telephone meetings may be necessary in some situations, particularly when new information must be urgently considered. In some settings, when the DMC has already had numerous meetings and the committee is very familiar with the trial and the analytical issues, telephone meetings may be sufficient. When telephone meetings are held, precautions may be needed to assure the confidentiality of the proceedings, and to prevent inadvertent access to conversations.

4.3.1.2. Meeting Structure

Attendance at meetings raises the same confidentiality issues as access to interim reports provided to the DMC. A central tenet of clinical trials is the importance of maintaining confidentiality of interim comparative data (see the Guidance for Industry, ICH E9, Statistical Principles for Clinical Trials, available at http://www.fda.gov/cder/ guidance/ICH_E9-fnl.pdf). Although FDA typically expects that confidentiality of the interim data will be maintained, the DMC may interact with the sponsor and/or trial lead investigators to clarify issues relating to the conduct of the trial, potential impact on the trial of external data, or other topics. In order to permit such interaction without compromising confidentiality, many DMC meetings include an "open" session in which information in the open report is discussed. These non-confidential data may include, for example, status of recruitment, baseline characteristics, ineligibility rate, accuracy and timeliness of data submissions, and other administrative data. Sponsors may also use open sessions to provide external data to the DMC that may be relevant to the study being monitored. Open session discussions might include representatives of the sponsor, steering committee, study investigators, FDA representatives, or others with trial responsibilities. There is a benefit to having a wider attendance at these sessions, since they provide an opportunity for those with the most intimate knowledge of the study to share their insights with the DMC and raise issues for the DMC to consider. The DMC generally considers the comparative interim data contained in the closed report in a "closed" session attended only by the DMC members and the statistician who prepared and is presenting the interim analyses to the DMC. Following the closed session, the DMC may meet again with the sponsor to relay any recommendations the DMC has made.

Section 6 describes the risks to study integrity when sponsor representatives have access to unblinded interim data and attend closed sessions of DMC meetings. In settings in which a sponsor chooses to permit its representatives or other non-DMC members to attend the closed session despite the risks of such arrangements, we recommend that the DMC have the option of conducting an "executive" session with no participants other than DMC members.

4.3.1.3. Initial Meeting

Scheduling the initial meeting of a DMC before the study is initiated has many advantages. At this meeting, the DMC can discuss the protocol and analytic plan, model informed consent form, data collection instruments and other important trial documents, and present any suggestions for modifications to the sponsor and/or steering committee. Regulatory considerations may also be discussed. Meeting participants typically discuss and complete plans for monitoring the safety and effectiveness data, including:

- Scheduling of meetings;

- Format for the interim reports to the DMC;

- Timing of the delivery of the report to the DMC members prior to the meeting;

- Definition of a "quorum" of DMC members, including representation of essential scientific and other disciplines;

- Handling of meeting minutes; and

• Other aspects of the process.

It is particularly important that the sponsor and the DMC agree on the data monitoring plan, including the approach to early termination.

4.3.1.4. Format of Interim Reports to the DMC and Use of Treatment Codes

It is important that the general format and content of interim reports to the DMC be acceptable to the DMC. This may be accomplished most efficiently if the sponsor proposes a template for these reports at its first meeting, so that changes requested by the DMC may be implemented before interim data are first presented. However, the templates may change during the course of the trial as experience is accrued. Further, the DMC will generally need easy and timely access to any additional data and analyses deemed important, and may request such additional material when needed.

We recommend that a DMC have access to the actual treatment assignments for each study group. Some have argued that DMCs should be provided only coded assignment information that permits the DMC to compare data between study arms, but does not reveal which group received which intervention, thereby protecting against inadvertent release of unblinded interim data and ensuring a greater objectivity of interim review. This approach, however, could lead to problems in balancing risks against potential benefits in some cases. For example, to maintain blinding of the actual treatment assignments, safety outcomes would have to be coded differently from effectiveness outcomes when adverse effects would reveal the assigned intervention. This would prevent the DMC from evaluating the balance of risks and benefits of the active interventions, its most critical responsibility.

Also, decisions about a trial are often asymmetric with respect to study arms; that is, a DMC may recommend termination of a study with a trend toward showing harm on the basis of data that, were they in the other direction, would not be considered strong enough to terminate early with a conclusion of benefit. Similarly, a trend suggesting a safety concern with a new intervention could be sufficient to suggest the need for trial modification, while a similar trend in the opposite direction (new intervention looks better than standard) might not.

A common approach is presentation of results in printed copy using codes (for example, Group A and Group B) to protect against inadvertent unblinding should a report be misplaced, with separate access to the actual study arm assignments provided to DMC members by the statistical group responsible for preparing DMC reports. To ensure the DMC's and sponsor's ability to address an emerging safety concern rapidly, they should establish a process to unblind treatment codes to DMC members in a timely fashion when needed (cf. 21 CFR 312.32(d); 21 CFR 812.46(b)(2)). For example, DMC members might routinely receive the unblinded treatment codes in a mailing separate from that containing the interim reports.

4.3.2. Statistical Methods

Statistical approaches to monitoring trials, and the principles involved in their implementation, are addressed in Guidance for Industry, ICH E9, Statistical Principles for Clinical Trials, available at http://www.fda.gov/cder/guidance/ICH_E9-fnl.pdf.[5] Planners of clinical trials most commonly use group sequential methods, in which interim analyses are performed at regular intervals based either on chronological time or amount of information accrued, but other approaches, such as those based on Bayesian methods, have been used as well. Statistical methods are also available to assess stopping for futility; that is, when the likelihood that the treatment effect being sought, based on the interim data, is very unlikely to be established. Other statistical strategies for monitoring may also be appropriate.

The sponsor or trial steering committee usually proposes the particular statistical approach to interim monitoring, but the DMC should generally review it before it is made final, to ensure that the DMC agrees to be guided in its actions by the planned approach. FDA will typically request that the sponsor submit a final monitoring plan once it has been put in place, and before the initiation of interim monitoring, as such a plan would typically be considered a critical component of the study protocol (see 21 CFR 312.23(a)(6)(iii)(g); 21 CFR 312.41(a); 21 CFR 812.150(b)(10)). Because statistical approaches based on classical hypothesis testing methods are by far the most common, the remaining discussion in this section will focus on issues within that framework. As noted earlier, other monitoring strategies may also be appropriate.

One of the major responsibilities of a DMC is to evaluate the relative treatment effects based on protocol-specified endpoints to determine if the trial is meeting its objectives. A major concern when data on group differences are assessed repeatedly as they accumulate is that the Type I error (false positive) rate may be inflated if adjustment is not made for the multiple looks at the data. Typically, the monitoring plan will specify a statistical approach that permits multiple interim reviews while maintaining the Type I error rate at the desired level. These approaches usually generate boundaries for interim estimates of benefit that indicate the magnitude of benefit needed to support stopping the trial at interim points prior to its planned completion, while maintaining the desired overall probability of Type I error. Such boundaries can serve as useful guidelines to the DMC in making recommendations regarding continued accrual to and conduct of the trial. The DMC will usually recommend termination when these hresholds are crossed, but it is not obligated to do so, since other aspects of the interim data may complicate the issue. For example, the data on effectiveness may be very strong, with a stopping boundary having been crossed, but emerging safety concerns may make the benefit-to-risk assessment non-definitive at that interim review. FDA expects the sponsor to direct the DMC to exercise its own judgment in such circumstances; the DMC can be flexible in assessing the data relative to the stopping boundaries. If the DMC recommends early termination for efficacy before a boundary is crossed, however, and this recommendation is implemented, the Type I error cannot be preserved and the study results may be difficult to interpret.

Statistical assessment may also suggest that early termination of a trial be considered on the basis of futility, as defined previously. In this case, a DMC may recommend early termination on the grounds that the trial is unlikely to meet its objectives and there is therefore no basis for continuing enrollment and/or follow-up. Before recommending that a trial be terminated due

---

[5] Although ICH documents are meant to provide guidance for drug and biologics sponsors, the statistical monitoring principles in ICH E9 could be used in the evaluation of medical devices as well.

to futility, a DMC will typically consider the Type II error, the chance of making a false negative conclusion. Stopping on the basis of futility does not raise concerns about Type I error in that trial, since the conclusions of the trial will not be positive. Nevertheless, protection of Type I error may be important even when there is a stated intention to stop early only for futility reasons since interim review of outcome data always raises the possibility that the DMC may find early results so persuasive that it would recommend early termination of the trial.

4.4. Potential DMC Responsibilities

4.4.1. Interim Monitoring

Most experience with DMCs has been in the setting of studies that address major outcomes such as mortality or serious irreversible morbidity. Although many such studies focus on short-term endpoints such as 30-day survival, other studies often use endpoints that require a substantial duration of follow-up after the intervention delivery has been completed. The need for monitoring in such studies often extends beyond the time when individuals are treated, since trends in survival or other serious outcomes may not become evident until some time during the follow-up period. Thus, the DMC's responsibility to monitor the study generally continues until the planned completion of follow-up, regardless of the duration of treatment.

4.4.1.1. Monitoring for Effectiveness

In studies with serious outcomes, all parties would wish that any major treatment advance be identified and made available as soon as possible. It is critical, however, that the study yield a valid and definitive result. Thus, tensions between ethical and scientific considerations may arise. Consider, for example, a placebo-controlled trial of a new product for a serious illness or condition for which there is no standard treatment. If the emerging data suggest that those receiving the treatment are doing better, one might expect that a DMC would consider whether the study should be terminated earlier than planned. Estimates of treatment effect, however, will be unstable at early points in a study, and the chance is substantial of observing a nominally statistically significant benefit (e.g., $p<0.05$) at one of multiple interim analyses during a study of an ineffective product (see Section 4.4.2). A DMC, guided by a pre-specified statistical monitoring plan acceptable to both the DMC and the study leadership, will generally be charged with recommending early termination on the basis of a positive result only when the data are truly compelling and the risk of a false positive conclusion is acceptably low.

A second type of consideration is whether the hypothesized benefit is likely ultimately to be achieved. If the interim data suggest that the new product is of no benefit—that is, there is no trend indicating superiority of the new product—or that accrual rates are too low or noncompliance too great to provide adequate power for identifying the specified benefit, a DMC may consider whether continuation of the study is futile and may recommend early termination on this basis. In this case, false negative conclusions are of concern; statistical procedures are available to guide such determinations (see Section 4.3.2).

4.4.1.2. Monitoring for Safety

There are several aspects to safety monitoring in long-term outcome studies. First, the primary efficacy endpoint itself often has safety implications. If individuals given the investigational intervention are found to be at higher risk for the outcome of interest (e.g., mortality, disease progression, loss of organ function) sooner than those given the control, the DMC may consider recommending early termination on safety grounds. Such assessments have potential implications for falsely concluding that there is an adverse effect, just as regular assessments of efficacy have the potential to lead to false positive conclusions about benefit. Statistical considerations for early stopping when the data are trending in the direction of harm are often different from the case of trends in the direction of benefit, however. It is usually appropriate to demand less rigorous proof of harm to justify early termination than would be appropriate for a finding of benefit. In some cases, however, it may be appropriate to establish a harmful effect more definitively—for example, if a positive effect on the primary endpoint has been demonstrated or appears to be emerging, a precise assessment of a negative trend on a potentially important safety endpoint may be required for benefit-to-risk considerations.

A second important aspect of safety monitoring in these trials is comparison of adverse event rates in each treatment arm. In some cases adverse events of particular concern can be identified in advance of the trial, and particular attention will be given to monitoring these events. For example, in a large trial of hormone replacement therapy, specific monitoring plans were established to detect a possible increase in breast cancer incidence in women taking active therapy. Because many types of adverse reactions cannot be anticipated prior to a large-scale study, the DMC should generally be provided with interim summaries by treatment arm of adverse events observed, not limited to those identified in advance. This is particularly important for serious events that may result from the disease being treated as well as the intervention itself, or occur at an observable background rate in the population under study. An effect of the drug on these events can only be detected by comparing the rates of the events in treatment and control groups.

To illustrate the process, consider acute myocardial infarctions (AMIs) occurring during a study of an antidiabetic therapy. Diabetics are at increased risk of AMI so that a specific AMI in a participant could not be attributed to the new drug. A DMC for such a trial, however, would regularly review the number of myocardial infarctions observed in each study arm. If an imbalance between groups emerges, concerns will arise that some of the myocardial infarctions may be due to the intervention rather than the disease itself. Since a potentially large number of adverse event categories may be observed and compared between the study arms, sensitivity on the part of the DMC to the issues of multiplicity, i.e., the elevated probability of "false positives" when performing multiple analyses, is warranted. Not all potential risks can be identified in advance of the trial, so pre-specifying risks of concern cannot always be done.

A third aspect of safety monitoring is consideration of individual events of particular concern. Although a DMC typically reviews summary adverse event data as discussed above, it will not usually review in detail every adverse event reported, or even every serious adverse event. This responsibility generally lies with the sponsor, who must assure review of such events promptly (see, e.g., 21 CFR 312.32(b); 21 CFR 812.42(d); 21 CFR 812.46(b)). The sponsor has the responsibility of reporting to FDA serious, unexpected adverse events in drugs and biologics trials under 21 CFR 312.32 and unanticipated adverse events in the case of device trials under 21 CFR 812.150(b)(1). For clinical trials involving drugs and biologics, we recommend that sponsors notify DMCs about any waivers granted by FDA for expedited reporting of certain serious events.

The involvement of a DMC in the review of individual adverse event reports will vary from case to case. In some studies, it may be important for the DMC to see detailed information on all deaths or other specified events, particularly events that are likely to have been caused by the product being tested (e.g., acute liver failure in a drug study). In other studies, where many deaths or other serious events are expected, the DMC may view only the summary tabulations and comparative statistics to determine whether there appears to be an excess of an important adverse event in one of the study arms. Simple listings of adverse outcomes, especially without treatment assignments, are rarely useful for DMC discussion.

The sponsor may ask the DMC to review any individual event thought to be of major significance by the study's medical monitor; such events would generally include deaths or other serious outcomes for which a causal connection with the intervention is plausible. We recommend that the DMC be informed in a timely manner of any cases for which unblinding of treatment code at the clinical site or by the treating clinician is thought to be necessary to provide an appropriate intervention, so that the DMC can assess the potential impact of such actions on the overall study blind. Review of individual cases by the DMC does not relieve the sponsor of the regulatory responsibilities, discussed above, regarding evaluation of these events and reporting as required to FDA.

Concerns about the extent and type of adverse events observed may lead to early termination of the trial when the DMC judges that the potential benefits of the intervention are unlikely to outweigh the risks. In other cases, a DMC may recommend measures short of termination that might reduce the risk of adverse events. For example, the DMC might recommend:

- Changing the eligibility criteria if the risks of the intervention seem to be concentrated in a particular subgroup.

- Altering the product dosage and/or schedule if the adverse events observed appear likely to be reduced by such changes.

- Instituting screening procedures that could identify those at increased risk of a particular adverse event.

- Informing current and future study participants of newly identified risks via changes in the consent form and, in some cases, obtaining reconsent of current participants to continued study participation.

4.4.1.3. Monitoring Study Conduct

The DMC typically shares responsibility for assessment of data related to study conduct with the sponsor, the study leadership (such as a steering committee), and to some extent with IRBs. A DMC will generally review data related to the conduct of the study (that is, the quality of the study and its ultimate ability to address the scientific questions of interest), in addition to data on effectiveness and safety outcomes. These data may include, among other items:

- Rates of recruitment, ineligibility, noncompliance, protocol violations and dropouts, overall and by study site;

- Completeness and timeliness of data;

- Degree of concordance between site evaluation of events and centralized review;

- Balance between study arms on important prognostic variables;

- Accrual within important subsets.

The DMC may issue recommendations to the sponsor regarding trial conduct when concerns arise that some aspects of trial conduct may threaten the safety of participants or the integrity of the study. For example, if the data presented to the DMC are not current, the DMC will not be able to meet its responsibility of ensuring that the study continues to be safe for its current and future participants. As another example, an excess of dropouts may endanger the ultimate interpretability of the study results.

4.4.1.4. Consideration of External Data

A DMC may be asked to consider the impact of external information on the study being monitored. Release of results of a related study may have implications for the design of the ongoing study, or even its continuation. In some cases, particularly when unexpected safety issues arise in related studies, the sponsor may bring external data to the attention of the DMC; in other cases, the data may be publicly reported. Such data may lead to recommendations ranging from termination of the study, termination of one or more study arms, changes in target population, dose and/or duration of the intervention, or use of concomitant treatments. The DMC may also recommend changes to the consent form or investigator's brochure, and/or letters from the sponsor to study participants describing the new results. The role of the DMC in considering interim changes to a study protocol or other aspects of study conduct in response to external information raises additional issues that merit consideration.

In many cases, access to the unblinded data will be essential to making the best decision regarding changes to an ongoing trial that are suggested by external data. For example, if external reports indicate that use of the study drug in a different indication raised serious, unexpected safety concerns, a decision about continuing the ongoing trial may depend on whether the interim data suggest important benefits that may make the newly found risks acceptable, or the extent to which the newly identified concerns are evident in the ongoing study. In some circumstances, DMCs of separate but closely related trials (e.g., trials of the same product in different patient populations) may consider sharing confidential interim data when unexpected safety issues arise in one trial and information from the two trials together may improve decision-making in both trials. Because such sharing limits the extent to which the trials can be considered independent, it should be pursued only in the rare situations when early stopping might be considered, but the issues leading to this consideration are ambiguous, for example, when a safety concern arises that appears biologically implausible. Both DMCs would typically require the express consent of the respective sponsors prior to sharing such information.

In some cases, however, significant involvement of the DMC in considerations of changes based on external data could have undesirable consequences precisely because the DMC is aware of the interim study results. Many kinds of trial modifications (e.g., changing endpoints, changing or adding to prespecified analysis subgroups) could, if made with knowledge of trial results, have significant effects on type I error and interpretation of final results. If it is perceived that emerging results could have influenced these types of interim protocol changes, the credibility of the trial may be severely damaged. In general, to minimize the potential for bias, the trial leadership, which is insulated from knowledge of the interim data, rather than the DMC, should be responsible for proposing potential changes other than those driven by safety considerations (cf. 21 CFR 314.126(b)(5), 21 CFR 860.7(f)(1)).

The principle that interim protocol changes should not be influenced by emerging results has implications for sponsors, who would initiate requests for protocol changes, and FDA staff, who would need to evaluate any such requests for protocol changes for INDs under 21 CFR 312.30 and for IDEs under 21 CFR 812.35. Sponsors who wish to have the ability to request interim protocol changes without raising concerns about biasing the study should establish procedures to minimize bias, such as ensuring that they are completely unaware of unblinded comparative data (see 21 CFR 314.126(b)(5), 21 CFR 860.7(f)(1)). If the study is performed with blinded treatment allocation, and access to unblinded data is limited to the DMC, making such changes as requested by the sponsor is straightforward. If treatment allocation is not blinded, it is more difficult to maintain confidentiality of interim comparative results, as sponsor staff such as medical monitors will be reviewing data on each case. In such circumstances it may be very advantageous for the sponsor to set up a "firewall" to ensure that those who would be proposing interim protocol changes based on external data are insulated from knowledge of interim comparative results. To avoid any influence of interim data on consideration of protocol changes, FDA staff will also generally remain blinded to the interim results. Under 21 CFR 312.41(a) (drugs) or 21 CFR 812.150(b)(10) (devices), we may request additional information or data to aid in FDA's review of protocol amendments and other aspects of clinical trials under an IND or IDE, respectively. Under these authorities, we will typically request that, once interim data have been seen by the sponsor, such data should also be available to FDA, provided such data form the basis for a request by the sponsor to amend a study protocol. It may be necessary for FDA to play a more active role regarding interim results in rare cases when there is an immediate need to evaluate a serious safety concern, especially when we may have important relevant information that may not otherwise be available to the DMC. Even in such cases, however, it will generally be preferable for FDA to provide such information to the DMC, where possible, rather than taking a direct role in interim evaluations.

4.4.1.5. Studies of Less Serious Outcomes

Many clinical trials evaluate interventions to relieve symptoms. These studies are generally short-term, evaluating treatment effect over periods of a few days to a few months. These studies tend to be smaller than major outcome studies and, therefore, are completed more quickly. Because the primary endpoints of such studies are not serious irreversible events, as in a major outcome study, the ethical issues for monitoring are different. In these studies, valuable secondary objectives such as characterization of the effect (i.e., magnitude, duration, time to response), assessment of the effect in population subsets, comparison of several doses and/or comparison of the new product to an active control can be ethically pursued even when the conclusion regarding the primary outcome is clear. Early termination for effectiveness is rarely appropriate in such studies. First, the study may be essentially completed by the time any interim analysis could be undertaken. Second, the effectiveness of an intervention to relieve symptoms would not generally be so compelling as to override the need to collect the full amount of safety data, or to collect other information of interest and importance that characterizes the effect, as noted above.

DMCs have not been commonly established for short-term studies of interventions to relieve symptoms. The need for an outside group to monitor data regularly to consider questions of early stopping for efficacy or protocol modification is usually not compelling in this situation. Such a group is probably warranted only when termination of the trial for efficacy, even at the expense of obtaining more complete safety information, would be indicated for ethical reasons.

For products intended solely to relieve symptoms, as opposed to curing or delaying progress of a serious disease or medical condition, an expert group to oversee all studies at all stages of development, monitor the developing safety database and make recommendations for design of successive studies based on early results may be useful. The sponsor or investigator could refer an unusual safety concern arising in any study to this type of external group for review, while maintaining its own primary role in monitoring the accumulating results. Such a group may be particularly valuable when the patient population is at relatively high risk of serious events; for example, in studies of drugs to control symptoms of angina, congestive heart failure, or chronic obstructive lung disease. The external group would independently evaluate individual events and overall event rates in ongoing studies and advise the sponsor about emerging concerns. Clearly, monitoring considerations of this type are more clinical than statistical. Sponsors frequently constitute internal groups to monitor these types of studies, and these may be satisfactory in most cases. Nevertheless, external advisors, who will be less committed to the existing development plan, may identify some problems more readily than internal reviewers. Thus, sponsors may find it valuable to augment such internal groups with one or more external advisors.

4.4.2. Early Studies

DMCs are not usually warranted in early studies such as Phase 1 or early Phase 2 studies, or pilot/feasibility studies, but formal monitoring groups may be useful for certain types of early clinical studies. While these formal monitoring groups will often consist of individuals internal to the sponsor and/or investigators, a DMC overseeing safety may be considered when risk to participants appears unusually high, e.g., with particularly novel approaches to treating a disease or condition. When the investigator is also the product manufacturer or IND/IDE sponsor, and thereby subject to potentially strong influences related to financial and/or intellectual incentives, a DMC could provide additional, independent oversight that would enhance safety of study participants and the credibility of the product development. Sponsors may therefore wish to consider establishing DMCs in such settings.

A DMC's role in early phase studies would be different from that in late Phase 2 or Phase 3 studies. Early studies are often exploratory in nature; they are frequently not randomized or controlled and therefore accumulating results are known to the investigators and sponsor. Issues regarding statistical interpretation of interim data, or confidentiality of interim data, are therefore generally less relevant in this setting. Nevertheless, for difficult situations in which the potential scientific gain from continuing a study must be evaluated in the context of ethical considerations for ensuring subjects' rights and welfare, particularly in settings such as those described above, DMCs may be helpful to investigators, sponsors and IRBs by providing independent, objective expert counsel. We expect, however, that the need for independent DMCs in early phase studies will be infrequent.

4.4.3. Other Responsibilities

4.4.3.1. Making Recommendations

A fundamental responsibility of a DMC is to make recommendations to the sponsor (and/or, as noted in the Introduction, a steering committee or other group delegated by the sponsor to make decisions about the trial) concerning the continuation of the study. Most frequently, a DMC's recommendation after an interim review is for the study to continue as designed. Other recommendations that might be made include study termination, study continuation with major or minor modifications, or temporary suspension of enrollment and/or study intervention until some uncertainty is resolved.

Because a DMC's actions potentially impact the safety of trial participants, it is important that a DMC express its recommendations very clearly to the sponsor. Both a written recommendation and oral communication, with opportunity for questions and discussion, can be valuable. Recommendations for modifications are best accompanied by the minimum amount of data required for the sponsor to make a reasoned decision about the recommendation, and the rationale for such recommendations should be as clear and precise as possible. Sponsors may wish to develop internal procedures to limit the interim data released by a DMC after a recommendation until a decision is made regarding acceptance or rejection of the recommendation, to facilitate maintaining confidentiality of the interim results should the trial continue. We recommend that a DMC document its recommendations, and the rationale for such recommendations, in a form that can be reviewed by the sponsor and then circulated, if and as appropriate, to IRBs, FDA, and/or other interested parties. Sections 5 and 7.2.1 address implications for reporting to FDA of DMC recommendations for major study changes such as early study termination.

4.4.3.2. Maintaining Meeting Records

We recommend that the DMC keep minutes of all meetings (see Guidance for Industry, ICH E6, Good Clinical Practice: Consolidated Guidance, Section 5.5 at 5.5.2, available at http://www.fda.gov/cder/guidance/959fnl.pdf). We also recommend that the DMC divide meeting minutes into two parts, according to whether they include discussion of confidential data (usually unblinded comparative data). The second part of the minutes will typically summarize discussion of the comparative unblinded outcome data and provide the rationale for the recommendations made to the sponsor. Generally, the DMC does not circulate this portion of the minutes or the interim study reports for the closed session outside the DMC membership until the trial is terminated.

We also recommend that after each meeting, the DMC issue a written report to the sponsor based on the meeting minutes. This report does not have to be extremely detailed, but should include sufficient information to explain the rationale for any recommended changes. Sponsors should establish procedures to minimize the potential for bias, such as requiring that reports to the sponsor include only those data generally available to the sponsor (e.g., number screened, number enrolled at each site) (see 21 CFR 314.126(b)(5) (drugs) and 21 CFR 860.7(f)(1) (devices)). If no changes are recommended, the report may be as simple as "The DMC recommends that the study continue as designed." We further recommend that the report to the sponsor include a summary of the discussion in any open session of the meeting and document any information provided orally to the sponsor that was not included in the written report. The sponsor may convey the relevant information in this report to other interested parties such as the study investigators, who should provide any such information, as appropriate, to participating IRBs. Of course, sponsors and/or investigators must report to participating IRBs, as well as to FDA, applicable changes in the protocol or study procedures made as a result of DMC recommendations (see 21 CFR 56.108(a)(3) and (4) and 312.30 and 312.66 for drugs and 21 CFR 812.40 for devices).

We recommend that the DMC or the group preparing the confidential interim reports to the DMC maintain all meeting records in order to best ensure continued confidentiality of interim data. We may request copies of these records when the study is completed (21 CFR 312.58 (drugs); 21 CFR 812.150(b)(10) (devices)). We may also request access to the electronic data sets used for each set of interim analysis. We therefore recommend that sponsors arrange for archiving such electronic data sets.

5. DMC Recommendations and Regulatory Reporting Requirements

All clinical trials conducted under an IND or IDE are subject to regulatory safety reporting requirements. These requirements include prompt reporting to FDA of certain serious and unexpected adverse events (see 21 CFR 312.32(c), 21 CFR 312.52, 21 CFR 812.46(b), 21 CFR 812.150(b)(1)). In general, for an event that is individually recognizable as a serious event potentially related to administration of a medical product (e.g., agranulocytosis, hepatotoxicity for drug studies), the sponsor (sometimes through a CRO managing that aspect of the trial, see 21 CFR 312.52) is responsible for notifying FDA (21 CFR 312.32, 21 CFR 812.150(b)(1)). The sponsor may make this notification with or without unblinding the individual case, as appropriate.

As discussed above in Section 4.4.1.2, evidence of a possible relationship between many serious adverse events and an investigational drug might be detectable only by comparison of rates in the two arms of a controlled trial and not by review of individual cases. For example, in a drug trial carried out in patients with coronary artery disease, in whom heart attacks and strokes would be expected to occur, an increased heart attack or stroke rate would not be recognized except by comparison to the rate in the control group; if such comparison demonstrated an increase in heart attack and stroke rate, it could be presumed that the increase in heart attack and stroke rate was drug-related. Such a finding involving a serious adverse event, conveyed to a sponsor by a DMC with a recommendation to change the trial (e.g., design, informed consent), could represent, on its face, a report of one or more serious unexpected adverse event(s). As required by 21 CFR 312.32(d)(1), the sponsor would need to investigate a DMC's recommendation relating to such events as potentially reportable to FDA under 21 CFR 312.32. If the sponsor concluded that the increased rate of serious unanticipated adverse events was "associated with the use of the drug," the finding, and support for it (which could include the DMC report, any analysis, and pertinent data) would need to be submitted as a serious unexpected adverse experience. These considerations would also apply to unanticipated adverse device effects under 21 CFR 812.50(b)(1).

Findings conveyed to a sponsor by a DMC as part of a recommendation to modify the trial could therefore mean that serious and unexpected events were occurring, and the sponsor would consequently be required to report an analysis of these events to FDA and to all study investigators according to 21 CFR 312.32(c)(1)(B)(ii) (drug trials) and 21 CFR 812.150(b)(1) (device trials). Study investigators are generally responsible for reporting such findings to their IRBs, according to 21 CFR 312.66 (drug trials) and 21 CFR 812.150(a)(1) and 21 CFR 812.40 (device trials), although direct reporting from sponsors to responsible IRBs may be arranged and may be preferable in some situations; for example, when a central IRB has been established. For a device trial, however, the sponsor is responsible for notifying all participating IRBs when an evaluation of an unanticipated adverse event is conducted (21 CFR 812.150(b)(1)).

The requirement to report DMC recommendations related to serious adverse events in an expedited manner in clinical trials of new drugs (21 CFR 312.32(c)) would not apply when the DMC recommendation is related to an excess of events not classifiable as serious. Nevertheless, we recommend that sponsors inform FDA about all recommendations related to the safety of the investigational product whether or not the adverse event in question meets the definition of "serious." Examples might be recommendations to lower the dose of a study agent because of excess toxicity, or to inform current and future trial participants of an emerging safety concern that had not been recognized at the start of the trial.

6. Independence of the DMC

Independence of a DMC depends on the relationships of its members to those sponsoring, organizing, conducting, and regulating the trial. Independence is greatest when members have no involvement in the design and conduct of the trial except through their role on the DMC, and have no financial or other important connections to the sponsor (other than their compensation for serving on the DMC) or other trial organizers that could influence (or be perceived to influence) their objectivity in evaluating trial data.

Independence is defined on a continuum. DMCs are rarely, if ever, entirely independent of the sponsor, as the sponsor generally selects the members, gives the committee its charge, and pays committee members for their expenses and services. Aside from being compensated for their duties as DMC members, however, we recommend that these members generally have no ongoing financial relationship with a trial's commercial sponsor and not be involved in the conduct of the trial in any role other than that of a DMC member.

A critical issue in planning and managing operations of a DMC is resolving the tension that can arise between having a maximally independent DMC and having a DMC that is well informed about the trial: its objectives, its design, and its conduct. To narrowly defining "independence" may result in eliminating from consideration the most knowledgeable researchers, who are likely to have had some past interaction with others sponsoring or performing research in their area of expertise. Additionally, while sponsor involvement in looking at comparative data threatens independence, sponsor representatives, study statisticians, and study investigators may contribute valuable perspectives regarding the trial that may not be available to the committee from more independent sources. With regard to sponsor/investigator involvement with the DMC, this tension is best resolved by permitting interaction with the committee in a carefully defined and limited manner, as described in Section 4.3.1.2. The involvement of such individuals with the DMC will typically be limited in terms of what interim data may be viewed, which sessions may be attended, what topics may be discussed, and what roles (e.g., observer, consultant, member), may be played. Some of the considerations in addressing these issues are discussed below.

6.1. Desirability of an Independent DMC

Independence of the DMC from the sponsor offers the following advantages:

- Independence from the sponsor helps ensure that sponsor interests do not unduly influence the DMC, promoting objectivity that benefits the subjects and the trial.

- Through enhancement of objectivity and reduction of the possibilities for bias, independence of the DMC increases the credibility of the trial's conclusions.

- Independence of the DMC and complete blinding of the sponsor to interim outcome data preserve the ability of the sponsor to make certain modifications to a trial in response to new external information without introducing bias.

- In a commercially sponsored trial, independence of the DMC may shield the sponsor (and thus the trial) from securities issues by maintaining the sponsor in a fully blinded situation.

6.2. Value of Sponsor Interaction with the DMC

A sponsor's decision to establish an independent DMC does not preclude interaction of the sponsor with the DMC. Sponsor involvement in an open part of the DMC meeting, during which data such as enrollment, compliance, and event rates may be viewed in aggregate but not separately by study arm, has significant advantages. The sponsor may provide important information to the DMC regarding the sponsor's goals, plans, and resources that the DMC can later integrate into its deliberation. Further, the review of interim comparative data may raise certain questions that the DMC might want to address to the sponsor. These interactions may improve the quality of the monitoring process and may also provide the sponsor with information relevant to the costs, timetable, and likely interpretability of the study that can be of significant value in planning future studies and/or other aspects of product development. The risk to the study of such sponsor involvement can be quite limited provided that (1) appropriate care is taken to ensure that the sponsor does not see outcome data separately by study arm and (2) the sponsor does not unduly influence the closed deliberations of the committee.

On the other hand, involvement by sponsor representatives and certain investigators in the portion of the DMC meeting where unblinded data are reviewed presents substantial disadvantages, as discussed in Sections 6.1 and 6.3. Even so, such involvement is not entirely without rationale. When a DMC is facing difficult decisions based on interim safety or efficacy data, the sponsor representatives, study statisticians, and study investigators may contribute valuable perspectives that may not be available from more independent sources. For example, such individuals might point out that unanticipated difficulties in collecting certain data may affect their reliability. In addition, such individuals might have detailed knowledge of other relevant information about the drug (or disease) gained from the trial in question or from other studies that could enhance the DMC's ability to monitor the current trial. To the extent such perspectives can be obtained through a combination of having independent DMC members who are very familiar with the drug, the disease, and trial and having sponsor involvement in open session only, some risks (see Section 6.3) will be minimized. When a trial's procedures are such that sponsor representatives or investigators do see unblinded data with the DMC, the DMC may wish to develop its recommendations in an executive session (see Section 4.3.1.2).

6.3. Risks of Sponsor Exposure to Interim Comparative Data

Sponsor exposure to unblinded interim data, through the DMC or otherwise, can present substantial risk to the integrity of the trial. One concern is that unblinding of the sponsor increases the risk of further unblinding, e.g., of participants, potential participants, or investigators, thereby potentially compromising objective safety monitoring, equipoise, recruitment, administration of the intervention, or other aspects of the trial. In some cases, this risk may be limited and manageable. However, even when unblinding is limited to a small group or a single individual within the sponsoring organization who maintains confidentiality of the results, it is possible that an individual with knowledge of interim data may reveal, or be perceived to reveal, information inadvertently, e.g., by facial expression or body language.

An additional problem arising from a sponsor's access to interim data is the diminution of the sponsor's ability to manage the trial without introducing bias. Many trials, particularly those with DMCs, take place over several years. During that time, it is not uncommon for scientific advancements, e.g., development of new tests, approval of new products, announcement of results of other trials, to significantly affect a given trial. Such developments may suggest a need for modifications of the experimental protocol, e.g., allowing certain concomitant treatments, changing endpoints. Non-scientific developments, such as new financial considerations, production problems, enrollment problems, and missing data, may also suggest the need for protocol changes. If the sponsor has had access to interim data, it may be impossible to avoid allowing that knowledge to influence decisions regarding modifications of the trial; it may also be impossible for outside evaluators to assess the impact of that influence. For example, if based on external developments, a sponsor were considering terminating accrual in one subgroup or changing an endpoint, knowledge of current results in that subgroup or with regard to that endpoint would introduce unavoidable, but unmeasurable bias. Thus, the sponsor who knows interim data may well find itself in a position where a protocol change that appears to be in the interest of the trial or even essential for continuing the trial, cannot be made without potentially introducing biases that can be neither quantified nor corrected. This may lead to major difficulties in interpreting the results of statistical comparisons.

In certain situations, exceptions to the strict maintenance of confidentiality of interim data will be warranted. For example, as noted earlier, in trials in which severe toxicity or other severe morbidity is expected and ongoing continual monitoring is required to ensure maximal protection of trial participants, the sponsor may need to be more actively involved in monitoring unblinded safety data despite the risks to confidentiality of the interim results.

6.4. Statisticians Conducting the Interim Analyses

As discussed in Section 2.1, DMCs add administrative complexity to a trial, adding complexity to the statistician's analyses of the trial. Traditionally, the primary trial statistician performs interim analyses and reports to the DMC. "Primary trial statistician" may refer to an individual statistician or statistical group responsible for designing the trial and managing its conduct in collaboration with the study chair and others in the trial leadership. This arrangement can be appealing because the primary trial statistician will be extremely knowledgeable about the study and will be able to provide the most informative interaction with the DMC.

Assigning the primary trial statistician the responsibility for interim analysis and reporting to the DMC can be problematic, however. When issues arise that might suggest changes to the trial design, a statistician performing both the primary collaborative and the interim analysis functions including reporting to the DMC, will probably be the only member of the trial management team with knowledge of the interim data. In considering possible changes, the statistician's objectivity will inevitably be compromised by the knowledge of the potential impact of such changes on the outcome of the trial. When statisticians with knowledge of interim data participate in trial management meetings in which potential changes to study size, entry criteria, or endpoints are discussed, perceptions of biased decision-making could arise. Even if the statisticians remain silent about the interim data, it is essentially impossible for any opinion they may express not to be influenced by knowledge of these data. When the statistician is present for such discussions and knows which of the alternative courses of action is more likely to result in the experimental intervention being shown effective, even unintentional non-verbal communication may reveal (or may be perceived to reveal) some of that knowledge. Furthermore, if the sponsor must make a decision with major financial implications and a statistician in the sponsor's employ possesses information critical to that decision, both may be placed in a very uncomfortable position in which the risk is high of verbal or non-verbal transmission of information regarding interim data.

For this reason, when there is a DMC and a formal mechanism for interim analyses, it is advantageous for the statistician performing the interim analysis to be uninvolved in managing the conduct of the trial, especially in regard to making decisions about design modifications. This could mean that a statistician other than the primary trial statistician would take responsibility for the interim analysis and reporting to the DMC.

Alternatively, the primary statistician could maintain this role, but forego further involvement in trial management once interim analysis was undertaken. Sponsors may identify and develop other approaches to reduce potential inappropriate influence of interim data on trial management.

Another important issue relating to the role of the statistician arises when the statistician is employed by the sponsor. Elsewhere in this guidance, we have described the concerns associated with sponsors being aware of the interim comparative data. For purposes of quality assurance, sponsors often wish to maintain control of the data and have their own statisticians perform the analyses, including the unblinded analyses for the DMC.

Typically and appropriately, such statisticians are instructed not to disclose interim data to others within the sponsoring organization. Questions can always arise, however, as to whether the statisticians are adequately separated from others within the sponsoring organization involved in managing the trial.

For these reasons, the integrity of the trial may be best protected when the statisticians preparing unblinded data for the DMC are external to the sponsor and uninvolved in discussions regarding potential changes in trial design while the trial is ongoing. This is an especially important consideration for critical studies intended to provide definitive evidence of effectiveness. Balanced against this concern, however, is the need for the statisticians reporting to the DMC to be very familiar with details of the study and have ample opportunity to assess the interim data. The primary trial statistician, whether or not employed by the sponsor, is usually best situated to play this role. We recognize that an external statistician contracted by the sponsor to perform interim analysis may not be entirely free from the types of pressures and concerns that may affect a statistician employed by the sponsor.

There has been substantial experience with the model in which the primary trial statistician also analyzes interim data and reports to the DMC. This arrangement has worked well, for the most part. In the admittedly infrequent situation in which interim protocol changes need to be considered, however, participation of statisticians with knowledge of interim data could complicate the interpretability of the data. Sponsors may wish to take the above considerations into account in establishing procedures for the operation of the DMC and the process of interim monitoring.

If the statistician reporting unblinded data to the DMC is not the primary trial statistician, it is particularly important that efforts be made to ensure both that the unblinded statistician is very familiar with the design, setting, and objectives of the trial, and has sufficient time and access to the data to provide insightful analyses responsive to the DMC's needs.

If the primary trial statistician takes on the responsibility for interim analysis and reporting to the DMC, we recommend that this statistician have no further responsibility for the management of the trial once interim analysis begins and have minimal contact with those who have such involvement. In this case, we recommend that sponsors establish and document procedures to ensure this separation, and designate a different statistician to advise on the management of the trial.

If the primary trial statistician takes on the responsibility for interim analysis and reporting to the DMC, and it appears infeasible or highly impractical for any other statistician to take over responsibilities related to trial management, we recommend that sponsors consider, develop and document procedures to minimize the risks of bias that are associated with such arrangements, as described above.

If the statistician responsible for interim analysis and reporting to the DMC is employed by the sponsor, special care should be taken to minimize the potential for bias, such as ensuring that confidential interim data are not revealed to anyone else within the sponsor organization (see 21 CFR 314.126(b)(5) (drugs) and 21 CFR 860.7(f)(1) (devices)).

While the sponsor or the DMC may suggest to the statistician the nature of the analyses and tables they wish reported to the DMC, we recommend the statistician also have the familiarity with the study and data access necessary to perform additional analyses that might be suggested by the accumulating data and/or requested by the DMC.

6.5. Sponsor Access to Interim Data for Planning Purposes

Often, sponsors wish to have access to unblinded interim data for the purpose of planning product development, e.g., designing/initiating further trials or making decisions regarding production facilities. This interest is understandable, but such access is problematic for reasons already discussed. In general, sponsors are advised to avoid seeking information about unblinded interim data because of the significant possibility that they may wind up impairing trial management or even making the trial results uninterpretable by doing so. Further, plans or decisions based on statistically imprecise interim data may often be suboptimal. Where the sponsor nonetheless has a compelling need to review such information, certain approaches may lessen, although they do not eliminate, risks to the trial:

- Discussion of such an action with FDA in advance. This is particularly advisable when the sponsor intends to use the study in support of a licensing or marketing application.

- Development of appropriate stopping rules and apportionment of type I error (α) before performing any unblinded interim analysis. This is important because any viewing of study arm-specific effectiveness data by the DMC and/or sponsor in a study of a serious illness raises the possibility that an unanticipated extreme finding of effectiveness might create an ethical imperative to stop the trial, and it would not be possible to quantitate the level of evidence provided by the data if the monitoring plan had not been established prior to data review.

- Determination of the minimum amount of information needed. For example, to assist in defining eligibility criteria for a subsequent trial, the sponsor may wish to know only whether estimates of treatment effect in a subgroup are less or greater than in the overall data set.

- Formulation of written questions, preferably with yes/no rather than numerical answers, that will elicit only that minimal required information and nothing more.

- Receiving only written information regarding the requested data (thereby documenting what was received and avoiding additional unnecessary communications) and abstaining from participation in closed DMC meetings or discussions of data with unblinded DMC members (except as otherwise requested by the DMC).

- Identification of those sponsor employees with a critical "need-to-know" and restriction of such information to those individuals only.

- Ensuring that individuals with access to the information avoid any subsequent role in the management of the trial and minimize interactions with others in that role.

- Ensuring that individuals who have access to such information make every effort to avoid taking actions that will assist others in inferring what the information is.

- Ensuring that reports of study findings describe any access to interim data by individuals involved with study management, and steps taken to prevent such access from potentially biasing the study results.

7. Sponsor Interaction with FDA Regarding Use and Operation of DMCs

There are many situations, several mentioned earlier, in which sponsor consultation with FDA on matters regarding a DMC is advisable.

7.1. Planning the DMC

In planning a clinical trial, a sponsor makes several decisions regarding use, types of membership, and operations of a DMC. Many of these can be critical to the success of the trial in meeting regulatory requirements. This guidance document is intended to provide general FDA guidance regarding those decisions, but each set of circumstances can raise unique considerations. Issues regarding use of DMCs are appropriate topics for FDA-sponsor meetings (in person or by telephone) at the sponsor's request.

7.2. Accessing Interim Data

As discussed above, accessing interim data by the sponsor carries many risks, not all of which may be fully appreciated by the sponsor. We recommend that sponsors contact FDA before initiating communication with the DMC regarding access to interim data from a trial likely to be an important part of a regulatory submission. While FDA permission is not required, a discussion regarding the potential risks and implications of that action and of methods to limit the risks may contribute to informed decision making.

7.2.1. DMC Recommendations to Terminate the Study

In almost all cases, a DMC is advisory to the sponsor; the sponsor decides whether to accept recommendations to discontinue a trial. FDA will rarely, if ever, tell a sponsor which decision to make. For trials that may be terminated early because a substantial benefit has been observed, however, consideration may still need to be given to the adequacy of data with regard to other issues such as safety, duration of benefit, outcomes in important subgroups and important secondary endpoints. We recommend that sponsors of trials that could potentially be terminated early for efficacy reasons discuss these issues with FDA prior to implementing the trial, when the statistical monitoring plan and early stopping boundaries are being developed. In these settings, consultation with FDA may provide the sponsor with important information regarding the regulatory and scientific implications of a decision and may lead to better decisions. Sponsors are encouraged to revisit these issues with FDA when considering DMC recommendations for early termination if new issues have arisen and/or if the regulatory implications of early termination were not adequately clarified at the outset of the trial.

For trials that may be terminated because of safety concerns, timely communication with FDA is often required (see, e.g., 21 CFR 312.56(d) (drugs); 21 CFR 812.150 (devices)). In such cases, we recommend that the sponsor initiate discussion as soon as possible about the appropriate course of action, for the trial in question as well as any other use of the investigational product.

We strongly recommend that sponsors initiate discussion with FDA prior to early termination of any trial implemented specifically to investigate a potential safety concern.

7.2.2. FDA Interaction with DMCs

In rare cases, we may wish to interact with a DMC of an ongoing trial to ensure that specific issues of urgent concern to FDA are fully considered by the DMC or to address questions to the DMC regarding the consistency of the safety data in the ongoing trial to that in the earlier trials, to optimize regulatory decisionmaking.

An example might be a situation in which FDA is considering a marketing application in which a safety issue is of some concern, and the sponsor has a second trial of the investigational agent ongoing. In such a situation, we might wish to be sure that the DMC for the ongoing trial is aware of the existing safety data contained in the application and is taking those data into consideration in evaluating the interim safety data from the ongoing trial. In such a case, we could request that the sponsor arrange for FDA to communicate with, or even meet with, the DMC (see 21 CFR 312.41(a); 21 CFR 812.150(b)(10)), and care should be taken to minimize the possibility of jeopardizing the integrity of the ongoing trial.

7.3. DMC Recommendations for Protocol Changes

A DMC may, in some instances, recommend changes to the study protocol, particularly in the context of their responsibilities for monitoring patient safety. Many protocol changes have little impact on the usefulness of a trial to gain regulatory approval. Certain types of changes to the protocol, however, such as changes in the primary endpoints, could have substantial impact on the validity of the trial and/or its ability to support the desired regulatory decision if they potentially could have been motivated by the interim data. We recommend that sponsors discuss proposed changes of the latter type with FDA before implementation.

8. Paperwork Reduction Act of 1995

This guidance contains information collection provisions that are subject to review by the Office of Management and Budget (OMB) under the Paperwork Reduction Act of 1995 (44 U.S.C.

3501-3520).

The time required to complete this information collection is estimated to average 11.75 hours per response, including the time to review instructions, search existing data resources, gather the data needed, and complete and review the information collection. Send comments regarding this burden estimate or suggestions for reducing this burden to:

Food and Drug Administration
Center for Biologics Evaluation and Research (HFM-99)
1401 Rockville Pike, Suite 200N
Rockville, MD 20852-1448

This guidance also refers to previously approved collections of information fund in FDA regulations. The collections of information in §§ 312.30, 312.32, 312.38, 312.55, and 312.56 have been approved under OMB Control No. 0910-0014; the collections of information in § 314.50 have been approved under OMB Control No. 0910-0001; and the collections of information in §§ 812.35 and 812.150 have been approved under OMB Control No. 0190-0078.

Note: *An agency may not conduct or sponsor, and a person is not required to respond to, a collection of information unless it displays a currently valid OMB Control No. The OMB Control No. for this information collection is 0910-0581 (Expires 6/30/2012).*

Data Retention When Subjects Withdraw from FDA-Regulated Clinical Trials

Data Retention When Subjects Withdraw from FDA-Regulated Clinical Trials

Guidance for Institutional Review Boards, Clinical Investigators, and Sponsors[1, 2]

U.S. Department of Health and Human Services
Food and Drug Administration
Office of the Commissioner (OC)
Good Clinical Practice Program (GCPP)

September 2008

Contains Nonbinding Recommendations

This guidance describes FDA's longstanding policy that already-accrued data, relating to individual who cease participating in a study, are to be maintained as part of the study data. This pertains to data from individuals who decide to discontinue participation in a study, who are withdrawn by their legally authorized representative, as applicable, or who are discontinued from participation by the clinical investigator.

I. Introduction

This guidance is intended for sponsors, clinical investigators and institutional review boards (IRBs). It describes the Food and Drug Administration's (FDA) longstanding policy that already-accrued data, relating to individuals who cease participating in a study, are to be maintained as part of the study data. This pertains to data from individuals who decide to discontinue participation in a study, who are withdrawn by their legally authorized representative, as applicable, or who are discontinued from participation by the clinical investigator. This policy is supported by the statutes and regulations administered by FDA as well as ethical and quality standards applicable to clinical research. Maintenance of these records includes, as with all study records, safeguarding the privacy and confidentiality of the subject's information.

---

[1] Available on the FDA website at: http://www.fda.gov/downloads/
RegulatoryInformation/Guidances/UCM126489.pdf

[2] This guidance document was developed by the Good Clinical Practice Program and the Office of the Chief Counsel (OCC), both in the Office of the Commissioner (OC), FDA.

FDA's guidance documents, including this guidance, do not establish legally enforceable responsibilities. Instead, guidances describe the Agency's current thinking on a topic and should be viewed only as recommendations, unless specific regulatory or statutory requirements are cited. The use of the word should in Agency guidances means that something is suggested or recommended, but not required.

II. Background

The Federal Food, Drug, and Cosmetic Act (the act) authorizes the study of an investigational product to develop safety and effectiveness data about the product. The act also requires the maintenance of records documenting these data and the submission of certain reports regarding this use to FDA.[3, 4]FDA (by delegation from the Secretary) implemented these provisions by issuing regulations relating to investigational drugs, the Investigational New Drug (IND) regulations at 21 CFR Part 312, and investigational devices, the Investigational Device Exemptions (IDE) regulations at 21 CFR Part 812. These regulations specify that data collection and maintenance are indispensable requirements when conducting a clinical investigation of an unapproved product.

For example, the IND regulations require investigators "… to prepare and maintain adequate and accurate case histories that record all observations and other data pertinent to the investigation on each individual administered the investigational drug or employed as a control in the investigation" (21 CFR 312.62(b)). Similarly, the IDE regulations require an investigator to maintain "Records of each subject's case history and exposure to the device" (21 CFR 812.140(a)(3)).

Additionally, the regulations relating to the submission of marketing applications require the submission of all relevant data in order for FDA to determine whether a product meets the standard for approval. A new drug application (NDA) must include a description and analysis of each clinical pharmacology study and controlled clinical study, a description of each uncontrolled trial, and an integrated summary of all available information about the safety of the drug product (21 CFR 314.50(d)(5)) and "copies of individual case report forms for each patient who died during a clinical study or who did not complete the study because of an adverse event, whether believed to be drug related or not, including patients receiving reference drugs or placebo" (21 CFR 314.50(f)(2)). Similarly, an application for premarket approval (PMA) for a device must include "safety and effectiveness data, adverse reactions and

---

[3] The investigational new drug provisions of the act condition use of such drugs upon, for example, "the establishment and maintenance of such records, and the making of such reports to the Secretary, by the manufacturer or the sponsor of the investigation of such drug, of data (including but not limited to analytical reports by investigators) obtained as the result of such investigational use of such drug, as the Secretary finds will enable him to evaluate the safety and effectiveness of such drug in the event of the filing of an application pursuant to subsection (b)." 21 USC 355(i)(1)(C).

[4] The investigational device provisions similarly require "that the person applying for an exemption for a device assure the establishment and maintenance of such records, and the making of such reports to the Secretary of data obtained as a result of the investigational use of the device during the exemption, as the Secretary determines will enable him to assure compliance with such conditions, review the progress of the investigation, and evaluate the safety and effectiveness of the device." 21 USC 360j(g)(2)(B)(ii).

complications, patient discontinuation, patient complaints, device failures and replacements, tabulations of data from all individual subject report forms and copies of such forms for each subject who died during a clinical investigation or who did not complete the investigation" (21 CFR 814.20(b)(6)(ii)). Likewise , an application for a biologics license application (BLA) must include data derived from nonclinical laboratory and clinical studies which demonstrate that the manufactured product is safe, pure, and potent (21 CFR 601.2(a)).

FDA law and regulations require the collection and maintenance of complete clinical study data. This includes information on subjects who withdraw from a clinical investigation, whether the subject decides to discontinue participation in the clinical trial (21 CFR 50.25(a)(8)) or is discontinued by the investigator because the subject no longer qualifies under the protocol (for example, due to a significant adverse event or due to failure to cooperate with study requirements). FDA recognizes that a subject may withdraw from a study; however, the withdrawal does not extend to the data already obtained during the time the subject was enrolled. FDA's longstanding policy has been that all data collected up to the point of withdrawal must be maintained in the database and included in subsequent analyses, as appropriate.[5]

III. Discussion

FDA law and regulations recognize that a complete and accurate risk/benefit profile of an investigational product depends upon the data from every subject's experience in the clinical trial. For example, if a subject's data could be withdrawn from a study, a sponsor would not have access to data on adverse events experienced by the subject and would be unable to evaluate whether changes to the protocol or the informed consent documents are needed to ensure the rights, safety, and welfare of other trial subjects.[6]

The validity of a clinical study would also be compromised by the exclusion of data collected during the study. There is long-standing concern with the removal of data, particularly when

---

[5] FDA previously addressed the topic of data withdrawal in the preamble to the 1996 final rule providing an exception from informed consent requirements for emergency research, 21 CFR 50.24. In response to a comment that a subject's legally authorized representative should be allowed to prevent the review of the subject's data, FDA stated: "FDA regulations (see, for example, Sec. 312.62 and Sec. 812.140(a)(3)) require investigators to prepare and maintain adequate case histories recording all observations and other data pertinent to the investigation on each individual treated with the drug or exposed to the device. The agency needs all such data in order to be able to determine the safety and effectiveness of the drug or device. The fact of having been in an investigation cannot be taken back. Also, if a subject were able to control the use (inclusion and exclusion) of his or her data, and particularly if the clinical investigation were not blinded, the bias potential would be immense. Thus, the agency rejects this comment because it could prevent FDA from learning of an important effect of the product and significantly bias the results of the investigation" ('see comment 95, 61 Federal Register 51498, 51519, October 2, 1996). It should be appreciated that FDA's response applies to the most potentially difficult situation, that is, studies involving an exception from the informed consent requirements in which subjects, due to a life threatening medical condition, are unable to provide informed consent to participate in the study. Subjects may subsequently withdraw from such studies, but the data collected up to withdrawal may not be removed.

[6] Such review of safety data by sponsors is required by 21 CFR 312.56 and 21 CFR 812.46.

---

removal is non-random, a situation called "informative censoring." FDA has long advised "intent-to-treat" analyses (analyzing data related to all subjects the investigator intended to treat), and a variety of approaches for interpretation and imputation of missing data have been developed to maintain study validity.[7] Complete removal of data, possibly in a non-random or informative way, raises great concerns about the validity of the study.

There is particular concern with a study's reliability when subjects withdraw their data in a non-random way because they are unhappy with their experience, either because they failed to obtain a desired effect or suffered an adverse event. Loss of these subjects' data could greatly distort effectiveness results and could hide important safety information (for example, toxicity) of a poorly tolerated treatment. Allowing subjects to withdraw data could even provide an opportunity for unscrupulous parties to "improve" study results by selectively encouraging certain subjects to withdraw from a study.

The importance of ensuring the scientific validity of clinical research is reflected not only in FDA's regulations but in international documents and published literature as well. The International Conference on Harmonization of Technical Requirements for Registration of Pharmaceuticals for Human Use (ICH), in which FDA participates, identifies as a principle of good clinical practice that "clinical trials should be scientifically sound."[8] Other international guidance documents include similar statements, such as the International Ethical Guidelines for Biomedical Research Involving Human Subjects, Guideline 1, which states "scientifically invalid research is unethical in that it exposes research subjects to risks without possible benefit …"[9]

Published literature on medical research ethics, dating back to the Nuremburg Code of 1947,[10] also emphasizes the importance of scientific validity. As maintained by Emanuel, et. al., "For a clinical research protocol to be ethical, the methods must be valid and practically feasible: the research must have a clear scientific objective; be designed using accepted principles, methods, and reliable practices; have sufficient power to definitively test the objective; and offer a plausible data analysis plan."[11] The importance of scientific validity to ethical research is also underscored in modern ethical documents, such as the Declaration of Helsinki[12] and the

---

[7] For a discussion of problems presented by missing data in the analysis of clinical trials, please see "Points to Consider on Missing Data" from the Committee for Proprietary Medicinal Products of the European Medicines Agency (EMEA), http://www.emea.europa.eu/pdfs/human/ewp/177699EN.pdf.

[8] http://www.fda.gov/cder/guidance/959fnl.pdf. ICH E6 Guidance for Industry, "Good Clinical Practice: Consolidated Guidance," adopted as official guidance by FDA, Section 2.5.

[9] http://www.cioms.ch/frame_guidelines_nov_2002.htm. These guidelines are published by the Counsel of International Organizations for Medical Sciences (CIOMS). Guideline 11 reiterates this principle.

[10] The Nuremberg Code: http://ohsr.od.nih.gov/guidelines/nuremberg.html

[11] Emanuel, EJ, Wendler, D, and Grady, C., "What Makes Clinical Research Ethical?" JAMA 283:20 (May 24/31, 2000 2701-11).

[12] "Medical research involving human subjects must conform to generally accepted scientific principles, be based on a thorough knowledge of the scientific literature, other relevant sources of information, and on adequate laboratory and, where appropriate, animal experimentation." The Declaration of Helsinki (2000) (as amended 2002, 2004) http://www.wma.net/e/policy/b3.htm.

Belmont Report, issued in 1979 by the National Commission for the Protection of Human Subjects of Biomedical and Behavioral Research.[13]

In summary, data collected on study subjects up to the time of withdrawal must remain in the trial database in order for the study to be scientifically valid. If a subject withdraws from a study, removal of already collected data would undermine the scientific, and therefore the ethical, integrity of the research. Such removal of data could also put enrolled subjects, future subjects, and eventual users of marketed products at an unreasonable risk. Finally, removal of data would fundamentally compromise FDA's ability to perform its mission, to protect public health and safety by ensuring the safety and effectiveness of regulated products.

IV. FDA Policy

Following are key points regarding FDA's policy on the withdrawal of subjects from a clinical investigation, whether the subject elects to discontinue further interventions or the clinical investigator terminates the subject's participation in further interventions:

- According to FDA regulations, when a subject withdraws from a study, the data collected on the subject to the point of withdrawal remains part of the study database and may not be removed.

- An investigator may ask a subject who is withdrawing whether the subject wishes to provide continued follow-up and further data collection subsequent to their withdrawal from the interventional portion of the study. Under this circumstance, the discussion with the subject would distinguish between study-related interventions and continued follow-up of associated clinical outcome information, such as medical course or laboratory results obtained through non-invasive chart review, and address the maintenance of privacy and confidentiality of the subject's information.

- If a subject withdraws from the interventional portion of the study, but agrees to continued follow-up of associated clinical outcome information as described in the previous bullet, the investigator must obtain the subject's informed consent for this limited participation in the study (assuming such a situation was not described in the original informed consent form). In accordance with FDA regulations, IRB approval of informed consent documents would be required (21 CFR 50.25, 56.109(b), 312.60, 312.66, 812.100).

- If a subject withdraws from the interventional portion of a study and does not consent to continued follow-up of associated clinical outcome information, the investigator must not access for purposes related to the study the subject's medical record or other confidential records requiring the subject's consent. However, an investigator may review study data

---

[13] The Belmont Report addresses the connection between scientific validity and ethics through the Principle of Beneficence. Beneficence has two complementary aspects: maximizing possible benefits and minimizing possible harms. The Report recognizes as one of the components of maximizing benefits, "In the case of scientific research in general, members of the larger society are obliged to recognize the longer term benefits and risks that may result from the improvement of knowledge and from the development of novel medical, psychotherapeutic, and social procedures." http://ohsr.od.nih.gov/ guidelines/belmont.html

related to the subject collected prior to the subject's withdrawal from the study, and may consult public records, such as those establishing survival status.

Financial Disclosure by Clinical Investigators

Financial Disclosure by Clinical Investigators, Guidance for Industry[1, 2]

U.S. Department of Health and Human Services
Food and Drug Administration
Center for Drug Evaluation and Research
Center for Biologics Evaluation and Research
Center for Devices and Radiological Health

March 20, 2001

Contains Nonbinding Recommendations

This guidance addresses questions received by FDA concerning the implementation of the final rule on financial disclosure (21 CFR 54). The financial disclosure regulations were intended to ensure that financial interests and arrangements of clinical investigators that could affect the reliability of data submitted to FDA were identified and disclosed to FDA by the applicant.

Note: *This guidance represents the Food and Drug Administration's (FDA's) current thinking on this topic. It does not create or confer any rights for or on any person and does not operate to bind FDA or the public. You can use an alternative approach if the approach satisfies the requirements of the applicable statutes and regulations. If you want to discuss an alternative approach, contact the appropriate FDA staff. If you cannot identify the appropriate FDA staff, call the appropriate number listed on the title page of this guidance.*

I. Introduction

On February 2, 1998, FDA published a final rule requiring anyone who submits a marketing application of any drug, biological product or device to submit certain information concerning the compensation to, and financial interests of, any clinical investigator conducting clinical studies covered by the rule. This requirement, which became effective on February 2, 1999, applies to any clinical study submitted in a marketing application that the applicant or FDA relies on to establish that the product is effective, and any study in which a single investigator makes a significant contribution to the demonstration of safety. This final rule requires applicants to certify to the absence of certain financial interests of clinical investigators or to

---

[1] Available on the FDA website at: http://www.fda.gov/downloads/
RegulatoryInformation/Guidances/ucm126832.pdf
[2] This guidance has been prepared by the Implementation Team for Financial Disclosure comprised of individuals in the Office of the Commissioner, the Center for Drug Evaluation and Research (CDER), Center Biologics Evaluation and Research (CBER) and Center Devices and Radiological Health (CDRH) at the Food and Drug Administration.

disclose those financial interests. If the applicant does not include certification and/or disclosure, or does not certify that it was not possible to obtain the information, the agency may refuse to file the application. On December 31, 1998, FDA published an amended final rule that reduced the need to gather certain financial information for studies completed before February 2, 1999. On October 26, 1999, FDA published a draft guidance to provide clarification in interpreting and complying with these regulations. The burden hours required for Section 21 CFR Part 54 are reported and approved under OMB Control Number 0910 0396.

II. Financial Disclosure Requirements

Under the applicable regulations (21 CFR Parts 54, 312, 314, 320, 330, 601, 807, 812, 814, and 860), an applicant is required to submit to FDA a list of clinical investigators who conducted covered clinical studies and certify and/or disclose certain financial arrangements as follows:

1. Certification that no financial arrangements with an investigator have been made where study outcome could affect compensation; that the investigator has no proprietary interest in the tested product; that the investigator does not have a significant equity interest in the sponsor of the covered study; and that the investigator has not received significant payments of other sorts; and/or

2. Disclosure of specified financial arrangements and any steps taken to minimize the potential for bias.

Disclosable Financial Arrangements:

A. Compensation made to the investigator in which the value of compensation could be affected by study outcome. This requirement applies to all covered studies, whether ongoing or completed as of February 2, 1999.

B. A proprietary interest in the tested product, including, but not limited to, a patent, trademark, copyright or licensing agreement. This requirement applies to all covered studies, whether ongoing or completed as of February 2, 1999.

C. Any equity interest in the sponsor of a covered study, i.e., any ownership interest, stock options, or other financial interest whose value cannot be readily determined through reference to public prices. This requirement applies to all covered studies, whether ongoing or completed;

D. Any equity interest in a publicly held company that exceeds $50,000 in value. These must be disclosed only for covered clinical studies that are ongoing on or after February 2, 1999. The requirement applies to interests held during the time the clinical investigator is carrying out the study and for 1 year following completion of the study; and

E. Significant payments of other sorts, which are payments that have a cumulative monetary value of $25,000 or more made by the sponsor of a covered study to the investigator or the investigators' institution to support activities of the investigator exclusive of the costs of conducting the clinical study or other clinical studies, (e.g., a grant to fund ongoing research, compensation in the form of equipment or retainers for ongoing consultation or honoraria) during the time the clinical investigator is carrying out the study and for 1 year following completion of the study. This requirement applies to payments made on or after February 2, 1999.

Agency Actions

If FDA determines that the financial interests of any clinical investigator raise a serious question about the integrity of the data, FDA will take any action it deems necessary to ensure the reliability of the data including:

Initiating agency audits of the data derived from the clinical investigator in question;

Requesting that the applicant submit further analyses of data, e.g., to evaluate the effect of the clinical investigator's data on the overall study outcome;

Requesting that the applicant conduct additional independent studies to confirm the results of the questioned study; and

Refusing to treat the covered clinical study as providing data that can be the basis for an agency action.

Definitions

Clinical Investigator - means any listed or identified investigator or subinvestigator who is directly involved in the treatment or evaluation of research subjects. The term also includes the spouse and each dependent child of the investigator.

Covered clinical study - means any study of a drug, biological product or device in humans submitted in a marketing application or reclassification petition that the applicant or FDA relies on to establish that the product is effective (including studies that show equivalence to an effective product) or any study in which a single investigator makes a significant contribution to the demonstration of safety. This would, in general, not include phase 1 tolerance studies or pharmacokinetic studies, most clinical pharmacology studies (unless they are critical to an efficacy determination), large open safety studies conducted at multiple sites, treatment protocols and parallel track protocols. An applicant may consult with FDA as to which clinical studies constitute "covered clinical studies" for purposes of complying with financial disclosure requirements.

Applicant - means the party who submits a marketing application to FDA for approval of a drug, device or biologic product or who submits a reclassification petition. The applicant is responsible for submitting the required certification and disclosure statements.

Sponsor of the covered clinical study - means the party providing support for a particular study at the time it was carried out.

III. Purpose

The financial disclosure regulations were intended to ensure that financial interests and arrangements of clinical investigators that could affect the reliability of data submitted to FDA are identified and disclosed by the applicant. FDA has received many questions concerning the implementation of this final rule. The agency is issuing this guidance to respond to these questions. FDA encourages applicants and sponsors to contact the agency for advice concerning specific circumstances that may raise concerns as early in the product development process as possible.

IV. Questions and Answers

1.Q. Why did FDA develop financial disclosure regulations?

A: In June 1991, the Inspector General of the Department of Health and Human Services submitted a management advisory report to FDA stating that FDA's failure to have a mechanism for collecting information on "financial conflicts of interest" of clinical investigators who study products that undergo FDA review could constitute a material weakness under the Federal Manager's Financial Integrity Act. Although FDA determined that a material weakness did not exist, the agency did conclude that there was a need to address this issue through rulemaking. During the rulemaking process, FDA also learned about potentially problematic financial arrangements through published newspaper articles, Congressional inquiries, public testimony, and comments. Based on the information gathered, FDA determined that it was appropriate to require the submission of certain financial information with marketing applications that include certain types of clinical data.

2.Q: Are applicants required to use FDA forms 3454 and 3455 in reporting this information?

A: Yes. The regulations require that the applicant submit one completed Form 3454 for all clinical investigators certifying to the absence of financial interests and arrangements. The applicant may append a list of investigator names to Form 3454 for those investigators certifying that those investigators hold none of the identifiable disclosable financial arrangements. For any clinical investigator for whom the applicant does not submit the certification, the applicant must submit a completed Form 3455 disclosing the financial interests and arrangements and steps taken to minimize the potential for bias.

Where an applicant cannot provide a blanket certification for all investigators because of the existence of disclosable financial arrangements for one or more investigators, an applicant should complete a disclosure form 3455 for each investigator having disclosable financial arrangements. The applicant should identify the specific covered clinical study (or studies) at issue and provide detailed information about the specific relationship that is being disclosed, (e.g., the nature of the contingent payment or the equity holdings of the investigator or the investigator's spouse or dependent child that exceeded the threshold). This disclosure needs to be linked to the specific covered clinical study (or studies) in which the investigators participated.

In those instances where the applicant cannot provide complete certification or disclosure on all arrangements (e.g., the applicant has information about 2 of the 4 requirements but has been unable to obtain the rest of the information), the applicant should certify that despite the applicant's due diligence in attempting to obtain the information, the applicant was unable to obtain the information and should include an explanation of how the applicant attempted to obtain the information and why the information was not obtainable.

3.Q: What does FDA mean by the term "due diligence"?

A: "Due diligence" is a measure of activity expected from a reasonable and prudent person under a particular circumstance. In complying with these rules, sponsors and applicants should use reasonable judgement in deciding how much effort needs to be expended to collect this information. If sponsors/applicants find it impossible to obtain the financial information in question, applicants should explain why this information was not obtainable and document attempts made in an effort to collect the information.

4.Q:Who, specifically, is responsible for signing the financial certification/disclosure forms?

A: The forms are to be signed and dated by a responsible corporate official or representative of the applicant (e.g., the chief financial officer)

5.Q. Where in a drug/biologic application should an applicant include certification that financial arrangements of concern do not exist or the disclosure of those arrangements that do exist? Where should the information be included in a device application?

For drugs and biological product applications, applicants should include the financial certification/disclosure forms as part of item 19 (other) of the application. See form 356h. FDA is revising the current form 356h and upon completion of this revision, financial certification and disclosure information will become number 19 and (other) will become number 20. For device applications, applicants should submit the financial certification/disclosure forms according to the format outlined in the appropriate submission checklist.

6.Q: What obligations do IND and IDE sponsors have regarding information collection prior to study start?

A: The regulations, 21 CFR 312.53 and 21 CFR 812.43, provide that before permitting an investigator to begin participation in an investigation, the IND/IDE sponsor shall obtain sufficient accurate financial information that will allow an applicant to submit complete and accurate certification or disclosure statements required under Part 54. The sponsor is also required to obtain the investigator's commitment to promptly update this information if any relevant changes occur during the course of the investigation and for one year following completion of the study. By collecting the information prior to study start, the sponsor will be aware of any potential problems, can consult with the agency early on, and take steps to minimize the potential for bias. See question and answer 7 for additional information.

7.Q: What is the responsibility of the IND/IDE sponsor for obtaining financial information from investigators at the IND/IDE stage when the IND/IDE sponsor is not the party who will be submitting a marketing application?

The term "sponsor" has somewhat different meanings in the regulations at 312.53/812.43 and 54.2. An applicant must report financial interests in the sponsor of the covered study. Under 21 CFR 54.2, "sponsor" is defined as the party "supporting a particular study at the time it was carried out." FDA interprets support to include those who provide "material support",e.g., monetary support or test product under study. The sponsor of an IND or IDE, as defined in 21 CFR 312 and 812 is the "party or parties who take responsibility for and initiate a clinical investigation". The term "sponsor" is also used in 312.53 and 812.43 to refer to the party who will be submitting a marketing application (who is also responsible for submitting the certification and disclosure statement required by Part 54).

In most cases, the IND/IDE sponsor, the sponsor of the covered study, and the applicant company are the same party, but there are times where they may be different. For example, when an academic or government institution or CRO conducts a covered study and is the IND/IDE sponsor (Part 312/812 sponsor), a drug or device company that provides funding or the test article used in the study is a Part 54 sponsor, and is likely to be the applicant if a marketing application is submitted to FDA. If the drug or device company that was a sponsor of the covered study sold the drug/device to another company, the applicant could be neither the IND/IDE sponsor nor a Part 54 sponsor.

The responsibility for reporting financial information to FDA falls upon the applicant; that is, the final rule (Part 54) requires that the applicant company submit financial information on clinical investigators at the time the marketing application is submitted to the agency. The information that the applicant must report, apart from compensation that may be affected by study outcome and proprietary interests is:

1. equity interests in a Part 54 sponsor of a covered study (e.g., any interest that cannot be valued through reference to public prices and interest in excess of $50,000 in a publicly held company), and

2. significant payments of other sorts by a Part 54 sponsor of a covered study.

Although reporting to the FDA is the responsibility of the applicant, the IND/IDE holder (part 312/812 sponsor) is required to collect the financial information before permitting an investigator to participate in a clinical study (312.53 and 812.43). The purpose of this requirement is two fold:

1. to alert the IND/IDE sponsor of the study to any potentially problematic financial interest as early in the drug development process as possible in order to minimize the potential for study bias and

2. to facilitate accurate collection of data that may be submitted many years later.

The IND/IDE sponsor, who is in contact with the investigator, is best placed to inquire as to the financial arrangements of investigators, and this obligation applies to any IND/IDE sponsor (e.g., commercial, government or CRO). The IND/IDE sponsor shall maintain complete and accurate records showing any financial interest as described in Section 54.4 (a) (3) (i-iv) in a sponsor of the covered study. The IND/IDE sponsor is responsible for ensuring that required financial information is collected and is made available to the applicant company, so that, the information can be included in the NDA/BLA/PMA submission.

8.Q. The applicant is obligated to disclose financial interests related only to >covered studies, specifically those relied upon to provide support for the effectiveness of a product and certain others. An IND holder (IND sponsor), acting much earlier, must inquire into investigator financial interests before the ultimate role of a study in the application is determined. How will the IND sponsor determine which studies will ultimately require certification/disclosure statements?

A: The IND sponsor will need to consider the potential role of a particular study based on study size, design and other considerations. Almost any controlled effectiveness study could, depending on outcome, become part of a marketing application, but other studies might be critical too, such as a pharmacodynamic study in a population subset or a bioequivalence study supporting a new dosage form.

9.Q. If a Contract Research Organization (CRO) is conducting a covered clinical study on behalf of another company, should the CRO collect the financial information from investigators? Is it necessary to collect financial information from investigators who have financial interests in CROs?

A: With regard to CRO and commercial sponsor arrangements, the same principles as articulated in answer 6 would apply. For example, if a CRO meets the definition of an IND/IDE sponsor or has contracted to collect financial information from investigators on behalf of an IND/IDE sponsor, the CRO must collect financial information on clinical investigators' interests in Part 54 sponsors (312.53, 812.43). If the CRO provides material support for a covered study, financial information on clinical investigators' interests in the CRO is to be collected. If another entity provided material support for the study, the CRO also would collect financial information relative to that entity.

10.Q: Suppose a public or academic institution conducts a study without any support from a commercial sponsor, but the study is then used by an applicant to support its marketing application. In that case, who is the "sponsor" of the study and what information should the applicant submit?

A: In this case, the Part 54 sponsor of the study is the public or academic institution. Because such institutions are not commercial entities, in many instances, there will not be relevant equity interests to report. However, any relevant interests under 54.4, such as any proprietary interest in the tested product, including but not limited to a patent, trademark, copyright or licensing agreement are to be reported.

11.Q: Does FDA have expectations about how the financial information should be collected? Will FDA consider it acceptable practice for a company to use a questionnaire to collect financial information from investigators rather than constructing an internal system to collect and report this information?

A: FDA has no preference as to how this information is collected from investigators. Under this rule, sponsors/applicants have the flexibility to collect the information in the most efficient and least burdensome manner that will be effective, for example, through questionnaires completed by the clinical investigators or by using information already available to the sponsor. FDA does not require sponsors to establish elaborate tracking systems to collect the information.

12.Q. What does FDA mean by the definition of clinical investigator and subinvestigator? Is it necessary to collect financial information on spouses and dependent children of subinvestigators?

A: The definition of "clinical investigator" in Part 54 is intended to identify the individuals who should be considered investigators for purposes of reporting under the rule, generally, the people taking responsibility for the study at a given study site. For drugs, biological products and devices, it should be noted that hospital staff, including nurses, residents, or fellows and office staff who provide ancillary or intermittent care but who do not make direct and significant contribution to the data are not meant to be included under the definition of clinical investigator. For purposes of this financial disclosure regulation, the term investigator also includes the spouse and each dependent child of the investigator and subinvestigator.

For drugs and biological products, clinical investigator means the individual(s) who actually conduct(s) and take(s) responsibility for an investigation, i.e. under whose immediate direction the drug or biologic is administered or dispensed to a subject or who is directly involved in the evaluation of research subjects. Where an investigation is directed by more than one person at a site, there may be more than one investigator who must report. For purposes of this rule, the terms investigators and subinvestigators include persons who fit any of these criteria: sign the Form FDA 1572, are identified as an investigator in initial submissions or protocol amendments under an IND, or are identified as an investigator in the NDA/BLA. For studies not conducted under an IND, the sponsor will need to identify the investigators and subinvestigators they consider covered by the rule in form 3454 and/or 3455. We expect that there will be at least one such person at each clinical site. If, however, there are other persons who are responsible for a study at a site, those persons should also be included as investigators.

For medical devices, clinical investigators are defined as individual(s), under whose immediate direction the subject is treated and the investigational device is administered, including follow-up evaluations and treatments. Where an investigation is conducted by a team of individuals, the investigator is the responsible leader of the team. In general, investigators and subinvestigators sign "investigator agreements" in accordance with 21 CFR 812.43(c) and it is these individuals whose interests should be reported. For studies not conducted under an FDA-approved IDE, (that is, a non-significant risk IDE or an exempt study), the sponsor would need to identify the investigators and subinvestigators they considered covered by the rule in form 3454 and 3455. We expect that there will be at least one such person at each site.

13.Q: Do the reporting requirements apply to efficacy studies that include large numbers of investigators and multiple sites? Will the agency consider a waiver mechanism to exempt applicants from collecting information from clinical investigators conducting these kinds of studies?

A:　　　Large multi-center efficacy studies with many investigators are considered covered clinical studies within the meaning of the final rule. See 21 CFR 54.2(c). Data from investigators having only a small percentage of the total subject population (in a study with large numbers of investigators and multiple sites) may still affect the overall study results. For example, if a sponsor submitted data collected during a large, multi-center, double blind study that included several thousand subjects and a single clinical investigator at one of the largest sites enrolled one percent of subjects, that investigator could still be responsible for a significant number of subjects. If the investigator fabricated data or otherwise affected the integrity of the data, remaining data for the drug may not meet the statistical criteria for efficacy as defined prospectively in the protocol.

Because the regulations (see 21 CFR 312.10, 812.10, 314.90 and 814.20) allow a sponsor to seek a waiver of certain requirements, applicants may seek waivers of the financial disclosure requirements. FDA believes it is highly unlikely, however, that any waivers will be justified for studies begun after February 2, 1999, because the sponsor should already have begun collecting the information on an ongoing basis. FDA will evaluate any request for waiver on a case-by-case basis.

14.Q: The rule requires that investigators provide information on financial interests during the course of the study and for one year after completion of the study (see 54.4(b))? What does "completion of the study" mean?

A:　　　Completion of the study means that all study subjects have been enrolled and follow up of primary endpoint data on all subjects has been completed in accordance with the clinical protocol. Many studies have more than one stage (e.g., a study could have a short term endpoint and a longer term follow up phase). Completion of the study here refers to that part of the study being submitted in the application. If there were a subsequent application based on longer term data, completion of the study would be defined similarly for the new data. It is not required that an applicant submit updated financial information to FDA after submission of the application, but applicants must retain complete records. Where there is more than one study site, the sponsor may consider completion of the study to be when the last study site is complete, or may consider each study site individually as it is completed.

15.Q: Do applicant companies need to collect information for a year after completion of the study? Who is responsible for collecting/providing this information?

A: According to the February 2, 1998 final rule, the investigator must provide updated information when the investigator holds any equity interest in a privately held company or if stock holdings in a publicly held company exceed $50,000 in value during the one year period following completion of the study. In addition, sponsors/applicants must keep records on file when significant payments of other sorts are paid by the sponsor of the covered study to the investigator or the investigator's institution to support activities of the investigator that have a cumulative monetary value of more than $25,000, exclusive of the costs of conducting the covered clinical studies, during the study and for one year following completion of the study. FDA specified the one-year time frame because anticipation of payments may be as influential as payments already received. Applicants need only report on these arrangements when the marketing application is submitted, but sponsors/applicants are responsible for keeping updated financial information from the investigators in company files.

16.Q: What information about a financial interest should be disclosed to the agency? For example, if an investigator owns more than $50,000 of stock in a publicly held company, can the applicant just disclose that there is an interest that exceeds the $50,000 threshold or is it necessary to disclose in written detail the arrangement in question?

A: The applicant must disclose specific details of the financial interest including the size and nature of the financial interest in question and any steps taken to minimize the potential for study bias that such an interest represents.

17.Q: Is the clinical investigator required to report all fluctuations above and below the $50,000 level during the course of the investigation and one year after completion of the study?

A: The rule requires sponsors/applicants to obtain financial information from clinical investigators and a commitment from clinical investigators to promptly update financial information, if any relevant changes occur during the course of the covered clinical study and for one year following the completion of the study [21 CFR 312.53(c)(4), 312.64(d), 812.43(c)(5), 812.110(d)]. In light of the potential volatility of stock prices, FDA recognizes that the dollar value of an investigator's equity holding in a sponsoring/applicant company is likely to fluctuate during the course of a trial. Clinical investigators should report an equity interest when the investigator becomes aware that the holding has exceeded the threshold and the investigator should use judgement in updating and reporting on fluctuations in equity interests exceeding $50,000. FDA does not expect the investigator to report when that equity interest fluctuates below that threshold.

18.Q: Are equity interests in mutual funds and 401K(s) reportable?

A: Because an investigator would not have control over buying or selling stocks in mutual funds, these would not be reportable. In most circumstances, interests in 401K(s) would not be reportable, although equity interest in a product over $50,000 would be reportable if it is a holding in a self directed 401K.

19.Q: Does the rule include ANDAs? Does the rule include 510(k)s that do not include clinical data?

A: The rule applies to any clinical study of a drug (including a biological product) or device submitted in a marketing application that the applicant or FDA relies on to establish that the product is effective, including studies of drugs that show equivalence to an approved product. This means that ANDAs are covered by the final rule. 510(k)s that do not include clinical data would not contain covered studies and therefore, no financial information from device manufacturers is needed for those applications.

20.Q: Do applicants need to provide information on investigators who participate in foreign studies?

A: Yes, applicants should include either a certification or disclosure of information for investigators participating in foreign covered studies. Where the applicant is unable to obtain the information despite acting with due diligence, the applicant may submit a statement documenting its efforts to obtain the information. In this case, it is unnecessary to submit a certification or disclosure form.

21.Q: Does the rule apply to studies in support of labeling changes?

A: The rule applies to studies submitted in a supplement when those studies meet the definition of a covered clinical study. It also applies to studies to support safety labeling changes where individual investigators make a significant contribution to the safety information.

22.Q: In the case where a subsidiary company of a larger parent company is conducting a covered clinical trial, is the applicant (subsidiary company) required to report information from clinical investigators about financial interests in only the subsidiary company, or is the applicant also required to report financial holdings, if any, of the investigator in the larger parent company?

A: If the subsidiary company meets the definition of sponsor of the covered study as defined under Part 54, the IND/IDE holder is required to collect from clinical investigators financial information related to the subsidiary company. The IND/IDE holder also must collect financial information related to the parent company if the parent company is a Part 54 sponsor of the study in question. If there are multiple companies providing material support for a covered study, the IND/IDE holder is responsible for collecting financial information from clinical investigators related to all companies providing that support. The applicant company is ultimately responsible for submitting financial information to the Agency at the time the marketing application is submitted.

23.Q: Do "actual use studies" to support a request to switch a drug product from prescription to over-the-counter (OTC) status fit the definition of covered clinical study?

A: Applicants who file supplements requesting that FDA approve a switch of a prescription drug to OTC status or who file a new drug application for direct OTC use often conduct "actual use studies." These may be intended to demonstrate that the product is safe and effective when used without the supervision of a licensed practitioner; in other cases, they may test labeling comprehension or other aspects of treatment. Actual use studies performed to support these applications would be considered covered clinical studies if they were used to demonstrate effectiveness in the OTC setting or if it is a safety study where any investigator makes a significant contribution.

24.Q: Are clinical investigators of in vitro diagnostics (IVDs) covered under this regulation since they often involve specimens, not human subjects?

A: Yes. Applicants who submit marketing applications for IVDs must include the appropriate financial certification or disclosure information. Under section 21 CFR 812.3(p), "subject" is defined as a "human who participates in an investigation, either as an individual on whom or on whose specimen an investigational device is used or as a control." Thus, an investigation of an IVD is considered a clinical investigation and, if it is used to support a marketing application, it would be subject to this regulation.

25.Q: How do significant payments of other sorts (SPOOS) relate to the variety of payments the sponsor might make to an individual or institution for various activities?

A: The term "significant payments of other sorts" was intended to capture substantial payments or other support provided to an investigator that could create a sense of obligation to the sponsor.

These payments do not include payments for the conduct of the clinical trial of the product under consideration or clinical trials of other products, under a contractual arrangement, but do include other payments made directly to the investigator or to an institution for direct support of the investigator. These payments would include honoraria, consulting fees, grant support for laboratory activities and equipment or actual equipment for the laboratory/clinic. This means that if an investigator were given equipment or money to purchase equipment for use in the laboratory/clinic, but not in relation to the conduct of the clinical trial, the payment would be considered a significant payment of other sorts and should be reported. If however, the investigator were provided with computer software or money to buy the software needed for use in the clinical trial, that would not need to be reported. Finally, payments made to the institution or to other nonstudy participating investigators that are not made on behalf of the investigator do not need to be reported.

26.Q: Are payments made to investigators to cover travel expenses (such as transportation, lodgings and meal expenses) trackable under significant payments of other sorts (SPOOS)?

A: Generally, reasonable payments made to investigators to cover reimbursable expenses such as transportation, lodgings and meals do not fall within the purview of SPOOS and, therefore, would not need to be tracked. Travel costs associated with transporting, providing lodgings and meals for family members of investigators are considered unnecessary and should be tracked as SPOOS. In addition, other payments that exceed reasonable expectations, (for example, an investigator is flown to a resort location for an extra week of vacation) are considered outside of normal reimbursable expenditures and are not considered expenses that are necessary to conduct the study. Therefore, these types of expenses are also reportable and should be tracked as SPOOS.

27.Q: Under what circumstances would FDA refuse to file an application?

A: FDA may refuse to file any marketing application that does not contain either a certification that no specified financial arrangement exists or a disclosure statement identifying the specified arrangements or a statement that the applicant has acted with due diligence to obtain the required information, and an explanation of why it was unable to do so. The agency does not anticipate that it will be necessary to use its refuse to file authority often in the context of this financial disclosure rule. Applicants are encouraged to discuss their concerns on particular matters about financial information with FDA.

28.Q: Who will review a disclosure of the specified financial arrangements when such information is submitted in a marketing application? How will the financial information be handled during the review of the application?

A: Applicants are required to disclose specified financial information and any steps taken to minimize the potential for bias in any drug, biological product or device marketing application submitted to the agency on or after February 2, 1999. (See 21 CFR 54.4(a)(3)).

FDA review staff, including project managers, consumer safety officers, medical officers and others in the supervisory chain will review this information on a case-by-case basis.

29.Q: Under what circumstances will FDA publicly discuss financial arrangements disclosed to the agency?

A: In the preamble to the final rule, FDA stated that certain types of financial information requested under the rule, notably clinical investigators' equity interests would be protected from public disclosure unless circumstances relating to the public interest clearly outweigh the clinical investigator's identified privacy interest. FDA cited the example of a financial arrangement so affecting the reliability of a study as to warrant its public disclosure during evaluation of the study by an advisory panel. FDA expects that only rarely would an investigator's privacy interest be outweighed by the public interest and thus warrant disclosure of the financial interest. It is difficult to predict all possible situations that may result in public disclosure of financial interests of a clinical investigator. The agency will carefully evaluate each circumstance on a case-by-case basis.

30.Q: Can FDA have access to documents related to financial disclosure or certification documents during an inspection?

A: Yes, FDA has the authority to have access to and to copy documents supporting an applicant's certification or disclosure statement submitted to the agency in a marketing application. Regulations implementing sections 505(i), 519, and 520(g) of the Act require sponsors to establish and maintain records of data (including but not limited to analytical reports by investigators) obtained during investigational studies of drugs, biological products, and devices, that will enable the Secretary to evaluate a product's safety and effectiveness. Under 54.6, applicants must retain certain information on clinical investigators' financial interests and permits FDA employees to have access to and copy them at reasonable times.

31.Q: What kind of documentation is necessary for manufacturers to keep in case questions about certification and/or disclosure arise?

A: To the extent that applicants have relied on investigators as the source of information about potentially disclosable financial interests in any of the four categories, the underlying documentation -- e.g., copies of executed questionnaires returned by investigators , correspondence on the subject of financial disclosure, mail receipts, etc. should be retained. Likewise, to the extent that applicants who did not sponsor a covered clinical study rely on information furnished by the sponsor, the underlying documentation, including all relevant correspondence with and reports from the sponsor should be retained. To the extent that applicants rely upon information available internally, all appropriate financial documentation regarding the financial interests or arrangements in question should be retained. For example, in the case of "significant payments of other sorts," sponsors should keep documentation including, but not limited to, check stubs, canceled checks, records of electronic financial transactions, certified mail deliver receipts, etc.

32.Q: Where are forms FDA 3454 and 3455 located on the Web?

The forms are located at the following Internet address:
http://www.fda.gov/opacom/morechoices/fdaforms/cder.html

33.Q: Who are the contact persons in each FDA Center to answer questions during this implementation phase?

The following persons may be contacted: Ms. Linda Carter in the Center for Drug Evaluation and Research, phone 301-594-6758, Dr. Joanne Less in the Center for Devices and Radiological Health, phone 301-594-1190, and Dr. Jerome Donlon in the Center for Biologics Evaluation and Research, phone 301-827-3028.

Financial Relationships and Interests in Research Involving Human Subjects

Financial Relationships and Interests in Research Involving Human Subjects: Guidance for Human Subject Protection[1]

Tommy G. Thompson
Secretary
Department of Health and Human Services

May 5, 2004

Contains Nonbinding Recommendations

This guidance document, developed at the department (DHHS) level, applies to all human subject research conducted or supported by HHS agencies or regulated by the Food and Drug Administration.

Note: This guidance represents the Food and Drug Administration's (FDA's) current thinking on this topic. It does not create or confer any rights for or on any person and does not operate to bind FDA or the public. You can use an alternative approach if the approach satisfies the requirements of the applicable statutes and regulations. If you want to discuss an alternative approach, contact the appropriate FDA staff. If you cannot identify the appropriate FDA staff, call the appropriate number listed on the title page of this guidance.

This document replaces the "HHS Draft Interim Guidance: Financial Relationships in Clinical Research: Issues for Institutions, Clinical Investigators, and IRBs to Consider when Dealing with Issues of Financial Interests and Human Subject Protection" dated January 10, 2001. This document is intended to provide guidance. It does not create or confer rights for or on any person and does not operate to bind the Department of Health and Human Services (HHS, or the Department), including the Food and Drug Administration (FDA), or the public. An alternative approach may be used if such approach satisfies the requirements of the applicable statutes and regulations.

I. Introduction

A. Purpose

In this guidance document, HHS raises points to consider in determining whether specific financial interests in research affect the rights and welfare of human subjects[2] and if so, what

---

[1] Available on the FDA website at: http://www.hhs.gov/ohrp/humansubjects/ finreltn/fguid.pdf

actions could be considered to protect those subjects. This guidance applies to human subjects research conducted or supported by HHS or regulated by the FDA. The consideration of financial relationships, as discussed in this document relates to human subject protection in research conducted under the HHS or FDA regulations (45 CFR part 46, 21 CFR parts 50, 56)[3] This document is nonbinding and does not change any existing regulations or requirements, and does not impose any new requirements.

Institutions and individuals involved in human subjects research may establish financial relationships related to or separate from particular research projects. Those financial relationships may create financial interests of monetary value, such as payments for services,

---

[2] Under the Public Health Service Act and other applicable law, HHS has authority to regulate institutions engaged in HHS conducted or supported research involving human subjects. For a description of what is meant by institutions engaged in research see the Office for Human Research Protections (OHRP) engagement policy at ttp://ohrp.osophs.dhhs.gov/humansubjects/ assurance/engage.htm. Under the Federal Food, Drug, and Cosmetic Act, FDA has the authority to regulate Institutional Review Boards (IRBs) and investigators involved in the review or conduct of FDA-regulated research.

[3] This document does not address HHS Public Health Service regulatory requirements that cover institutional management of the financial interests of individual investigators who conduct Public Health Service (PHS) supported research (42 CFR part 50, subpart F, and 45 CFR part 94). This document also does not address FDA regulatory requirements that place responsibilities on sponsors to disclose certain financial interests of investigators to FDA in marketing applications (21 CFR part 54). Guidelines interpreting the application of the PHS regulations to research conducted or supported by the National Institutes of Health (NIH) that involve human subjects are available at http://grants.nihgov/grants/guide/notice-files/NOT-OD-00-040.html. Guidance interpreting the provisions of the FDA regulations appears at http://www.fda.gov/oc/guidance/financialdis.html.

The PHS regulations require grantee institutions and contractors to designate one or more persons to review investigators' financial disclosure statement describing their significant financial interests and ensure that conflicting financial interests are managed, reduced, or eliminated before expenditure of funds (42 CFR 50.604(b), 45 CFR 94.4(b)). The PHS threshold for significant financial interest is $10,000 per year income or equity interests over $10,000 and 5 percent ownership in a company (42 CFR 50.603, 45 CFR 94.3). The regulations give several examples of methods for managing investigators' financial conflicts of interest (42 CFR 50.605(a), 54 CFR 94.5(a)).

Sponsors are required to disclose certain financial interests of clinical investigators to FDA in marketing approval applications under the Federal Food, Drug and Cosmetic Act (FD&C Act) (21 CFR part 54). FDA regulations at 21 CFR part 54 address requirements for the disclosure of certain financial interests held by clinical investigators. The purpose of these regulations is to provide additional information to allow FDA to assess the reliability of the clinical data (21 CFR 54.1). The FDA regulations require sponsors seeking marketing approval for products to certify that investigators do not have certain financial interests, or to disclose those interests to FDA (21 CFR 54.4). These regulations require sponsors to report (1) financial arrangements between the sponsor and the investigator whereby the value of the investigator's compensation could be influenced by the outcome of the trial, (2) any proprietary interest in the product studied held by the investigator; (3) significant payments of other sorts over $25,000 beyond costs of the study; or (4) any significant equity interest in the sponsor of a covered study (21 CFR 54.4).

Note that when the PHS regulations were promulgated, the National Science Foundation (NSF) Investigator Financial Disclosure Policy was revised to match closely the PHS regulations. The NSF conflict of interest policy appears at http://www.nsf.gov/bfa/cpo/gpm95/ch5.htm#ch5.

---

equity interests, or intellectual property rights. A financial interest related to a research study may be a conflicting financial interest. The Department recognizes that some conflicting financial interests in research may affect the rights and welfare of human subjects. This document provides some possible approaches to consider in assuring that human subjects are adequately protected. Institutional review boards (IRBs), institutions, and investigators engaged in human subjects research each have appropriate roles in ensuring that financial interests do not compromise the protection of research subjects.[4]

B. Target Audiences

The principal target audiences include investigators, IRB members and staffs, institutions engaged in human subjects research and their officials, and other interested members of the research community.

C. Underlying Principles

The regulations protecting human research subjects are based on the ethical principles described in the Belmont report:[5] respect for persons, beneficence, and justice. The Belmont principles should not be compromised by financial relationships. Openness and honesty are indicators of respect for persons, characteristics that promote ethical research and can only strengthen the research process.

D. Basis for This Document

The HHS human subject protection regulations (45 CFR part 46) require that institutions performing HHS conducted or supported non-exempt research involving human subjects have the research reviewed and approved by an IRB whose goal is to help ensure that the rights and welfare of human subjects are protected. The comparable FDA regulations (21 CFR parts 50 and 56) require that FDA regulated research involving human subjects is reviewed and approved by such an IRB. Under these regulations, IRBs are responsible for, among other things, determining that:

- Risks to subjects are minimized (45 CFR 46.111(a)(1), 21 CFR 56.111(a)(1));

- Risks to subjects are reasonable in relation to anticipated benefits, if any, to subjects (45 CFR 46.111(a)(2), 21 CFR 56.111(a)(2));

- Selection of subjects is equitable (45 CFR 46.111(a)(3), 21 CFR 56.111(a)(3));

- Informed consent will be sought from each prospective subject (45 CFR 46.111(a)(4), 21 CFR 56.111(a)(4)); and,

- The possibility of coercion or undue influence is minimized (45 CFR 46.116, 21 CFR 50.20).

---

[4] The Department recognizes that some non-financial conflicting interests related to research also may affect the rights and welfare of human subjects. However, non-financial interests are beyond the scope of this guidance document.
[5] http://ohrp.osophs.dhhs.gov/humansubjects/guidance/belmont.htm

In addition the IRB may

- Require that additional information be given to subjects "when in the IRB's judgment the information would meaningfully add to protection of the rights and welfare of subjects" (45 CFR 46.109(b), 21 CFR 56.109(b)).

For HHS conducted or supported research, the funding agency may impose additional conditions as necessary for the protection of human subjects (45 CFR 46.124).

IRBs are also responsible for ensuring that members who review research have no conflicting interest. 45 CFR 46.107(e) directly addresses conflicts of interest by requiring that "no IRB may have a member participate in the IRB's initial or continuing review of any project in which the member has a conflicting interest, except to provide information requested by the IRB." FDA regulations include identical language at 21 CFR 56.107(e).

Concerns have grown that financial conflicts of interest in research, derived from financial relationships and the financial interests they create, may affect the rights and welfare of human subjects in research. Financial interests are not prohibited, and not all financial interests cause conflicts of interest or affect the rights and welfare of human subjects. HHS recognizes the complexity of the relationships between government, academia, industry and others, and recognizes that these relationships often legitimately include financial relationships. However, to the extent financial interests may affect the rights and welfare of human subjects in research, IRBs, institutions, and investigators need to consider what actions regarding financial interests may be necessary to protect those subjects.

In May 2000, HHS announced five initiatives to strengthen human subject protection in clinical research. One of these was to develop guidance on financial conflict of interest that would serve to further protect research participants. As part of this initiative, HHS held a conference on the topic of human subject protection and financial conflict of interest on August 15-16, 2000. A draft interim guidance document, "Financial Relationships in Clinical Research: Issues for Institutions, Clinical Investigators, and IRBs to Consider when Dealing with Issues of Financial Interests and Human Subject Protection," based on information obtained at and subsequent to that conference was made available to the public for comment on January 10, 2001.[6] This document replaces that draft interim guidance. The Department notes that other organizations have also addressed financial interests in human research via reports, guidance and recommendations.[7] Many of these contain strong and sound ideas for

---

[6] http://ohrp.osophs.dhhs.gov/humansubiects/finreltn/finguid.htm.

[7] Recent Federal and Private Sector Activities: In addition to the HHS initiative, several Federal organizations have examined the issues related to financial relationships in human subjects research:

* The National Bioethics Advisory Commission (NBAC), in a comprehensive examination of the "Ethical and Policy Issues in Research Involving Human Participants," in Chapter 3 recommended development of federal, institutional, and sponsor policies and guidance to ensure that research subjects' rights and welfare are protected from the effects of conflicts of interest (http://www.georgetown.edu/ research/nrcbl/nbac/human/overvol1.pdf).

* The HHS Office of the Inspector General (OIG) has issued a series of reports examining regulation and activities of IRBs. A June 2000 OIG report addressed recruitment practices and found that about one quarter

actions to deal with potential financial conflicts of interest on the part of institutions, investigators and IRBs.

II. Guidance for Institutions, IRBs and Investigators

A. General Approaches to Address Financial Relationships and Interests in Research Involving Human Subjects

The Department recommends that in particular, IRBs, institutions, and investigators consider whether specific financial relationships create financial interests in research studies that may

---

of the surveyed IRBs consider financial arrangements with sponsors of research as part of their protocol review (http://oig.hhs.gov/oei/reports/oei-01-97-00195.pdf).

* The National Human Research Protections Advisory Committee (NHRPAC) offered advice to HHS regarding the content and finalization of the HHS Draft Interim Guidance in August, 2001 (http://ohrp.osophs.dhhs.gov/ nhrpac/documents/aug01a.pdf).

* In December 2001, the General Accounting Office released report 02-89 "Biomedical Research: HHS Direction Needed to Address Financial Conflicts of Interest." The report recommended that the Secretary of Health and Human Services develop specific guidance or regulations concerning institutional financial conflicts of interest (http://www.gao.gov/).

* A number of nongovernmental organizations recently have addressed financial interests in reports and issued new or updated policies or guidelines of varying scope and specificity, including the Association of American Universities, October 2001 (http://www.aau.edu/research/COI.01.pdf), the Association of American Medical Colleges, December 2001 and October 2002 (http://www.aamc.org/ members/coitf/firstreport.pdf and http://www.aamc.org/ members/coitf/2002coireport.pdf), the International Committee of Medical Journal Editors October 2001 (http://www.icmje.org/sponsor.htm), the American Medical Association, January 2002 (http://jama.ama-assn.org/cgi/content/short/287/1/78), and opinions E-8.0315 Managing Conflicts of Interest in the Conduct of Clinical Trials (http://www.ama-assn.org/ama/pub/category/8471.html) and E-8031 Conflicts of Interest: Biomedical Research (http://www.ama-assn.org/ama/pub/category/8470.html), the American Society of Gene Therapy, April 2000 (http://www.asgt.org/policy/index.html), the American Society of Clinical Oncology, June 2003 (http://www.jco.org/cgi/content/full/21/12/2394), and the Institute of Medicine, October 2002, report "Responsible Research: A Systems Approach to Protecting Research Participants" (http://www.nap.edu/books/0309084881/html/).

* Two accrediting bodies for human subject protection programs have included elements addressing individual and institutional conflicts of interest in their accreditation evaluations, the Association for the Accreditation of Human Research Protection Programs (http://www.aahrpp.org/ images/Evaluation_Instrument_1.pdf) and the National Committee for Quality Assurance, http://www.ncqa.org/ Programs/QSG/VAHRPAP/vahrpapfindstds.pdf).

Internationally, the World Medical Association's revision in 2000 of the Declaration of Helsinki, (http://www.wma.net/e/policv/17-c_e.html) principle 22, includes "sources of funding" among the items of information to be provided to subjects. A number of individual institutions also have developed policies for their own situations, as noted in the NIH Guide Notice issued in June 2000 (http://grants.nih.grants/guide/notice-files/NOT-OD-00-040.html). Some of these policies involve conflicts of interest management methods and address institutional financial interests as well as individual interests.

---

adversely affect the rights and welfare of subjects. These entities may find it useful to include the following questions in their deliberations:

- What financial relationships and resulting financial interests could cause potential or actual conflicts of interest?

- At what levels should those potential or actual financial conflicts of interest be managed or eliminated?

- What procedures would be helpful, including those to

  - collect and evaluate information regarding financial relationships related to research,

  - determine whether those relationships potentially cause a conflict of interest, and

  - determine what actions are necessary to protect human subjects and ensure that those actions are taken?

- Who should be educated regarding financial conflict of interest issues and policies?

- What entity or entities would examine individual and/or institutional financial relationships and interests?

B. Points for Consideration

Financial interests determined to create a conflict of interest may be managed by eliminating them or mitigating their impact. A variety of methods or combinations of methods may be effective. Some methods may be implemented by institutions engaged in the conduct of research, and some methods may be implemented by IRBs or investigators. Some of those may apply before research begins, and some may apply during the conduct of the research.

In establishing and implementing methods to protect the rights and welfare of human subjects from conflicts of interest created by financial relationships of parties involved in research, the Department recommends that IRBs, institutions engaged in research, and investigators consider the questions below. Additional questions may be appropriate. The Department's intent is not to be exhaustive, but to suggest ways to examine the issues so that appropriate actions can be taken to protect the rights and welfare of human research subjects. The Department recognizes that a number of institutions currently address such issues in their consideration of financial interests of parties involved in human subject research.

- Does the research involve financial relationships that could create potential or actualconflicts of interest?

  - How is the research supported or financed?

  - Where and by whom was the study designed?

  - Where and by whom will the resulting data be analyzed?

- What interests are created by the financial relationships involved in the situation?

  - Do individuals or institutions receive any compensation that may be affected by the study outcome?

- Do individuals or institutions involved in the research:

  -- have any proprietary interests in the product, including patents, trademarks, copyrights, or licensing agreements?

  -- have an equity interest in the research sponsor and, if so, is the sponsor a publicly held company or non-publicly held company?

  -- receive significant payments of other sorts? (e.g., grants, compensation in the form of equipment, retainers for ongoing consultation, or honoraria)

  -- receive payment per participant or incentive payments, and are those payments reasonable?

- Given the financial relationships involved, is the institution an appropriate site for the research?

- How should financial relationships that potentially create a conflict of interest be managed?

- Would the rights and welfare of human subjects be better protected by any or a combination of the following:

  - reduction of the financial interest?

  - disclosure of the financial interest to prospective subjects?

  - separation of responsibilities for financial decisions and research decisions?

  - additional oversight or monitoring of the research?

  - an independent data and safety monitoring committee or similar monitoring body?

  - modification of role(s) of particular research staff or changes in location for certain research activities, e.g., a change of the person who seeks consent, or a change of investigator?

  - elimination of the financial interest?

C. Specific Points for Consideration

1. Institutions

The Department recommends that institutions engaged in HHS conducted or supported human subjects research consider whether the following actions or other actions would help ensure that financial interests do not compromise the rights and welfare of human research subjects.

Actions to consider:

- Establishing the independence of institutional responsibility for research activities from the management of the institution's financial interests.

- Establishing conflict of interest committees (COICs)[8] or identifying other bodies or persons and procedures to

  - deal with individuals' or institutional financial interests in research or verify the absence of such interests and

  - address institutional financial interests in research.

- Establishing criteria to determine what constitutes an institutional conflict of interest, including identifying leadership positions for which the individual's financial interests are such that they may need to be treated as institutional financial interests.

- Establishing clear channels of communication between COICs and IRBs.

- Establishing policies on providing information, recommendations, or findings from COIC deliberations to IRBs.

- Establishing measures to foster the independence of IRBs and COICs.

- Determining whether particular individuals should report financial interests to the COIC.

These individuals could include IRB members and staff and appropriate officials of the institution, along with investigators, among those who report financial interests to COICs.

- Establishing procedures for disclosure of institutional financial relationships to COICs.

- Providing training to appropriate individuals regarding financial interest requirements.

- Using independent organizations to hold or administer the institution's financial interest.

- Including individuals from outside the institution in the review and oversight of financial interests in research.

- Establishing policies regarding the types of financial relationships that may be held by parties involved in the research and circumstances under which those financial relationships and interests may or may not be held.

2. IRB Operations

The Department recommends that institutions engaged in human subjects research and IRBs that review HHS conducted or supported human subjects research or FDA regulated human subjects research consider whether establishing policies and procedures addressing IRB member potential and actual conflicts of interest as part of overall IRB policies and procedures would help ensure that financial interests do not compromise the rights and welfare of human research subjects. As noted, 45 CFR 46.107(e) and 21 CFR 56.107(e) prohibit an IRB member with a conflicting interest in a project from participating in the IRB's initial or continuing review, except to provide information as requested by the IRB.

---

[8] The acronym COIC will be used to represent the body or person(s) designated to review financial interests.

Policies and procedures to consider:

- Reminding members of conflict of interest policies at each meeting and documenting any actions taken regarding IRB member conflicts of interest related to particular protocols.

- Developing educational materials for IRB members to ensure their awareness of federal regulations and institutional policies regarding financial relationships and interests in human subjects research.

3. IRB Review

The Department recommends that IRBs reviewing HHS conducted or supported human subjects research or FDA regulated human subjects research consider whether the following actions, or other actions related to conduct or oversight of research, would help ensure that financial interests do not compromise the rights and welfare of human research subjects.

Actions to consider:

- Determining whether methods used for management of financial interests of parties involved in the research adequately protect the rights and welfare of human subjects.

- Determining whether other actions are necessary to minimize risks to subjects.

- Determining the kind, amount, and level of detail of information to be provided to research subjects regarding the source of funding, funding arrangements, financial interests of parties involved in the research, and any financial interest management techniques applied.

4. Investigators

The Department recommends that investigators conducting human subjects research consider the potential effects that a financial relationship of any kind might have on the research or on interactions with research subjects, and what actions to take.

Actions to consider:

- Including information in the informed consent document, such as

  - the source of funding and funding arrangements for the conduct and review of research, or

  - information about a financial arrangement of an institution or an investigator and how it is being managed.

- Using special measures to modify the informed consent process when a potential or actual financial conflict exists, such as

  - having a another individual who does not have a potential or actual conflict of interest involved in the consent process, especially when a potential or actual conflict of interest could influence the tone, presentation, or type of information presented during the consent process.

- Using independent monitoring of the research.

Investigator Responsibilities—Protecting the Rights, Safety, and Welfare of Study Subjects

Investigator Responsibilities—Protecting the Rights, Safety, and Welfare of Study Subjects, Guidance for Industry[1, 2]

U.S. Department of Health and Human Services
Food and Drug Administration
Center for Drug Evaluation and Research (CDER)
Center for Biologics Evaluation and Research (CBER)
Center for Devices and Radiological Health (CDRH)

October 2009

Contains Nonbinding Recommendations

This guidance provides an overview of the responsibilities of a person who conducts a clinical investigation of a drug, biological product, or medical device (an investigator as defined in 21 CFR 312.3(b) and 21 CFR 812.3(i)). It is intended to clarify for investigators and sponsors FDA's expectations concerning the investigator's responsibility (1) to supervise a clinical study in which some study tasks are delegated to employees or colleagues of the investigator or other third parties and (2) to protect the rights, safety, and welfare of study subjects.

Note: *This guidance represents the Food and Drug Administration's (FDA's) current thinking on this topic. It does not create or confer any rights for or on any person and does not operate to bind FDA or the public. You can use an alternative approach if the approach satisfies the requirements of the applicable statutes and regulations. If you want to discuss an alternative approach, contact the appropriate FDA staff. If you cannot identify the appropriate FDA staff, call the appropriate number listed on the title page of this guidance.*

I. Introduction

This guidance provides an overview of the responsibilities of a person who conducts a clinical investigation of a drug, biological product, or medical device (an investigator as defined in 21 CFR 312.3(b) and 21 CFR 812.3(i)). The goal of this guidance is to help investigators better

---

[1] Available on the FDA website at: http://www.fda.gov/downloads/Drugs/
GuidanceComplianceRegulatoryInformation/Guidances/UCM187772.pdf
[2] This guidance has been prepared by the Investigator Responsibilities Working Group, which includes representatives from the Office of the Commissioner, the Center for Drug Evaluation and Research (CDER), the Center for Biologics Evaluation and Research (CBER), and the Center for Devices and Radiological Health (CDRH) at the Food and Drug Administration.

meet their responsibilities with respect to protecting human subjects and ensuring the integrity of the data from clinical investigations. This guidance is intended to clarify for investigators and sponsors FDA's expectations concerning the investigator's responsibility (1) to supervise a clinical study in which some study tasks are delegated to employees or colleagues of the investigator or other third parties and (2) to protect the rights, safety, and welfare of study subjects.

FDA's guidance documents, including this guidance, do not establish legally enforceable responsibilities. Instead, guidances describe the Agency's current thinking on a topic and should be viewed only as recommendations, unless specific regulatory or statutory requirements are cited. The use of the word should in Agency guidances means that something is suggested or recommended, but not required.

II. Overview of Investigator Responsibilities

In conducting clinical investigations of drugs, including biological products, under 21 CFR part 312 and of medical devices under 21 CFR part 812, the investigator is responsible for:

- Ensuring that a clinical investigation is conducted according to the signed investigator statement for clinical investigations of drugs, including biological products, or agreement for clinical investigations of medical devices, the investigational plan, and applicable regulations

- Protecting the rights, safety, and welfare of subjects under the investigator's care

- Controlling drugs, biological products, and devices under investigation (21 CFR 312.60, 21 CFR 812.100)

Although specific investigator responsibilities in drug and biologics clinical trials are not identical to the investigator responsibilities in medical device clinical trials, the general responsibilities are essentially the same. This guidance discusses the general investigator responsibilities that are applicable to clinical trials of drugs, biologics, and medical devices.

An investigator's responsibilities in conducting clinical investigations of drugs or biologics are provided in 21 CFR Part 312. Many of these responsibilities are included in the required investigator's signed statement, Form FDA-1572 (see Attachment A) (hereinafter referred to as 1572). Note that although the 1572 specifically incorporates most of the requirements directed at investigators in part 312, not all requirements are listed in the 1572. Investigators and sponsors should refer to 21 CFR Parts 11, 50, 54, 56, and 312 for a more comprehensive listing of FDA's requirements for the conduct of drug and biologics studies.[3]

An investigator's responsibilities in conducting clinical investigations of a medical device are provided in 21 CFR Part 812, including the requirement that there be a signed agreement between the investigator and sponsor (see 21 CFR 812.43(c)(4) and 812.100). The medical

---

[3] As a reminder, some investigators may be responsible for submitting certain clinical trial information to the National Institutes of Health clinical trials data bank under 42 U.S.C 282(j), 402(j) of the Public Health Service Act. Although not all investigators will be expected to meet this requirement, go to www.clinicaltrials.gov for further information about potential responsibilities.

device regulations do not require use of a specific form for an investigator's statement; and there are additional requirements not listed above (see Attachment B). Investigators and sponsors should refer to 21 CFR Parts 11, 50, 54, 56, and 812 for a more comprehensive listing of FDA's requirements for the conduct of device studies.

Nothing in this guidance is intended to conflict with recommendations for investigators contained in the International Conference on Harmonisation (ICH) guidance for industry, E6 Good Clinical Practice: Consolidated Guidance (Good Clinical Practice Guidance).[4]

III. Clarification of Certain Investigator Responsibilities

This section of the guidance clarifies the investigator's responsibility to supervise the conduct of the clinical investigation and to protect the rights, safety, and welfare of participants in drug and medical device clinical trials.

A. Supervision of the Conduct of a Clinical Investigation

As stated above, investigators who conduct clinical investigations of drugs, including biological products, under 21 CFR Part 312, commit themselves to personally conduct or supervise the investigation. Investigators who conduct clinical investigations of medical devices, under 21 CFR Part 812, commit themselves to supervise all testing of the device involving human subjects. It is common practice for investigators to delegate certain study-related tasks to employees, colleagues, or other third parties (individuals or entities not under the direct supervision of the investigator). When tasks are delegated by an investigator, the investigator is responsible for providing adequate supervision of those to whom tasks are delegated. The investigator is accountable for regulatory violations resulting from failure to adequately supervise the conduct of the clinical study.

In assessing the adequacy of supervision by an investigator, FDA focuses on four major areas: (1) whether individuals who were delegated tasks were qualified to perform such tasks, (2) whether study staff received adequate training on how to conduct the delegated tasks and were provided with an adequate understanding of the study, (3) whether there was adequate supervision and involvement in the ongoing conduct of the study, and (4) whether there was adequate supervision or oversight of any third parties involved in the conduct of a study to the extent such supervision or oversight was reasonably possible.

1. What Is Appropriate Delegation of Study-Related Tasks?

The investigator should ensure that any individual to whom a task is delegated is qualified by education, training, and experience (and state licensure where relevant) to perform the delegated task. Appropriate delegation is primarily an issue for tasks considered to be clinical or medical in nature, such as evaluating study subjects to assess clinical response to an investigational therapy (e.g., global assessment scales, vital signs) or providing medical care to subjects during the course of the study. Most clinical/medical tasks require formal medical

---

[4] Guidances, including ICH guidances, are available on the Agency's Web page. See the Web addresses on the second title page of this guidance.

training and may also have licensing or certification requirements. Licensing requirements may vary by jurisdiction (e.g., states, countries). Investigators should take such qualifications/licensing requirements into account when considering delegation of specific tasks. In all cases, a qualified physician (or dentist) should be responsible for all trial-related medical (or dental) decisions and care.[5]

During inspections of investigation sites, FDA has identified instances in which study tasks have been delegated to individuals lacking appropriate qualifications. Examples of tasks that have been inappropriately delegated include:

- Screening evaluations, including obtaining medical histories and assessment of inclusion/exclusion criteria

- Physical examinations

- Evaluation of adverse events

- Assessments of primary study endpoints

- Obtaining informed consent

The investigator is responsible for conducting studies in accordance with the protocol (see 21 CFR 312.60, Form FDA-1572, 21 CFR 812.43 and 812.100). In some cases a protocol may specify the qualifications of the individuals who are to perform certain protocol-required tasks (e.g., physician, registered nurse), in which case the protocol must be followed even if state law permits individuals with different qualifications to perform the task (see 21 CFR 312.23(a)(6) and 312.40(a)(1)). For example, if the state in which the study site is located permits a nurse practitioner or physician's assistant to perform physical examinations under the supervision of a physician, but the protocol specifies that physical examinations must be done by a physician, a physician must perform such exams.

The investigator should maintain a list of the appropriately qualified persons to whom significant trial-related duties have been delegated.[6] This list should also describe the delegated tasks, identify the training that individuals have received that qualifies them to perform delegated tasks (e.g., can refer to an individual's CV on file), and identify the dates of involvement in the study. An investigator should maintain separate lists for each study conducted by the investigator.

2. What Is Adequate Training?

The investigator should ensure that there is adequate training for all staff participating in the conduct of the study, including any new staff hired after the study has begun to meet unanticipated workload or to replace staff who have left. The investigator should ensure that staff:

- Are familiar with the purpose of the study and the protocol

---

[5] Guidance for industry, E6 Good Clinical Practice: Consolidated Guidance, section 4.3.1.
[6] Ibid, section 4.1.5

- Have an adequate understanding of the specific details of the protocol and attributes of the investigational product needed to perform their assigned tasks

- Are aware of regulatory requirements and acceptable standards for the conduct of clinical trials and the protection of human subjects

- Are competent to perform or have been trained to perform the tasks they are delegated

- Are informed of any pertinent changes during the conduct of the trial and receive additional training as appropriate

If the sponsor provides training for investigators in the conduct of the study, the investigator should ensure that staff receive the sponsor's training, or any information (e.g., training materials) from that training that is pertinent to the staff's role in the study.

3. What Is Adequate Supervision of the Conduct of an Ongoing Clinical Trial?

For each study site, there should be a distinct individual identified as an investigator who has supervisory responsibility for the site. Where there is a subinvestigator at a site, that individual should report directly to the investigator for the site (i.e., the investigator should have clear responsibility for evaluating the subinvestigator's performance and the authority to terminate the subinvestigator's involvement with the study) and the subinvestigator should not be delegated the primary supervisory responsibility for the site.

The investigator should have sufficient time to properly conduct and supervise the clinical trial. The level of supervision should be appropriate to the staff, the nature of the trial, and the subject population. In FDA's experience, the following factors may affect the ability of an investigator to provide adequate supervision of the conduct of an ongoing clinical trial at the investigator's site:

- Inexperienced study staff

- Demanding workload for study staff

- Complex clinical trials (e.g., many observations, large amounts of data collected)

- Large number of subjects enrolled at a site

- A subject population that is seriously ill

- Conducting multiple studies concurrently

- Conducting a study from a remote (e.g., off-site) location

- Conducting a study at multiple sites under the oversight of a single investigator, particularly where those sites are not in close proximity

The investigator should develop a plan for the supervision and oversight of the clinical trial at the site. Supervision and oversight should be provided even for individuals who are highly qualified and experienced. A plan might include the following elements, to the extent they apply to a particular trial:

- Routine meetings with staff to review trial progress, adverse events, and update staff on any changes to the protocol or other procedures

- Routine meetings with the sponsor's monitors

- A procedure for the timely correction and documentation of problems identified by study personnel, outside monitors or auditors, or other parties involved in the conduct of a study

- A procedure for documenting or reviewing the performance of delegated tasks in a satisfactory and timely manner (e.g., observation of the performance of selected assessments or independent verification by repeating selected assessments)

- A procedure for ensuring that the consent process is being conducted in accordance with 21 CFR Part 50 and that study subjects understand the nature of their participation and the risks

- A procedure for ensuring that source data are accurate, contemporaneous, and original

- A procedure for ensuring that information in source documents is accurately captured on the case report forms (CRFs)

- A procedure for dealing with data queries and discrepancies identified by the study monitor

- Procedures for ensuring study staff comply with the protocol and adverse event assessment and reporting requirements

- A procedure for addressing medical and ethical issues that arise during the course of the study in a timely manner

4. What Are an Investigator's Responsibilities for Oversight of Other Parties Involved in the Conduct of a Clinical Trial?

a. Study Staff Not in the Direct Employ of the Investigator

Staff involved directly in the conduct of a clinical investigation may include individuals who are not in the direct employ of the investigator. For example, a site management organization (SMO) may hire an investigator to conduct a study and provide the investigator with a study coordinator or nursing staff employed by the SMO. In this situation, the investigator should take steps to ensure that the staff not under his/her direct employ are qualified to perform delegated tasks (see section III.A.1) and have received adequate training on carrying out the delegated tasks and on the nature of the study (see section III.A.2), or the investigator should provide such training. The investigator should be particularly cautious where documentation needed to comply with the investigator's regulatory responsibilities is developed and maintained by SMO staff (e.g., source documents, CRFs, drug storage and accountability records, institutional review board correspondence). A sponsor who retains an SMO shares responsibility for the quality of the work performed by the SMO.

The investigator is responsible for supervising the study tasks performed by this staff, even though they are not in his/her direct employ during the conduct of the study (see section III.A.3). This responsibility exists regardless of the qualifications and experience of staff members. In the event that the staff's performance of study-related tasks is not adequate and cannot be made satisfactory by the investigator, the investigator should document the observed deficiencies in writing to the staff member's supervisor(s) and inform the sponsor. Depending on the severity of the deficiencies, the clinical trial may need to be voluntarily suspended until personnel can be replaced.

b. Parties Other than Study Staff

There are often critical aspects of a study performed by parties not involved directly in patient care or contact and not under the direct control of the clinical investigator. For example, clinical chemistry testing, radiologic assessments, and electrocardiograms are commonly done by a central independent facility retained by the sponsor. Under these arrangements, the central facility usually provides the test results directly to the sponsor and to the investigator. Because the activities of these parties are critical to the outcome of the study and because the sponsor retains the services of the facility, the sponsor is responsible for ensuring that these parties are competent to fulfill and are fulfilling their responsibilities to the study.

Less frequently, a study may require that investigators arrange to obtain information critical to the study that cannot be obtained at the investigator's site. For example, if the study protocol requires testing with special equipment or expertise not available at the investigator's site, the investigator might make arrangements for an outside facility to perform the test. In this case, the results are usually provided directly to the investigator, who then submits the information to the sponsor. If the investigator retains the services of a facility to perform study assessments, the investigator should take steps to ensure that the facility is adequate (e.g., has the required certification or licenses). The investigator may also institute procedures to ensure the integrity of data and records obtained from the facility providing the information (e.g., a process to ensure that records identified as coming from the facility are authentic and accurate). Procedures are particularly important when assessments are crucial to the evaluation of the efficacy or safety of an intervention or to the decision to include or exclude subjects who would be exposed to unreasonable risk.

Investigators should carefully review the reports from these external sources for results that are inconsistent with clinical presentation. To the extent feasible, and considering the specifics of study design, investigators should evaluate whether results appear reasonable, individually, and in aggregate, and they should document the evaluation. If investigators detect possible errors or suspect that results from a central laboratory or testing facility might be questionable, the investigator should contact the sponsor immediately.

c. Special Considerations for Medical Device Studies

Field clinical engineers (device sponsor employees) have traditionally played a role in some investigational device procedures (e.g., cardiology, orthopedics, and ophthalmology) by providing technical assistance to the device investigator. The field clinical engineer should be supervised by the investigator because the field clinical engineer's presence or activities may have the potential to bias the outcome of studies, may affect the quality of research data, and/or may compromise the rights and welfare of human subjects. The field clinical engineer's activities should be described in the protocol. If the field engineer has face-to-face contact with subjects or if the activities of the field engineer directly affect the subject, those activities should also be described in the informed consent.

B. Protecting the Rights, Safety, and Welfare of Study Subjects

Investigators are responsible for protecting the rights, safety, and welfare of subjects under their care during a clinical trial (21 CFR 312.60 and 812.100). This responsibility should include:

- Providing reasonable medical care for study subjects for medical problems arising during participation in the trial that are, or could be, related to the study intervention

- Providing reasonable access to needed medical care, either by the investigator or by another identified, qualified individual (e.g., when the investigator is unavailable, when specialized care is needed)

- Adhering to the protocol so that study subjects are not exposed to unreasonable risks

The investigator should inform the subject's primary physician about the subject's participation in the trial if the subject has a primary physician and the subject agrees to the primary physician being informed.

1. Reasonable Medical Care Necessitated by Participation in a Clinical Trial

During a subject's participation in a trial, the investigator (or designated subinvestigator) should ensure that reasonable medical care is provided to a subject for any adverse events, including clinically significant laboratory values, related to the trial participation. If the investigator does not possess the expertise necessary to provide the type of medical care needed by a subject, the investigator should make sure that the subject is able to obtain the necessary care from a qualified practitioner. For example, if the study involves placement of a carotid stent by an interventional neuroradiologist and the subject suffers a cerebral stroke, the neuroradiologist should assess the clinical status of the subject and arrange for further care of the subject by a neurologist. Subjects should receive appropriate medical evaluation and treatment until resolution of any emergent condition related to the study intervention that develops during or after the course of their participation in a study, even if the follow-up period extends beyond the end of the study at the investigative site.

The investigator should also inform a subject when medical care is needed for conditions or illnesses unrelated to the study intervention or the disease or condition under study when such condition or illness is readily apparent or identified through the screening procedures and eligibility criteria for the study. For example, if the investigator determines that the subject has had an exacerbation of an existing condition unrelated to the investigational product or the disease or condition under study, the investigator should inform the subject. The subject should also be advised to seek appropriate care from the physician who was treating the illness prior to the study, if there is one, or assist the subject in obtaining needed medical care.

2. Reasonable Access to Medical Care

Investigators should be available to subjects during the conduct of the trial for medical care related to participation in the study. Availability is particularly important when subjects are receiving a drug that has significant toxicity or abuse potential. For example, if a study drug has potentially fatal toxicity, the investigator should be readily available by phone or other electronic communication 24 hours a day and in reasonably close proximity to study subjects (e.g., not in another state or on prolonged travel). Study subjects should be clearly educated on the possible need for such contact and on precisely how to obtain it, generally by providing pertinent phone numbers, e-mail addresses, and other contact information, in writing. Prior to undertaking the conduct of a study, prospective investigators should consider whether they can be available to the extent needed given the nature of the trial.

During any period of unavailability, the investigator should delegate responsibility for medical care of study subjects to a specific qualified physician who will be readily available to subjects during that time (in the manner a physician would delegate responsibility for care in clinical practice). If the investigator is a non-physician, the investigator should make adequate provision for any necessary medical care that the investigator is not qualified to provide.

3. Protocol Violations that Present Unreasonable Risks

There are occasions when a failure to comply with the protocol may be considered a failure to protect the rights, safety, and welfare of subjects because the non-compliance exposes subjects to unreasonable risks. For example, failure to adhere to inclusion/exclusion criteria that are specifically intended to exclude subjects for whom the study drug or device poses unreasonable risks (e.g., enrolling a subject with decreased renal function in a trial in which decreased function is exclusionary because the drug may be nephrotoxic) may be considered failure to protect the rights, safety, and welfare of the enrolled subject. Similarly, failure to perform safety assessments intended to detect drug toxicity within protocol-specified time frames (e.g., CBC for an oncology therapy that causes neutropenia) may be considered failure to protect the rights, safety, and welfare of the enrolled subject. Investigators should seek to minimize such risks by adhering closely to the study protocol.

Attachment A: Copy of Form 1572

DEPARTMENT OF HEALTH AND HUMAN SERVICES FOOD AND DRUG ADMINISTRATION STATEMENT OF INVESTIGATOR *(TITLE 21, CODE OF FEDERAL REGULATIONS (CFR) PART 312)* (See instructions on reverse side.)	Form Approved: OMB No. 0910-0014. Expiration Date: May 31, 2009. *See OMB Statement on Reverse* NOTE: No investigator may participate in an investigation until he/she provides the sponsor with a completed, signed Statement of Investigator, Form FDA 1572 (21 CFR 312.53(c)).

1. NAME AND ADDRESS OF INVESTIGATOR

2. EDUCATION, TRAINING, AND EXPERIENCE THAT QUALIFIES THE INVESTIGATOR AS AN EXPERT IN THE CLINICAL INVESTIGATION OF THE DRUG FOR THE USE UNDER INVESTIGATION. ONE OF THE FOLLOWING IS ATTACHED.

☐ CURRICULUM VITAE   ☐ OTHER STATEMENT OF QUALIFICATIONS

3. NAME AND ADDRESS OF ANY MEDICAL SCHOOL, HOSPITAL OR OTHER RESEARCH FACILITY WHERE THE CLINICAL INVESTIGATION(S) WILL BE CONDUCTED

4. NAME AND ADDRESS OF ANY CLINICAL LABORATORY FACILITIES TO BE USED IN THE STUDY.

5. NAME AND ADDRESS OF THE INSTITUTIONAL REVIEW BOARD (IRB) THAT IS RESPONSIBLE FOR REVIEW AND APPROVAL OF THE

6. NAMES OF THE SUBINVESTIGATORS (*e.g., research fellows, residents, associates*) WHO WILL BE ASSISTING THE INVESTIGATOR IN THE CONDUCT OF THE INVESTIGATION(S).

7. NAME AND CODE NUMBER, IF ANY, OF THE PROTOCOL(S) IN THE IND FOR THE STUDY(IES) TO BE CONDUCTED BY THE INVESTIGATOR.

8. ATTACH THE FOLLOWING CLINICAL PROTOCOL INFORMATION:

☐ FOR PHASE 1 INVESTIGATIONS, A GENERAL OUTLINE OF THE PLANNED INVESTIGATION INCLUDING THE ESTIMATED DURATION OF
THE STUDY AND THE MAXIMUM NUMBER OF SUBJECTS THAT WILL BE INVOLVED

☐ FOR PHASE 2 OR 3 INVESTIGATIONS, AN OUTLINE OF THE STUDY PROTOCOL INCLUDING AN APPROXIMATION OF THE NUMBER OF SUBJECTS TO BE TREATED WITH THE DRUG AND THE NUMBER TO BE EMPLOYED AS CONTROLS, IF ANY, THE CLINICAL USES TO BE INVESTIGATED, CHARACTERISTICS OF SUBJECTS BY AGE, SEX AND CONDITION, THE KIND OF CLINICAL OBSERVATIONS AND LABORATORY TESTS TO BE CONDUCTED, THE ESTIMATED DURATION OF THE STUDY, AND COPIES OR A DESCRIPTION OF CASE REPORT FORMS TO BE USED

9. COMMITMENTS:

I agree to conduct the study(ies) in accordance with the relevant, current protocol(s) and will only make changes in a protocol after notifying the sponsor, except when necessary to protect the safety, rights, or welfare of subjects.

I agree to personally conduct or supervise the described investigation(s).

I agree to inform any patients, or any persons used as controls, that the drugs are being used for investigational purposes and I will ensure
that the requirements relating to obtaining informed consent in 21 CFR Part 50 and institutional review board (IRB) review and approval in 21 CFR Part 56 are met.

I agree to report to the sponsor adverse experiences that occur in the course of the investigation(s) in accordance with 21 CFR 312.64.

I have read and understand the information in the investigator's brochure, including the potential risks and side effects of the drug.

I agree to ensure that all associates, colleagues, and employees assisting in the conduct of the study(ies) are informed about their obligations
in meeting the above commitments.

I agree to maintain adequate and accurate records in accordance with 21 CFR 312.62 and to make those records available for inspection in accordance with 21 CFR 312.68.

I will ensure that an IRB that complies with the requirements of 21 CFR Part 56 will be responsible for the initial and continuing review and approval of the clinical investigation. I also agree to promptly report to the IRB all changes in the research activity and all unanticipated
problems involving risks to human subjects or others. Additionally, I will not make any changes in the research without IRB approval except where necessary to eliminate apparent immediate hazards to human subjects.

I agree to comply with all other requirements regarding the obligations of clinical investigators and all other pertinent requirements in 21 CFR
Part 312.

INSTRUCTIONS FOR COMPLETING FORM FDA 1572
STATEMENT OF INVESTIGATOR:

1. Complete all sections. Attach a separate page if additional space is needed.

2. Attach curriculum vitae or other statement of qualifications as described in Section 2.

3. Attach protocol outline as described in Section 8.

4. Sign and date below.

5. FORWARD THE COMPLETED FORM AND ATTACHMENTS TO THE SPONSOR. The sponsor will incorporate this information along with other technical data into an Investigational New Drug Application (IND). INVESTIGATORS SHOULD NOT SEND THIS FORM DIRECTLY TO THE FOOD AND DRUG ADMINISTRATION.

10. SIGNATURE OF INVESTIGATOR	11. DATE

(WARNING: A willfully false statement is a criminal offense. U.S.C. Title 18, Sec. 1001.)

Public reporting burden for this collection of information is estimated to average 100 hours per response, including the time for reviewing instructions, searching existing data sources, gathering and maintaining the data needed, and completing and reviewing the collection of information. Send comments regarding this burden estimate or any other aspect of this collection of information, including suggestions for reducing this burden to:

Department of Health and Human Services	Department of Health and Human Services	"An agency may not conduct or
Food and Drug Administration	Food and Drug Administration	sponsor, and a person is not
Center for Drug Evaluation and Research	Center for Biologics Evaluation and Research (HFM-99)	required to respond to, a
Central Document Room	1401 Rockville Pike	collection of information unless it
5901-B Ammendale Road	Rockville, MD 20852-1448	displays a currently valid OMB
Beltsville, MD 20705-1266		control number."

Please DO NOT RETURN this application to this address.

Attachment B: Investigator Responsibilities for Significant Risk Device Investigations

This document is intended to assist investigators in identifying and complying with their responsibilities in connection with the conduct of clinical investigations involving medical devices. Although this guidance primarily addresses duties imposed upon clinical investigators by regulations of the Food and Drug Administration (FDA), investigators should be cognizant of additional responsibilities that may derive from other sources (such as the study protocol itself, the investigator agreement, any conditions of approval imposed by FDA or the governing institutional review board, as well as institutional policy and state law).

General Responsibilities of Investigators (21 CFR 812.100)

1. Ensuring that the investigation is conducted according to the signed agreement, the investigational plan, and applicable FDA regulations

2. Protecting the rights, safety, and welfare of subjects under the investigator's care

3. Controlling devices under investigation

4. Ensuring that informed consent is obtained from each subject in accordance with 21 CFR Part 50 and that the study is not commenced until FDA and IRB approvals have been obtained.

Specific Responsibilities of Investigators (21 CFR 812.110)

1. Awaiting IRB approval and any necessary FDA approval before requesting written informed consent or permitting subject participation

2. Conducting the investigation in accordance with:

   a. The signed agreement with the sponsor

   b. The investigational plan

   c. The regulations set forth in 21 CFR Part 812 and all other applicable FDA regulations

   d. Any conditions of approval imposed by an IRB or FDA

3. Supervising the use of the investigational device. An investigator shall permit an investigational device to be used only with subjects under the investigator's supervision. An investigator shall not supply an investigational device to any person not authorized under 21 CFR Part 812 to receive it.

4. Disposing of the device properly. Upon completion or termination of a clinical investigation or the investigator's part of an investigation, or at the sponsor's request, an investigator shall return to the sponsor any remaining supply of the device or otherwise dispose of the device as the sponsor directs.

Maintaining Records (21 CFR 812.140)

An investigator shall maintain the following accurate, complete, and current records relating to the investigator's participation in an investigation:

1. Correspondence with another investigator, an IRB, the sponsor, a monitor, or FDA

2. Records of receipt, use or disposition of a device that relate to:

   a. The type and quantity of the device, dates of receipt, and batch numbers or code marks

   b. Names of all persons who received, used, or disposed of each device

   c. The number of units of the device returned to the sponsor, repaired, or otherwise disposed of, and the reason(s) therefore

3. Records of each subject's case history and exposure to the device, including:

   a. Documents evidencing informed consent and, for any use of a device by the investigator without informed consent, any written concurrence of a licensed physician and a brief description of the circumstances justifying the failure to obtain informed consent

   b. All relevant observations, including records concerning adverse device effects (whether anticipated or not), information and data on the condition of each subject upon entering, and during the course of, the investigation, including information about relevant previous medical history and the results of all diagnostic tests;

   c. A record of the exposure of each subject to the investigational device, including the date and time of each use, and any other therapy.

4. The protocol, with documents showing the dates of and reasons for each deviation from the protocol

5. Any other records that FDA requires to be maintained by regulation or by specific requirement for a category of investigations or a particular investigation

Inspections (21 CFR 812.145)

Investigators are required to permit FDA to inspect and copy any records pertaining to the investigation including, in certain situations, those which identify subjects.

Submitting Reports (21 CFR 812.150)

An investigator shall prepare and submit the following complete, accurate, and timely reports:

1. To the sponsor and the IRB:

- Any unanticipated adverse device effect occurring during an investigation. (Due no later than 10 working days after the investigator first learns of the effect.)

- Progress reports on the investigation. (These reports must be provided at regular intervals, but in no event less often than yearly. If there is a study monitor, a copy of the report should also be sent to the monitor.)

- Any deviation from the investigational plan made to protect the life or physical well-being of a subject in an emergency. (Report is due as soon as possible but no later than 5 working days after the emergency occurs. Except in emergency situations, a protocol deviation requires prior sponsor approval; and if the deviation may affect the scientific soundness of the plan or the rights, safety, or welfare of subjects, prior FDA and IRB approval are required.)

- Any use of the device without obtaining informed consent. (Due within 5 working days after such use.)

- A final report. (Due within 3 months following termination or completion of the investigation or the investigator's part of the investigation. For additional guidance, see the discussion under the section entitled "Annual Progress Reports and Final Reports.")

- Any further information requested by FDA or the IRB about any aspect of the investigation.

2. To the Sponsor:

- Withdrawal of IRB approval of the investigator's part of an investigation. (Due within 5 working days of such action).

Investigational Device Distribution and Tracking

The IDE regulations prohibit an investigator from providing an investigational device to any person not authorized to receive it (21 CFR 812.110(c)). The best strategy for reducing the risk that an investigational device could be improperly dispensed (whether purposely or inadvertently) is for the sponsor and the investigators to closely monitor the shipping, use, and final disposal of devices. Upon completion or termination of a clinical investigation (or the investigator's part of an investigation), or at the sponsor's request, an investigator is required to return to the sponsor any remaining supply of the device or otherwise to dispose of the device as the sponsor directs (21 CFR 812.110(e)). Investigators must also maintain complete, current, and accurate records of the receipt, use, or disposition of investigational devices (21 CFR 812.140(a)(2)). Specific recordkeeping requirements are set forth at 21 CFR 812.140(a).

Prohibition of Promotion and Other Practices (21 CFR 812.7)

The IDE regulations prohibit the promotion and commercialization of a device that has not been first cleared or approved for marketing by FDA. This prohibition is applicable to sponsors and investigators (or any person acting on behalf of a sponsor or investigator) and encompasses the following activities:

1. Promotion or test marketing of the investigational device

2. Charging subjects or investigators for the device a price larger than is necessary to recover the costs of manufacture, research, development, and handling

3. Prolonging an investigation beyond the point needed to collect data required to determine whether the device is safe and effective

4. Representing that the device is safe or effective for the purposes for which it is being investigated

Monitoring Clinical Investigations

Monitoring Clinical Investigations

Guidance for Industry - Guideline for the Monitoring of Clinical Investigations[1]

U.S. Department of Health and Human Services
Food and Drug Administration
Office of Regulatory Affairs

January 1988

(Minor editorial and formatting changes November 1998)

Contains Nonbinding Recommendations

This guidance addresses issues pertaining to monitoring of clinical investigations.

Note: *This guidance represents the Food and Drug Administration's (FDA's) current thinking on this topic. It does not create or confer any rights for or on any person and does not operate to bind FDA or the public. You can use an alternative approach if the approach satisfies the requirements of the applicable statutes and regulations. If you want to discuss an alternative approach, contact the appropriate FDA staff. If you cannot identify the appropriate FDA staff, call the appropriate number listed on the title page of this guidance.*

Purpose

The purpose of this guideline is to present acceptable approaches to monitoring clinical investigations. Existing requirements for sponsors of clinical investigations involving new drugs for human and animal use (including biological products for human use) and medical devices under 21 CFR Parts 312 and 511, and 812 and 813, respectively, require that a sponsor monitor the progress of a clinical investigation. The monitoring functions may be delegated to a contract research organization as defined under 21 CFR 312.3. Proper monitoring is necessary to assure adequate protection of the rights of human subjects and the safety of all subjects involved in clinical investigations and the quality and integrity of the resulting data submitted to the Food and Drug Administration (FDA).

Introduction

This guideline, issued under 21 CFR 10.90, reflects principles recognized by the scientific community as desirable approaches to monitoring clinical research involving human and animal subjects. These principles are not legal requirements but represent a standard of

---

[1] Available on the FDA website at: http://www.fda.gov/RegulatoryInformation/Guidances/ucm126400.htm

practice that is acceptable to FDA. A sponsor may rely upon this guideline or may develop different procedures. A sponsor who selects different procedures for monitoring a clinical investigation may, but is not required to, submit those procedures to FDA for review and comment to avoid the possibility of employing monitoring procedures that FDA might later determine to be inadequate. Sponsors wishing to obtain such a review should contact FDA's Bioresearch Program Coordinator (HFC-230), Food and Drug Administration, 5600 Fishers Lane, Rockville, MD 20857.

FDA may amend this guideline from time to time on the basis of comments submitted by interested persons or information obtained from agency inspections of sponsors, monitors, and investigators.

A. Selection of a Monitor

A sponsor may designate one or more appropriately trained and qualified individuals to monitor the progress of a clinical investigation. Physicians, veterinarians, clinical research associates, paramedical personnel, nurses, and engineers may be acceptable monitors depending on the type of product involved in the study. A monitor need not be a person qualified to diagnose and treat the disease or other condition for which the test article is under investigation, but somewhere in the direct line of review of the study data there should be a person so qualified.

For any given study, the factors that should be considered in determining the number of monitors and the education, training, or expertise necessary should include:

- The number of investigators conducting the study.

- The number and location of the facilities in which the study is being conducted.

- The type of product involved in the study (i.e., drug for human use, drug for animal use, medical device).

- The complexity of the study.

- The nature of the disease or other condition under study.

B. Written Monitoring Procedures

A sponsor should establish written procedures for monitoring clinical investigations to assure the quality of the study and to assure that each person involved in the monitoring process carries out his or her duties. A single written monitoring procedure need not be developed for each clinical investigation. Rather, a standardized, written procedure, sufficiently detailed to cover the general aspects of clinical investigations, may be used as a basic monitoring plan and supplemented by more specific or additional monitoring procedures tailored to the individual clinical investigation.

C. Preinvestigation Visits

A sponsor is responsible for assuring, through personal contact between the monitor and each investigator, that the investigator clearly understands and accepts the obligations incurred in undertaking a clinical investigation.

Prior to the initiation of a clinical investigation, the monitor should visit the site of the clinical investigation to assure that the investigator:

- Understands the investigational status of the test article and the requirements for this accountability.

- Understands the nature of the protocol or investigational plan.

- Understands the requirements for an adequate and well-controlled study.

- Understands and accepts his or her obligations to obtain informed consent in accordance with 21 CFR Part 50. The monitor should review a specimen of each consent document to be used by the investigator to assure that reasonably foreseeable risks are adequately explained.

- Understands and accepts his or her obligation to obtain IRB review and approval of a clinical investigation before the investigation may be initiated and to ensure continuing review of the study by the IRB in accordance with 21 CFR Part 56, and to keep the sponsor informed of such IRB approval and subsequent IRB actions concerning the study.

- Has access to an adequate number of suitable subjects to conduct the investigation.

- Has adequate facilities for conducting the clinical investigation.

- Has sufficient time from other obligations to carry out the responsibilities to which the investigator is committed by applicable regulations.

D. Periodic Visits

A sponsor is responsible for assuring throughout the clinical investigation that the investigator's obligations, as forth in applicable regulations, are being fulfilled and that the facilities used in the clinical investigation continue to be acceptable. The most effective way to achieve this assurance is to maintain personal contact between the monitor and the investigator throughout the clinical investigation. The monitor should visit the investigator at the site of the investigation frequently enough to assure that:

- The facilities used by the investigator continue to be acceptable for purposes of the study.

- The study protocol or investigational plan is being followed.

- Changes to the protocol have been approved by the IRB and/or reported to the sponsor and the IRB.

- Accurate, complete, and current and current records are being maintained.

- Accurate, complete, and timely reported are being made to the sponsor and IRB.

- The investigator is carrying out the agreed-upon activities and has not delegated them to other previously unspecified staff.

E. Review of Subject Records

A sponsor is responsible for assuring that the data submitted to FDA in support of the safety and effectiveness of a test article are accurate and complete. The most effective way to assure the accuracy of the data submitted to FDA is to review individual subject records and other supporting documents and compare those records with the reports prepared by the investigator for submission to the sponsor. Therefore, during a periodic visit, the monitor should compare a representative number of subject records and other supporting documents with the investigator's reports to determine that:

- The information recorded in the investigator's report is complete, accurate, and legible.

- There are no omissions in the reports of specific data elements such as the administration to any subject of concomitant test articles or the development of an intercurrent illness.

- Missing visits or examinations are noted in the reports.

- Subjects failing to complete the study and the reason for each failure are noted in the reports.

- Informed consent has been documented in accordance with 21 CFR Parts 50 and 56.

F. Record of On-Site Visits

The monitor or the sponsor should maintain a record of the findings, conclusions, and action taken to correct deficiencies for each on-site visit to an investigator. Such a record may enable FDA to determine that a sponsor's obligations in monitoring the progress of a clinical investigation are being fulfilled. The record may include such elements as:

- The date of the visit.

- The name of the individual who conducted the visit.

- The name and address of the investigator visited.

- A statement of the findings, conclusions and any actions taken to correct any deficiencies noted during the visit.

Patient-Reported Outcome Measures: Use in Medical Product Development to Support Labeling Claims

Guidance for Industry

Patient-Reported Outcome Measures: Use in Medical Product Development to Support Labeling Claims[1], [2]

U.S. Department of Health and Human Services
Food and Drug Administration
Center for Drug Evaluation and Research (CDER)
Center for Biologics Evaluation and Research (CBER)
Center for Devices and Radiological Health (CDRH)

December 2009

Clinical/Medical

Contains Nonbinding Recommendations

This guidance describes how FDA reviews and evaluates existing, modified, or newly created patient-reported outcome (PRO) instruments used to support claims in approved medical product labeling. A PRO instrument (i.e., a questionnaire plus the information and documentation that support its use) is a means to capture PRO data used to measure treatment benefit or risk in medical product clinical trials. This guidance does not address the use of PRO instruments for purposes beyond evaluation of claims made about a medical product in labeling nor disease-specific issues.

Note: *This guidance represents the Food and Drug Administration's (FDA's) current thinking on this topic. It does not create or confer any rights for or on any person and does not operate to bind FDA or the public. You can use an alternative approach if the approach satisfies the requirements of the applicable statutes and regulations. If you want to discuss an alternative approach, contact the appropriate FDA staff. If you cannot identify the appropriate FDA staff, call the appropriate number listed on the title page of this guidance.*

---

[1] Available on the FDA website at: http://www.fda.gov/downloads/Drugs/ GuidanceComplianceRegulatoryInformation/Guidances/UCM193282.pdf

[2] This guidance has been prepared by the Center for Drug Evaluation and Research (CDER) in cooperation with the Center for Biologics Evaluation and Research (CBER) and the Center for Devices and Radiological Health (CDRH) at the Food and Drug Administration.

I. Introduction

This guidance describes how the Food and Drug Administration (FDA) reviews and evaluates existing, modified, or newly created *patient-reported outcome (PRO) instruments* used to support *claims* in approved medical product labeling.[3] A PRO instrument (i.e., a *questionnaire* plus the information and documentation that support its use) is a means to capture PRO data used to measure *treatment benefit* or risk in medical product clinical trials. This guidance does not address the use of PRO instruments for purposes beyond evaluation of claims made about a medical product in labeling. This guidance also does not address disease-specific issues. Guidance on clinical trial endpoints for specific diseases can be found on various FDA Web sites.[4]

By explicitly addressing the review issues identified in this guidance, sponsors can increase the efficiency of their discussions with the FDA during the medical product development process, streamline the FDA's review of PRO instrument adequacy and resultant PRO data collected during a clinical trial, and provide optimal information about the patient perspective for use in making conclusions about treatment effect at the time of medical product approval. PRO instrument development is an iterative process and we recognize there is no single correct way to develop a PRO instrument. Different strategies and methods can be used to address FDA review issues.

The Glossary defines many of the terms used in this guidance. Words or phrases found in the Glossary appear in *bold italics* at first mention. Specifically, we encourage sponsors to familiarize themselves with the terms *conceptual framework of a PRO instrument*, *endpoint model*, and *content validity*.

FDA's guidance documents, including this guidance, do not establish legally enforceable responsibilities. Instead, guidances describe the Agency's current thinking on a topic and should be viewed only as recommendations, unless specific regulatory or statutory requirements are cited. The use of the word *should* in Agency guidances means that something is suggested or recommended, but not required.

---

[3] Labeling, as used in this guidance, refers to the information about an FDA-approved medical product intended for the clinician to use in treating patients. See 21 CFR 201.56 and 201.57 for regulations pertaining to prescription drug (including biological drug) labeling. Section 201.56 specifically describes the need for labeling that is not false or misleading. See 21 CFR part 801 for medical device labeling. See 21 CFR 606.122 for blood and blood products for transfusion.

[4] See the following FDA Web sites: http://www.fda.gov/Drugs/ GuidanceComplianceRegulatoryInformation/Guidances/default.htm (CDER), http://www.fda.gov/BiologicsBloodVaccines/GuidanceComplianceRegulatoryInformation/default.htm (CBER), and http://www.fda.gov/MedicalDevices/DeviceRegulationandGuidance/GuidanceDocuments/default.htm (CDRH).

II. Background

A PRO is any report of the status of a patient's health condition that comes directly from the patient, without interpretation of the patient's response by a clinician or anyone else. The outcome can be measured in absolute terms (e.g., severity of a *symptom*, *sign*, or state of a disease) or as a change from a previous measure. In clinical trials, a PRO instrument can be used to measure the effect of a medical intervention on one or more *concepts* (i.e., the *thing* being measured, such as a symptom or group of symptoms, effects on a particular function or group of functions, or a group of symptoms or functions shown to measure the severity of a health condition).

Generally, findings measured by a well-defined and reliable PRO instrument in appropriately designed investigations can be used to support a claim in medical product labeling if the claim is consistent with the instrument's documented measurement capability. The amount and kind of evidence that should be provided to the FDA is the same as for any other labeling claim based on other data. Use of a PRO instrument is advised when measuring a concept best known by the patient or best measured from the patient perspective. A PRO instrument, like physician-based instruments, should be shown to measure the concept it is intended to measure, and the FDA will review the evidence that a particular PRO instrument measures the concept claimed. The concepts measured by PRO instruments that are most often used in support of labeling claims refer to a patient's symptoms, signs, or an aspect of functioning directly related to disease status. PRO measures often represent the effect of disease (e.g., heart failure or asthma) on health and functioning from the patient perspective.

Claims generally appear in either the Indications and Usage or Clinical Studies section of labeling, but can appear in any section. Regardless of the labeling section, PRO instrument evaluation principles described here apply.

III. Evaluation of a Pro Instrument

The evaluation of a PRO instrument to support claims in medical product labeling includes the following considerations:

- The population enrolled in the clinical trial
- The clinical trial objectives and design
- The PRO instrument's conceptual framework
- The PRO instrument's *measurement properties*

Because the purpose of a PRO measure is to capture the patient's experience, an instrument will not be a credible measure without evidence of its usefulness from the target population of patients. Sponsors should provide documented evidence of patient input during instrument development and of the instrument's performance in the specific application in which it is used (i.e., population, condition). An existing instrument can support a labeling claim if it can be shown to reliably measure the claimed concept in the patient population enrolled in the clinical trial.

A. Endpoint Model

Sponsors should define the role a PRO *endpoint* is intended to play in the clinical trial (i.e., a primary, key secondary, or exploratory endpoint) so that the instrument development and performance can be reviewed in the context of the intended role, and appropriate statistical methods can be planned and applied. It is critical to plan these approaches in what can be called an endpoint model.

Figures 1 and 2 show examples of endpoint models. In Figure 1, a PRO symptom assessment is a secondary endpoint with a physiologic measure as the primary endpoint intended to support an indication for the treatment of Disease X. In this case, the clinical trial would need to succeed on the physiologic endpoint before success could be attained on the secondary endpoints. In Figure 2, a PRO symptom assessment is the primary clinical trial endpoint intended to support an indication for the treatment of symptoms associated with Disease Y and the physical performance and limitation measures would be the key secondary endpoints. PRO instrument adequacy depends on its role and relationships with other clinical trial endpoints as depicted in the endpoint model. The endpoint model explains the exact demands placed on the PRO instrument to attain the evidence to meet the clinical trial objectives and support the targeted claims corresponding to the concepts measured.

Figure 1. Endpoint Model: Treatment of Disease X

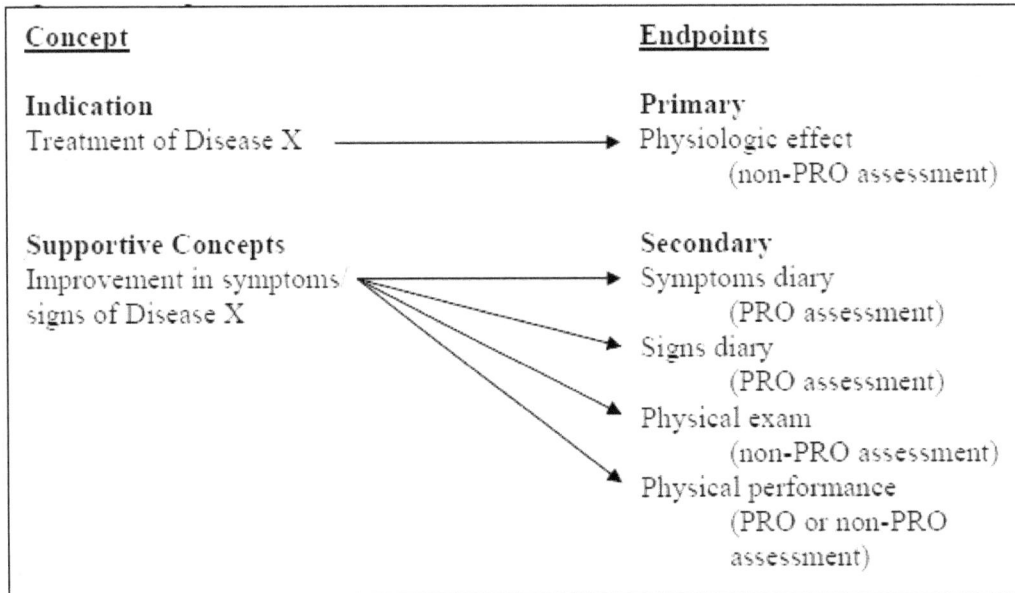

Figure 2. Endpoint Model: Treatment of Symptoms Associated with Disease Y

```
Concept                               Endpoints

Indication                            Primary
Treatment of symptoms ─────────▶      Total Disease Y symptoms score
of Disease Y                               (PRO assessment)


Supportive Concepts                   Secondary
Other treatment benefit ──────▶       Physical performance
                               ╲           (PRO or non-PRO
                                ╲           assessment)
                                 ▶     Disease Y-related physical
                                           limitations (PRO
                                           assessment)
```

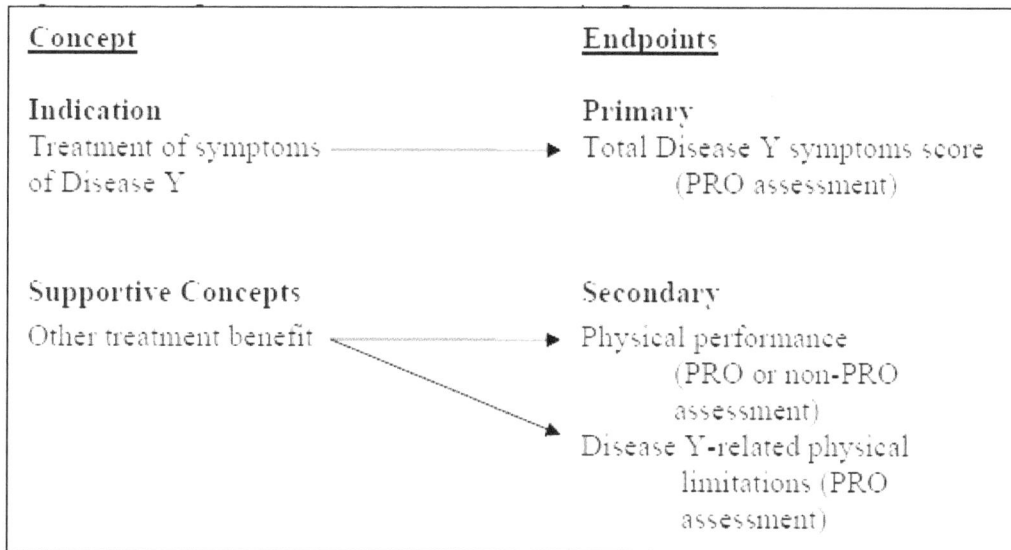

To help specify potential labeling claims and to facilitate communication with the FDA about the specific clinical trials designed to assess the planned concepts, sponsors can use a *target product profile (TPP)*, which is a clinical development program summary in the context of prescribing information goals (i.e., targeted labeling claims).[5, 6]

B. Choice of PRO Instrument

Early in medical product development, sponsors planning to use a PRO instrument in support of a labeling claim are encouraged to determine whether an adequate PRO instrument exists to assess and measure the concepts of interest. If it does not, a new PRO instrument can be developed. In some situations, the new instrument can be developed by modifying an existing instrument.

The adequacy of any PRO instrument, whether existing, modified, or newly developed, as a measure to support medical product labeling claims depends on whether its characteristics (see this section), conceptual framework (see section III.C.), content validity (see section III.D.), and other measurement properties (see section III.E.) are satisfactory. The FDA will review documentation of PRO instrument development and testing in conjunction with clinical trial results to determine whether a labeling claim is substantiated. The Appendix lists the type of PRO information sponsors should provide to the FDA to facilitate instrument review.

Characteristics of PRO instruments that are reviewed by the FDA include the following:

─────────────────────

[5] See the draft guidance for industry and review staff *Target Product Profile — A Strategic Development Process Tool.* When final, this guidance will represent the FDA's current thinking on this topic. For the most recent version of a guidance, check the FDA Drug guidance Web page at http://www.fda.gov/Drugs/GuidanceComplianceRegulatoryInformation/Guidances/default.htm.
[6] Although the TPP process is used for drug and biologic approvals, the concept of beginning with the desired claims and designing the clinical trials to assess these claims is similar for medical devices.

- Concepts being measured

- Number of *items*

- Conceptual framework of the instrument

- Medical condition for intended use

- Population for intended use

- Data collection method

- Administration mode

- Response options

- *Recall period*

- Scoring

- Weighting of items or *domains*

- Format

- Respondent burden

- Translation or cultural adaptation availability

We encourage instrument developers to make their instruments and related development history available and accessible publicly. When development history is not available, sponsors generally should provide documentation of content validity with an application (i.e., evidence that the instrument measures what it is intended to measure), including open-ended patient input from the appropriate population. Content validity is discussed in more detail in section III.D., Content Validity. In addition, we anticipate empiric evidence of an instrument's other measurement properties, discussed in more detail in section III.E., Reliability, Other Validity, and Ability to Detect Change.

We suggest that an instrument's measurement properties be well established before enrollment begins for confirmatory clinical trials. Therefore, sponsors should begin instrument development and evaluation early in medical product development, and engage the FDA in a discussion about a new or unique PRO instrument before confirmatory clinical trial protocols are finalized.

Requests for FDA input should be addressed to the review division responsible for the medical product in question. For the FDA to provide useful early input, sponsors should provide their labeling goals, a hypothesized PRO instrument conceptual framework, and the relationship of the PRO endpoints to other clinical trial endpoints in preliminary endpoint models for the planned confirmatory trials.

If the measurement goal is to support a complex, multidomain concept, PRO instruments that measure a simple concept may not be adequate to substantiate the complex claim. For example, PRO-based evidence of improved symptoms alone will only support claims specific to improvement of the symptoms and would not support a general claim related to improvement in a patient's ability to function or the patient's psychological state. In addition, a complex, multidomain claim cannot be substantiated by instruments that do not adequately measure the individual component domain concepts adequately.

PRO instruments can be used to measure important safety concerns if those concerns represent symptoms or signs that are best captured from the patient perspective. The principles for PRO instrument development are not different for this application.

Claims representing general concepts often are not supported, even though the PRO instrument was developed to measure the general concepts, because the instrument may not distinguish adverse side effects of treatment that affect the general concept and that may not be known at the time the clinical trials are designed. If adverse effects are captured, PRO instruments should aim to measure the adverse consequences of treatment separately from the effectiveness of treatment. As with any clinical trial evaluating FDA-regulated medical products, all adverse events detected with a PRO instrument should be included in the clinical trial report.

Figure 3 summarizes the iterative process used in developing a PRO instrument for use in clinical trials. FDA review of the developmental process documentation is discussed in more detail in section III.C., Conceptual Framework of a PRO Instrument, through section III.G., PRO Instruments Intended for Specific Populations.

Figure 3. Development of a PRO Instrument: An Iterative Process

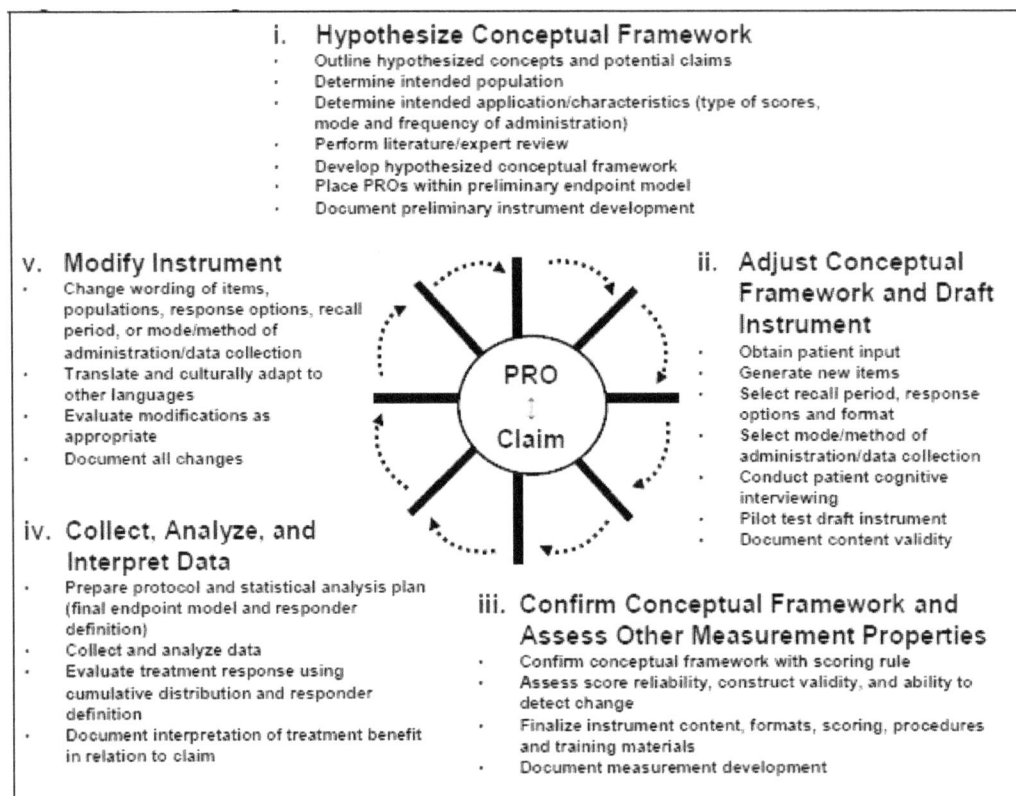

i. Hypothesize Conceptual Framework
- Outline hypothesized concepts and potential claims
- Determine intended population
- Determine intended application/characteristics (type of scores, mode and frequency of administration)
- Perform literature/expert review
- Develop hypothesized conceptual framework
- Place PROs within preliminary endpoint model
- Document preliminary instrument development

v. Modify Instrument
- Change wording of items, populations, response options, recall period, or mode/method of administration/data collection
- Translate and culturally adapt to other languages
- Evaluate modifications as appropriate
- Document all changes

ii. Adjust Conceptual Framework and Draft Instrument
- Obtain patient input
- Generate new items
- Select recall period, response options and format
- Select mode/method of administration/data collection
- Conduct patient cognitive interviewing
- Pilot test draft instrument
- Document content validity

iv. Collect, Analyze, and Interpret Data
- Prepare protocol and statistical analysis plan (final endpoint model and responder definition)
- Collect and analyze data
- Evaluate treatment response using cumulative distribution and responder definition
- Document interpretation of treatment benefit in relation to claim

iii. Confirm Conceptual Framework and Assess Other Measurement Properties
- Confirm conceptual framework with scoring rule
- Assess score reliability, construct validity, and ability to detect change
- Finalize instrument content, formats, scoring, procedures and training materials
- Document measurement development

PRO
Claim

C. Conceptual Framework of a PRO Instrument

The adequacy of a proposed instrument to support a claim depends on the conceptual framework of the PRO instrument. The conceptual framework explicitly defines the concepts measured by the instrument in a diagram that presents a description of the relationships between items, domain (subconcepts), and concepts measured and the *scores* produced by a PRO instrument.

1. Concepts Measured

One fundamental consideration in the review of a PRO instrument is the adequacy of the item generation process to support the final conceptual framework of the instrument. In some cases, the question of what to measure may be obvious given the condition being treated. For example, to assess the effect of treatment on pain, patients from the target population are queried about pain severity using a single-item PRO instrument. Generally, when it is not obvious, instrument developers initially can hypothesize a conceptual framework to support the measurement of the concept of interest drafting the domains and items to be measured based on literature reviews and expert opinion. Subsequently, patient interviews, focus groups, and qualitative *cognitive interviewing* ensures understanding and completeness of the concepts contained in the items. (See section III.D.1., Item Generation.)

The conceptual framework of a PRO instrument will evolve and be confirmed over the course of instrument development as a sponsor gathers empiric evidence to support item grouping and scores. When used in a clinical trial, the PRO instrument's conceptual framework should again be confirmed by the observed relationships among items and domains.

Documentation of the instrument development process should reveal the means by which the items and domains were identified. The exact words used to represent the concepts measured by domain or total scores should be derived using patient input to ensure the conclusions drawn using instrument scores are valid.

For measures of general concepts, we intend to review how individual items are thought to be associated with each other, how items are associated with each domain, and how domains are associated with each other and the general concept of interest based on the conceptual framework of the PRO instrument. The diagram in Figure 4 depicts a generic example of a conceptual framework of a PRO instrument where Domain 1, Domain 2, and General Concept each represent related but separate concepts. Items in this diagram are aggregated into domains. The final framework is derived and confirmed by measurement property testing.

Figure 4. Diagram of the Conceptual Framework of a PRO Instrument

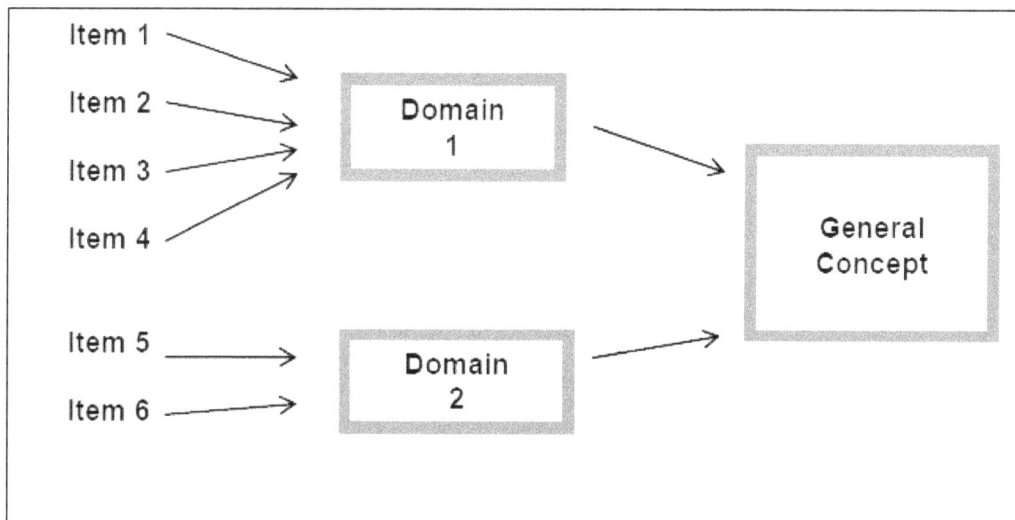

The conceptual framework of a PRO instrument may be straightforward if a single item is a reliable and valid measure of the concept of interest (e.g., pain intensity). If the concept of interest is general (e.g., physical function), a single-item PRO instrument does not provide a useful understanding of the treatment's effect because a stand-alone single item does not capture the domains of the general concept. For this reason, single-item questions about general concepts that include multiple items or domains rarely provide sufficient evidence to support claims about that general concept. For example, in clinical trials of functional disorders defined by clusters of specific symptoms and signs, a PRO instrument consisting of a single-item global question usually would be inadequate as an endpoint to support labeling claims and would be uninformative about the effects on each specific symptom and sign. Instead, the effect of treatment on each of the appropriate symptoms and signs should be adequately measured.

The conceptual framework for PRO instruments intended to measure a general concept will be complex because identifying all of the appropriate domains and items of the general concept can be difficult. Multidomain PRO instruments can be used to support claims about a general concept if the PRO instrument has been developed to measure the important and relevant domains of the general concept contained in the claim. However, the complex nature of multidomain PRO instruments often raises significant questions about how to interpret and report results in a way that is not misleading. For example, if improvement in a score for a general concept (e.g., symptoms associated with a certain condition) is driven by a single responsive item (e.g., pain intensity improvement) whereas other important items (e.g., other symptoms) did not show a response, a general claim about the general concept (e.g., improvements in symptoms associated with the condition) cannot be supported. However, that single responsive item or domain may support a claim specific to that item or domain.

We intend to examine the final version of an instrument in light of its development history, including documentation of the complete list of items generated and the reasons for deleting or modifying items, as illustrated in Table 1. We will determine from empiric evidence provided whether the PRO instrument's final conceptual framework (e.g., the hypothesized relationships among items, domains, and concepts measured) is confirmed in the appropriate study population and is consistent with the endpoint model of the planned clinical trials.

Table 1. Common Reasons for Changing Items during PRO Instrument Development

Item Property	Reason for Change or Deletion
Clarity or relevance	• Reported as not relevant by a large segment of the target population • Generates an unacceptably large amount of missing data points • Generates many questions or requests for clarification from patients as they complete the PRO instrument • Patients interpret items and responses in a way that is inconsistent with the PRO instrument's conceptual framework
Response range	• A high percent of patients respond at the floor (response scale's worst end) or ceiling (response scale's optimal end) • Patients note that none of the response choices applies to them • Distribution of item responses is highly skewed
Variability	• All patients give the same answer (i.e., no variance) • Most patients choose only one response choice • Differences among patients are not detected when important differences are known
Reproducibility	• Unstable scores over time when there is no logical reason for variation from one assessment to the next
Inter-item correlation	• Item highly correlated (redundant) with other items in the same

Item Property	Reason for Change or Deletion
	concept of interest
Ability to detect change	• Item is not sensitive (i.e., does not change when there is a known change in the concepts of interest)
Item discrimination	• Item is highly correlated with measures of concepts other than the one it is intended to measure • Item does not show variability in relation to some known population characteristics (i.e., severity level, classification of condition, or other known characteristic)
Redundancy	• Item duplicates information collected with other items that have equal or better measurement properties
Recall period	• The population, disease state, or application of the instrument can affect the appropriateness of the recall period

2. Intended Population

Using documentation of the process described in Figure 3 and of the measurement properties as described in Table 2, we plan to compare the patient population studied in the PRO instrument development process to the population enrolled in the clinical trial to determine whether the instrument is applicable for that population. See the Appendix for a description of the types of information sponsors should provide for FDA discussion and review of PRO instruments.

Specific measurement considerations posed by pediatric, cognitively impaired, or seriously ill patients are discussed in section III.G., PRO Instruments Intended for Specific Populations.

Table 2. Measurement Properties Considered in the Review of PRO Instruments Used in Clinical Trials

Measurement Property	Type	What Is Assessed?	FDA Review Considerations
Reliability	Test-retest or intra-interviewer reliability (for interviewer-administered PROs only)	Stability of scores over time when no change is expected in the concept of interest	• Intraclass correlation coefficient • Time period of assessment
	Internal consistency	• Extent to which items comprising a scale measure the same concept • Intercorrelation of items that contribute to a score • Internal consistency	• Cronbach's alpha for summary scores • Item-total correlations
	Inter-interviewer reliability (for interviewer-	Agreement among responses when the PRO is administered by two or more	• Interclass correlation coefficient

Measurement Property	Type	What Is Assessed?	FDA Review Considerations
	administered PROs only)	different interviewers	
Validity	Content validity	Evidence that the instrument measures the concept of interest including evidence from qualitative studies that the items and domains of an instrument are appropriate and comprehensive relative to its intended measurement concept, population, and use. Testing other measurement properties will not replace or rectify problems with content validity.	• Derivation of all items • Qualitative interview schedule • Interview or focus group transcripts Items derived from the transcripts • Composition of patients used to develop content • Cognitive interview transcripts to evaluate patient understanding
	Construct validity	Evidence that relationships among items, domains, and concepts conform to a priori hypotheses concerning logical relationships that should exist with measures of related concepts or scores produced in similar or diverse patient groups	• Strength of correlation testing a priori hypotheses (discriminant and convergent validity) • Degree to which the PRO instrument can distinguish among groups hypothesized a priori to be different (known groups validity)
Ability to detect change		Evidence that a PRO instrument can identify differences in scores over time in individuals or groups (similar to those in the clinical trials) who have changed with respect to the measurement concept	• Within person change over time • Effect size statistic

D. Content Validity

Content validity is the extent to which the instrument measures the concept of interest. Content validity is supported by evidence from qualitative studies that the items and domains of an instrument are appropriate and comprehensive relative to its intended measurement concept, population, and use. Content validity is specific to the population, condition, and treatment to be studied. For PRO instruments, items, domains, and general scores reflect what is important to patients and comprehensive with respect to patient concerns relevant to the concept being assessed. Documentation of patient input in item generation as well as evaluation of patient understanding through cognitive interviewing can contribute to evidence of content validity. Evidence of other types of validity (e.g., **construct validity**) or reliability (e.g., consistent scores) will not overcome problems with content validity because we evaluate instrument adequacy to measure the concept represented by the labeling claim. It is important to establish content validity before other measurement properties are evaluated.

When evaluating the utility of an existing instrument or developing a new PRO instrument, sponsors are encouraged to support the adequacy of the instrument's content validity by documenting the following development processes and instrument attributes.

1. Item Generation

Item generation should include input from the target patient population to establish the items that reflect the concept of interest and contribute to its evaluation. The population will help generate item wording, evaluate the completeness of item coverage, and perform initial assessment of clarity and readability. PRO instrument items can be generated from literature reviews, transcripts from focus groups, or interviews with patients, clinicians, family members, researchers, or other sources. We may review whether appropriate individuals and sources were used and how information gleaned from those sources was used in the PRO instrument development process. We will also review whether open-ended patient interviews provide a full understanding of the patient's perspective of the concept of interest.

Item generation generally incorporates the input of a wide range of patients with the condition of interest to represent variations in severity and in population characteristics such as age, sex, ethnicity, and language groups in accordance with the anticipated clinical trial design.

Without adequate documentation of patient input, a PRO instrument's content validity is likely to be questioned. We will review documentation to determine that the items cover all aspects of the concept important to patients, and that **saturation** has been reached. Saturation is reached at the point when no new relevant or important information emerges and collecting additional data will not likely add to the understanding of how patients perceive the concept of interest and the items in the questionnaire.

Documentation provided to the FDA to support content validity should include all item generation techniques used, including any theoretical approach; the populations studied; source of items; selection, editing, and reduction of items; cognitive interview summaries or transcripts; pilot testing; importance ratings; and quantitative techniques for item evaluation. Table 1 lists common reasons for changing items.

If items are not generated in all language groups included in the clinical trials, the appropriateness of the content should be addressed in cognitive interviewing in each language group tested. An **item tracking matrix** may be helpful to document the changes or deletions in items and the reasons for those changes.

With existing instruments, it cannot be assumed that the instrument has content validity if patients were not involved in instrument development. New qualitative work similar to that conducted when developing a new instrument can provide documentation of content validity for existing instruments if patient interviews or focus groups are conducted using open-ended methods to elicit patient input. Such qualitative testing of existing instruments is particularly important if a review of the instrument content gives cause for concern. For example, if symptoms known to be common to the population to be studied in the clinical trial are missing from a measure meant to capture important symptoms in that population, we will question the instrument's content validity. We cannot provide recommendations for the number or size of the individual patient interviews or focus groups for establishing content validity. The sample size depends on the completeness of the information obtained from analysis of the transcripts. Generally, the number of patients is not as critical as interview quality and patient diversity included in the sample in relation to intended clinical trial population characteristics.

Items that ask patients to respond hypothetically may cause patients to respond on the basis of their desired condition rather than on their actual condition and therefore are not recommended for clinical trials. For example, in assessing the concept *ability to perform daily activities*, it is more appropriate to ask whether or not the patient performed specific activities (and if so, with how much difficulty) than whether or not the patient perceived that he or she can perform daily activities, because patients may report they are able to perform a task even when they never do the task.

When using multi-item instruments, it is important that all items be relevant to most of the patients in the clinical trial. Using the example in the previous paragraph, it would be severely disadvantageous to use a measure with items that include activities most of the clinical trial patients would not perform. Doing so would yield a *bias toward the null*, or a tendency to show no effect of treatment, even if the treatment were effective. In such cases, a negative response (or indication of little to no activity) is not useful. Use of *not applicable* response options creates problems with scoring. Skip patterns may create difficulties in administration.

2. Data Collection Method and Instrument Administration Mode

Sponsors should consider the data collection method and all procedures and protocols associated with the instrument administration mode, including instructions to interviewers, instructions for self-administration, or instructions for supervising self-administration. We will review data quality control procedures specific to the data collection method or instrument administration mode along with case report forms or screen shots of electronic PRO instruments. Administration modes can include self-administration, interview, or a combination of both. Data collection methods can include paper-based, computer-assisted, and telephone-based assessments. We intend to review the comparability of data obtained when using multiple data collection methods or administration modes within a single clinical trial to determine whether the treatment effect varies by method or mode. If a patient diary or some other form of unsupervised data entry is used, we plan to review the clinical trial protocol to determine what steps are taken to ensure that patients make entries according to the clinical trial design and not, for example, just before a clinic visit when their reports will be collected.

3. Recall Period

Sponsors should also evaluate the rationale and the appropriateness of the recall period for a PRO instrument. To this end, it is important to consider patient ability to validly recall the information requested. The choice of recall period that is most suitable depends on the instrument's purpose and intended use; the variability, duration, frequency, and intensity of the concept measured; the disease or condition's characteristics; and the tested treatment. When evaluating PRO-based claims, we intend to review the clinical trial protocol to determine what steps were taken to ensure that patients understood the instrument recall period. In many cases, what is of real interest is not the integrated effect over a short time period (e.g., 2-week period), but the effect at regular intervals (e.g., 2, 4, and 6 weeks), similar to how measurements might be made every 2 weeks in a blood pressure trial. In that case, patients can be asked to report on recent status. Note also that any problems created by differential recall are likely to add noise and obscure treatment effects.

PRO instruments that call for patients to rely on memory, especially if they must recall over a long period of time, compare their current state with an earlier period, or average their response over a period of time, are likely to undermine content validity. Response is likely to be influenced by the patient's state at the time of recall. For these reasons, items with short recall periods or items that ask patients to describe their current or recent state are usually preferable. If detailed recall of experience over a period of time is necessary, we recommend the instrument use appropriate methods and techniques for enhancing the validity and reliability of retrospectively reported data (e.g., ask patients to respond based on their worst (or best) experience over the recall period or make use of a diary for data collection).

4. Response Options

It is also important to consider whether the response options for each item are consistent with its purpose and intended use. Table 3 describes some of the various types of item response options that are typically seen in PRO instruments.

Table 3. Response Option Types

Type	Description
Visual analog scale (VAS)	A line of fixed length (usually 100 mm) with words that anchor the scale at the extreme ends and no words describing intermediate positions. Patients are instructed to indicate the place on the line corresponding to their perceived state. The mark's position is measured as the score.
Anchored or categorized VAS	A VAS that has the addition of one or more intermediate marks positioned along the line with reference terms assigned to each mark to help patients identify the locations between the scale's ends (e.g., half-way).
Likert scale	An ordered set of discrete terms or statements from which patients are asked to choose the response that best describes their state or experience.
Rating scale	A set of numerical categories from which patients are asked to choose the category that best describes their state or experience. The ends of rating scales are anchored with words but the categories are numbered rather than labeled with words.
Recording of events as they occur	Specific events are recorded as they occur using an event log that can be included in a patient diary or other reporting system (e.g., interactive voice response system).
Pictorial scale	A set of pictures applied to any of the other response option types. Pictorial scales are often used in pediatric questionnaires but also have been used for patients with cognitive impairments and for

Type	Description
	patients who are otherwise unable to speak or write.
Checklist	Checklists provide a simple choice between a limited set of options, such as Yes, No, and Don't know. Some checklists ask patients to place a mark in a space if the statement in the item is true. Checklists are reviewed for completeness and nonredundancy.

Item response options generally are considered appropriate when:

- Wording used in responses is clear and appropriate (e.g., anchoring a *scale* using the term *normal* assumes that patients understand what is normal for the general population).

- The item response options are appropriate for the intended population. For example, patients with visual impairment may find a VAS difficult to complete.

- Responses offer a clear distinction between choices (e.g., patients may not distinguish between *intense* and *severe* if both are offered as response choices to describe their pain).

- Instructions to patients for completing items and selecting responses for the items are adequate.

- The number of response options is justified empirically (e.g., using qualitative research, initial instrument testing, or existing literature).

- Responses for an item are appropriately ordered and represent similar intervals.

- Responses for items avoid potential ceiling or floor effects (e.g., it may be necessary to introduce more responses to capture worsening or improvement so that fewer patients respond at the response continuum top or bottom).

- Responses do not bias the direction of responses (e.g., bias exists if possible responses are weighted toward the severity spectrum's mild end with two severity options for mild and only one each for *moderate* and *severe*).

5. Instrument Format, Instructions, and Training

Results obtained using a PRO instrument can vary according to the instructions given to patients or the training given to the interviewer or persons supervising PRO data collection during a clinical trial. Sponsors should consider all PRO instrument instructions and procedures contained in publications and user manuals provided by developers, including procedures for reviewing completed questionnaires and procedures used to avoid missing data or clarify responses.

It is important that the PRO instrument format used in the clinical trial be consistent with the format that is used during the instrument development process. *Format* refers to the exact questionnaire, diary, or interview script appearance used to collect the PRO data. Format is specific to the administration mode and the data collection method. We plan to review the specific format used in the clinical trial including the order and numbering of items, the presentation of response options in single response or grid formats, the grouping of items, patterns for skipping questions, and all instructions to interviewers or patients.

We recommend that the user manual provided by a developer during the PRO instrument development process specify how to incorporate the instrument into a clinical trial in a way that minimizes administrator burden, patient burden, missing data, and poor data quality. The user manual should explain to investigators and interviewers critical principles of PRO administration.

6. Patient Understanding

When the initial and subsequent drafts of an instrument are prepared, sponsors are encouraged to examine all items and procedures in a pilot test of whether patients understand the items and instructions included in the PRO instrument. This examination should include documentation that the concepts represented in the PRO instrument's conceptual framework are confirmed, that the response options and recall period are appropriately comprehended, and that the instrument's readability is adequate for the intended population. The FDA's evaluation of these procedures is likely to include a review of a cognitive interviewing report containing the script used in patient cognitive interviews, the interview transcripts, the readability test used (if applicable), the *usability testing* process description (if applicable), the cognitive interviews analysis, and the actions taken to delete or modify items, response scales, or patient instructions in response to the cognitive interview or pilot test results. Evidence from the patient cognitive interview studies (i.e., the interview schedule, transcript, and listing of all concepts elicited by a single item) can be used to determine when a concept is adequately captured. Repeating cognitive interviews can help confirm content validity.

7. Scoring of Items and Domains

For each item, numerical scores generally should be assigned to each answer category based on the most appropriate scale of measurement for the item (e.g., nominal, ordinal, interval, or ratio scales). We will review the distribution of item responses to ensure that the response choices represent appropriate intervals.

A scoring algorithm creates a single score from multiple items. We will review the evidence that the summary score is appropriate. Equally weighted scores for each item are appropriate when the responses to the items are independent. If two items are dependent, their collected information is less than two independent items and they are over-weighted when they are treated as two equally weighted items. Over-weighting also may be a concern when the number of response options or the values associated with response options varies by item. The same weighting concerns apply with added complexity when combining domain scores into a single general score. Using qualitative research or defined statistical techniques, sponsors should justify the method chosen to combine items to create a score or to combine domain scores to create a general score.

When empirically determined patient preference ratings are used to weight items or domains, we intend to review the composition of samples and the process used to determine the preference weights. Because preference weights are often developed for use in resource allocation (e.g., as in cost-effectiveness analysis that may use predetermined community weights), it is tempting to use those same weights in the clinical trial setting to demonstrate treatment benefit. However, this practice is discouraged unless the preference weights' relationship to the intended clinical trial population is known and found adequate and appropriate.

Total scores combining multiple domains should be supported by evidence that the total score represents a single albeit complex concept. As described earlier in section III.C., Conceptual Framework of a PRO Instrument, the instrument's final conceptual framework documents the concept represented by each score. If a score is intended to support a targeted claim, the concept measured will match the targeted claim language. Generally, we discourage claims expressed in terms of domain or instrument titles because they often do not represent the concept measured.

8. Respondent and Administrator Burden

Undue physical, emotional, or cognitive strain on patients generally decreases the quality and completeness of PRO data. Factors that can contribute to respondent burden include the following:

- Length of questionnaire or interview

- Formatting

- Font size too small to read easily

- New instructions for each item

- Requirement that patients consult records to complete responses

- Privacy of the setting in which the PRO is completed (e.g., not providing a private space for patients to complete questionnaires containing sensitive information about their sexual performance or substance abuse history)

- Inadequate time to complete questionnaires or interviews

- Literacy level too high for population

- Questions that patients are unwilling to answer

- Perception by patients that the interviewer wants or expects a particular response

- Need for physical help in responding (e.g., turning pages, holding a pen, assistance with a telephone or computer keyboard)

The degree of respondent burden that is tolerable for instruments in clinical trials depends on the frequency and timing of PRO assessments in a protocol and on patient cognition, illness severity, or treatment toxicity. For example, if the questionnaire contains instructions to skip one or more questions based on response to a previous question, respondents may fail to understand what to do and make errors in responding or find the assessment too complicated to complete. Sponsors should consider missing data and the refusal rate as possible indications of inappropriate respondent burden or inappropriate items or response options.

E. Reliability, Other Validity, and Ability to Detect Change

Once the instrument's content validity has been established, we intend to consider the following additional measurement properties during FDA review of a PRO instrument: reliability, construct validity, and ability to detect change. We plan to review the measurement properties that are specific to the documented PRO instrument's conceptual framework, confirmed scoring algorithm, administration procedures, and questionnaire format in light of the clinical trial's objectives, design, enrolled population, and statistical analysis plan (SAP). We also plan to review whether the population and medical conditions included in any sample used to develop or test a PRO instrument are appropriate for the planned clinical trials.

In addition, an adequate study to evaluate any specific measurement property of a PRO instrument should be designed to test a prespecified hypothesis. For example, if the study compares a new PRO measure to an existing measure of the same concept administered during the same interview or within a short time of each other to establish construct validity, the study should be designed to test the hypothesized level of correlation and the results should be discussed in light of that hypothesis.

1. Reliability

Because clinical trials measure change over time, the adequacy of a PRO instrument for use in a clinical trial depends on its reliability or ability to yield consistent, reproducible estimates of true treatment effect.

We will review documentation of tests to determine if reproducibility (e.g., test-retest reliability) has been demonstrated. Test-retest is most informative when the time interval chosen between the test and retest is long enough in stable patients to minimize memory effects. The time interval chosen depends on the variability of the state or experience being evaluated and on the potential for change in the condition or population over time that reflects actual change in the condition rather than variability in stable patients. Test-retest reliability can be tested over a variety of periods to satisfy different study protocols or even in different intervals between visits in the same protocol. We acknowledge that for remitting and relapsing or episodic diseases, test-retest reliability may be difficult or impossible to establish.

Internal consistency reliability tests (e.g., Cronbach's alpha) to determine agreement among responses to different questions, in the absence of test-retest reliability, may not constitute sufficient evidence of reliability for clinical trial purposes. However, as is true for other imperfections in testing, in general, flaws in reliability tend to increase the beta (Type II) error, and instruments demonstrating poor reliability are unlikely to give a false positive result.

When PRO instruments are interviewer-administered, we will review inter-interviewer reproducibility. Inter-interviewer reproducibility depends on instrument administration standardization and interviewer training on this standard.

2. Other Validity

In addition to content validity (discussed in section III.D., Content Validity), we will evaluate evidence of construct validity, and if appropriate, *criterion validity*.

Construct validity is determined by evidence that relationships among items, domains, and concepts conform to a *priori* hypotheses concerning logical relationships that should exist with other measures or characteristics of patients and patient groups.

We will review the construct validity of an instrument to determine whether the documented relationships between results gathered using the instrument and results gathered using other measures are consistent with pre-existing hypotheses concerning those relationships (i.e., discriminant and convergent validity). We will also review evidence that the instrument can differentiate between clinically distinct groups (i.e., known groups validity).

As stated earlier, single-item questions about general concepts are not useful to support claims; however, they can be useful to help assess the construct validity of multi-item measures of the same concept and to determine whether important items or domains of a general concept are missing. For example, when results using single-item general questions do not correlate with results using a multi-item questionnaire of the same general concept, this may be evidence that the questionnaire is not capturing all the important domains of the general concept.

Criterion validity is the extent to which the scores of a PRO instrument are related to a known *gold standard* measure of the same concept. In rare cases, we will also review the criterion validity of an instrument if a criterion measure is purported for the PRO concept assessed (e.g., comparing a new sleep scale to a clinical measure of polysomnography). However, for most PROs, criterion validity testing is not possible because the nature of the concept to be measured does not allow for a criterion measure to exist. This is true for any symptom measure where the symptom is known only to the patient. If a criterion measure is used, sponsors should provide rationale and support for that criterion. We will review the extent to which the PRO measure is correlated with the criterion measure as well as the sensitivity, specificity, and predictive value of the criterion measure.

3. Ability to Detect Change

We will review an instrument's ability to detect change using data that compare change in PRO scores to change in other similar measures that indicate that the patient's state has changed with respect to the concept of interest. A review of the ability to detect change includes evidence that the instrument is equally sensitive to gains and losses in the measurement concept and to change at all points within the entire range expected for the clinical trial population.

When patient experience of a concept is predicted to change, the values for the PRO instrument measuring that concept should change. If there is clear evidence that patient experience relative to the concept has changed, but the PRO scores do not change, then either the ability to detect change is inadequate or the PRO instrument's validity should be questioned. If there is evidence that PRO scores are affected by changes that are not specific to the concept of interest, the PRO instrument's validity may be questioned.

The ability of an instrument to detect change influences the sample size for evaluating the effectiveness of treatment. The extent to which the PRO instrument's ability to detect change varies by important patient subgroups (e.g., sex, race, or age) can affect clinical trial results. If important subgroup differences in ability to detect change are known, these documented differences can be taken into account in assessing results. In general, an inability to detect change tends to support the null hypothesis of no treatment effect.

F. Instrument Modification

The adequacy of an instrument's development and testing is specific to its intended application in terms of population, condition, and other aspects of the measurement context for which the instrument was developed. When a PRO instrument is modified, sponsors generally should provide evidence to confirm the new instrument's adequacy. That is **not** to say that every small change in application or format necessitates extensive studies to document the final version's measurement properties. Additional qualitative work may be adequate depending on the type of modification made. Examples of changes that can alter the way that patients respond to the same set of questions include:

- Changing an instrument from paper to electronic format

- Changing the timing of or procedures for PRO instrument administration within the clinic visit

- Changing the application to a different setting, population, or condition

- Changing the order of items, item wording, response options, or recall period or deleting portions of a questionnaire

- Changing the instructions or the placement of instructions within the PRO instrument

A small nonrandomized study may be adequate to compare the distribution of responses between versions of a questionnaire with different formats (e.g., changing a response scale from vertical to horizontal). If the PRO instrument will be used in a significantly different patient population (e.g., a different disease or age group), we may recommend using qualitative studies to confirm content validity in the new population. A small randomized study to ascertain the measurement properties in the new population may minimize the risk that the instrument will not perform adequately.

G. PRO Instruments Intended for Specific Populations

As previously mentioned, if multiple versions of an instrument will be used in a clinical trial, documentation should exist that the content validity and other measurement properties of those versions are similar to each other. Measurement of PRO concepts in children and adolescents, in patients who have cognitive impairment or are unable to communicate because of serious illness, and across culture or language groups introduces challenges in addition to those already mentioned. These challenges are discussed below.

1. Children and Adolescents

In general, the review issues related to the development process for pediatric PRO instruments are similar to the issues detailed for adults. Additional review issues for PRO instruments applied in children and adolescents include age-related vocabulary, language comprehension, comprehension of the health concept measured, and duration of recall. Instrument development within fairly narrow age groupings is important to account for developmental differences and to determine the lower age limit at which children can understand the questions and provide reliable and valid responses that can be compared across age categories. We discourage ***proxy-reported outcome*** measures for this population (i.e., reports by someone who is not the patient responding as if that person were the patient). For patients who cannot respond for themselves (e.g., infant patients), we encourage observer reports that include only those events or behaviors that can be observed. For example, observers cannot validly report an infant's pain intensity but can report infant behavior thought to be caused by pain.

2. Patients Cognitively Impaired or Unable to Communicate

We discourage proxy-reported outcome measures for this population. For patients who cannot respond for themselves (e.g., cognitively impaired), we encourage observer reports that include only those events or behaviors that can be observed.

3. Culture or Language Subgroups

Because many development programs are multinational, application of PRO instruments to multiple cultures or languages is common in clinical trials. Regardless of whether the instrument was developed concurrently in multiple cultures or languages or whether a fully developed instrument was adapted or translated to new cultures or languages, we recommend that sponsors provide evidence that the content validity and other measurement properties are adequately similar between all versions used in the clinical trial. We will review the process used to translate and culturally adapt the instrument for populations that will use them in the trial.

IV. Clinical Trial Design

The same clinical trial design principles that apply to other endpoint measures also apply to PROs. Therefore, this section is not a comprehensive overview of those principles but rather focuses primarily on issues unique to PRO endpoints.

A. General Protocol Considerations

If the PRO measurement goal is to support labeling claims, PRO concept measurement should be stated as a specific clinical trial objective or hypothesis. It is important that the case report form in the protocol include the exact format and version of the specific PRO instrument to be administered. If an electronic version of the instrument will be used, the protocol can include screen shots or other similar instrument representations. In the process of considering the new drug application (NDA)/biologics license application (BLA)/medical device premarket approval (PMA) or NDA/BLA/PMA supplement, we intend to compare both the planned and actual PRO instrument used and its analysis.

1. Blinding and Randomization

Open-label clinical trials, where patients and investigators are aware of assigned therapy, are rarely adequate to support labeling claims based on PRO instruments. Patients who know they are in an active treatment group may overestimate benefit whereas patients who know they are not receiving active treatment may underreport any improvement actually experienced. For the same reasons, to prevent influencing patient perceptions, PRO instruments administered during a clinic visit should be administered before other clinical assessments or procedures.

In blinded clinical trials, patients should be blinded to treatment assignment throughout the trial. If the treatment has obvious effects, such as adverse events, the clinical trial may be at risk for unintentional unblinding. In these situations, sponsors can use PRO instrument administration techniques that may minimize the effects of possible unblinding, such as using response options that ask for current status, not giving patients access to previous responses, and using instruments that include many items about the same concept.

Suspicion of inadvertent unblinding can be a problematic review consideration for the FDA when assessing PRO endpoints. Therefore, when PRO instruments are included in a clinical trial, we encourage sponsors to include a single item during or at the end of the trial to ask patients to identify the clinical trial arm in which they believe they participated.

The effect of intentional unblinding is important to consider in the interpretation of clinical trial results. There are certain situations, such as in the evaluation of some medical devices or administration of identifiable treatment regimens, where blinding is not feasible and other situations where there is no reasonable control group (and therefore no randomization). When a PRO instrument appears useful in assessing patient benefit in those situations, we encourage sponsors to confer with the appropriate review division.

2. Clinical Trial Quality Control

The quality of a clinical trial can be optimized at the design stage by specifying in the protocol procedures to minimize inconsistencies in trial conduct. We recommend a standardized order by which PRO and other clinical assessments are administered. Other examples of standardized instructions and processes that can appear in the protocol include:

- Training and instructions to patients for self-administered PRO instruments

- Interviewer training and interview format for PRO instruments administered in an interview format

- Instructions for the clinical investigators regarding patient supervision, timing and order of questionnaire administration during or outside the office visit, processes and rules for questionnaire review for completeness, and documentation of how and when data are filed, stored, and transmitted to or from the clinical trial site

- Plans for confirmation of the instrument's measurement properties using clinical trial data

3. Handling Missing Data

Sometimes patients fail to report for visits, fail to complete questionnaires, or withdraw from a clinical trial before its planned completion. The resulting missing data can introduce bias and interfere with the ability to compare effects in the test group with the control group because only a subset of the initial randomized population contributes, and these patient groups may no longer be comparable. Missing data is a major challenge to the success and interpretation of any clinical trial. The clinical trial protocol should describe how missing data will be handled in the analysis.

The protocol can increase the likelihood that a clinical trial will still be informative by establishing backup plans for gathering all treatment-related reasons for patients failing to report at scheduled times or withdrawing from a treatment or the clinical trial and by trying to minimize patient dropouts before trial completion. Patients should remain in the clinical trial, even if they have discontinued treatment, and should continue to provide PRO data. The protocol should also establish a process by which PRO measurement is obtained before or shortly after patient withdrawal from treatment should early withdrawal be unpreventable.

B. Frequency of Assessments

The frequency of PRO assessment should correspond with the specific research questions being addressed, length of recall asked by the instrument's response options, demonstrated instrument measurement properties, the disease or condition's natural history, the treatment's nature, and planned data analysis. Some diseases, conditions, or clinical trial designs may necessitate more than one baseline assessment and several PRO assessments during treatment.

C. Clinical Trial Duration

The duration of PRO assessment depends on the PRO research questions being posed. It is important to consider whether the clinical trial's duration is of adequate length to support the proposed claim and assess a durable outcome in the disease or condition being studied. Generally, duration of follow-up with a PRO assessment should be the same as for other measures of effectiveness. However, the clinical trial duration appropriate for the PRO-related objective may not be the same duration as for other endpoints.

D. Design Considerations for Multiple Endpoints

A single hierarchy of endpoints as diagrammed in an endpoint model (see Figures 1 and 2 in section III.A., Endpoint Model) is determined by the trial's stated objectives and the clinical relevance and importance of each specific measure independently and in relationship to each other. We consider any endpoints that are not part of the prespecified hierarchy of primary and key secondary endpoints to be exploratory. Endpoints included for economic evaluation that are not intended for labeling claims should be designated as such, and will be regarded as exploratory. A PRO measurement can be the clinical trial's primary endpoint measure, a co-primary endpoint measure in conjunction with another PRO measure, other clinical endpoints or physician-rated measurements, or a secondary endpoint measure whose analysis is considered according to a hierarchical sequence. It is critical that the clinical trial protocol define the endpoint measures and the criteria for the statistical analysis and interpretation of results, including a specification of the conditions for a positive clinical trial conclusion, because determination of these criteria and conditions after data are unblinded will not be credible. Sponsors should avoid separate consideration of PRO endpoints from the clinical trial's primary objectives in terms of clinical trial design or data analysis. Sponsors also should avoid *cherry picking* or post hoc selective picking of PRO endpoint results for inclusion in proposed labeling.

E. Planning for Clinical Trial Interpretation Using a Responder Definition

Regardless of whether the primary endpoint for the clinical trial is based on individual responses to treatment or the group response, it is usually useful to display individual responses, often using an *a priori* **responder definition** (i.e., the individual patient PRO score change over a predetermined time period that should be interpreted as a treatment benefit). The responder definition is determined empirically and may vary by target population or other clinical trial design characteristics. Therefore, we will evaluate an instrument's responder definition in the context of each specific clinical trial.

The empiric evidence for any responder definition is derived using anchor-based methods. Anchor-based methods explore the associations between the targeted concept of the PRO instrument and the concept measured by the anchors. To be useful, the anchors chosen should be easier to interpret than the PRO measure itself. For example, the number of incontinence episodes collected in incontinence diaries has been used to determine a responder definition for PRO instruments assessing the annoyance of incontinence. A 50 percent reduction in incontinence episodes might be proposed as the anchor for defining a responder on the PRO instrument. Confirmation of this anchor approach in early clinical trials can provide the basis for the proposed responder definition in the confirmatory trials.

Another anchor-based approach to defining responders makes use of patient ratings of change administered at different periods of time or upon exit from a clinical trial. These numerical ratings range from *worse* to *the same* and *better*. The difference in the PRO score for persons who rate their condition *the same* and *better* or *worse* can be used to define responders to treatment. Patient ratings of change are less useful as anchors when patients are not blinded to treatment assignment.

Another set of approaches to defining a responder are distribution-based methods that use, for example, the between-person standard deviation or the standard error of measurement to define a meaningful change on a scale. Distribution-based methods can be used to categorize these changes as small, moderate, and large and often can be combined with anchor-based estimates to provide confidence in the responder definition. Distribution-based methods for determining clinical significance of particular score changes should be considered as supportive and are not appropriate as the sole basis for determining a responder definition.

Alternatively, it is possible to present the entire distribution of responses for treatment and control group, avoiding the need to pick a responder criterion. Whether the individual responses are meaningful represents a judgment, but that problem is present with almost all endpoints except survival. Such cumulative distribution displays show a continuous plot of the percent change from baseline on the X-axis and the percent of patients experiencing that change on the Y-axis. This display type may be preferable to attempting to provide categorical definitions of *responders*. A variety of responder definitions can be identified along the cumulative distribution of response curve.

Guidance on interpretation considerations for a clinical trial's SAP is found in section V.E., Interpretation of Clinical Trial Results.

F. Specific Concerns When Using Electronic PRO Instruments

When PRO instruments are used, sponsors must ensure that FDA regulatory requirements are met for sponsor and investigator record keeping, maintenance, and access.[7] These responsibilities are independent of the method used to record clinical trial data and, therefore, apply to all types of PRO data including electronic PRO data. Sponsors are responsible for providing investigators with all information to conduct the investigation properly, for monitoring the investigation, for ensuring that the investigation is conducted in accordance with the investigational plan, and for permitting the FDA to access, copy, and verify records and reports relating to the investigation.

The principal record keeping requirements for clinical investigators include the preparation and maintenance of adequate and accurate case histories (including the case report forms and supporting data), record retention, and provision for the FDA to access, copy, and verify records (i.e., source data verification). The investigator's responsibility to control, access, and maintain source documentation can be satisfied easily when paper PRO instruments are used, because the patient usually returns the diary to the investigator who either retains the original or a certified copy as part of the case history. The use of electronic PRO instruments, however, may pose a problem if direct control over source data is maintained by the sponsor or the contract research organization and not by the clinical investigator. We consider the investigator to have met his or her responsibility when the investigator retains the ability to control and provide access to the records that serve as the electronic source documentation for the purpose of an FDA inspection. The clinical trial protocol, or a separate document, should specify how the electronic PRO source data will be maintained and how the investigator will meet the regulatory requirements.

In addition, the FDA has previously provided guidance to address the use of computerized systems to create, modify, maintain, archive, retrieve, or transmit clinical data to the FDA and to clarify the requirements and application of 21 CFR part 11.[8, 9] Because electronic PRO data (including data gathered by personal digital assistants or phone-based interactive voice recording systems) are part of the case history, electronic PRO data should be consistent with the data standards described in that guidance. Sponsors should plan to establish appropriate system and security controls, as well as cyber-security and system maintenance plans that address how to ensure data integrity during network attacks and software updates.

Sponsors also should avoid the following:

- Direct PRO data transmission from the PRO data collection device to the sponsor, clinical investigator, or other third party without an electronic audit trail that documents all changes to the data after it leaves the PRO data collection device.

---

[7] For the principal record keeping requirements for clinical investigators and sponsors developing drugs and biologics, see 21 CFR 312.50, 312.58, 312.62, and 312.68. For medical devices, see 21 CFR 812.140 and 812.145.

[8] See the guidance for industry Computerized Systems Used in Clinical Investigations (http://www.fda.gov/Drugs/GuidanceComplianceRegulatoryInformation/Guidances/default.htm).

[9] See the guidance for industry Part 11, Electronic Records; Electronic Signatures — Scope and Application (http://www.fda.gov/Drugs/GuidanceComplianceRegulatoryInformation/ Guidances/default.htm).

- Source document control by the sponsor exclusively.

- Clinical investigator inability to maintain and confirm electronic PRO data accuracy. The data maintained by the clinical investigator should include an audit trail to capture any changes made to the electronic PRO data at any point in time after it leaves the patient's electronic device.

- The existence of only one database without backup (i.e., risk of data corruption or loss during the trial with no way to reconstitute or verify the data).

- Ability of any entity other than the investigator (and/or site staff designated by the investigator) to modify the source data.

- Loss of adverse event data.

- Premature or unplanned access to unblinded data.

- Inability of an FDA investigator to inspect, verify, and copy the data at the clinical site during an inspection.

- An insecure system where records are easily altered.

- Direct PRO data transmission of important safety information to sponsors, clinical research organizations, and/or third parties, without ensuring the timely transmission of the data to the clinical investigator responsible for the patients.

V. Data Analysis

Incorporating PRO instruments as clinical trial endpoint measures introduces challenges in the analysis of clinical trial data. The most important of these challenges are discussed in the following sections.

A. General Statistical Considerations

The statistical analysis considerations for PRO endpoints are not unlike statistical considerations for any other endpoint used in medical product development.[10] Every protocol should describe the principal data analysis features in the statistical section with a detailed elaboration of the analysis in an SAP. We intend to determine the adequacy of clinical trial data to support claims in light of the prespecified method for endpoint analysis. We usually view unplanned or post hoc statistical analyses conducted after unblinding as exploratory and, therefore, unable to serve as the basis of a labeling claim of effectiveness.

B. Statistical Considerations for Using Multiple Endpoints

PROs in a clinical trial, like non-PRO clinical endpoints, can be primary or secondary endpoints. Primary endpoints are those endpoints on which the main benefit of a clinical trial's test treatment is judged rigorously. Primary endpoints are used to determine the clinical trial

---

[10] See the ICH guidance for industry E9 Statistical Principles for Clinical Trials (http://www.fda.gov/Drugs/GuidanceComplianceRegulatoryInformation/Guidances/default.htm).

sample size and are the endpoints that will be tested statistically. They are clinically meaningful but may not be the most important endpoint because choice of clinical trial endpoints is a complex evolution of expected effect size, expected number of events, and other factors.

There are often multiple endpoints that would be of clinical interest. Analysis of multiple endpoints, where an effect on any of the endpoints will be considered evidence of effectiveness, can inflate the probability of false positive findings known as the Type I error rate, an inflation that can be controlled by a prospectively planned multiplicity adjustment. It is common to analyze secondary endpoints only after success on a primary endpoint. This can be done using a sequential analysis, testing additional endpoints in a defined sequence each at the usual alpha = 0.05 level of statistical significance. The analyses cease when a failure occurs. It is important that the clinical trial protocol specify all primary and secondary endpoints. The SAP should describe the planned primary analysis in detail noting whether the endpoint will be analyzed as a continuous variable (mean scores), dichotomous variable (success or failure), or some graded response; the primary and secondary endpoints; adjustments for multiplicity to control the overall Type I error rate; and the specific statistical methods planned. Sponsors should provide the FDA with the clinical trial's SAP for review.

Cases arise in clinical trials where a clinically meaningful treatment benefit depends on having two or more primary endpoints achieving statistical significance at a specified alpha level (e.g., alpha = 0.05). For example, a clinical trial may identify two endpoints with a decision rule that each should show that the treatment is better than control. Such a decision rule does not require multiplicity adjustment because the maximum Type I error rate (alpha) is actually reduced. However, this type of decision rule will increase the Type II error. Therefore, we recommend sizing the trial carefully for this situation.

There is no single best statistical procedure for multiplicity adjustment because the choice of procedure depends upon the clinical trial's objectives, the most important endpoints, the decision rule for declaring treatment benefit, and other considerations. Some of the statistical procedures that can be useful for a more efficient analysis approach include methods that prespecify a sequence or order of testing or a hierarchy of comparisons that should first be satisfied before others are considered for testing as described above. These methods can be less conservative than the conventional nonhierarchical type methods, such as Bonferroni, the step-down or step-up tests, and prospective alpha allocations schemes, which ignore the hierarchy of comparisons or their families. These conventional type methods should be used when a restriction on the order of testing is not warranted.

A multidomain PRO measure may successfully support a labeling claim based on one or a subset of the domains measured if an a priori analysis plan prespecifies the domains that will be targeted as endpoints and the method of analysis that will adjust for the multiplicity of tests for the specific claim. The use of domain subsets as clinical trial endpoints presupposes that the PRO instrument was adequately developed and validated to measure the subset of domains independently from the other domains.

C. Statistical Considerations for Composite Endpoints

For a PRO instrument with multiple domains, combining the scores to calculate a general score creates a composite endpoint. Composite endpoints have a few advantages (e.g., they can reduce multiplicity problems), but their use for confirmatory clinical trials for specific claims of treatment benefit poses many difficulties and challenges.

Rules for interpretation of composite endpoints depend on substantial experience with the measure in the clinical trial setting. Therefore, development of a composite endpoint at the time the confirmatory clinical trial protocol is written depends on special considerations and substantial empirical evidence of the following: the components are of similar importance to patients, the more important and less important components are equally likely to occur with similar frequency, and the components are likely to have roughly similar treatment effects. Therefore, we discourage the use of a composite endpoint for confirmatory clinical trials when large variations are predicted to exist between its components.

Multiplicity problems arise when the multiple individual components of a composite endpoint are intended as possible claims. In general, individual components of a composite endpoint will not be adequate to support a labeling claim for the components unless the components are prespecified in the protocol as separate endpoints and all prespecified components are reported in labeling as suggested in current guidance.[11] The components of a composite endpoint will be shown in labeling to convey what drove a favorable result. Sequential testing approaches can be used to test the components of a composite. The components are tested only when there is a statistically significant treatment benefit for the composite.

D. Statistical Considerations for Patient-Level Missing Data

When the amount of missing data becomes large, clinical trial results can be inconclusive. As described in section IV., Clinical Trial Design, we encourage prespecified procedures in the clinical trial protocol to avoid missing data. We also encourage prespecified procedures for obtaining data on each patient at the time of early withdrawal from the clinical trial. If a measurement is taken at the time of withdrawal, this information can be handled according to rules established in the SAP. In clinical trials of terminal illness, it is critical to plan ahead for how missing data because of death will be handled. Missing data may occur because of the treatment received or the underlying disease and can introduce bias in the analysis of treatment differences and conclusions about treatment effect.

Even with the best planning, data may be missing at the end of the clinical trial. The SAP should address plans for how the statistical analyses will handle missing data when evaluating treatment benefit and when considering patient success or patient response.

1. Missing Items within Domains

At a specific patient visit, a domain measurement may be missing some, but not all, items. One approach to handling this type of missing data is to define rules that specify the number of items that can be missing and still consider the domain as adequately measured. Rules for handling missing data should be specific to each PRO instrument and usually should be determined during the instrument development process. The SAP should specify all rules for

---

[11] See the guidance for industry Clinical Studies Section of Labeling for Human Prescription Drug and Biological Products — Content and Format (http://www.fda.gov/Drugs/ GuidanceComplianceRegulatoryInformation/Guidances/default.htm).

handling missing data. For example, the SAP can specify the proportion of items that can be missing before a domain is treated as missing.

2. Missing Entire Domains or Entire Measurements

We will consider a variety of statistical strategies to deal with missing data because of a patient's early termination before planned completion of a trial. No single method is generally considered as preferred. All of these strategies are imperfect, as they involve strong or weak assumptions about what caused data to be missing, assumptions that usually cannot be verified from the data. Methods of missing data imputation should take the patient population, disease progression, and respondent burden into account. How to impute the missing data for a PRO endpoint and any related supportive endpoints should be addressed in the protocol and the SAP. In addition, the sensitivity analyses in analyzing the PRO endpoints should be proposed in the protocol and the SAP to assess the robustness of statistical estimation for endpoints with the missing data imputed. We recommend that in the protocol the sponsor propose two or more sensitivity analyses with different methods for missing data imputation.

E. Interpretation of Clinical Trial Results

Because statistical significance can sometimes be achieved for small changes in PRO measures that may not be clinically meaningful (i.e., do not indicate treatment benefit), we encourage sponsors to avoid proposing labeling claims based on statistical significance alone.

To demonstrate treatment benefit, we find it informative to examine the cumulative distribution function (CDF) of responses between treatment groups to characterize the treatment effect and examine the possibility that the mean improvement reflects different responses in patient subsets. To interpret the CDF, sponsors can apply the responder definition along the CDF curve at each level of response (see section IV.E., Planning for Clinical Trial Interpretation Using a Responder Definition).

Interpretation of PRO endpoints follows similar considerations as for all other endpoint types used to evaluate treatment benefit of a medical product.

Glossary

Ability to detect change — Evidence that a PRO instrument can identify differences in scores over time in individuals or groups who have changed with respect to the measurement concept.

Claim — A statement of treatment benefit. A claim can appear in any section of a medical product's FDA-approved labeling or in advertising and promotional labeling of prescription drugs and devices.

Cognitive interviewing — A qualitative research tool used to determine whether concepts and items are understood by patients in the same way that instrument developers intend. Cognitive interviews involve incorporating follow-up questions in a field test interview to gain a better understanding of how patients interpret questions asked of them. In this method, respondents are often asked to think aloud and describe their thought processes as they answer the instrument questions.

Concept — The specific measurement goal (i.e., the thing that is to be measured by a PRO instrument). In clinical trials, a PRO instrument can be used to measure the effect of a medical intervention on one or more concepts. PRO concepts represent aspects of how patients function or feel related to a health condition or its treatment.

Conceptual framework of a PRO instrument — An explicit description or diagram of the relationships between the questionnaire or items in a PRO instrument and the concepts measured. The conceptual framework of a PRO instrument evolves over the course of instrument development as empiric evidence is gathered to support item grouping and scores. We review the alignment of the final conceptual framework with the clinical trial's objectives, design, and analysis plan.

Construct validity — Evidence that relationships among items, domains, and concepts conform to a priori hypotheses concerning logical relationships that should exist with other measures or characteristics of patients and patient groups.

Content validity — Evidence from qualitative research demonstrating that the instrument measures the concept of interest including evidence that the items and domains of an instrument are appropriate and comprehensive relative to its intended measurement concept, population, and use. Testing other measurement properties will not replace or rectify problems with content validity.

Criterion validity — The extent to which the scores of a PRO instrument are related to a known gold standard measure of the same concept. For most PROs, criterion validity cannot be measured because there is no gold standard.

Domain — A subconcept represented by a score of an instrument that measures a larger concept comprised of multiple domains. For example, psychological function is the larger concept containing the domains subdivided into items describing emotional function and cognitive function.

Endpoint — The measurement that will be statistically compared among treatment groups to assess the effect of treatment and that corresponds with the clinical trial's objectives, design, and data analysis. For example, a treatment may be tested to decrease the intensity of symptom Z. In this case, the endpoint is the change from baseline to time T in a score that represents the concept of symptom Z intensity.

Endpoint model — A diagram of the hierarchy of relationships among all endpoints, both PRO and non-PRO, that corresponds to the clinical trial's objectives, design, and data analysis plan.

Health-related quality of life (HRQL) — HRQL is a multidomain concept that represents the patient's general perception of the effect of illness and treatment on physical, psychological, and social aspects of life. Claiming a statistical and meaningful improvement in HRQL implies: (1) that all HRQL domains that are important to interpreting change in how the clinical trial's population feels or functions as a result of the targeted disease and its treatment were measured; (2) that a general improvement was demonstrated; and (3) that no decrement was demonstrated in any domain.

Instrument — A means to capture data (i.e., a questionnaire) plus all the information and documentation that supports its use. Generally, that includes clearly defined methods and instructions for administration or responding, a standard format for data collection, and well-documented methods for scoring, analysis, and interpretation of results in the target patient population.

Item — An individual question, statement, or task (and its standardized response options) that is evaluated by the patient to address a particular concept.

Item tracking matrix — A record of the development (e.g., additions, deletions, modifications, and the reasons for the changes) of items used in an instrument.

Measurement properties — All the attributes relevant to the application of a PRO instrument including the content validity, construct validity, reliability, and ability to detect change. These attributes are specific to the measurement application and cannot be assumed to be relevant to all measurement situations, purposes, populations, or settings in which the instrument is used.

Patient-reported outcome (PRO) — A measurement based on a report that comes directly from the patient (i.e., study subject) about the status of a patient's health condition without amendment or interpretation of the patient's response by a clinician or anyone else. A PRO can be measured by self-report or by interview provided that the interviewer records only the patient's response.

Proxy-reported outcome — A measurement based on a report by someone other than the patient reporting as if he or she is the patient. A proxy-reported outcome is not a PRO. A proxy report also is different from an observer report where the observer (e.g., clinician or caregiver), in addition to reporting his or her observation, may interpret or give an opinion based on the observation. We discourage use of proxy-reported outcome measures particularly for symptoms that can be known only by the patient.

Quality of life — A general concept that implies an evaluation of the effect of all aspects of life on general well-being. Because this term implies the evaluation of nonhealth-related aspects of life, and because the term generally is accepted to mean what the patient thinks it is, it is too general and undefined to be considered appropriate for a medical product claim.

Questionnaire — A set of questions or items shown to a respondent to get answers for research purposes. Types of questionnaires include diaries and event logs.

Recall period — The period of time patients are asked to consider in responding to a PRO item or question. Recall can be momentary (real time) or retrospective of varying lengths.

Reliability — The ability of a PRO instrument to yield consistent, reproducible estimates of true treatment effect.

Responder definition — A score change in a measure, experienced by an individual patient over a predetermined time period that has been demonstrated in the target population to have a significant treatment benefit.

Saturation — When interviewing patients, the point when no new relevant or important information emerges and collecting additional data will not add to the understanding of how patients perceive the concept of interest and the items in a questionnaire.

Scale — The system of numbers or verbal anchors by which a value or score is derived for an item. Examples include VAS, Likert scales, and rating scales.

Score — A number derived from a patient's response to items in a questionnaire. A score is computed based on a prespecified, validated scoring algorithm and is subsequently used in statistical analyses of clinical trial results. Scores can be computed for individual items, domains, or concepts, or as a summary of items, domains, or concepts.

Sign — Any objective evidence of a disease, health condition, or treatment-related effect. Signs are usually observed and interpreted by the clinician but may be noticed and reported by the patient.

Symptom — Any subjective evidence of a disease, health condition, or treatment-related effect that can be noticed and known only by the patient.

Target product profile (TPP) — A clinical development program summary in the context of labeling goals where specific types of evidence (e.g., clinical trials or other sources of data) are linked to the targeted labeling claims or concepts.

Treatment benefit — The effect of treatment on how a patient survives, feels, or functions. Treatment benefit can be demonstrated by either an effectiveness or safety advantage. For example, the treatment effect may be measured as an improvement or delay in the development of symptoms or as a reduction or delay in treatment-related toxicity. Measures that do not directly capture the treatment effect on how a patient survives, feels, or functions are surrogate measures of treatment benefit.

Usability testing — A formal evaluation with documentation of respondents' abilities to use the instrument, as well as comprehend, retain, and accurately follow instructions.

Appendix: Information On A Pro Instrument Reviewed By The FDA

The following topics represent areas that should be addressed in PRO documents provided to the FDA for review. The extent of background information provided in each section will vary depending upon the PRO instrument used. Some sections may be less relevant for a particular PRO instrument application than others, or may be less complete for discussions in early stages of medical product development. Refer to the content of this guidance for additional information concerning the types of evidence needed in each of the following areas.

If the PRO information is provided electronically, it should be placed in section 5.3.5.3 of the electronic common technical document.[12]

---

[12] See the ICH guidance for industry M2 eCTD: Electronic Common Technical Document Specification (http://www.fda.gov/Drugs/GuidanceComplianceRegulatoryInformation/ Guidances/default.htm).

I. Instrument (review cannot begin without a copy of the proposed instrument):

    A. Exact version of the instrument proposed or used in the clinical trial (protocol) under review and all instructions for use. Include screen shots or interviewer scripts, if relevant.

    B. Prior versions, if relevant.

    C. Instructions for use: An instrument user manual can be provided as Appendix A and referenced here.

        1. Administration timing, method (e.g., paper or pencil, electronic), and mode (e.g., self-, clinician-, or interviewer-administered)

        2. The scoring algorithm

        3. Training method and materials used for questionnaire administration

            a. Patient training — summarize here and include a copy of all materials in Appendix A1

            b. Investigator training — summarize here and include a copy of all materials in Appendix A2

            c. Other training — summarize here and include a copy of all materials in Appendix A3

II. Targeted Claims or Target Product Profile (TPP)[13]

Include language describing all specific targeted labeling claims related to all clinical trial endpoint measures, both PRO and non-PRO, and specific to:

- Disease or condition with stage, severity, or category, if relevant

- Intended population (e.g., age group, sex, other demographics)

- Data analysis plan

III. Endpoint Model

    A. Relationships (known and hypothesized) among all clinical trial endpoints, both PRO and non-PRO. These endpoints can include physiologic/lab/physical, caregiver, or clinician-reported measures in addition to PROs.

    B. Hierarchy of all PRO and non-PRO endpoints intended to support claims corresponding with the planned data analyses.

---

[13] See the draft guidance for industry and review staff Target Product Profile — A Strategic Development Process Tool. When final, this guidance will represent the FDA's current thinking on this topic. For the most recent version of a guidance, check the FDA Drug guidance Web page at http://www.fda.gov/Drugs/GuidanceComplianceRegulatoryInformation/Guidances/default.htm.

IV. The PRO Instrument's Conceptual Framework

Diagram of hypothesized (proposed) or final PRO instrument conceptual framework showing relationship of items to domains and domains to total score. Ensure that the PRO instrument's conceptual framework corresponds to the clinical trial endpoints described in the clinical trial protocol and proposed as labeling claims.

V. Content Validity Documentation

Evidence that instrument captures all of the most clinically important concepts and items, and that items are complete, relevant (appropriate), and understandable to the patient. This evidence applies to both existing and newly created instruments and is specific to the planned clinical trial population and indication. Documentation includes:

A. Literature review and documentation of expert input

B. Qualitative study protocols, interview guides, and summary of results for:

    1. Focus group testing (include transcripts in Appendix C1)

    2. Open-ended patient interviews (include transcripts in Appendix C2)

    3. Cognitive interviews (include transcripts in Appendix C3)

C. Origin and derivation of items with chronology of events for item generation, modification, and finalization

Item tracking matrix for versions tested with patients showing items retained and items deleted providing evidence of saturation. Summarize here and include complete materials under Appendix B.

D. Qualitative study summary that supports content validity for:

    1. Item content

    2. Response options

    3. Recall period

    4. Scoring

E. Summary of qualitative studies demonstrating how item pool was generated, reduced, and finalized. Specify type of study (i.e., focus group, patient interview, or cognitive interview) and characteristics of study population. Include full transcripts and datasets in Appendix C.

VI. Assessment of Other Measurement Properties

Assuming content validity is established in the intended population and application, evidence that the instrument is reliable, valid, and able to detect change. The same version of the instrument to be used in the clinical trial should be used to assess measurement properties.

   A.   Protocols for instrument testing

   B.   Summary of testing results for each domain or summary score proposed as support for claims:

      1.   Reliability (internal; test-retest)

      2.   Construct validity (convergent, discriminant, known-groups)

      3.   Ability to detect change

VII. Interpretation of Scores

   A.   Summary of the logic and methods used to interpret the clinical meaningfulness of clinical trial results

   B.   Responder definition (i.e., definition of meaningful within-person change specific to the clinical trial population)

VIII. Language Translation and Cultural Adaptation

   A.   Process used to translate and culturally adapt the instrument for populations that will use them in the trial

   B.   Description of patient testing, language- or culture-specific concerns, and rationale for decisions made to create new versions.

   C.   Copies of translated or adapted versions

   D.   Evidence that content validity and other measurement properties are comparable between the original and new instruments

IX. Data Collection Method

   A.   Process used to develop data collection methods (e.g., electronic, paper) intended for use in the clinical trial

      If electronic data collection is used to assess PRO endpoints, evidence that procedures for maintenance, transmission, and storage of electronic source documents comply with regulatory requirements.

   B.   Evidence that content validity and other measurement properties are comparable among all data collection methods

   C.   User manual for each additional data collection method

X. Modifications

      Any change in the original instrument (e.g., wording of items, response options, recall period, use in a new population or indication)

   A.   Rationale for and process used to modify the instrument

B. Copy of original and new instruments

C. Evidence that content validity and other measurement properties are comparable between the original and modified instruments (including use in a new indication or population)

XI. PRO-Specific Plans Related to Clinical Trial Design and Data Analysis

A. Clinical trial protocol. Ensure in the protocol that:

- Each PRO endpoint is stated as a specific clinical trial objective and multiplicity concerns are addressed

- The clinical trial will be adequately blinded

- Procedures for training are well-described for:
    - Patients
    - Interviewers
    - Clinical investigators

- Plans for instrument administration are consistent with instrument's user manual

- Plans for PRO instrument scoring are consistent with those used during instrument development

- Procedures include assessment of PRO endpoint before or shortly after a patient withdraws from the clinical trial

- Frequency and timing of PRO assessments are appropriate given patient population, clinical trial design and objectives, and demonstrated PRO measurement properties

- Clinical trial duration is adequate to support PRO objectives

- Plans are included for handling missing data

- Plans are included for a cumulative distribution function comparison among treatment groups

- Data collection, data storage, and data handling and transmission of procedures, including electronic PROs, are specified

B. Statistical analysis plan (SAP). Ensure the SAP includes:

- Plans for multiplicity adjustment

- Plans for handling missing data at both the instrument and patient level

- Description of how between-group differences will be portrayed (e.g., cumulative distribution function)

XII. Key References

List and attach all relevant published and unpublished documents

Appendix A — User Manual

A1: Patient training

A2: Investigator training

A3: Other training

Appendix B — Item Tracking Matrix

Appendix C — Transcripts

C1: Focus groups

C2: Open-ended patient interviews

C3: Cognitive interviews

Guidance for Industry Pharmacogenomic Data Submissions

Guidance for Industry Pharmacogenomic Data Submissions, Guidance for Industry[1,2]

U.S. Department of Health and Human Services
Food and Drug Administration
Center for Drug Evaluation and Research (CDER)
Center for Biologics Evaluation and Research (CBER)
Center for Devices and Radiological Health (CDRH)

March 2005

Procedural

Contains Nonbinding Recommendations

The guidance provides recommendations to sponsors holding investigational new drug applications (INDs), new drug applications (NDAs), and biologics license applications (BLAs) on what pharmacogenomic data to submit to the agency during the drug development process, the format of submissions, and how the data will be used in regulatory decision making. The guidance is intended to facilitate scientific progress in the area of pharmacogenomics.

Note: *This guidance represents the Food and Drug Administration's (FDA's) current thinking on this topic. It does not create or confer any rights for or on any person and does not operate to bind FDA or the public. You can use an alternative approach if the approach satisfies the requirements of the applicable statutes and regulations. If you want to discuss an alternative approach, contact the appropriate FDA staff. If you cannot identify the appropriate FDA staff, call the appropriate number listed on the title page of this guidance.*

I. Introduction

This guidance is intended to facilitate scientific progress in the field of pharmacogenomics and to facilitate the use of pharmacogenomic data in drug development. The guidance provides recommendations to sponsors holding investigational new drug applications (INDs), new drug applications (NDAs), and biologics license applications (BLAs) on (1) when to submit

---

[1] Available on the FDA website at: http://www.fda.gov/downloads/RegulatoryInformation/ Guidances/UCM126957.pdf

[2] This guidance has been prepared by the Center for Drug Evaluation and Research (CDER) and the Center for Biologics Evaluation and Research (CBER), in cooperation with the Center for Devices and Radiological Health (CDRH) at the Food and Drug Administration.

pharmacogenomic data to the Agency during the drug or biological drug product[3] development and review processes, (2) what format and content to provide for submissions, and (3) how and when the data will be used in regulatory decision making. Key information, including examples of when pharmacogenomic data submissions would be required and when voluntary genomic data submissions (VGDSs) would be welcome are provided in a separate companion document (Pharmacogenomic Data Submissions, Attachment: Examples of Voluntary Submissions or Submissions Required Under 21 CFR 312, 314, or 601).

For the purposes of this guidance, the term pharmacogenomics is defined as the use of a pharmacogenomic or pharmacogenetic test (see glossary for definitions) in conjunction with drug therapy. Pharmacogenomics does not include the use of genetic or genomic techniques for the purposes of biological product characterization or quality control (e.g., cell bank characterization, bioassays). The FDA plans to provide guidance on those uses at a future time. Pharmacogenomics also does not refer to data resulting from proteomic or metabolomic techniques. This document is not meant to provide guidance on pharmacoproteomics or multiplexed protein analyte based technologies. However, the voluntary submission process described in this guidance may be used to submit such data if so desired.

FDA's guidance documents, including this guidance, do not establish legally enforceable responsibilities. Instead, guidances describe the Agency's current thinking on a topic and should be viewed only as recommendations, unless specific regulatory or statutory requirements are cited. The use of the word should in Agency guidances means that something is suggested or recommended, but not required.

II. Background

The promise of pharmacogenomics lies in its potential to help identify sources of inter-individual variability in drug response (both effectiveness and toxicity); this information will make it possible to individualize therapy with the intent of maximizing effectiveness and minimizing risk.

However, the field of pharmacogenomics is currently in early developmental stages, and such promise has not yet been realized. The Agency has heard that pharmaceutical sponsors have been reluctant to embark on programs of pharmacogenomic testing during FDA-regulated phases of drug development because of uncertainties in how the data will be used by FDA in the drug application review process. This guidance is intended to help clarify FDA policy in this area.

Sponsors submitting or holding INDs, NDAs, or BLAs are subject to FDA requirements for submitting to the Agency data relevant to drug safety and effectiveness (including 21 CFR 312.22, 312.23, 312.31, 312.33, 314.50, 314.81, 601.2, and 601.12). Because these regulations were developed before the advent of widespread animal or human genetic or gene expression testing, they do not specifically address when such data must be submitted. The FDA has

---

[3] For the purposes of this guidance, the term drug or drug product includes human drug and biological products.

received numerous inquiries about what these regulations require of sponsors who are conducting such testing.

From a public policy perspective, a number of factors should be considered when interpreting how these regulations apply to the developing field of pharmacogenomics. Because the field of pharmacogenomics is rapidly evolving, in many circumstances, the experimental results may not be well enough established scientifically to be suitable for regulatory decision making. For example:

- Laboratory techniques and test procedures may not be well validated. In addition, test systems may vary so that results may not be consistent or generalizable across different platforms. A move to standardize assays is underway, and much more information should be available within the next several years.

- The scientific framework for interpreting the physiologic, toxicologic, pharmacologic, or clinical significance of certain experimental results may not yet be well understood.

- The findings from a specific study often cannot be extrapolated across species or to different study populations (e.g., various human subpopulations with different genetic backgrounds).

- The standards for transmission, processing, and storage of the large amounts of highly dimensional data generated from microarray technology have neither been well defined nor widely tested.

Despite these concerns, some pharmacogenetic tests — primarily those related to drug metabolism — have well-accepted mechanistic and clinical significance and are currently being integrated into drug development decision making and clinical practice.

It is important for FDA to have a role in the evaluation of pharmacogenomic tests, both to ensure that evolving FDA policies are based on the best science and to provide public confidence in the field. The FDA developed this guidance to facilitate the use of pharmacogenomic tests during drug development and encourage open and public sharing of data and information on pharmacogenomic test results.

To this end, the Agency has undertaken a process for obtaining input from the scientific community and the public. On May 16 and 17, 2002, the Agency held a workshop, cosponsored by pharmaceutical industry groups, to identify key issues associated with the application of pharmacogenetics and pharmacogenomics to drug development. Subsequently, on April 8, 2003, a public presentation was made to the FDA Science Board. This presentation contained a proposal for developing guidance on the submission of information on pharmacogenomic tests and a potential algorithm for deciding whether submission of such data is voluntary or required. The Science Board endorsed moving forward with both of these proposals. In November 2003, FDA published a draft version of this guidance and received public comment on the draft guidance. The Agency also has developed internal policy related to pharmacogenomics and voluntary submissions.[4]

---

[4] A charter has been developed outlining the organization, principles, and function of the inter-center Interdisciplinary Pharmacogenomics Review Group (IPRG) (MaPP 4180.2). In addition, policy has been

The policies and processes outlined in this final guidance are intended to take the above factors into account and to assist in advancing the field in a manner that will benefit both drug development programs and the public health.

III. Submission Policy

A. General Principles

The FDA recognizes that its pharmacogenomic data submission policies must be consistent with the relevant codified regulatory submission requirements for investigational and marketing application submitters and holders. At present, many pharmacogenomic results are not well enough established scientifically to be appropriate for regulatory decision making.[5] This guidance interprets FDA's regulations for investigational and marketing submissions, with the goal of clarifying FDA's current thinking about when the regulations require pharmacogenomic data to be submitted and when the submission of such data would be welcome on a voluntary basis. In some cases, complete reports of pharmacogenomic studies suffice, while in others, an abbreviated report or synopsis should or must be submitted.[6]

Because FDA regulations establish different requirements for investigational applications, unapproved marketing applications, and approved marketing applications, this guidance sets out different submission algorithms for each of these categories. The guidance also clarifies how the Agency currently intends to use such data in regulatory decision making — that is, when the data will be considered sufficiently reliable to serve as the basis for regulatory decision making; when it will be considered only supportive to a decision; and when the data will not be used in regulatory decision making.

This guidance also makes a distinction between pharmacogenomic tests that may be considered either probable or known *valid biomarkers*, which may be appropriate for regulatory decision making, and other less well-developed tests that are either observational or exploratory biomarkers that, alone, are insufficient for making regulatory decisions. Although, currently, most pharmacogenomic measurements are not considered valid biomarkers, certain markers (e.g., for drug metabolism) are well established biomarkers with clear clinical significance. Undoubtedly, the distinction between what tests are appropriate for regulatory decision making and those that are not will change over time as the science evolves. Throughout the development of these tests, as appropriate, FDA will continue to seek public

---

developed for Agency staff, explaining how voluntary genomic data submissions (VGDSs) will be received and reviewed in the Agency (MaPP 4180.3; SOPP 8114).

[5] For purposes of this document, the term *regulatory decision making*, as defined here, applies to decisions that FDA may make in the evaluation of pharmacogenomic information used to establish the dosing, safety, or effectiveness of a drug or biological product. FDA regulatory decisions occur throughout the investigational stages of product development, during premarket review, and during postmarket regulation.

[6] For further information on when abbreviated study reports can be submitted in NDAs and BLAs, see the guidance for industry *Submission of Abbreviated Reports and Synopses in Support of Marketing Applications*, developed under section 118 of the Food and Drug Administration Modernization Act.

comment as we evaluate whether a biomarker is a *valid biomarker* (e.g., via discussions at Advisory Committee meetings).

For the purposes of this guidance, a pharmacogenomic test result may be considered a *valid biomarker* if (1) it is measured in an analytical test system with well-established performance characteristics and (2) there is an established scientific framework or body of evidence that elucidates the physiologic, pharmacologic, toxicologic, or clinical significance of the test results. For example, the consequences for drug metabolism of genetic variation in the human enzymes CYP2D6 and thiopurine methyltransferase are well understood in the scientific community and are reflected in certain approved drug labels. The results of genetic tests that distinguish allelic variants of these enzymes are considered to be well established and, therefore, valid biomarkers.

This guidance makes an additional distinction between known valid biomarkers that have been accepted in the broad scientific community and probable valid biomarkers that appear to have predictive value for clinical outcomes, but may not yet be widely accepted or independently verified by other investigators or institutions (see Glossary). When a sponsor generates, or possesses, data sufficient to establish a significant association between a pharmacogenomic test result and clinical outcomes, the test result represents a probable valid biomarker. It would be expected that this biomarker would meet criteria (1) and (2) above, and its association with a meaningful outcome would have been demonstrated in more than one experiment.

The algorithms described below for investigational and marketing application holders describe when to submit to FDA data on known valid biomarkers. Data on probable valid biomarkers need not be submitted to the IND unless they are used by a sponsor to make decisions regarding specific animal safety studies or clinical trials (e.g., using biomarker data as inclusion or exclusion criteria, assessment of treatment-related prognosis, or stratifying patients by dose) or are a probable valid biomarker in human safety studies (see section IV.A).[7] However, we recommend that sponsors or applicants submit reports on all probable valid biomarkers to new (i.e., unapproved) NDAs or BLAs according to the algorithm in section IV.B.

Many pharmacogenomic testing programs implemented by pharmaceutical sponsors or by scientific organizations are intended to develop the knowledge base necessary to establish the validity of new genomic biomarkers. During such a period of scientific exploration, test results are not useful in making regulatory judgments pertaining to the safety or effectiveness of a drug and are not considered known or probable valid biomarkers. However, scientific development of this sort is highly desirable for advancing the understanding of relationships between genotype or gene expression and responses to drugs and, therefore, should be encouraged and facilitated. For these reasons, although submission of exploratory pharmacogenomic data is not required under the regulations, FDA is encouraging voluntary submission of such data, as described below.

---

[7] For the purposes of this guidance, the phrase decision making by the sponsor, as defined here, refers to study- or trial-specific decisions that a sponsor might make in the development of a drug, but not to overall strategies related to drug development or portfolio management.

B. Specific Uses of Pharmacogenomic Data in Drug Development and Labeling

As the field of pharmacogenomics advances, it is likely (and desirable) that sponsors will begin to use pharmacogenomic tests to support drug development and/or to guide therapy. Sponsors may choose to submit pharmacogenomic data that have not achieved the status of a valid biomarker to an investigational or marketing application to support scientific contentions related to dosing and dosing schedule, safety, or effectiveness. For example, a sponsor may wish to provide supportive data demonstrating that changes in drug-induced gene expression differ between species that have different toxicologic responses to a drug, thus correlating changes in certain gene expression patterns with a specific toxicity. Or, a pharmacogenomic test result might also be used to stratify patients in a clinical trial or to identify patients at higher risk for an adverse event to correlate test results with clinical outcome.

When pharmacogenomic results affect the design of a specific animal safety trial, or human safety or efficacy trial, the submission algorithms described below suggest that full information on the test system must be submitted to the IND (§§ 312.30(b) and 312.31). In contrast, results from earlier feasibility studies done under the same IND (or outside the IND) to establish the potential usefulness of the pharmacogenomic test (e.g., from samples taken during a doseresponse study) are not a required submission, but would be encouraged as a voluntary submission. However, a plan to perform any invasive test, including phlebotomy, with the possible intent to conduct pharmacogenomic testing on a sample, must be noted both in the protocol and the informed consent document (§§ 312.23(a)(6), 312.30(b), and 50.25).

If a pharmacogenomic test shows promise for enhancing the dose selection, safety, or effectiveness of a drug, a sponsor may wish to fully integrate pharmacogenomic data into the drug development program. This integration could occur in two ways:

1. The pharmacogenomic data may be intended to be included in the drug labeling in an informational manner.

   For example, such data might be used to describe the potential for dose adjustment by drug metabolism genotype (e.g., CYP2D6*5) or to mention the possibility of a side effect of greater severity or frequency in individuals of a certain genotype or gene expression profile. In such cases, the pharmacogenomic test result would be considered a known valid biomarker. However, an FDA-approved pharmacogenomic test may not be available or required to be available, or a commercial pharmacogenomic test may not be widely available. Given this level of complexity, at the current time, sponsors should consult the relevant FDA review division for advice on how to proceed in a specific case. However, whenever a sponsor intends to include pharmacogenomic data in the drug label, complete information on the test and results must be submitted to the Agency as described under §§ 314.50 and 601.2.

2. The pharmacogenomic data and resulting test or tests may be intended to be included in the drug labeling to choose a dose and dose schedule, to identify patients at risk, or to identify patient responders. Inclusion of a pharmacogenomic test in the labeling would be contingent upon its performance characteristics. For example:

- Patients will be tested for drug metabolism genotype and dosed according to the test results.

- Patients will be selected as potential responders for an efficacy trial (or deselected because of a high risk) based on genotype (e.g., of either the patient or the patient's tumor) or gene expression profile.

- Patients will be excluded from a clinical trial based on genotype or gene expression profile (e.g., biomarker for risk of an adverse event).

In all of these cases, FDA recommends co-development of the drug and the pharmacogenomic tests, if they are not currently available, and submission of complete information on the test/drug combination to the Agency. The FDA plans to issue further guidance on co-development of pharmacogenomic tests and drugs.

The Office of In Vitro Diagnostics in CDRH, appropriate review divisions in CBER, and the Clinical and Clinical Pharmacology Review divisions in CBER or CDER are willing to meet jointly with sponsors to discuss both scientific and regulatory issues with regard to new pharmacogenomic tests. The CDRH has both formal (IDE) and informal (pre- IDE) processes to evaluate protocols for pharmacogenomic test development.

C. Benefits of Voluntary Submissions to Sponsors and FDA

At the current time, most pharmacogenomic data are of an exploratory or research nature, and FDA regulations do not require that these data be submitted to an IND, or that complete reports be submitted to an NDA or BLA. However, voluntary submissions can benefit both the industry and FDA in a general way by providing a means for sponsors to ensure that regulatory scientists are familiar with and prepared to appropriately evaluate future genomic submissions. The FDA and industry scientists alike would benefit from an enhanced understanding of relevant scientific issues, such as the following:

- The types of genetic loci or gene expression profiles being explored by the pharmaceutical industry for pharmacogenomic testing

- The test systems and techniques being employed

- The problems encountered in applying pharmacogenomic tests to drug development

- The ability to transmit, store, and process large amounts of complex pharmacogenomic data streams with retention of fidelity

- The scientific rationale for standardizing naming and characterization of the genes used on different genomic analysis platforms and for developing bioinformatics software programs used to evaluate pharmacogenomic data

- Facilitate identification of predictors of safety, effectiveness, or toxicity

A greater understanding of the issues surrounding the use of pharmacogenomic data may prevent delays in reviews of future submissions where genomics are an integral part of specific studies in a drug development program.

Therefore, FDA is requesting that sponsors conducting such programs consider providing pharmacogenomic data to the Agency *voluntarily*, when such data are not otherwise required under the regulations. To facilitate VGDSs, FDA has established a cross-center Interdisciplinary Pharmacogenomic Review Group (IPRG) to review VGDSs, to work on policy development, and, upon request, to advise review divisions on interpretation and evaluation of pharmacogenomic data.

For sponsors, voluntary submission of genomic data offers a number of specific potential benefits:

- Meet *informally* with FDA and receive peer review assessments of scientific data from pharmacogenomic experts at the Agency

- Obtain insight into the evolving regulatory decision making process as it relates to genetic and genomic information

- Familiarize FDA scientists with novel pharmacogenomic experiments, data analysis, and interpretation approaches at an early stage

- Conserve time and resources by obtaining feedback from FDA on a VGDS that might highlight unaddressed issues that could prove time consuming or costly later during product development

- Identify new opportunities for drug development (e.g., feedback from FDA might help reach new strategic decisions). For example, a shelved product may be continued when new tools such as genotyping assays become available to demonstrate effectiveness in a subpopulation.

- Make a contribution to the VGDS data repository to facilitate advancement of pharmacogenomics and development of rational, data-based policies and guidances

IV. Submission Of Pharmacogenomic Data

The FDA's regulations establish different requirements for INDs, new (i.e., unapproved) NDAs and BLAs, and approved NDAs and BLAs. For this reason, there are different submission algorithms for the submission of pharmacogenomic data.

A. Submission of Pharmacogenomic Data During the IND Phase

Section 312.23 describes information submission requirements for an IND, including data generated or available during the IND phase. Section 312.23(a)(8) contains the requirements for pharmacology and toxicology information: "Adequate information about pharmacologic and toxicological studies of the drug involving laboratory animals or in vitro, *on the basis of which* the sponsor has concluded that it is reasonably safe to conduct the proposed clinical investigations" (emphasis added). The in vitro and animal studies needed to establish a basis for proceeding with human trials of various types are well established internationally. Therefore, pharmacogenomic data relevant to, or derived from, animal or in vitro studies must ordinarily be submitted according to § 312.23(a)(8) when the sponsor wishes to use these data to make a scientific case, or when the pharmacogenomic test is a known valid biomarker.

Section 312.23(a)(9) sets forth the requirements for submitting previous human experience with an investigational drug. The application must include a summary of trials or human experience relevant to an evaluation of the safety or effectiveness of a drug. Therefore, sponsors must submit human data of known relevance (e.g., known valid pharmacogenomic biomarkers). In addition, sponsors or applicants must submit "any other information that would aid evaluation of the proposed clinical investigations with respect to their safety or their design and potential as controlled clinical trials to support the marketing of the drug" (§ 312.23(a)(10)(iv)). Sponsors may possess human data that suggest that a particular biomarker is a probable valid biomarker for evaluating the safety of the drug being evaluated. In these cases, information on the biomarker must be submitted to the IND because it could potentially aid in evaluation of the safety of the investigations per the regulations.

In addition, section 312.23(a)(11) states that a sponsor must submit "if requested by FDA, any other relevant information needed for review of the application." Therefore, during the IND review, FDA may request pharmacogenomic information the Agency considers relevant (e.g., information related to the mechanism of action of the drug).

Sponsors holding INDs who generate or possess pharmacogenomic data related to an investigational drug can comply with FDA requirements using the following algorithm:

Pharmacogenomic data must be submitted to the IND under § 312.23 if ANY of the following apply:

1. The test results are used for making decisions pertaining to a specific clinical trial, or in an animal trial used to support safety (e.g., the results will affect dose and dose schedule selection, entry criteria into a clinical trial safety monitoring, or subject stratification).

2. A sponsor is using the test results to support scientific arguments pertaining to, for example, the pharmacologic mechanism of action, the selection of drug dosing and dosing schedule, or the safety and effectiveness of a drug.

3. Test results constitute a known valid biomarker for physiologic, pathophysiologic, pharmacologic, toxicologic, or clinical states or outcomes in humans, or the test is a known valid biomarker for a safety outcome in animal studies. If the information on the biomarker (example, human CYP2D6 status) is *not* being used for purposes 1 or 2 above, the information can be submitted to the IND as an abbreviated report.

Submission to an IND is NOT required, but voluntary submission is encouraged (i.e., information does not meet the criteria of § 312.23) if

4. Information is from exploratory studies or is research data, such as from general gene expression analyses in cells/animals/humans, or single-nucleotide polymorphism (SNP) analysis of trial participants.

5. Information consists of results from test systems where the validity of the biomarker is not established.

Although submission of such data in cases 4 and 5 is not required under the regulations, FDA would welcome voluntary submission of the data in a VGDS. See Appendix A for additional guidance on assessing whether to submit pharmacogenomic data to an IND.

Note: Regardless of requirements for submission, the fact that samples will be collected for potential analysis must be noted in any clinical protocol (§ 312.23(a)(6)) and informed consent documents (§ 50.25).

Data from a VGDS submission concerning a product under an IND will not be used for regulatory decision making. However, after the sponsor submits a VGDS, if additional information becomes available that triggers the requirements for submission under §§ 312, 314, or 601, the sponsor must submit the data to the relevant application and should follow the appropriate algorithm.

B. Submission of Pharmacogenomic Data to a New NDA, BLA, or Supplement

Section 314.50 outlines the NDA submission requirements; section 601.2 generally outlines BLA submission requirements. As the introduction to § 314.50 states, "the [NDA] application is required to contain reports of all investigations of the drug product sponsored by the applicant, and all other information about the drug product pertinent to an evaluation of the application that is received or otherwise obtained by the applicant from any source." Therefore, to comply with these regulations, sponsors must provide reports of certain pharmacogenomic investigations in their NDAs, and to permit a thorough analysis of a biologics application, a sponsor must submit such a report in its BLA. However, the extent and format of such reports will depend on the relevance and application of the information.

Subsequent paragraphs of § 314.50 outline the submission requirements in specific disciplines. Nonclinical pharmacology and toxicology submission requirements are described in § 314.50(d)(2); human pharmacokinetics and bioavailability requirements in § 314.50(d)(3); and clinical data requirements in § 314.50(d)(5).

Section 601.2 generally outlines the BLA submission requirements. Section 601.2 states that the BLA manufacturer shall submit data derived from nonclinical laboratory and clinical studies that demonstrate that the manufactured product meets prescribed requirements of safety. Like NDA sponsors, BLA sponsors must provide reports of certain pharmacogenomic investigations in their BLAs. However, the extent and format of such reports will depend on the relevance and application of the information.

Sponsors who have generated or possess pharmacogenomic data related to a drug can comply with the regulations' requirements using the algorithm below describing what kind of report to submit:

1. Provide full (complete) reports on pharmacogenomic investigations intended by the sponsor to be used in the drug label or as part of the scientific database being used to support approval as complete submissions (not in the form of an abbreviated report, synopsis, or VGDS), including information about test procedures and complete data, in the relevant sections of the NDA or BLA. If the pharmacogenomic test is already approved by FDA or is the subject of an application submitted to the Agency, information on the test itself can be provided by cross reference.

The following examples would fit this category.

- Pharmacogenomic test results from clinical trials used to support scientific arguments made by the sponsor about selecting drug doses, assessing safety, selecting patients for treatment, or monitoring the beneficial responses

- Pharmacogenomic test results that the sponsor proposes to describe in the drug labeling

- Pharmacogenomic tests that are essential to achieving the dosing, safety, or effectiveness described in the drug labeling

2. Submit reports of pharmacogenomic test results that constitute known valid biomarkers for physiologic, pathophysiologic, pharmacologic, toxicologic, or clinical states or outcomes in the relevant species, but that the sponsor is not relying on or mentioning in the label, to the Agency as an abbreviated report (not in the form of a synopsis or VGDS). (If a pharmacogenomic test of this type was conducted as part of a larger overall study, the reporting of the pharmacogenomic test results can be incorporated into the larger study report.)

3. Submit reports of pharmacogenomic tests that represent probable valid biomarkers for physiologic, pathophysiologic, pharmacologic, toxicologic, or clinical states or outcomes in the relevant species to the NDA or BLA as an abbreviated report. (If the pharmacogenomic testing of this type was conducted as part of a larger study, the abbreviated report can be appended to the report of the overall study.)

4. There is no need to submit detailed reports of general exploratory or research information, such as broad gene expression screening, collection of sera or tissue samples, or results of pharmacogenomic tests that are not known, or probable valid biomarkers to the NDA or BLA. Because the Agency does not view such studies as germane in determining the safety or effectiveness of a product, the submission requirements in §§ 314.50 or 601.2 will be satisfied by the submission of a synopsis of the study. However, the Agency encourages the voluntary submission of the data from such a study in a VGDS.

See Appendix B for additional guidance on how to assess whether to submit pharmacogenomic data to an unapproved NDA or BLA.

C. Submission to a Previously Approved NDA or BLA

The requirements for submitting new scientific information to a previously approved NDA or BLA are outlined in §§ 314.81(b)(2) and 601.12. Results of nonclinical or clinical pharmacogenomic investigations on known or probable valid biomarkers must be submitted in the annual report as synopses or abbreviated reports (§ 314.81(b)(2)).

Pharmacogenomic study results of other types do not meet the submission requirements outlined in the regulations (§ 314.81(b)(2)). However, such reports can be voluntarily submitted to the NDA or BLA as a VGDS.

Pharmacogenomic data collected in pharmacoepidemiologic and observational studies can be submitted as a VGDS by the applicant in accordance with the recommendations in this guidance (see Section VI).

D. Compliance with 21 CFR Part 58

Questions have been raised about the need for pharmacogenomic studies to comply with the requirements of 21 CFR part 58, which describes good laboratory practices (GLPs) for nonclinical laboratory studies that support INDs and NDAs. Section 58.3(d) (21 CFR 58.3(d)) defines nonclinical laboratory studies as "in vivo or in vitro experiments in which test articles are studied prospectively in test systems under laboratory conditions to determine their safety.

The term does not include studies using human subjects or clinical studies or field trials in animals. The term does not include basic exploratory studies carried out to determine whether a test article has any potential utility...."

The requirements of part 58 apply to nonclinical studies submitted to support safety findings, including nonclinical pharmacogenomic studies intended to support regulatory decision making. If full compliance with 21 CFR Part 58 cannot be met, a sponsor must clearly indicate in the study report the areas in which such data do not comply with Part 58 (§§ 312.23(a)(8)(iii) and 314.50(d)(2)(v)). Any studies eligible to be submitted in an abbreviated report, synopsis, or VGDS under the algorithms discussed above do not fall under part 58.

The FDA recognizes that it may not be feasible to conduct separate, long-term, non-GLP preclinical studies. For this reason, FDA encourages sampling of tissues from GLP studies for investigational purposes. Removal of tissue samples and the reason for removal (e.g., exploratory, mechanistic study, tissue banking) should be specified in the protocol. Removal of specimens for investigational purposes from a study does not invalidate the GLP status of the main toxicology study, if otherwise acceptable. If the tissue samples are subsequently analyzed, the results should be reported to the NDA as a synopsis. The FDA would also be interested in receiving these data in a VGDS. If findings from these studies are considered by the sponsor to be relevant to the safety of the compound under study (e.g., related to a known valid biomarker), the findings must be reported to the application, as is necessary for any other relevant nonclinical study findings 312.23(a)(8), 312.32(c)(1)(i)(B), 314.50(d)(2)).

E. Submission of Voluntary Genomic Data from Application-Independent Research

The FDA will also accept pharmacogenomic data from investigators who may not have an active IND, NDA, or BLA , but who wish to provide the information voluntarily to FDA, according to the process described in Section VI of this guidance.

We recommend that all VGDSs be prominently marked as **VGDS**, or **VOLUNTARY SUBMISSIONS**, on the cover letter that accompanies the submission (see Appendix E).

V. Format and Content of a VGDS

The FDA invites submission of exploratory pharmacogenomic data on drugs or candidate drugs whether or not the molecules are currently the subject of an active IND, NDA, or BLA. Exploratory genomic data may result from, for example, microarray expression profiling experiments, genotyping or single-nucleotide polymorphism (SNP) profiling experiments, or from other studies using evolving methodologies that are intended to facilitate global analysis of gene functions, but not specific claims pertaining to drug dosing, safety assessments, or effectiveness evaluations. Currently, consensus standards do not exist for presenting and exchanging genomic data, although such standards are evolving. Therefore, this guidance does not recommend a specific data format for the VGDS.

We recommend that, to achieve the goals of the VGDS process as delineated in Section III(C), the content of a VGDS, and the level of detail, be sufficient for the Agency to interpret the information and independently analyze the data, verify results, and explore possible genotypephenotype correlations across studies. We do not, however, want the submission of a VGDS to be overly burdensome and time-consuming for sponsors. Therefore, VGDS could be submitted in a number of forms:

- As an article submitted to a peer-reviewed scientific journal with raw or processed data submitted electronically

- As an evolving public standard for specific types of experiments, such as the Minimum Information About a Microarray Experiment (MIAME) standard for microarray expression data.[8] Using an approach similar in content to MIAME one can format a VGDS containing genotyping or other genomic data derived from technology platforms other than nucleic acid hybridization arrays.

- As a full report on a gene expression microarray experiment, the content could contain the following analytical, preclinical and/or clinical information, for example:

  - Title page
  - Table of contents
  - Background and scientific rationale
  - Primary and secondary study goals
  - Synopses and summary of findings
  - Study design and sample collection
  - Array design and description
    - Sample processing and preparation

---

[8] Brazma, A., et al., Nature Genetics, 29, 365-371, 2001 and http://www.mged.org/workgroups/miame.html.

- Demonstration of quality of RNA or DNA

- Hybridization procedures and parameters

- Measures of performance of hybridization such as spike-in control

- Measurements and quantification

- Normalization controls

- Number of repeats (array hybridized), number of biological assays performed

— Data Analysis

- Statistical analysis

- Bioinformatics tools and software used. Source of gene annotation

— Results and conclusions, including, for example, data visualization (e.g., scatter plots, principle component analysis (PCA), hierarchical clustering (heat maps)), correlation between expression profiles and outcomes, and appropriate information about relevant co-factors

— References

• Additional Study Information related to mircroarray studies might include the following:

— Confirmation of SNP analysis by sequencing or other assays

- Confirmation of gene expression by other conventional assays (e.g., Northern blot, RT-PCR (real time polymerase chain reaction)). As much as possible, all genes of importance should be confirmed with secondary assays. However, if the genomic profile is of importance, it may be appropriate to sample a selected subpopulation of affected genes

- Alternative approaches that examine endpoints other than gene expression changes may also be appropriate under certain circumstances (e.g., immunohistochemistry or Western blot, if reagents available).

VI. Process for Submitting Pharmacogenomic Data

Using the decision trees (see Appendices A-C), sponsors should submit genomic data according to the following recommendations.

• For required submissions, complete reports, abbreviated reports, or synopses of pharmacogenomic studies should be submitted to INDs, NDAs, or BLAs in the usual manner.

• For candidate drugs or stand alone voluntary submissions (submissions not related to any application), sponsors should submit the package clearly labeled as **VOLUNTARY GENOMIC DATA SUBMISSION (VGDS)**. A voluntary submission cover sheet that can be used is included in Appendix E. For VGDSs related to an existing IND, NDA, or BLA, please include the reference number on the voluntary submission cover sheet.

VII. Agency Review of Voluntary Genomic Data Submissions

The FDA has received many questions about the use of pharmacogenomic data in the application review process. Questions reflect the concern that the Agency will raise new questions and require additional data based on findings from exploratory pharmacogenomic studies, that new studies will be required or suggested based on preliminary human pharmacogenomic data, that indicated populations will be narrowed or restricted based on the pharmacogenomic results in subpopulations, or that new studies in subpopulations will be required after retrospective analysis suggests differential responses based on pharmacogenomic subgrouping. There is also concern about the availability of staff who are experts in interpretation of such data.

The FDA will not use genomic information submitted through the voluntary process for regulatory decision making on INDs, BLAs, or NDAs.

VGDSs will be reviewed by the Interdisciplinary Pharmacogenomic Review Group (IPRG). The review process is intended to ensure that scientific staff experienced in the evaluation of genomics studies participate first-hand in analysis and review of the data. Any data evaluation will be conducted for scientific and informational purposes — not for regulatory decision making. If additional information becomes available after a sponsor submits a VGDS that triggers the submission requirements under §§ 312, 314, or 601, the sponsor must resubmit the data to the investigational or marketing application and should follow the appropriate algorithm described in this guidance for a required submission. Also, a review division may consult the IPRG when pharmacogenomic data are submitted as part of an IND, NDA, or BLA.

The animal and in vitro toxicology database needed to support human trials at various stages of the IND process and to support marketing of short- or long-term use drugs is well established. Any proposals for the substitution or addition of new animal genomic safety tests will ordinarily be the product of a public process involving the international scientific and drug development communities. If FDA becomes aware that a particular pharmacogenomic test has taken on great significance based upon cumulative experience (e.g., from evaluating results across submissions, and/or obtaining input from Advisory Committees), the Agency will notify sponsors about its findings.

Currently, as discussed above, only a few pharmacogenetic tests for certain drug metabolizing enzymes are considered known valid biomarkers in humans. Considerable concern has been expressed about how FDA will evaluate newer types of pharmacogenomic data (e.g., results that may predict increased risk of adverse events, or point to an enhanced probability of effectiveness response). The FDA has considerable experience dealing with these issues in other contexts. Examples of how pharmacogenomic studies fit into this experience include the following.

- Descriptions of drug metabolizing phenotypes and discussion of their effects on dosing are common in drug labels. Extrapolation of this information to pharmacogenetic testing is straightforward.

- There are many conditions or co-factors that may increase an individual's susceptibility to an adverse event (e.g., co-morbid conditions, metabolic susceptibilities such as hepatic failure, or concomitant drug therapies) or the probability of a beneficial response.

The FDA's usual approach in such cases has been to request that information be added to the drug labeling that describes the possible interaction and relevant co-factors and advises on precautions. If a sponsor discovers a new pharmacogenomic test that could possibly distinguish patients at greater risk for a serious adverse event, it is likely that both the sponsor and the Agency would have great interest in exploring the correlation in the appropriate populations. However, if the sponsor also moved forward on developing the drug in the overall indicated population, FDA would evaluate the safety database on its merits. If the sponsor decided to develop the drug solely in populations from which certain patients were excluded based on pharmacogenomic testing, FDA would recommend co-development of the pharmacogenomic test (as a diagnostic) and the drug because FDA would be unable to approve a drug for which the risk or benefit was predicated on a pharmacogenomic test that was unavailable.

It is most likely that, in the near future, pharmacogenomic biomarkers that predict drug toxicity will be identified and developed on a path parallel with overall drug development. In other words, a drug would be developed in a conventional manner with a parallel effort to identify appropriate predictors of toxicity. If the drug's risk-benefit profile were acceptable in the entire target population, the drug could be approved prior to the completion of efforts to refine and develop the relevant pharmacogenomic tests. When and if a test's predictive values were to be established and the test were to become commercially available (either as an approved device or as a service), the drug label could be changed to reflect the data.

- The FDA has similar experience with tests used to target populations likely to respond to therapy.

Several decades ago, broad indications for use were described in labels. Over time, as more exact diagnoses were developed, narrower indications were sought by sponsors, based on the clinical trials conducted. A similar evolution occurred in the field of anti-HIV therapies as drug resistance testing became available. We encourage sponsors to continue to develop pharmacogenomic tests that are predictive of subpopulations with enhanced response to therapy. However, if overall drug development is pursued in the larger population, the effectiveness and risk-benefit will be evaluated in that population, and approval decisions will be based on the overall database.

Much of the concern about FDA actions in this area is based on the perception that pharmacogenomic testing is likely to give definitive answers about the probability of safety and effectiveness in subpopulations. Such specificity may occur occasionally (e.g., where a product is designed to inhibit a specific molecular target), and in such cases, rapid development of a diagnostic test is highly encouraged. However, this is unlikely to be the ordinary case. In most instances, a genotype or particular gene expression profile is likely to be one of a number of factors that affects the probability of an adverse event or a favorable response. For this reason, pharmacogenomic biomarkers can ordinarily be handled like other non-genomic predictive markers in the clinical arena.

Glossary

The following definitions are for use in the processes outlined in this guidance and are not intended to be broadly applicable to the entire field.

Biological marker (biomarker): A characteristic that is objectively measured and evaluated as an indicator of normal biologic processes, pathogenic processes, or pharmacologic responses to a therapeutic intervention.[9]

Pharmacogenetic test: An assay intended to study interindividual variations in DNA sequence related to drug absorption and disposition (pharmacokinetics) or drug action (pharmacodynamics), including polymorphic variation in the genes that encode the functions of transporters, metabolizing enzymes, receptors, and other proteins

Pharmacogenomic test: An assay intended to study interindividual variations in whole-genome or candidate gene, single-nucleotide polymorphism (SNP) maps, haplotype markers, or alterations in gene expression or inactivation that may be correlated with pharmacological function and therapeutic response. In some cases, the pattern or profile of change is the relevant biomarker, rather than changes in individual markers.

Valid biomarker: A biomarker that is measured in an analytical test system with wellestablished performance characteristics and for which there is an established scientific framework or body of evidence that elucidates the physiologic, toxicologic, pharmacologic, or clinical significance of the test results. The classification of biomarkers is context specific. Likewise, validation of a biomarker is context-specific and the criteria for validation will varywith the I ntended use of the biomarker. The clinical utility (e.g., predict toxicity, effectiveness or dosing) and use of epidemiology/population data (e.g., strength of genotype-phenotype associations) are examples of approaches that can be used to determine the specific context and the necessary criteria for validation.

- ***Known valid biomarker:*** A biomarker that is measured in an analytical test system with well-established performance characteristics and for which there is widespread agreement in the medical or scientific community about the physiologic, toxicologic, pharmacologic, or clinical significance of the results

- ***Probable valid biomarker:*** A biomarker that is measured in an analytical test system with well-established performance characteristics and for which there is a scientific framework or body of evidence that appears to elucidate the physiologic, toxicologic, pharmacologic, or clinical significance of the test results. A probable valid biomarker may not have reached the status of a known valid marker because, for example, of any one of the following reasons:

  - The data elucidating its significance may have been generated within a single company and may not be available for public scientific scrutiny.

---

[9] Biomarkers Definitions Working Group, "Biomarkers and Surrogate Endpoints: Preferred Definitions and Conceptual Framework," Clinical Pharm. & Therapeutics, vol. 69, N. 3, March 2001.

- The data elucidating its significance, although highly suggestive, may not be conclusive.

- Independent verification of the results may not have occurred.

Voluntary genomic data submission (VGDS): The designation for pharmacogenomic data submitted voluntarily to FDA.

Appendix A: Submission of Pharmacogenomic (PG) Data to an IND

Reports of pharmacogenomic investigations should be submitted to the IND in accordance with the decision tree below and in the formats indicated here or in the body of the guidance:

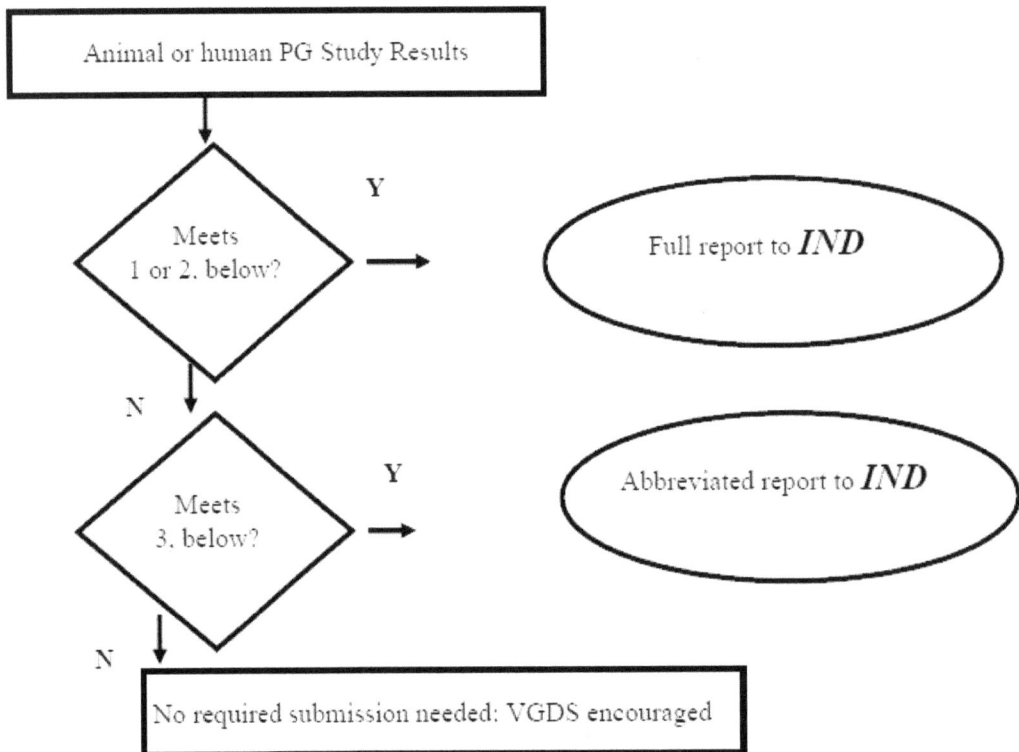

Pharmacogenomic data must be submitted to the IND under § 312.23 if ANY of the following apply:

1. The test results are used for making decisions pertaining to a specific clinical trial, or in an animal trial used to support safety (e.g., the results will affect dose selection, entry criteria into a clinical trial safety monitoring, or subject stratification).

2. A sponsor is using the test results to support scientific arguments pertaining to, for example, the pharmacologic mechanism of action, the selection of drug dosing or the safety and effectiveness of a drug.

3. The test results constitute a known, valid biomarker for physiologic, pathophysiologic, pharmacologic, toxicologic, or clinical states or outcomes in humans, or is a known valid biomarker for a safety outcome in animal studies or a probable valid biomarker in human safety studies. If the information on the biomarker (example, human CYP2D6 status) is not being used for purposes 1 or 2 above, the information can be submitted to the IND as an abbreviated report.

Submission to an IND is NOT required, but voluntary submission is encouraged (i.e., information does not meet the criteria of § 312.23) if

4. Information is from exploratory studies or is research data, such as from general gene expression analyses in cells/animals/humans, or singlenucleotide polymorphism (SNP) analysis of trial participants.

5. Information consists of results from test systems where the validity of the biomarker is not established.

Appendix B: Submission of Pharmacogenomic (PG) Data to a New NDA, BLA, or Supplement

Reports of pharmacogenomic investigations should be submitted to the NDA in accordance with the decision tree below and in the formats indicated here or in the body of the guidance:

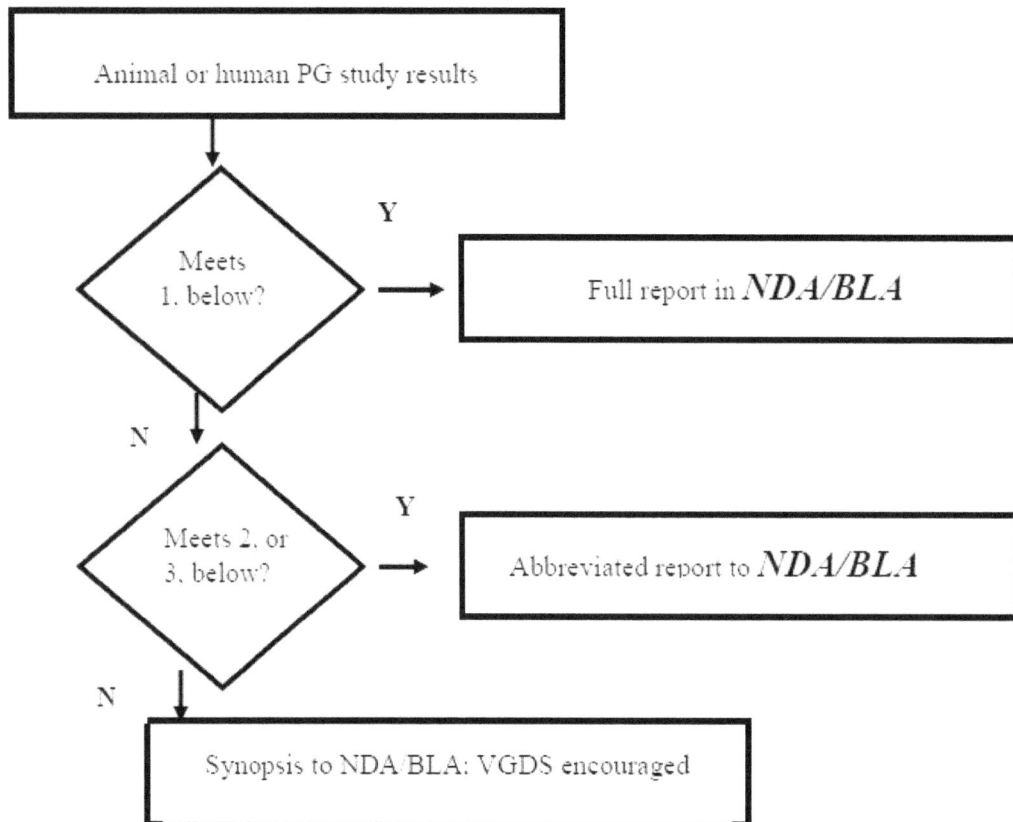

1. The sponsor will use the test results in the drug labeling or as part of the scientific database being used to support approval as complete submissions (not in the form of an abbreviated report, synopsis, or VGDS), including information about test procedures and complete data, in the relevant sections of the NDA or BLA. If the pharmacogenomic test is already approved by FDA or is the subject of an application filed with the Agency, information on the test itself can be provided by cross reference.

   The following examples would fit this category.

   – Pharmacogenomic test results that are being used to support scientific arguments made by the sponsor about drug dosing, safety, patient selection, or effectiveness

   – Pharmacogenomic test results that the sponsor proposes to describe in the drug label

   – Pharmacogenomic tests that are essential to achieving the dosing, safety, or effectiveness described in the drug label

2. The test results are known valid biomarkers for physiologic, pathophysiologic, pharmacologic, toxicologic, or clinical states or outcomes in the relevant species, but the sponsor is not relying on or mentioning this in the label. Submit to the Agency as an abbreviated report (not as a synopsis or VGDS). If a pharmacogenomic test of this type was conducted as part of a larger overall study, the reporting of the pharmacogenomic test results can be incorporated into the larger study report.

3. The test results represent probable valid biomarkers for physiologic, pathophysiologic, pharmacologic, toxicologic, or clinical states or outcomes in the relevant species. Submit to the Agency as an abbreviated report. If the pharmacogenomic testing of this type was conducted as part of a larger study, the abbreviated report can be appended to the report of the overall study.

4. Information from general exploratory or research studies, such as broad gene expression screening, collection of sera or tissue samples, or results of pharmacogenomic tests that are not known or probable valid biomarkers to the NDA or BLA are not required to be submitted. Because the Agency does not view these studies as germane in determining the safety or effectiveness of a drug, the submission requirements in §§ 314.50 or 601.2 will be satisfied by the submission of a synopsis of the study. However, the Agency encourages the voluntary submission of the data from the study in a VGDS submitted to the NDA or BLA.

Appendix C: Submission of Pharmacogenomic (PG) Data to an Approved NDA, BLA, or Supplement

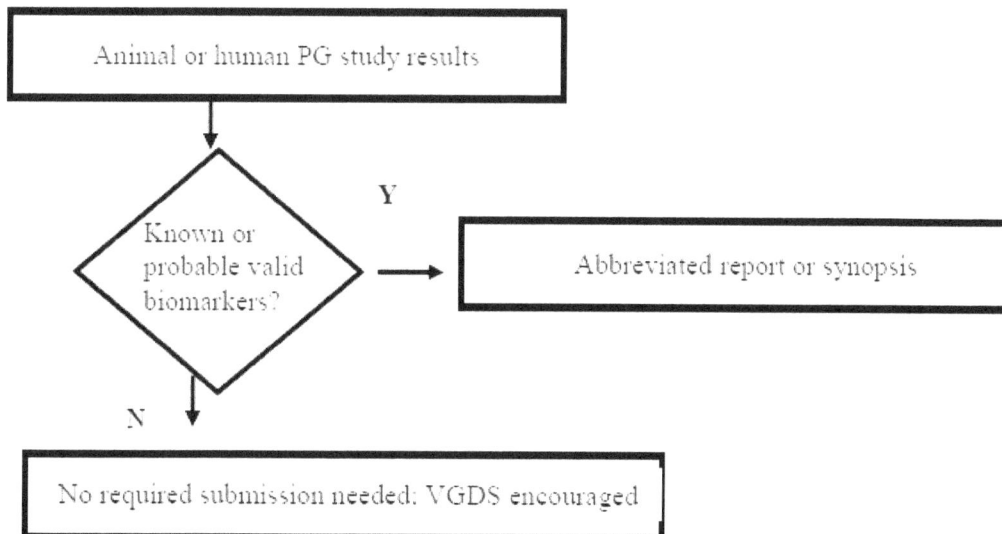

Appendix D: Quick Reference on Pharmacogenomic Submissions

Submitting data to an:	IND	New (Unapproved) NDA, BLA, or Supplement	Previously Approved NDA or BLA
Known Valid Biomarker	Must be submitted, pursuant to 21 CFR 312.23 (a) (8), (9), (10) (iv) or (11).	Must be submitted, pursuant to 21 CFR 314.50 and 601.2. See section IV.B. of the guidance.	Must be submitted pursuant to 21 CFR 314.81 in annual report and should be submitted pursuant to § 601.12 as synopses or abbreviated reports.
Probable Valid Biomarker	Does not need to be submitted.[10] The FDA welcomes voluntary submission of such data in a VGDS.	The FDA recommends submission, using algorithm in section IV.B. of the guidance.	Must be submitted pursuant to 21 CFR 314.81 in annual report and should be submitted pursuant to § 601.12 as synopses or abbreviated reports.
Exploratory or Research Pharmacogenomic Data	The FDA welcomes voluntary submission of such data in a VGDS.	The FDA recommends submission, using algorithm in section IV.B. of the guidance. The FDA welcomes voluntary submission of such data in a VGDS.	The FDA welcomes voluntary submission of such data in a VGDS.

---

[10] Except if used in human safety studies.

Appendix E: Voluntary Submission Cover Sheet

Send all CDER voluntary genomic data submissions to the following address accompanied by this coversheet:

FDA/CDER
Central Document Room (CDR)
5901-B Ammendale Road
Beltsville, MD 20705-1266

Attention!

This is a

Voluntary
Genomic Data Submission

Application number _____ (leave blank if this is the first submission for a stand-alone VGDS)

_____ Initial Submission

_____ Subsequent Submission

Please route directly to the IPRG (HFD-850)
After processing in the CDR!

Collection of Race and Ethnicity Data in Clinical Trials

Collection of Race and Ethnicity Data in Clinical Trials, Guidance for Industry[1,2]

U.S. Department of Health and Human Services
Food and Drug Administration
Center for Drug Evaluation and Research (CDER)
Center for Biologics Evaluation and Research (CBER)
Center for Devices and Radiologic Health (CDRH)
Office of the Commissioner (OC)

September 2005

Clinical Medical

Contains Nonbinding Recommendations

This guidance recommends using a standardized approach for collecting and reporting race and ethnicity information in clinical trials conducted in the United States and abroad for certain FDA regulated products. The recommended standardized approach was developed by the Office of Management and Budget (OMB). The guidance lists the OMB categories for race and ethnicity and describes FDA's reasons for recommending the use of these categories. In addition, this guidance recommends a format for race and ethnicity data within study data that are submitted in standardized data sets such as the Study Data Tabulation Model or in the electronic Common Technical Document (eCTD).

Note: *This guidance represents the Food and Drug Administration's (FDA's) current thinking on this topic. It does not create or confer any rights for or on any person and does not operate to bind FDA or the public. You can use an alternative approach if the approach satisfies the requirements of the applicable statutes and regulations. If you want to discuss an alternative approach, contact the appropriate FDA staff. If you cannot identify the appropriate FDA staff, call the appropriate number listed on the title page of this guidance.*

I. Introduction

This guidance recommends using a standardized approach for collecting and reporting race and ethnicity information in clinical trials conducted in the United States and abroad for certain

---

[1] Available on the FDA website at: http://www.fda.gov/RegulatoryInformation/Guidances/ucm126340.htm

[2] This guidance has been developed by the Agency-wide Race and Ethnicity Working Group from the Office of the Commissioner (OC), the Center for Biologics Evaluation and Research (CBER), the Center for Drug Evaluation and Research (CDER), and the Center for Devices and Radiological Health (CDRH) of the Food and Drug Administration (FDA).

FDA regulated products. The recommended standardized approach was developed by the Office of Management and Budget (OMB). The guidance lists the OMB categories for race and ethnicity and describes FDA's reasons for recommending the use of these categories. In addition, this guidance recommends a format for race and ethnicity data within study data that are submitted in standardized data sets such as the Study Data Tabulation Model[3] or in the electronic Common Technical Document (eCTD).[4]

This document is intended to provide guidance on meeting the requirements set forth in the 1998 final rule on investigational new drug (IND) applications and new drug applications (NDAs)[5] (Demographic Rule). The Demographic Rule requires IND holders to tabulate in their annual report the number of subjects enrolled in clinical studies of drugs and biologic products by age, race, and gender, and sponsors of NDAs to include summaries of effectiveness and safety data for important demographic subgroups, including racial subgroups.[6] This guidance is also intended to help applicants in preparing biologics license applications (BLAs).

Although the regulations governing medical devices do not include requirements for the collection of demographic data comparable to those for INDs and NDAs, for those cases in which race and ethnicity data are relevant to determining the safety and effectiveness of a device, FDA encourages sponsors to collect the data in accordance with the OMB recommendations and the information collection standards discussed in this guidance document. Sponsors are also encouraged to discuss any race or ethnicity issues with the appropriate review division in the Office of Device Evaluation, Center for Devices and Radiological Health, when developing their study protocols.

This guidance does not address the level of participation of racial and ethnic groups in clinical trials. For questions related to the level of participation or the size of a study sponsors should consult with the review division prior to the start of a study.

FDA's guidance documents, including this guidance, do not establish legally enforceable responsibilities. Instead, guidances describe the Agency's current thinking on a topic and should be viewed only as recommendations, unless specific regulatory or statutory requirements are cited. The use of the word should in Agency guidances means that something is suggested or recommended, but not required.

II. Background

FDA regulations require sponsors of NDAs to present a summary of safety and effectiveness data by demographic subgroups (age, gender, race), as well as an analysis of whether

[3] See CDISC standardized Study Data Tabulation Model (SDTM), Analysis Data Model (ADaM), Operational Data Model (ODM) at http://www.cdisc.org.
[4] See FDA's guidance for industry titled M4 Common Technical Document for the Registration of Pharmaceuticals for Human Use (eCTD guidance) for the submission file location of the table format presented in section V of this guidance. Available at http://www.fda.gov/cder/regulatory/ersr/ectd.htm.
[5] FR 6854 (February 11, 1998) (codified at 21 CFR 312.33(a)(2) and 21 CFR 314.50(d)(5)).
[6] 21 CFR 312.33(a)(2). See also 21 CFR 314.50(d)(5)(v) and (vi)(a) regarding demographic data submission in NDAs.

modifications of dose or dosage intervals are needed for specific subgroups (21 CFR 314.50 (d)(5)(v) and (vi)(a)).[7] One consideration in such summaries is the identification of a subject's race or ethnicity.

In 1997, OMB issued its revised recommendations for the collection and use of race and ethnicity data by Federal agencies (Policy Directive 15).[8] FDA is now recommending the use of the standardized OMB race and ethnicity categories for data collection in clinical trials for two reasons. First, the use of the recommended OMB categories will help ensure consistency in demographic subset analyses in applications submitted to FDA (21 CFR 314.50(d)(5)(v) and (vi)(a), 312.120, 314.106(b), and 601.2) and in data collected by other government agencies. Second, consistency in these categories may make the demographic subset analysis more useful in evaluating potential differences in the safety and efficacy of pharmaceutical products among population subgroups. To assess potential subgroup differences in a meaningful way, it is important to use uniform, standard methods of defining racial and ethnic subgroups.

A. Relevance of Population Subgroup Studies

Differences in response to medical products have already been observed in racially and ethnically distinct subgroups of the U.S. population.[9] These differences may be attributable to intrinsic factors (e.g., genetics, metabolism, elimination), extrinsic factors (e.g., diet, environmental exposure, sociocultural issues), or interactions between these factors. For example, in the United States, Whites[10] are more likely than persons of Asian and African heritage to have abnormally low levels of an important enzyme (CYP2D6) that metabolizes drugs belonging to a variety of therapeutic areas, such as antidepressants, antipsychotics, and beta blockers (Xie 2001). Other studies have shown that Blacks respond poorly to several classes of antihypertensive agents (beta blockers and angiotensin converting enzyme (ACE) inhibitors) (Exner 2001 and Yancy 2001). Racial differences in skin structure and physiology that can affect response to dermatologic and topically applied products have been noted (Taylor 2002). Clinical trials have demonstrated lower responses to interferon-alpha used in the

---

[7] Under 21 CFR 314.101(d)(3), the Agency may refuse to file an NDA if it is incomplete because it does not contain information required by 21 CFR 314.50. Thus, if there is an inadequate evaluation for safety and/or effectiveness of the population intended to use the drug, including pertinent subsets, such as gender, age, and racial subsets the Agency may refuse to file the application. See FDA's guidance for industry titled New Drug Evaluation Guidance Document: Refusal to File, available on the Internet at http://www.fda.gov/cder/guidance/index.htm.

[8] Statistical Policy Directive No. 15, Race and Ethnic Standards for Federal Statistics and Administrative Reporting, 1997 (reprinted in Appendix 2). See also the OMB guidance entitled Provisional Guidance on the Implementation of the 1997 Standards for Federal Data on Race and Ethnicity (2000), available at http://www.whitehouse.gov/omb/bulletins/b00-02.html.

[9] In fact, in June of 2005, FDA approved BiDil, the first drug approved by the Agency to treat a disease in patients identified by race. The drug was approved for the treatment of heart failure in black patients. The company conducted two trials in the general population that failed to show a benefit, but suggested a benefit of BiDil in black patients. The company then studied the drug in 1,050 self-identified black patients and it was shown to be safe and effective.

[10] The terms used in this guidance to describe the various racial and ethnic groups are those used by OMB.

---

treatment of hepatitis C among Blacks when compared with other racial subgroups (McHutchison 2000 and Reddy 1999).

B. FDA Decision to Recommend Use of the OMB Categories

The OMB stated that its race and ethnicity categories were not anthropologic or scientifically based designations, but instead were categories that described the sociocultural construct of our society. The Department of Health and Human Services (HHS) chose to adopt these standardized categories for its agencies that report statistics because the categories are relevant to assessing various health related data, including public health surveillance and research. FDA believes that the use of the current OMB categories and any future revisions will facilitate comparisons across clinical studies analyzed by FDA and data collected by other agencies. Collection of data using standard categories can enhance patient safety by helping FDA evaluate potential differences in drug response among subpopulations. Using standard categories may also facilitate analyses seeking to identify differences in response.

Although FDA has long requested race and ethnicity data on subjects in certain clinical trials, the Agency has not previously made explicit recommendations on the categories to use when collecting and reporting the data. In 1998, the Agency issued the Demographic Rule, which reflected growing recognition within the Agency and the health community that (1) different subgroups of the population may respond differently to specific drug products, and (2) although an effort should be made to look for differences in effectiveness and adverse reactions among such subgroups, that effort is not being made consistently.[11] In the Demographic Rule, FDA discussed the importance of collecting data in clinical trials (and of presenting those data in applications to the Agency) on population subgroups organized by gender, race, age, and other relevant categories. The Agency recommended that sponsors ask subjects in certain clinical trials to identify their racial group and, if desired, that sponsors use OMB categories when collecting race and ethnicity data.[12]

During the past two decades, efforts have been under way in a number of Federal organizations to collect race and ethnicity data in Federal programs in a standardized way. (See Appendix 1 for a summary of those efforts). In 1997, HHS issued a document entitled Policy Statement on Inclusion of Race and Ethnicity in DHHS Data Collection Activities.[13] In this policy statement, HHS adopted the revised OMB categories for including race and ethnicity in HHS funded and sponsored data collection and reporting systems. The HHS policy states that

---

[11] 63 FR 6854 at 6855, February 11, 1998.

[12] In the preamble to the final rule, FDA stated that it did not believe it was necessary to define specific racial categories in the rule itself because drug sponsors have been successful in identifying the relevant racial categories to examine safety and efficacy profiles of drugs (63 FR 6854 at 6859). However, FDA now believes that using uniform categories will enhance the consistency and comparability of data across studies submitted in marketing applications and other government reported statistics.

[13] Memorandum issued by HHS Sec. Donna Shalala on October 24, 1997 reaffirming HHS' commitment to the inclusion of data on minority groups in research, services and related activities. Effective as of November 1,1997. Available at http://aspe.hhs.gov/datacncl/inclusn.htm.

the categories described in revised OMB Directive 15 and its future revisions should be used when collecting and reporting data in HHS data systems or reporting HHS funded statistics.[14]

The Agency recommends that sponsors use the categories outlined in this guidance when collecting race and ethnicity data in clinical studies for FDA-regulated products conducted in United States and abroad. More detailed race and ethnicity data can be collected when appropriate to the study or locale, but we recommend that these more detailed race and ethnicity data be related to the identified OMB categories of all clinical trial participants when submitting such data to the Agency.

III. Collecting Race and Ethnicity Data in Clinical Trials

The recommendations in this section reflect the Agency's interest in more consistent data collection. For studies conducted in the United States, the Agency recommends that a two-question format be used, and that trial participants self-report their racial and ethnic ancestral origins. Based on the current OMB Directive, the Agency provides the following recommendations for the collection of the data:

1. We recommend using the two-question format for requesting race and ethnicity information, with the ethnicity question preceding the question about race.

2. We recommend that study participants self-report race and ethnicity information whenever feasible, and that individuals be permitted to designate a multiracial identity. When the collection of self-reported designations is not feasible (e.g., because of the subject's inability to respond), we recommend that the information be requested from a first-degree relative or other knowledgeable source.

3. For ethnicity, we recommend the following minimum choices be offered:

    *   Hispanic or Latino

    *   Not Hispanic or Latino

4. When race and ethnicity information is collected separately, we recommend the following minimum choices be offered for race[15]

    *   American Indian or Alaska Native

    *   Asian

    *   Black or African American

    *   Native Hawaiian or Other Pacific Islander

---

[14] OMB directed these activities to begin by January 1, 2003, in all Federal programs, including HHS. Although in the past FDA sought and received a variance from OMB exempting the Agency from reporting data using the Directive 15 categories, FDA now recommends the use of the categories to enhance data consistency. To view the policy memorandum see: http://www.hhs.gov/oirm/infocollect/nclusion.html

[15] To identify ancestral origins for each of the named categories see OMB Directive 15 (Appendix 2).

- White

5. In certain situations, as recommended in OMB Directive 15, more detailed race and ethnicity information may be desired (e.g., White can reflect origins in Europe, the Middle East, or North Africa; Asian can reflect origins from areas ranging from India to Japan). If more detailed characterizations of race or ethnicity are collected to enhance data consistency, we recommend these characterizations be traceable to the five minimum designations for race and two designations for ethnicity (five and two) listed in numbers 3 and 4. When more detailed characterizations are desired, the use of Race and Ethnicity vocabulary tables located within Health Level Seven's[16] Reference Information Model Structural Vocabulary Tables is recommended. These tables provide the OMB characterizations traceable to more detailed characterizations and concept ID code sets to help ensure that traceability is consistent. Where gaps exist in the representation of race or ethnicity categories, sponsors are encouraged to discuss the race or ethnicity issue with the appropriate review division.

IV. Clinical Trials Conducted Outside of the United States

To assist in assessing the relevance of foreign study population data to U.S. populations, we recommend that sponsors use the OMB standardized categories when collecting data from study participants in clinical trials conducted outside of the United States. However, FDA recognizes that the recommended categories for race and ethnicity were developed in the United States and that these categories may not adequately describe racial and ethnic groups in foreign countries. Therefore, for studies conducted outside the United States, we recommend using more detailed categories to provide sponsors the flexibility to adequately characterize race and ethnicity. If sponsors choose to use more detailed characterizations of race and ethnicity, it is important for analytical purposes that the data can be traced back to the recommended categories described below. When more detailed characterizations are desired, the use of the Race and Ethnicity vocabulary tables located within Health Level Seven's Reference Information Model Structural Vocabulary Tables is recommended[17]. These tables provide the five and two OMB characterizations traceable to more detailed characterizations and concept ID code sets and their use will help ensure that traceability is consistent. Where gaps exist in the representation of race or ethnicity categories, sponsors are encouraged to discuss the issue with the appropriate review division.

1. For ethnicity, we recommend the following categories or the use of categories that are mappable to the two categories listed below:

- Hispanic or Latino

---

[16] Health Level Seven (HL7), an American National Standards Institute (ANSI) accredited organization, has been designated by the Department as a Standards Development Organization for the development of interoperability standards for health care and health related information, and is available at http://hl7.org.
[17] These tables are located at http://hl7.org. To locate the tables from HL7's home page click on HL7 Standards under Resources, then RIMS, HL7 Reference Model Structured Vocabulary Tables in HTML, 2.1 HL7 Vocabulary Domain Values and select either the Race or Ethnicity table from the list.

- Not Hispanic or Latino

2. When race is collected separately in clinical studies conducted in foreign countries, we recommend that the categories be modified to reflect the following, as appropriate:

- American Indian or Alaska Native

- Asian

- Black

- Native Hawaiian or Other Pacific Islander

- White

Note that the ethnic and racial categories for studies inside and outside the United States are the same, except for one racial designation: the racial designation is African American in the United States, whereas it is Black for studies conducted in foreign countries.

V. Presentation of Demographic Tabulations in the eCTD

For INDs, NDAs, BLAs and relevant device submissions we recommend the submission of tabulated demographic data based on the Demographic Rule for all clinical studies using the characterizations of race and ethnicity described in this guidance.

For submitting an electronic application not in FDA's typical application format (i.e., when using the ICH document for submitting a marketing application to FDA and regulatory agencies in Japan and Europe, the eCTD) presentation of demographic data is described in ICH M4E eCTD Guidance (section 2.7.4.1.3 and table 2.7.4.2), which suggests a tabular display of demographic characteristics (e.g., age, gender, race) by treatment group (e.g., active drug, placebo). The document suggests specific kinds of demographic information to be collected as a part of a clinical trial, but does not provide rigid specifications on how the data should be presented, noting, for example, that "if relative exposure of demographic groups in the controlled trials differ from the overall exposure, it may be useful to provide separate tables." Choices of how best to summarize demographic data depend on the nature of the data to be conveyed. For some trials, it may be useful to show the distribution of one demographic characteristic within a second demographic (e.g., the age distribution of men and women enrolled in a set of controlled trials).

With regard to the description of race and ethnicity, the categories that are suggested previously in this document (sections III and IV) are preferable to those suggested in ICH M4E eCTD guidance.

Appendix 1: History of Federal Efforts in Data Collection on Race and Ethnicity and Other Subpopulations

For more than 20 years, a number of U.S. Government initiatives have tried to address questions related to whether to and how to collect race and ethnicity data. Major initiatives are reviewed briefly here.

Office of Management and Budget (OMB) Initiatives

In May 1977, OMB issued "Statistical Policy Directive No. 15, Race and Ethnic Standards for Federal Statistics and Administrative Reporting." The standards were developed in response to the need to enforce civil rights laws in education. These classifications were not to be interpreted as being scientific or anthropological in nature, or to be viewed as determinants of eligibility for participation in any Federal program. They were developed in response to needs expressed by both the Executive Branch and the Congress to provide for the collection and use of compatible, nonduplicated, exchangeable race and ethnicity data by Federal agencies. This Directive specified four categories for race:

- American Indian or Alaskan Native
- Asian or Pacific Islander
- Black
- White

And two categories for ethnicity:

- Hispanic
- Not of Hispanic origin

The OMB Directive specified two questionnaire formats for data collection: (1) a format combining race and ethnicity, and (2) a preferred format with two separate questions for race and ethnicity.

Since 1993, efforts have been under way to standardize the collection of race and ethnicity data to foster comparability across data collection and reporting systems. In 1997, OMB published Directive 15, "Revisions to the Standards for the Classification of Federal Data on Race and Ethnicity" (see Appendix 2). These revisions specified the minimum racial and ethnic diversity categories to be used when race and ethnicity are included in data collection and reporting for Federal programs. The Directive does not require that race and ethnicity be included in data collection and reporting; rather, it specifies what formats and categories to use when collecting this kind of data.

The revised OMB standards made the following changes:

- Introduced the option of reporting more than one race for multiracial persons
- Divided the Asian or Pacific Islander category into two — one labeled Asian, the other Native Hawaiian or Other Pacific Islander
- Changed Hispanic to Hispanic or Latino
- Changed Black to Black or African-American
- Strongly encouraged the use of self-identification
- Maintained the two-question format for race and Hispanic ethnicity when self-identification is used (the Hispanic origin question should precede the race question)

The revised categories were described in an OMB guidance entitled Implementation of the 1997 Standards for Federal Data on Race and Ethnicity (2000) as sociopolitical and intended for use in the collection of health data, among other types of statistics.

Department of Health and Human Services Initiatives

In 1999, the Department of Health and Human Services (HHS) issued a report, Improving the Collection and Use of Racial and Ethnic Data in HHS. The report describes HHS policy on collecting and reporting data on race and ethnicity for HHS programs. The report asks for the inclusion of race and ethnicity categories in HHS funded and sponsored data collection and reporting systems in all HHS programs, including in both health and human services. This policy clearly states that the minimum standard categories in OMB Directive 15 and revisions should be used when collecting and reporting data in HHS data systems or reporting HHS funded statistics. The policy was developed to (1) help monitor HHS programs, (2) determine whether Federal funds are being used in a nondiscriminatory manner, and (3) promote the availability of standard race and ethnicity data across various agencies to facilitate HHS responses to major health and human services issues.

National Institutes of Health Initiatives

In 1993, the National Institutes of Health (NIH) Revitalization Act directed NIH to establish guidelines for including women and minorities in NIH-sponsored clinical research. NIH was directed to ensure that women and minorities were included as subjects, unless their exclusion was justified due to circumstances specified by NIH guidelines. Furthermore, clinical trials were to be designed and carried out in a manner that would elicit information about individuals of both genders and diverse racial and ethnic groups to examine differential effects on such groups. NIH guidelines stipulate that when proposing a Phase 3 clinical trial, evidence must be reviewed to establish whether or not there are potentially clinically important gender- and minority-based differences in the anticipated effects of the intervention. If previous studies support the existence of significant differences, the primary questions and design of the study must specifically accommodate this. For example, if men and women are thought to respond differently to an intervention, then the Phase 3 clinical trial must be designed to answer two separate primary questions, one for men, and the other for women. When prior studies support no significant differences for either gender or minorities with a given intervention, then gender and minority status will not be required as subject selection criteria, although the inclusion and analysis of both genders and minorities is strongly encouraged. When prior studies neither support nor negate significant differences, then the design of the Phase 3 clinical trial will be required to support sufficient representation of both genders and minorities to allow for valid analysis of the intervention effects across all groups. However, the trial will not be required to provide high statistical power for these comparisons.

Food and Drug Administration Initiatives

Beginning in the 1980s, FDA grew concerned about possible differences in drug safety and efficacy among different population subgroups. Because the origins of subpopulation issues stem from the identification of differences in response in women and geriatric populations, references to those initiatives are included below. In 1983, the Agency initiated development of guidance on the study of drugs to be used in geriatric patients. FDA's Guideline for the Study of Drugs Likely to be Used in the Elderly was issued in 1989.

The first regulation specifying the analysis of population subsets appeared in 1985 in 21 CFR 314.50, which called for evidence to support the "dosage and administration section of the labeling, including support for the dosage and dose interval recommended," and modifications for specific subgroups (e.g., pediatrics, geriatrics, patients with renal failure) (21 CFR 314.50(d)(5)(v)).

In 1988, the Agency issued guidance describing elements of a new drug application's analysis of clinical study data. Guideline for the Format and Content of the Clinical and Statistical Sections of New Drug Applications emphasized the importance of conducting subset analyses on data from clinical studies submitted in new drug applications (NDAs). This guidance specified race and ethnicity as types of population subsets for which separate analyses of data from clinical studies should be conducted for assessments of product safety and effectiveness.

In July 1993, FDA published a guidance on the study of drugs in both genders entitled Guideline for the Study and Evaluation of Gender Differences in the Clinical Evaluation of Drugs. The guidance specifically called for analyzing trials by gender and for evaluating pharmacokinetics in women. In the Federal Register notice announcing the guidance, FDA also abandoned the policy explained in a 1977 guidance, excluding women of childbearing potential from participation in the earliest phases of clinical trials.

In 1993, FDA also published New Drug Evaluation Guidance Document: Refusal to File, on the Agency's use of the refusal-to-file (RTF) option if certain analyses were not performed. The guidance states that the Agency can exercise its RTF authority under 21 CFR 314.101(d)(3) if there is "inadequate evaluation for safety and/or effectiveness of the population intended to use the drug, including pertinent subsets, such as gender, age, and racial subsets."

In the Food and Drug Administration Modernization Act of 1997 (the Modernization Act), Congress directed FDA to examine issues related to the inclusion of racial and ethnic groups in clinical trials of new drugs. Section 115(b) of the Modernization Act required the Secretary, "in consultation with the Director of the National Institutes of Health and with representatives of the drug manufacturing industry, [to] review and develop guidance, as appropriate, on the inclusion of women and minorities in clinical trials. . . ." (codified at 21 U.S.C. 355(b)(1)). In response, FDA established the Women and Minorities Working Group to review and implement this section of the Modernization Act. In a report issued on July 20, 1998, the Working Group concluded that the Agency would implement procedures to enhance its ability to gather and evaluate demographic data, and then decide whether additional guidance should be developed in the future.

In 1998, the Agency published the Demographic Rule, which amended the language in 21 CFR 312.33(a)(2) and 314.50(d)(5), requiring sponsors to (1) tabulate the numbers of participants in clinical trials by age group, gender, and race in investigational new drug application (IND) annual reports, (2) characterize the data in NDAs according to these same subgroups, and (3) when appropriate, present safety data from other subgroups of the population of patients, such as for patients with hepatic or renal failure or patients with different levels of severity of the disease.

In 1999, a guidance for industry entitled Population Pharmacokinetics made recommendations on the use of population pharmacokinetics in the drug development process to help identify differences in drug safety and efficacy among population subgroups, including race and ethnicity. This guidance recommended that industry conduct clinical studies in subjects representative of the population to be treated with the drug.

In 2002, the Best Pharmaceuticals for Children Act (Public Law 107-109, January 4, 2002) directed FDA to monitor the racial and ethnic designations of children participating in clinical studies for pharmaceutical products.

ICH E5 - Guidance on Ethnic Factors in the Acceptability of Foreign Clinical Data

In 1999, as part of an international effort by Japan, the European Union, and the United States to harmonize technical requirements for pharmaceutical drug development and regulation (the International Conference on Harmonization (ICH)), the FDA published a guidance entitled E5 Guidance on Ethnic Factors in the Acceptability of Foreign Clinical Data (63 FR 31790, June 10, 1999), to permit the clinical data collected in one region to be used in the registration or approval of a drug or biological product in another region, while allowing for the influence of ethnic factors. The E5 guidance defines ethnic factors that affect response in terms of both intrinsic and extrinsic issues. Because differences in ethnic factors have the potential to affect responses in some subpopulations, the E5 guidance provides a general framework for evaluating medicines with regard to their sensitivity to ethnic factors.

Appendix 2: Revised Directive 15

OMB Standards for Maintaining, Collecting, and Presenting Federal Data on Race and Ethnicity

(Adopted on October 30, 1997)

This classification provides a minimum standard for maintaining, collecting, and presenting data on race and ethnicity for all Federal reporting purposes. The categories in this classification are social-political constructs and should not be interpreted as being scientific or anthropological in nature. They are not to be used as determinants of eligibility for participation in any Federal program. The standards have been developed to provide a common language for uniformity and comparability in the collection and use of data on race and ethnicity by Federal agencies.

The standards have five categories for data on race: American Indian or Alaska Native, Asian, Black or African American, Native Hawaiian or Other Pacific Islander, and White. There are two categories for data on ethnicity: Hispanic or Latino, and Not Hispanic or Latino.

1. Categories and Definitions

The minimum categories for data on race and ethnicity for Federal statistics, program administrative reporting, and civil rights compliance reporting are defined as follows:

American Indian or Alaska Native. A person having origins in any of the original peoples of North and South America (including Central America), and who maintains tribal affiliation or community attachment.

Asian. A person having origins in any of the original peoples of the Far East, Southeast Asia, or the Indian subcontinent, including, for example, Cambodia, China, India, Japan, Korea, Malaysia, Pakistan, the Philippine Islands, Thailand, and Vietnam.

Black or African American. A person having origins in any of the black racial groups of Africa. Terms such as "Haitian" or "Negro" can be used in addition to "Black or African American."

Hispanic or Latino. A person of Cuban, Mexican, Puerto Rican, South or Central American, or other Spanish culture or origin, regardless of race. The term, "Spanish origin," can be used in addition to "Hispanic or Latino."

Native Hawaiian or Other Pacific Islander. A person having origins in any of the original peoples of Hawaii, Guam, Samoa, or other Pacific Islands.

White. A person having origins in any of the original peoples of Europe, the Middle East, or North Africa.

Respondents shall be offered the option of selecting one or more racial designations. Recommended forms for the instruction accompanying the multiple response question are "Mark one or more" and "Select one or more."

2. Data Formats

The standards provide two formats that may be used for data on race and ethnicity. Self-reporting or self-identification using two separate questions is the preferred method for collecting data on race and ethnicity. In situations where self-reporting is not practicable or feasible, the combined format may be used.

In no case shall the provisions of the standards be construed to limit the collection of data to the categories described above. The collection of greater detail is encouraged; however, any collection that uses more detail shall be organized in such a way that the additional categories can be aggregated into these minimum categories for data on race and ethnicity.

With respect to tabulation, the procedures used by Federal agencies shall result in the production of as much detailed information on race and ethnicity as possible. However, Federal agencies shall not present data on detailed categories if doing so would compromise data quality or confidentiality standards.

a. Two-question format

To provide flexibility and ensure data quality, separate questions shall be used wherever feasible for reporting race and ethnicity. When race and ethnicity are collected separately, ethnicity shall be collected first. If race and ethnicity are collected separately, the minimum designations are:

Race:

- American Indian or Alaska Native
- Asian
- Black or African American
- Native Hawaiian or Other Pacific Islander
- White

Ethnicity:

- Hispanic or Latino
- Not Hispanic or Latino

When data on race and ethnicity are collected separately, provision shall be made to report the number of respondents in each racial category who are Hispanic or Latino. When aggregate data are presented, data producers shall provide the number of respondents who marked (or selected) only one category, separately for each of the five racial categories. In addition to these numbers, data producers are strongly encouraged to provide the detailed distributions, including all possible combinations of multiple responses to the race question. If data on multiple responses are collapsed, at a minimum the total number of respondents reporting "more than one race" shall be made available.

b. Combined format

The combined format may be used, if necessary, for observer-collected data on race and ethnicity. Both race (including multiple responses) and ethnicity shall be collected when appropriate and feasible, although the selection of one category in the combined format is acceptable. If a combined format is used, there are six minimum categories:

- American Indian or Alaska Native
- Asian
- Black or African American
- Hispanic or Latino
- Native Hawaiian or Other Pacific Islander
- White

When aggregate data are presented, data producers shall provide the number of respondents who marked (or selected) only one category, separately for each of the six categories. In addition to these numbers, data producers are strongly encouraged to provide the detailed distributions, including all possible combinations of multiple responses. In cases where data on multiple responses are collapsed, the total number of respondents reporting "Hispanic or Latino and one or more races" and the total number of respondents reporting "more than one race" (regardless of ethnicity) shall be provided.

3. Use of the Standards for Record Keeping and Reporting

The minimum standard categories shall be used for reporting as follows:

a. Statistical reporting

    These standards shall be used at a minimum for all federally sponsored statistical data collections that include data on race and/or ethnicity, except when the collection involves a sample of such size that the data on the smaller categories would be unreliable, or when the collection effort focuses on a specific racial or ethnic group. Any other variation will have to be specifically authorized by the OMB through the information collection clearance process. In those cases where the data collection is not subject to the information collection clearance process, a direct request for a variance shall be made to OMB.

b. General program administrative and grant reporting

    These standards shall be used for all Federal administrative reporting or record keeping requirements that include data on race and ethnicity. Agencies that cannot follow these standards must request a variance from OMB. Variances will be considered if the agency can demonstrate that it is not reasonable for the primary reporter to determine racial or ethnic background in terms of the specified categories, that determination of racial or ethnic background is not critical to the administration of the program in question, or that the specific program is directed to only one or a limited number of racial or ethnic groups.

c. Civil rights and other compliance reporting

    These standards shall be used by all Federal agencies in either the separate or combined format for civil rights and other compliance reporting from the public and private sectors and all levels of government. Any variation requiring less detailed data or data which cannot be aggregated into the basic categories must be specifically approved by OMB for executive agencies. More detailed reporting which can be aggregated to the basic categories may be used at the agencies' discretion.

4. Presentation of Data on Race and Ethnicity

Displays of statistical, administrative, and compliance data on race and ethnicity shall use the categories listed above. The term "nonwhite" is not acceptable for use in the presentation of Federal Government data. It shall not be used in any publication or in the text of any report. In cases where the standard categories are considered inappropriate for presentation of data on particular programs or for particular regional areas, the sponsoring agency may use:

a. The designations "Black or African American and Other Races" or "All Other Races" as collective descriptions of minority races when the most summary distinction between the majority and minority races is appropriate;

b. The designations "White," "Black or African American," and "All Other Races" when the distinction among the majority race, the principal minority race, and other races is appropriate; or

c. The designation of a particular minority race or races, and the inclusion of "Whites" with "All Other Races" when such a collective description is appropriate. In displaying detailed information that represents a combination of race and ethnicity, the description of the data being displayed shall clearly indicate that both bases of classification are being used.

When the primary focus of a report is on two or more specific identifiable groups in the population, one or more of which is racial or ethnic, it is acceptable to display data for each of the particular groups separately and to describe data relating to the remainder of the population by an appropriate collective description.

5. Effective Date

The provisions of these standards are effective immediately for all new and revised record keeping or reporting requirements that include racial and/or ethnic information. All existing record keeping or reporting requirements shall be made consistent with these standards at the time they are submitted for extension, or not later than January 1, 2003.

References

Exner, D., D. Dries, M. Donamski, and J. Cohn, 2001, "Lesser Response to Angiotensin-Converting-Enzyme Inhibitor Therapy in Black as Compared with White Patients With Left Ventricular Dysfunction," N Engl J Med, 344: 1351-1357.

Freedman, L.S., R.M. Simon, M.A. Foulkes, L.M. Freedman, N.L. Gordon, et al., 1995, "Inclusion of Women and Minorities in Clinical Trials and the NIH Revitalization Act — the Perspective of NIH Clinical Trialists," Control Clin Trials, 16(5):277-285, 310-312.

McHutchison, J., T. Poynard, S. Pianko, S. Gordon, A. Reid, et al., 2000, "The Impact of Interferon Alpha Plus Ribavirin on Response to Therapy in Black Patients with Chronic Hepatitis C. The International Hepatitis Interventional Therapy Group," Gastroenterology, 119: 1317-23.

Reddy, K. R., J. Hoofnagle, M. Tong, W. Lee, P. Pockros, et al., 1999, "Racial Differences in Responses to Therapy with Interferon in Chronic Hepatitis C. Consensus Interferon Study Group," Hepatology, 30: 787-93.

Taylor, S., 2002, "Skin of Color: Biology, Structure, Function, and Implications for Dermatologic Disease," Journal of American Academy of Dermatology, 46:S41-62.

Xie, H., R. Kim, A. Wood, and C. Stein, 2001, "Molecular Basis of Ethnic Differences in Drug Disposition and Response," Annu Rev Pharmacol Toxicol, 41:815-50.

Yancy, C., M. Fowler, W. Colucci, E. Gilbert, M. Bristow, et al., 2001, "Race and the Response to Adrenergic Blockade with Carvedilol in Patients with Chronic Heart Failure," N Engl J Med, 344: 1358-1365.

Bibliography

DHHS Policy and Reports

Department of Health and Human Services (DHHS), 1997, Policy For Improving Race and Ethnicity Data, October 24, 1997, http://www.hhs.gov/oirm/infocollect/nclusion.html.

DHHS, 1999, Improving the Collection and Use of Racial and Ethnic Data in HHS, joint report of the HHS Data Council Working Group on Racial and Ethnic Data and the Data Work Group of the HHS Initiative to Eliminate Racial and Ethnic Disparities in Health, December 1999, http://aspe.hhs.gov/datacncl/racerpt/index.htm.

FDA Regulations

Food and Drug Administration (FDA), 1998, "Investigational New Drug Applications and New Drug Applications," Final Rule, Federal Register (63 FR 6854, February 11, 1998).

FDA, Content and Format for Human Prescription Drugs; Addition of "Geriatric Use" Subsection in the Labeling, Final Rule, Federal Register (62 FR 45313, August 27, 1997).

FDA Guidance for Industry

Clinical Studies Section of Labeling for Prescription Drugs and Biologics (Draft)[18]

Content and Format for Geriatric Labeling

Content and Format of the Adverse Reactions Section of Labeling for Human Prescription Drugs and Biologics (Draft)

General Considerations for Pediatric Pharmacokinetic Studies for Drugs and Biological Products (Draft)

General Considerations for the Clinical Evaluation of Drugs

---

[18] Draft guidances are included for completeness only. As draft documents, they are not intended to be implemented until published in final form.

Format and Content of the Clinical and Statistical Sections of an Application

Study and Evaluation of Gender Differences in the Clinical Evaluation of Drugs

Study of Drugs Likely to be Used in the Elderly

In Vivo Drug Metabolism/Drug Interaction Studies — Study Design, Data Analysis, and Recommendations for Dosing and Labeling

Pharmacokinetics in Patients with Impaired Hepatic Function: Study Design, Data Analysis, and Impact on Dosing and Labeling

Pharmacokinetics in Patients with Impaired Renal Function: Study Design, Data Analysis and Impact on Dosing and Labeling

Population Pharmacokinetics

Refusal to File

ICH Guidances

ICH, E4 Dose Response Information to Support Drug Registration

ICH, E5 Ethnic Factors in the Acceptability Of Foreign Data

ICH, E7 Studies in Support of Special Populations: Geriatrics

ICH, E11 Clinical Investigation of Medicinal Products in the Pediatric Population

ICH, M4 Common Technical Document for the Registration of Pharmaceuticals for Human Use

Other Sources

Office of Management and Budget, 1994, Standards for the Classification of Federal Data on Race and Ethnicity, http://www.whitehouse.gov/omb/fedreg/notice_15.html.

General Accounting Office, FDA Needs to Ensure More Study of Gender Differences in Prescription Drug Testing, GAO/HFD-93-17.

NIH Revitalization Act of 1993, (PL 103-43).

NIH Guidelines on the Inclusion of Women and Minorities as Subjects in Clinical Research-Updated August 1, 2000. Available at http://grants.nih.gov/grants/funding/women_min/women_min.htm.

Food and Drug Administration Modernization Act of 1997 (Public Law 105-115).

Best Pharmaceuticals Act for Children of 2002 (Public Law 107-109).

Part II

Selected FDA GCP/Clinical Trial Guidance Documents: Institutional Review Boards (IRBs) and Informed Consent

Adverse Event Reporting to IRBs — Improving Human Subject Protection

Guidance for Clinical Investigators, Sponsors, and IRBs:
Adverse Event Reporting to IRBs — Improving Human Subject Protection[1, 2]

U.S. Department of Health and Human Services
Food and Drug Administration
Office of the Commissioner (OC)
Center for Drug Evaluation and Research (CDER)
Center for Biologics Evaluation and Research (CBER)
Center for Devices and Radiological Health (CDRH)
Good Clinical Practice Program (GCPP)

January 2009

Procedural

Contains Nonbinding Recommendations

This guidance is intended to provide assistance to the research community in interpreting requirements for submitting reports of unanticipated problems, including certain adverse events reports, to the IRB.

Note: *This guidance represents the Food and Drug Administration's (FDA's) current thinking on this topic. It does not create or confer any rights for or on any person and does not operate to bind FDA or the public. You can use an alternative approach if the approach satisfies the requirements of the applicable statutes and regulations. If you want to discuss an alternative approach, contact the appropriate FDA staff. If you cannot identify the appropriate FDA staff, call the appropriate number listed on the title page of this guidance.*

I. Introduction

This guidance is intended to assist the research community in interpreting requirements for submitting reports of *unanticipated problems*, including certain adverse events reports, to the institutional review board (IRB) under Title 21 of the Code of Federal Regulations (21 CFR) part 56 (Institutional Review Boards), part 312 (Investigational New Drug Application), and part 812 (Investigational Device Exemptions). Specifically, the guidance provides

---

[1] Available on the FDA website at: http://www.fda.gov/downloads/RegulatoryInformation/ Guidances/UCM126572.pdf

[2] This guidance has been prepared by the Office of the Commissioner, the Center for Drug Evaluation and Research (CDER), the Center for Biologics Evaluation and Research (CBER), the Center for Devices and Radiological Health (CDRH), and the Good Clinical Practice Program (GCPP) at the Food and Drug Administration.

recommendations for sponsors and investigators conducting investigational new drug (IND) trials to help them differentiate between those adverse events that are unanticipated problems that must be reported to an IRB and those that are not. The guidance also makes suggestions about how to make communicating adverse events information to IRBs more efficient.

FDA developed this guidance in response to concerns raised by the IRB community, including concerns raised at a March 2005 public hearing,[3] that increasingly large volumes of individual adverse event reports submitted to IRBs—often lacking in context and detail—are inhibiting, rather than enhancing, the ability of IRBs to protect human subjects.

FDA regulations use different terms when referring to an *adverse event*. For example, *adverse effect* is used in 21 CFR 312.64; *adverse experience* is used in § 312.32; and *unanticipated problems* is used in § 312.66. For the purposes of this guidance, the term *adverse event* is used, except when quoting specific regulations. For device studies, part 812 uses the term *unanticipated adverse device effect*, which is defined in 21 CFR 812.3(s).

FDA's guidance documents, including this guidance, do not establish legally enforceable responsibilities. Instead, guidances describe the Agency's current thinking on a topic and should be viewed only as recommendations, unless specific regulatory or statutory requirements are cited. The use of the word *should* in Agency guidances means that something is suggested or recommended, but not required.

II. Background

FDA regulates clinical studies authorized under sections 505(i) (drugs and biologics) and 520(g) (devices) of the Federal Food, Drug, and Cosmetic Act. All such clinical studies must be reviewed and approved by an IRB before the study is initiated, in accordance with the requirements of 21 CFR part 50 (Protection of Human Subjects), part 56 (Institutional Review Boards), and either part 312 (Investigational New Drug Application) or part 812 (Investigational Device Exemptions) (see §§ 50.1, 56.101, 312.23(a)(1)(iv), 312.40(a), 812.2(b)(1)(ii), 812.2(c) and 812.62(a)).[4] After the initial review and approval of a clinical study, an IRB must conduct continuing review of the study at intervals appropriate to the degree of risk presented by the study, but at least annually (§ 56.109(f)). The primary purpose of both initial and continuing review of the study is "to assure the protection of the rights and welfare of the human subjects" (§ 56.102(g)). To fulfill its obligations during the conduct of a clinical study, an IRB must have, among other things, information concerning unanticipated problems involving risk to human subjects in the study, including adverse events (AEs) that are considered unanticipated problems (§§ 56.108(a)(3), (4), (b)).[5]

---

[3] *Federal Register,* "Reporting of Adverse Events to Institutional Review Boards; Public Hearing," (70 FR 6693, March 21, 2005).

[4] As described below, there are some differences between the requirements for investigational new drug and investigational device exemption studies, as they concern obligations to report to a reviewing IRB.

[5] Unanticipated problems may be adverse events or other types of problems, i.e., adverse events are a subset of unanticipated problems.

For clinical investigations of drug and biological products conducted under an investigational new drug (IND) application, information about adverse events[6] must be communicated among investigators, sponsors, and IRBs as follows:

- Investigators are required to report promptly "to the sponsor any adverse effect that may reasonably be regarded as caused by, or probably caused by, the drug. If the adverse effect is alarming, the investigator shall report the adverse effect immediately" (§312.64(b)).

- Sponsors are specifically required to notify all participating investigators (and FDA) in a written IND safety report of "any adverse experience associated with the use of the drug that is both serious and unexpected" and "any finding from tests in laboratory animals that suggests a significant risk for human subjects" (§ 312.32(c)(1)(i)(A),(B)). And, more generally, sponsors are required to "keep each participating investigator informed of new observations discovered by or reported to the sponsor on the drug, particularly with respect to adverse effects and safe use" (§ 312.55(b)).

-

Investigators are required to report promptly "to the IRB… all unanticipated problems involving risks to human subjects or others," including adverse events that should be considered unanticipated problems (§§ 56.108(b)(1), 312.53(c)(1)(vii), and 312.66).

A critical question for studies conducted under part 312 is what adverse events should be considered unanticipated problems that merit reporting to an IRB. In the years since the IRB and IND regulations issued, changes in the conduct of clinical trials (e.g., increased use of multi-center studies, international trials) have complicated the reporting pathways for adverse event information described in the regulations. In particular, the practice of local investigators reporting individual, unanalyzed events to IRBs, including reports of events from other study sites that the investigator receives from the sponsor of a multi-center study—often with limited information and no explanation of how the event represents an unanticipated problem—has led to the submission of large numbers of reports to IRBs that are uninformative. IRBs have expressed concern that the way in which investigators and sponsors of IND studies typically interpret the regulatory requirement to inform IRBs of all "unanticipated problems" does not yield information about adverse events that is useful to IRBs and thus hinders their ability to ensure the protection of human subjects. This guidance is intended to help differentiate those adverse events that should be considered unanticipated problems (and thus reported to the IRB) from those that should not, thereby helping to ease the burden on IRBs and make the adverse events information they receive more informative and useful.

---

[6] The IND regulations use the term adverse effect (§ 312.64) and adverse experience (§ 312.32). These terms are interchangeable with adverse event.

III. Reporting AEs to IRBs in Clinical Trials of Drug and Biological Products Conducted Under IND Regulations

A. How to Determine If an AE is an Unanticipated Problem that Needs to Be Reported

In general, an AE observed during the conduct of a study should be considered an unanticipated problem involving risk to human subjects, and reported to the IRB, *only* if it were unexpected, serious, and would have implications for the conduct of the study (e.g., requiring a significant, and usually safety-related, change in the protocol such as revising inclusion/exclusion criteria or including a new monitoring requirement, informed consent, or investigator's brochure). An individual AE occurrence *ordinarily* does not meet these criteria because, as an isolated event, its implications for the study cannot be understood.

Many types of AEs generally require an evaluation of their relevance and significance to the study, including an aggregate analysis of other occurrences of the same (or similar) event, before they can be determined to be an unanticipated problem involving risk to human subjects. For example, an aggregate analysis of a series of AEs that are commonly associated with the underlying disease process that the study intervention is intended to treat (e.g., deaths in a cancer trial), or that are otherwise common in the study population independent of drug exposure (e.g., cardiovascular events in an elderly population) may reveal that the event rate is higher in the drug treatment group compared to the control arm. In this case, the AE would be considered an unanticipated problem. In the absence of such a finding, the event is uninterpretable.

The major exceptions to the general rule that an isolated event is not informative are serious AEs that are uncommon and strongly associated with drug exposure, such as angioedema, agranulocytosis, anaphylaxis, hepatic injury, or Stevens Johnson syndrome. In most cases, a single, unexpected occurrence of this type of event would be considered an unanticipated problem involving risk to human subjects and, thus, must be reported to the IRB. Similarly, one or a small number of serious events that are not commonly associated with drug exposure, but are otherwise uncommon in the study population (e.g., tendon rupture, progressive multifocal leukoencephalopathy) should be considered an unanticipated problem involving risk to human subjects.

Because they have been previously observed with a drug, the AEs listed in the investigator's brochure would, by definition,[7] not be considered unexpected and thus would not be

---

[7] An unexpected adverse drug experience is defined as "[a]ny adverse drug experience, the specificity or severity of which is not consistent with the current investigator brochure; or, if an investigator brochure is not required or available, the specificity or severity of which is not consistent with the risk information described in the general investigational plan or elsewhere in the current application, as amended. For example, under this definition, hepatic necrosis would be unexpected (by virtue of greater severity) if the investigator brochure only referred to elevated hepatic enzymes or hepatitis. Similarly, cerebral thromboembolism and cerebral vasculitis would be unexpected (by virtue of greater specificity) if the investigator brochure only listed cerebral vascular accidents. Unexpected, as used in this definition, refers to an adverse drug experience that has not been previously observed (e.g., included in the investigator

unanticipated problems. Possible exceptions would include situations in which the specificity or severity of the event is not consistent with the description in the investigator's brochure, or it can be determined that the observed rate of occurrence for a serious, expected AE in the clinical trial represents a clinically important increase in the expected rate of occurrence.

Therefore, FDA recommends that there be careful consideration of whether an AE is an unanticipated problem that must be reported to IRBs. In summary, FDA believes that only the following AEs should be considered as unanticipated problems that must be reported to the IRB.

- A single occurrence of a serious, unexpected event that is uncommon and strongly associated with drug exposure (such as angiodema, agranulocytosis, hepatic injury, or Stevens-Johnson syndrome).

- A single occurrence, or more often a small number of occurrences, of a serious, unexpected event that is not commonly associated with drug exposure, but uncommon in the study population (e.g., tendon rupture, progressive multifocal leukoencephalopathy).

- Multiple occurrences of an AE that, based on an aggregate analysis, is determined to be an unanticipated problem. There should be a determination that the series of AEs represents a signal that the AEs were not just isolated occurrences and involve risk to human subjects (e.g., a comparison of rates across treatment groups reveals higher rate in the drug treatment arm versus a control). We recommend that a summary and analyses supporting the determination accompany the report.

- An AE that is described or addressed in the investigator's brochure, protocol, or informed consent documents, but occurs at a specificity or severity that is inconsistent with prior observations. For example, if transaminase elevation is listed in the investigator's brochure and hepatic necrosis is observed in study subjects, hepatic necrosis would be considered an unanticipated problem involving risk to human subjects. We recommend that a discussion of the divergence from the expected specificity or severity accompany the report.

- A serious AE that is described or addressed in the investigator's brochure, protocol, or informed consent documents, but for which the rate of occurrence in the study represents a clinically significant increase in the expected rate of occurrence (ordinarily, reporting would only be triggered if there were a credible baseline rate for comparison). We recommend that a discussion of the divergence from the expected rate accompany the report.

- Any other AE or safety finding (e.g., based on animal or epidemiologic data) that would cause the sponsor to modify the investigator's brochure, study protocol, or informed consent documents, or would prompt other action by the IRB to ensure the protection of human subjects. We recommend that an explanation of the conclusion accompany the report.

---

brochure), rather than from the perspective of such experience not being anticipated from the pharmacological properties of the pharmaceutical product." (21 CFR 312.32(a))

B. How to Report Unanticipated Problems to IRBs

In a multicenter study, it is clear that individual investigators must rely on the sponsor to provide them information about AEs occurring at other study sites. It is also clear that the sponsor receives AE information from all study sites and typically has more experience and expertise with the study drug than an investigator. Accordingly, the sponsor is in a better position to process and analyze the significance of AE information from multiple sites and—when the determination relies on information from multiple study sites or other information not readily accessible to the individual investigators (e.g., a sponsor's preclinical data that supports the determination)—to make a determination about whether an AE is an unanticipated problem. Furthermore, the regulations require the sponsor of an IND to promptly review all information relevant to the safety of the drug and to consider the significance of the report within the context of other reports (§ 312.32)[8]

The regulations state that for studies conducted under 21 CFR part 312, investigators must report all "unanticipated problems" to the IRB (§§ 312.66, 312.53(c)(1)(vii), and 56.108(b)(1)). However, as discussed above, we recognize that for multicenter studies, the sponsor is in a better position to process and analyze adverse event information for the entire study and to assess whether an adverse event occurrence is both *unanticipated* and a *problem* for the study.

Accordingly, to satisfy the investigator's obligation to notify the IRB of unanticipated problems, an investigator participating in a multicenter study may rely on the sponsor's assessment and provide to the IRB a report of the unanticipated problem prepared by the sponsor. In addition, if the investigator knows that the sponsor has reported the unanticipated problem directly to the IRB, because the investigator, sponsor, and IRB made an explicit agreement for the sponsor to report directly to the IRB,[9] and because the investigator was copied on the report from the sponsor to the IRB, FDA intends to exercise its enforcement discretion and would not expect an investigator to provide the IRB with a duplicate copy of the report received from the sponsor.

IV. Reporting AEs to IRBs in Clinical Trials of Devices Under the IDE Regulations

The investigational device exemption (IDE) regulations define an unanticipated adverse device effect (UADE) as "any serious adverse effect on health or safety or any life-threatening problem or death caused by, or associated with, a device, if that effect, problem, or death was not previously identified in nature, severity, or degree of incidence in the investigational plan or application (including a supplementary plan or application), or any other unanticipated serious problem associated with a device that relates to the rights, safety, or welfare of subjects" (21

---

[8] Section 312.32(c)(1)(ii) requires a sponsor preparing an IND safety report to, among other things, "analyze the significance of the adverse experience in light of previous, similar reports." Section 312.32(b) requires the sponsor to "promptly review all information relevant to the safety of the drug obtained or otherwise received by the sponsor from any source . . . ."

[9] Note that such an agreement would be required to be incorporated into the IRB's written procedures (21 CFR 56.108(b)(1), 56.115(a)(6)).

CFR 812.3(s)). UADEs must be reported by the clinical investigator to the sponsor and the reviewing IRB, as described below:

- For device studies, investigators are required to submit a report of a UADE to the sponsor and the reviewing IRB as soon as possible, but in no event later than 10 working days after the investigator first learns of the event (§ 812.150(a)(1)).

- Sponsors must immediately conduct an evaluation of a UADE and must report the results of the evaluation to FDA, all reviewing IRBs, and participating investigators within 10 working days after the sponsor first receives notice of the effect (§§ 812.46(b), 812.150(b)(1)).

The IDE regulations, therefore, require sponsors to submit reports to IRBs in a manner consistent with the recommendations made above for the reporting of unanticipated problems under the IND regulations.

V. Conclusion

The receipt of a large volume of individual AE reports without analysis of their significance to a clinical trial rarely supports an IRB's efforts to ensure human subject protection. Sponsors can assess the implications and significance of AE reports promptly and are required to report serious, unexpected events associated with the use of a drug or device, including analyses of such events, to investigators and to FDA. In addition, sponsors are required to report analyses of unexpected adverse device experiences to IRBs. FDA encourages efforts by investigators and sponsors to ensure that IRBs receive meaningful AE information. The ultimate goal is to provide more meaningful information to IRBs, particularly when sponsor analysis (including an analysis of the significance of the adverse event, with a discussion of previous similar events where appropriate) is made available to IRBs.

Using a Centralized IRB Process in Multicenter Clinical Trials

Guidance for Industry on Using a Centralized IRB Process in Multi center Clinical Trials [†, 1]

U.S. Department of Health and Human Services
Food and Drug Administration
Good Clinical Practice Program, Office of the Commissioner (OC)
Center for Drug Evaluation and Research (CDER)
Center for Biologics Evaluation and Research (CBER)
Office of Regulatory Affairs (ORA)

March 2006

Procedural

Contains Nonbinding Recommendations

This guidance is intended to assist sponsors, institutions, institutional review boards (IRBs), and clinical investigators involved in multicenter clinical research in meeting the requirements of 21 CFR part 56 by facilitating the use of a centralized IRB review process (use of a single central IRB), especially in situations where centralized review could improve efficiency of IRB review.

Note: *This guidance represents the Food and Drug Administration's (FDA's) current thinking on this topic. It does not create or confer any rights for or on any person and does not operate to bind FDA or the public. You can use an alternative approach if the approach satisfies the requirements of the applicable statutes and regulations. If you want to discuss an alternative approach, contact the appropriate FDA staff. If you cannot identify the appropriate FDA staff, call the appropriate number listed on the title page of this guidance.*

I. Introduction

This guidance is intended to assist sponsors, institutions, institutional review boards (IRBs), and clinical investigators involved in multicenter clinical research in meeting the requirements of 21 CFR part 56 by facilitating the use of a centralized IRB review process (use of a single central IRB), especially in situations where centralized review could improve efficiency of IRB review.

[†] Available on the FDA website at: http://www.fda.gov/downloads/RegulatoryInformation/Guidances/UCM127013.pdf

[1] This guidance has been prepared by the Center for Drug Evaluation and Research (CDER), the Center for Biologics Evaluation and Research (CBER), the Good Clinical Practice Program in the Office of the Commissioner (OC), and the Office of Regulatory Affairs (ORA) at the Food and Drug Administration.

The guidance (1) describes the roles of the participants in a centralized IRB review process, (2) offers guidance on how a centralized IRB review process might consider the concerns and attitudes of the various communities participating in a multicenter clinical trial, (3) makes recommendations about documenting agreements between a central IRB and the IRBs at institutions involved in the centralized IRB review process concerning the respective responsibilities of the central IRB and each institution's IRB, (4) recommends that IRBs have procedures for implementing a centralized review process, and (5) makes recommendations for a central IRB's documentation of its reviews of studies at clinical trial sites not affiliated with an IRB. This guidance applies to clinical investigations conducted under 21 CFR part 312 (investigational new drug application, or IND regulations).

FDA's guidance documents, including this guidance, do not establish legally enforceable responsibilities. Instead, guidances describe the Agency's current thinking on a topic and should be viewed only as recommendations, unless specific regulatory or statutory requirements are cited. The use of the word *should* in Agency guidances means that something is suggested or recommended, but not required.

II. Background

Clinical investigations that are subject to the requirements of IND regulations must be reviewed and approved by an IRB in accordance with the requirements of 21 CFR part 56. The IRB requirements evolved at a time when most clinical trials were conducted at a single study site or at a small number of sites. In the intervening years, there has been substantial growth in the amount of clinical research generally, the number of multicenter trials, and the size and complexity of late-stage clinical trials. These changes have placed considerable burdens on IRBs and on sponsors and clinical investigators who are seeking IRB review for multicenter trials.[2, 3]

For example, sometimes the IRB at each center of a multicenter trial conducts a complete review of the protocol and informed consent. Such multiple reviews by multiple IRBs can result in unnecessary duplication of effort, delays, and increased expenses in the conduct of multicenter clinical trials.[4, 5, 6] Greater reliance on a centralized IRB review process, in appropriate circumstances, could reduce IRB burdens and delays in the conduct of multicenter trials.

---

[2] Department of Health and Human Services (DHHS), Office of the Inspector General Report, *Institutional Review Boards: A Time for Reform*, June 1998.

[3] Burman WE, RR Randall, DL Cohn, RT Schooley, Breaking the Camel's Back: Multicenter clinical trials and the local institutional review boards, *Ann Intern Med*, 134(2) : 152-157, 2001.

[4] Burman W, P Breese, S Weis, N Bock, J Bernardo, A Vernon, The Effects of local review on informed consent documents; from a multicenter clinical trials consortium, *Controlled Clin Trials*, 24(2003) 245-255.

[5] Silverman H, S Chandros Hull, J Sugarman, Variability among institutional review boards' decisions within the context of a multicenter trial, *Crit Care Med* 29(2), 235-241, 2001.

[6] MeWilliams R, J I-loover-Fong, A Hamosh, S Beck, T Beatty, G Cutting, Problematic Variation in Local Institutional Review of a Multicenter Genetic Epidemiology Study, *JAMA*, 290(3), 360-361, 2003.

Use of a centralized IRB review process is consistent with the requirements of existing IRB regulations. Section 56.114 (21 CFR 56.114, Cooperative Research) provides that, "institutions involved in multi-institutional studies may use joint review, reliance upon the review of another qualified IRB, or similar arrangements aimed at avoidance of duplication of effort." When this rule was proposed, the preamble to the proposed rule indicated that the purpose of this section is "to explicitly reduce duplicative review of multi-institutional studies."[7] The preamble to the final rule also stated that "the purpose of this section is to assure IRBs that FDA will accept reasonable methods of joint review."[8] An IRB that is at a different location from the research site can review the research, provided that the IRB is competent to understand the local context of the research. As stated in 21 CFR 56.107(a), this would require sensitivity to community attitudes and the ability to ascertain the acceptability of proposed research in terms of institutional commitments and regulations, applicable law, and standards of professional conduct and practice (see Section IV).

A centralized IRB review process involves an agreement under which multiple study sites in a municeriter trial rely in whole or in part on the review of an IRB other than the IRB affiliated with the research site. Because the goal of the centralized process is to increase efficiency and decrease duplicative efforts that do not contribute to meaningful human subject protection, it will usually be preferable that a central IRB take responsibility for all aspects of IRB review at each site partfcipating in the centralized review process. Other approaches may be appropriate as well. For example, an institution may permit a central IRB to be entirely responsible for initial and continuing review of a study, or apportion IRB review responsibilities between the central IRB and its own IRB.

III. Roles in Ensuring IRB Review

The following sections describe the roles and responsibilities of the principal parties as they relate to a centralized IRB review process .

A. Institution

Under 21 CFR 56 .114, institutions that participate in multicenter studies can use joint review, rely on the review of another qualified IRB, or establish other arrangements aimed at reducing duplicative efforts. For example, the decision could be made in a multicenter trial to rely primarily on the review of a central IRB while establishing an agreement that all site-specific IRBs will review their informed consent documents for local concerns. Institutions should develop policies for determining when and which studies conducted in the institution would be appropriate for centralized review and how a centralized review would be conducted for such studies.

---

[7] *Federal Register*, Vol. 40, pp. 47688, 47700, August 14, 1979.
[8] *Federal Register*, Vol. 46, pp. 8958, 8970, January 27, 1981.

B. Institution's IRB

An *institution's IRB* is the IRB designated or formed by an institution for the purpose of reviewing research conducted at the institution or with institutional support. For multicenter studies, an institution's IRB can serve as a central IRB ; an institution's IRB can rely on the review of a centralized IRB (in whole or in part) in place of its own IRB review of the study; or it can conduct its own review of the study. The institution's policies will dictate under what circumstances the institution's IRB can participate in a centralized review process and the role of the institution's IRB in that process.

C. Sponsor

For studies conducted under an IND, 21 CFR part 312 provides that a *sponsor* is responsible for obtaining a commitment from each investigator that he or she will ensure that requirements in part 56 relating to IRB review and approval are met with respect to the research conducted by the investigator (21 CFR 312 .53(c)(1)(vi)(d)). Sponsors can also initiate plans for use of a centralized IRB review process and facilitate agreements and other necessary communications among the parties involved.

D. Investigator

Under 21 CFR part 312, an *investigator* is responsible for ensuring that there will be initial and continuing review by a qualified IRB of research conducted by that investigator (21 CFR 312.53(c)(1)(vi)(d); 312.66). If the investigator is conducting clinical research as part of a multiceriter study at an institution with its own IRB and is subject to the policies of that institution, those policies would dictate how the investigator will ensure IRB review. Those policies may provide that the investigator's responsibility can be met by ensuring review through a centralized IRB review, through the institution's IRB, or through apportionment of IRB review responsibilities between a centralized IRB and the institution's IRB.

E. Central IRB

For multicenter studies, the *central IRB* is the IRB that conducts reviews on behalf of all study sites that agree to participate in the centralized review process. For sites at institutions that have an IRB that would ordinarily review research conducted at the site, the central IRB should reach agreement with the individual institutions participating in centralized review and those institutions' IRBs about how to apportion the review responsibilities between local IRBs and the central IRB (21 CFR 56 .114).

IV. Addressing Local Aspects of IRB Review

The implementation of a centralized IRB review process involves addressing a number of issues related to the communities where the research will take place. The requirements for IRB membership in 21 CFR 56.107(a) specify that the membership of an IRB must have sufficient experience, expertise, and diversity to promote respect for its advice and counsel in safeguarding the rights and welfare of human subjects. This requirement was intended to implement a recommendation of the National Commission for the Protection of Human

Subjects of Biomedical and Behavioral Research that IRB members be "men and women of diverse backgrounds and sufficient maturity, experience, and competence to assure that the Board will be able to discharge its responsibilities and that its determinations will be accorded respect by investigators and the community served by the institution or in which it is located."[9] In addition, IRB members must "be able to ascertain the acceptability of the proposed research in terms of institutional commitments and regulations, applicable law, and standards or professional conduct and practice" (21 CFR 56 .107(a)). Thus, IRB review, through its membership, is intended to provide meaningful consideration of various local factors in assessing research activities, including the cultural backgrounds (e.g., ethnicity, educational level, religious affiliations) of the population from which research subjects will be drawn, community attitudes"[10] about the nature of the proposed research, and the capacity of the institution to conduct or support the proposed research. Inter-community differences could influence, among other things, assessments of whether mechanisms of subject selection will be equitable, whether adequate provision is made to minimize risks to vulnerable populations, and the adequacy of the informed consent process.

The preamble to the final rule indicates that where a centralized IRB review process is used (21 CFR 56.114), the review should consider the ethical standards of the local community."[11] Therefore, a centralized IRB review process should include mechanisms to ensure meaningful consideration of these relevant local factors. Possible mechanisms include:

- Provision of relevant local information to the central IRB in writing by individuals or organizations familiar with the local community, institution, and/or clinical research

- Participation of consultants with relevant expertise, or IRB members from the institution's own IRB, in the deliberations of the central IRB

- Limited review of a central IRB-reviewed study by the institution's own IRB, with that Limited review focusing on issues that are of concern to the local community

Other mechanisms may also be appropriate.[12] IRB meeting minutes or other records should document how relevant community issues were considered in the review (21 CFR 56.115(a)) (see section V).

---

[9] *Federal Register*, Vol. 44, p. 47699, August 14, 1979, and *Federal Register*, Vol. 43, p. 56174, November 30, 1978 .

[10] *Community attitudes* is usually interpreted to refer to the attitudes of the local community where research will be conducted. However, it could also refer to a community of other individuals, such as a community of individuals with the same disease. For purposes of a discussion of special issues that arise in the context of central IRB review of multicenter research, when we refer to *community attitudes*, we are referring to any considerations that may be unique to the various communities from which research subjects will be drawn.

[11] *Federal Register*, Vol. 46, p. 8966, January 27, 1981

[12] Guidance issued by the Department of Health and Human Services, Office of Human Research Protections (OHRP) (IRB *Knowledge of Local Research Context*) discusses mechanisms for ensuring adequate consideration of local factors by 1RBs in review of clinical research that is supported by DHHS funding. Although the guidance applies only to DHI3S funded research, it may be also helpful to off-site IRBs seeking to provide meaningful consideration of relevant local factors for non-DHHS funded clinical research.

V. IRB Records - Documenting Agreements and Procedures

IRBs and institutions are required to prepare and maintain adequate documentation of IRB activities (21 CFR 56.115(a)). IRBs are also required to follow written procedures for the conduct of initial and continuing review of clinical research and for reporting their findings and actions to the investigator and the institution (21 CFR section 56.108(a), 56.115(a)(6)). The following recommendations should help IRBs fulfill these requirements.

A. Documenting Agreements

If an institution, its IRB, and a central IRB agree (under 21 CFR 56.114) to participate in a centralized IRB review process, they should document that action in an agreement signed by the parties. IRBs should report this action to the investigator and the institution, for example, by providing copies of the agreement to the investigator, and the institution.[13] If the agreement apportions IRB review responsibilities between a central IRB and the institution's IRB, the agreement should delineate the specific responsibilities of the central IRB and the institution's IRB for the initial and continuing review of the study.

B. Written Procedures

When an institution and an institution's IRB rely on review by a central IRB, both IRBs must have wriitten procedures in place to implement the centralized IRB review process (21 CFR 56.108, 56.1 14). For example, procedures should address the following:

- How the institution's IRB determines that the central IRB is qualified to review research conducted at the institution

- How the central IRB intends to communicate with relevant institutions, the institutions' IRBs, and investigators regarding its review

- How the central IRB ensures that it provides meaningful consideration of relevant local factors for communities from which research subjects will be drawn (see Section IV)

- How the central IRB assesses the ability of a geographically remote site to participate in a study (e.g., whether the site has medical services appropriate to the complexity of the study)

When an institution, an institution's IRB, and a central IRB agree to apportion IRB review responsibilities between the two IRBs, each IRB must have written procedures describing how it implements its responsibilities under the agreement (21 CFR 56.108, 56.115(a)(6)).

---

[13] When research covered by a Federal wide assurance (FWA) approved by the Office for Human Research Protections (OHRP) is to be reviewed by a central IRB, the central IRB must be designated under the FWA (45 CFR 46.103(b)(2)) . Procedures for respective responsibilities for IRB review activities must be documented (45 CFR 46.103(b)(4)). OHRP has a sample IRB Authorization Agreement on its Website atwww.hhs .gov/ohrpihumansubjects/assurance/iprotsup.rtf that may be useful to allocate responsibilities between IRBs, or the institutions may develop their own agreement .

VI. Using a Central IRB at Unaffiliated Sites

At clinical sites that are not already affiliated with an IRB, investigators and sponsors typically rely on the review and oversight of a central IRB. In this situation, the central IRB should document in meeting minutes or other records how it considered relevant local factors for the various communities from which research subjects are to be drawn (see Section N). The central IRB must also document its action in agreeing to conduct IRB review for the site (21 CFR 56.115), and must have written procedures in place that describe how it will perform its initial and continuing review responsibilities at remote sites (21 CFR 56.108, 56.115(a)(6)) (see Section V).

VII. Examples of Cooperative IRB Review Models

There are a variety of mechanisms that have been used to distribute IRB review responsibilities between an institution's IRB and a central IRB. This guidance is not intended to endorse any particular mechanism. These examples are provided only to illustrate possible mechanisms.

A. Trial in Which Multiple Sites Rely on a Central IRB

The primary model contemplated by this guidance is a centralized IRB review process used for a single multicenter trial performed by a commercial or publicly funded sponsor. Under 21 CFR 56.114, IRBs affiliated with the study sites could enter into agreements with a central IRB to rely on all or some of the review findings of the central IRB, or could decline to participate in a centralized IRB review (i.e., do their own complete review). Study sites not already affiliated with an IRB would rely on a central IRB for all IRB review responsibilities.

B. Central IRB Formed to Review Multicenter Trials in a Therapeutic Category

A central IRB can be formed to review multicenter trials in a therapeutic category. For example, the National Cancer Institute (NCI) has created a freestanding central IRB (NCI central IRB) to provide the option for centralized IRB review for the many multicenter cancer trials conducted by NCI. This NCI central IRB is a standing body with subject matter expertise that reviews all NCI-sponsored phase 3 trials in adults with cancer. The IRBs affiliated with the study sites have the option of accepting the review of the NCI central IRB, or doing their own complete review of the protocol and informed consent.[14]

C. Regional and Nonregional Cooperatives

IRBs at some academic medical centers have entered into ongoing cooperative agreements in which their IRBs have the option of accepting reviews by IRBs at other centers when both centers are participating in a multicenter trial.

---

[14] See http://www.ncicirb.arg/Div_Responsibilities1.pdf.

VIII. Conclusion

The Agency hopes that sponsors, institutions, institutional review boards (IRBs), and clinical investigators involved in multicenter clinical research will consider the use of a single central IRB (centralized IRB review process), especially if using centralized review could improve the efficiency of IRB review.

IRB Review of Stand-Alone HIPAA Authorizations Under FDA Regulations

Guidance for Industry:
IRB Review of Stand-Alone HIPAA Authorizations Under FDA Regulations[1]

U.S. Department of Health and Human Services
Food and Drug Administration
Office of the Commissioner

October 21, 2003

Final Guidance

Contains Nonbinding Recommendations

This guidance provides clarification for IRBs of their responsibilities for reviewing and approving stand-alone authorizations under the HIPAA Privacy Rule.

Note: *This guidance represents the Food and Drug Administration's (FDA's) current thinking on this topic. It does not create or confer any rights for or on any person and does not operate to bind FDA or the public. You can use an alternative approach if the approach satisfies the requirements of the applicable statutes and regulations. If you want to discuss an alternative approach, contact the appropriate FDA staff. If you cannot identify the appropriate FDA staff, call the appropriate number listed on the title page of this guidance.*

I. Introduction

This guidance provides the current recommendations of the Food and Drug Administration (FDA) concerning Institutional Review Boards' (IRB) review and approval under 2 1 C.F.R. Part 56 of stand-alone authorizations that are created by covered entities (or by third parties), pursuant to the Health Insurance Portability and Accountability Act of 1996 (HIPAA) Privacy Rule, and that are provided to research subjects prior to enrolling in clinical investigations after April 14, 2003, to obtain their permission to use and/or disclose their health information for research. A stand-alone HIPAA authorization (for research) is a document that is used to obtain permission from an individual for a covered entity to use and/or disclose the individual's identifiable health information for a research study, and that is not combined with an informed consent document to participate in the research study itself. The Privacy Rule refers to a HIPAA authorization that has been combined with an informed consent document as a "compound authorization." IRBs would be required to review the HIPAA authorization in

---

[1] Available on the FDA website at: http://www.fda.gov/downloads/RegulatoryInformation/
Guidances/UCM126577.pdf

a "compound authorization" because IRBs are required, with certain exceptions, to review and approve informed consent documents. See 21 C.F.R. Part 56.

Because issues addressed in this Level 1 guidance require immediate resolution, it is not feasible or appropriate for FDA to seek comments before implementing it. See 21 C.F.R. §§ 10.115(g)(2) and 10.115(g)(3). This guidance is intended to encourage IRBs to permit the enrollment of subjects in clinical investigations without prior IRB review and/or approval of stand-alone HIPAA authorizations, since such review and/or approval is not required by the HIPAA Privacy Rule.

II. Background

To improve the efficiency and effectiveness of the health care system, Congress passed the Health Insurance Portability and Accountability Act of 1996 (HIPAA), Public Law 104- 19 1, which included "Administrative Simplification" provisions that required, among other things, the Department of Health and Human Services (HHS) to adopt national standards for certain electronic health care transactions. At the same time, Congress recognized that advances in electronic technology could erode the privacy of health information. Consequently, Congress incorporated, into HIPS, provisions that mandated the adoption of Federal privacy protections for certain individually identifiable health information.

In response to the HIPS mandate, HHS published a final regulation, "Standards for Privacy of Individually Identifiable Health Information," generally known as the Privacy Rule, in December 2000. HHS subsequently amended the Privacy Rule on August 14, 2002. See 45 C.F.R. Part 160 and Part 164, Subparts A and E. This Privacy Rule sets minimum national standards for the protection of certain individually identifiable health information. The Privacy Rule, however, applies only to three types of entities, known as covered entities: health plans, health care clearinghouses, and health care providers who transmit health information in electronic form in connection with a transaction for which HHS has adopted a standard. As of April 14, 2003 (April 14, 2004, for small health plans), covered entities have been required to comply with the standards to protect and guard against the misuse and improper disclosure of individually identifiable health information. Failure to comply with these standards may, under certain circumstances, trigger the imposition of civil or criminal penalties. For a more complete description of the Privacy Rule, see www.hhs.gov/ocr/hipaa.

Health care providers who transmit health information electronically in connection with a transaction for which HHS has adopted a standard, such as most hospitals, are covered entities. In addition, clinical investigators who are health care providers but do not transmit health information in electronic form in connection with a transaction for which HHS has adopted a standard, but who are nonetheless part of the covered entities' workforce, must comply with the Privacy Rule in their work at or for the covered entity - unless they are workforce members of hybrid entities, and they are not part of the designated health care component. See 45 C.F.R. 164.105. Lastly, clinical investigators or other researchers who are not in either of the above categories are not covered by the Privacy Rule, but they should be aware of the Rule and its restrictions on the use and disclosure of protected health information.

Most IRBs are not covered entities under the Privacy Rule because they do not meet the Privacy Rule's definitions of health plan, health care clearinghouse, or health care provider who transmits any health information in electronic form in connection with a transaction for which HHS has adopted a standard. See 45 C.F.R. 160.103. However, Irks interact with clinical investigators who might be, or who are employed by or are otherwise part of the workforce of covered entities under the Privacy Rule. In addition, IRBs may review research in which individually identifiable health information will be obtained from covered entities, and therefore, they need to understand how the Privacy Rule interacts with Federal informed consent requirements, including FDA human subjects protection regulations and regulations on institutional review boards (see 21 C.F.R. Parts 50 and 56).

The Privacy Rule provides that, unless the use or disclosure is otherwise permitted or required by the rule, the use or disclosure of the protected health information of an individual, such as a research subject, is permitted only if the individual signs an authorization for the use or disclosure. See 45 C.F.R. 164.508. For example, in the context of a clinical investigation conducted by a covered entity, a valid and properly executed HIPAA authorization is a permission from the subject for the covered entity to use and/or disclose the subject's protected health information for the clinical investigation. A HIPAA authorization is different than a subject's informed consent. A HIPAA authorization, when executed, is the subject's permission for his/her identifiable health information to be used and/or disclosed for a research purpose. An informed consent document, on the other hand, apprises potential research subjects of the possible risks and benefits associated with participating in the clinical investigation and, when executed, indicates their willingness to participate. The Privacy Rule permits, but does not require, clinical investigators to combine a HIPAA authorization with informed consent documents; this combined form is known as a compound authorization under the Privacy Rule. See 45 C.F.R. 164.508(b)(3).

III. Discussion

Prior to the April 14th, 2003 Privacy Rule compliance date, FDA and the Office for Civil Rights (OCR), which are both components of HHS, received requests for clarification of IRBs' responsibilities with respect to reviewing and approving stand-alone HIPAA authorizations under the Privacy Rule, Federal regulations governing human subjects protection and IRBs, including 21 C.F.R Parts 50 and 56, and international guidelines (see, for example, International Conference on Harmonisation (ICH) Good Clinical Practice: Consolidated Guidance (E6).'[2] These requests expressed concern that, after the Privacy Rule's compliance date, clinical investigations might be impeded because IRBs could be backlogged with requests to review thousands of stand-alone HIPAA authorizations. The requests further stated that some IRBs would halt clinical investigation enrollment pending their review of these stand-alone HIPAA authorizations.

---

[2] The IHC Good Clinical Practice: Consolidated Guideline (E6) states, for example, "Before initiating a trial, the investigator/institution should have written and dated approval/favourable opinion from the IRB/IEC for the trial protocol, written informed consent form, consent form updates, subject recruitment procedures (e.g., advertisements), *and any other written information to be provided to subjects.*" (Emphasis added.) (See ICH E6 4.4.1.).

On April 15, 2003, OCR issued guidance entitled, "Privacy Guidance about Authorizations for Research and Institutional Review Boards" (OCR Guidance, available at www.hhs.gov/ocr/hipaa/privguideresearch.pdf) addressing these issues, OCR clarified that IRBs are not required to review and approve stand-alone HIPAA authorizations under the Privacy Rule or the HHS Human Subjects Protection Regulations at 45 C.F.R. Part 46. Furthermore, pursuant to FDA's permission, OCR clarified that IRBs are not required to review stand-alone authorizations under the FDA regulations at 2 1 C.F.R. Part 56, so long as an IRB's written procedures, adopted pursuant to 2 1 C.F.R. 56.108(a), do not require such review and approval.

With FDA's permission, OCR also addressed, in further detail, IRBs' responsibilities under the ICH guideline entitled, "Good Clinical Practice: Consolidated Guideline" (1996), which some IRBs have misunderstood as requiring them to review all written materials provided to subjects, including stand-alone HIPAA authorizations. The OCR guidance clarified that the ICH guidelines are not legal requirements subject to enforcement by U.S. authorities. Furthermore, in adopting and publishing the ICH guideline, FDA's "Good Clinical Practice: Consolidated Guideline" states, as is true of all guidance, that "[i]t does not create or confer any rights for or on any person and does not operate to bind FDA or the public. An alternative approach may be used if such approach satisfies the requirements of the applicable statutes, regulations, or both." See 62 FR 25692, May 9, 1997. FDA is issuing this guidance to clarify that use of a stand-alone HIPAA authorization that an entity other than an IRB, such as an investigator or sponsor, has determined meets the requirements of FDA's "Good Clinical Practice: Consolidated Guideline" would be an acceptable alternative, so long as it is permitted by the IRB's written procedures. Since the Privacy Rule does not require IRB review or approval of HIPAA stand-alone authorizations, the Privacy Rule would not affect the acceptability of this alternative.

FDA regulations governing IRBs require, in pertinent part, that IRBs follow written procedures they have adopted for reviewing clinical research. See 21 C.F.R. 56.108(a). Pursuant to this provision, IRBs that have written procedures requiring them to review all written materials provided to potential research subjects would have to review and approve stand-alone HIPAA authorizations, even though such review is not otherwise required under the Privacy Rule, FDA regulations governing IRBs, or international guidelines. Accordingly, if IRBs are backlogged in their review of stand-alone HIPAA authorizations, these IRBs might believe they have to halt enrollment in clinical investigations in order to complete their review in accordance with their written procedures.[3]

In order to ensure the continued enrollment of subjects in clinical investigations, and to encourage IRB flexibility with respect to handling possible backlogs, FDA is announcing its intention to exercise ongoing enforcement discretion with respect to the requirements of 2 1 C.F.R. 56.108(a) to the extent that IRBs' written procedures require the review and/or approval of standalone HIPAA authorizations because those written procedures require them to review all written materials provided to potential research subjects. FDA believes that enrollment in well-designed and well-conducted clinical investigations should not be

---

[3] Alternatively, IFWs could modify their written procedures to exclude a requirement for review and/or approval of stand-alone HIPAA authorizations.

interrupted for the purpose of IRB review and approval of stand-alone HIPAA authorizations. In exercising its enforcement discretion, FDA does not intend to take enforcement actions against IRBs that decide not to review stand-alone HIPAA authorization even where an IRB's written procedures would otherwise require this review and/or approval. FDA's exercise of ongoing enforcement discretion with respect to 2 1 C.F.R. 56.108(a), along with OCR's clarification that IRBs are not required to review and approve stand-alone HIPAA authorizations under the Privacy Rule, HHS Human Subject Protection Regulations, or international guidelines, should allow important studies to proceed in the best interest of the public health.

Frequently Asked Questions — IRB Registration

Frequently Asked Questions — IRB Registration, Guidance for Institutional Review Boards (IRBs)[1]

U.S. Department of Health and Human Services
Food and Drug Administration
Office of the Commissioner
Office of Science and Health Coordination
Good Clinical Practice Program

July 2009

Contains Nonbinding Recommendations

This guidance is intended to assist IRBs in complying with the requirement for IRB registration under amended 21 CFR 56.106, effective July 14, 2009. Registration is accomplished through a modified version of the Internet-based registration system used by OHRP for registration of IRBs that are designated by institutions under FWAs. This guidance document addresses basic information, such as why FDA issued a new rule requiring registration, which IRBs are subject to the regulation, the type of information to be provided when registering, and implications of non-compliance.

Note: *This guidance represents the Food and Drug Administration's (FDA's) current thinking on this topic. It does not create or confer any rights for or on any person and does not operate to bind FDA or the public. You can use an alternative approach if the approach satisfies the requirements of the applicable statutes and regulations. If you want to discuss an alternative approach, contact the appropriate FDA staff. If you cannot identify the appropriate FDA staff, call the appropriate number listed on the title page of this guidance.*

This guidance is intended to assist institutional review boards (IRBs) in complying with the new requirement for IRB registration. (See 74 FR 2358 (Jan. 15, 2009))[2] This requirement is an amendment to Part 56, Institutional Review Boards, (21 CFR 56.106), that requires each IRB in the United States (U.S.) that reviews FDA-regulated studies to register. IRB registration information is entered into an Internet-based registration system maintained by the Department of Health and Human Services (HHS). This system is a modification of the one used by the Office for Human Research Protections (OHRP) for registration of IRBs that are designated by institutions under Federalwide Assurances (FWAs). OHRP has issued a similar

---

[1] Available on the FDA website at: http://www.fda.gov/downloads/RegulatoryInformation/Guidances/UCM171256.pdf

[2] The Federal Register notice can be found at http://edocket.access.gpo.gov/2009/E9-682.htm

rule requiring IRBs designed under FWAs to register or update their registration information at this modified site. (See 74 FR 2399 (Jan. 15, 2009))[3]

FDA's guidance documents, including this guidance, do not establish legally enforceable responsibilities. Instead, guidances describe the Agency's current thinking on a topic and should be viewed only as recommendations, unless specific regulatory or statutory requirements are cited. The use of the word should in Agency guidances means that something is suggested or recommended, but not required.

I. Rationale for the New Requirement

1. Why is FDA requiring all IRBs that review FDA-regulated studies to register?

Because our information at the present time is derived from research and marketing applications, FDA (we) cannot be certain that we have current information about IRBs that review FDA-regulated studies. For example, some drug and device studies are exempt from the Investigational New Drug (IND, 21 CFR Part 312) and Investigational Device Exemptions (IDE, 21 CFR 812) submission requirements and are conducted without FDA involvement. In addition, many device studies (e.g., non-significant risk and many in vitro diagnostic (IVD) studies) are conducted with only IRB approval. We, therefore, do not have real-time information about these studies or the IRBs that review them.

In addition, several reports from the HHS Office of Inspector General (OIG) regarding our oversight of the conduct of clinical studies[4] recommended IRB registration, stressing the importance of a comprehensive listing of all IRBs that review FDA-regulated research. The 2001 OIG report also expressed concern about our ability to assure an equivalent level of human subject protection in clinical studies of FDA-regulated products conducted outside of the U.S. as compared to those conducted in the U.S. While registration of non-U.S. IRBs (often referred to as Independent or Research Ethics Committees – IECs/RECs) is voluntary, information we receive from them will be helpful in addressing this concern.

2. Why does FDA believe this information is necessary?

The new rule will provide FDA, as well as other interested parties (e.g., the IRB community, sponsors, and clinical investigators), with a comprehensive listing of all U.S. IRBs that review FDA-regulated research. It will also provide information about non-U.S. IRBs/IECs/RECs that review FDA-regulated research and choose to voluntarily register. This more complete knowledge of IRBs that actively review FDA-regulated studies will:

---

[3] This Federal Register notice can be found at http://edocket.access.gpo.gov/2009/E9-588.htm

[4] "Institutional Review Boards – A Time for Reform" (1998) available at http://www.oig.hhs.gov/oei/reports/oei-01-97-00193.pdf;; "The Globalization of Clinical Trials" (2001) available at http://oig.hhs.gov/oei/reports/oei-01-00-00190.pdf; and "The Food and Drug Administration's Oversight of Clinical Trials (2007) available at http://www.oig.hhs.gov/oei/reports/oei-01-06-00160.pdf.

- Facilitate our sharing of educational and other information with IRBs. We believe that the lack of an accurate, complete, and regularly updated listing of IRBs involved with the review of FDA-regulated studies limits our outreach and educational efforts;

- Assist us in scheduling and conducting IRB inspections under our bioresearch monitoring (BIMO) inspection program, by assuring up-to-date contact information; and

- Help us to prioritize IRB inspections.

3. Does registration imply that an IRB is in full compliance with 21 CFR Part 56 or is otherwise meeting a particular standard of competence or expertise?

No. IRB registration is not a form of accreditation or certification by FDA that the IRB is in full compliance with 21 CFR Part 56. While a U.S. IRB that reviews FDA-regulated studies must register to be in compliance with 21 CFR Part 56.106(a), IRB registration does not address issues regarding an IRB's competence or expertise nor does it require IRBs to meet a particular standard in order to conduct a review.

II. Registration Requirements

4. Who must register?

Each IRB in the U.S. that either:

a. reviews clinical investigations regulated by FDA under sections 505(i) (21 U.S.C. 355(i)) or 520(g) (21 U.S.C. 360j(g)) of the Federal Food, Drug, and Cosmetic Act (the Act);

or

b. reviews clinical investigations that are intended to support applications for research or marketing permits for FDA-regulated products. [See 21 CFR 56.106(a)]

5. May IRBs other than those required under 21 CFR 56.106(a), including those located outside of the U.S., register if they wish?

Yes, any IRB may choose to register voluntarily.

6. How does an IRB submit an initial registration?

IRBs that are not already in the registration system must submit an initial registration. IRBs can submit this registration electronically through http://ohrp.cit.nih.gov/efile. If your IRB lacks the ability to register electronically, it must send its registration information, in writing, to the Good Clinical Practice Program (HF-34), Office of Science and Health Coordination, Food and Drug Administration, 5600 Fishers Lane, Rockville, MD 20857.

As noted above, we are utilizing a modified version of the Internet-based system OHRP has employed for registration of IRBs designated under FWAs. Both OHRP and FDA will be using this same modified system.

The electronic registration system provides instructions to assist you in providing the appropriate information, depending on whether your IRB is subject to regulation by only OHRP, only FDA, or both OHRP and FDA. 7. What if my IRB is already registered in the OHRP system?

If your IRB is already registered in the OHRP system, the registration information must be updated to include all of the information required by FDA (see # 11 below). For IRBs that are currently reviewing FDA-regulated research, the additional information must be added to the database by September 14, 2009. For IRBs that are not currently reviewing FDA-regulated research, this must be done before any research involving FDA-regulated products is reviewed once the compliance date has passed. (See below for more information on the effective and compliance dates.)

8. What is the effective date of the final rule and, by what date, must IRBs complete an initial registration or submit additional information as required by the FDA rule?

This rule is effective July 14, 2009. However, in order to allow IRBs to submit their initial registrations or the additional information required by the rule, FDA will not enforce the rule until September 14, 2009, the compliance date. Therefore, as stated above, IRBs that are not already in the OHRP system must submit an initial registration by September 14, 2009. For IRBs that are already in the OHRP system and are reviewing FDA-regulated research, the required additional information must be added to the database by this compliance date. After September 14, 2009, if an IRB that is not reviewing FDA-regulated research decides to do so, this additional information must be submitted before the IRB reviews research involving FDA-regulated products.

9. Is assistance available if my IRB encounters technical problems when attempting to register electronically?

If your IRB encounters technical problems with the electronic registration system, it should contact an OHRP IRB Coordinator listed at http://www.hhs.gov/ohrp/daqi-staff.html.

10. Will my IRB receive confirmation that its registration was completed?

If your IRB registers electronically, it will receive a notification that registration was accepted by HHS, sent to the electronic e-mail address that it provided as part of the registration process.

If your IRB submits written information, as described in #6 above, an electronic notification will be sent once the information is successfully entered into the system and accepted by HHS.

11. What information does the final rule require from each IRB in the U.S. that reviews FDA-regulated studies?

The final rule requires the following information [see 21 CFR 56.106(b)]:

(1) The name, mailing address, and street address (if different from the mailing address) of the institution operating the IRB and the name, mailing address, phone number, facsimile number, and electronic mail address of the senior officer of that institution who is responsible for overseeing activities performed by the IRB;

(2) The IRB's name, mailing address, street address (if different from mailing address), phone number, facsimile number, and electronic mail address; each IRB chairperson's name, phone number, and electronic mail address; and the name, mailing address, phone number, facsimile number, and electronic mail address of the contact person providing the registration information;

(3) The approximate number of active protocols involving FDA-regulated products reviewed. For purposes of this rule, an "active protocol" is any protocol for which an IRB conducted an initial review or a continuing review at a convened meeting or under an expedited review procedure during the preceding 12 months; and

A description of the types of FDA-regulated products (such as biological products, color additives, food additives, human drugs, or medical devices) involved in the protocols that the IRB reviews.

III. Revisions to Information

12. Once my IRB is registered or its existing information is updated to comply with this rule, is registration permanent unless there is a change in required information?

No, an IRB is required to renew its registration and verify the required information every 3 years from the date of the last entry/change made to the registration information.

13. Once registered, when is an IRB required to revise its registration information?

The final rule specifies the following circumstances that require a revision to the registration information [see 21 CFR 56.106(e)]:

- If an IRB's contact or chairperson information changes, the IRB must revise its registration information within 90 days of the change.

- If an IRB decides to review new types of FDA-regulated products (e.g., to review device studies if it only reviewed drug studies previously) or to discontinue reviewing clinical investigations regulated by FDA, it must report this within 30 days of the change.

• If an IRB decides to disband, this must be reported within 30 days of permanent cessation of the IRB's review of research.

14. Do IRBs need to update the number of active protocols under review when changes occur?

No, we realize that the number of active protocols may change quite frequently and that continuously updating this information would be burdensome. Therefore, for changes in the number of active protocols, as well as any other additions or changes not specifically covered in 21 CFR 56.106(e) (see previous response), IRBs may wait to modify the information until it is necessary to change any of the information specified in #13 or time to renew its registration.

IV. Consequences of failure to register

15. What are the consequences of an IRB failing to register as required by the final rule?

An IRB that fails to register could be considered noncompliant with these regulations. As part of our inspectional activities, FDA may conduct an inspection of an IRB to verify compliance with regulatory requirements, including the requirement for an IRB to register under 21 CFR 56.106(a).

Sponsors and clinical investigators are required by FDA regulations governing the conduct of clinical studies (21 CFR Part 312 for drugs and biologics and 21 CFR Part 812 for devices) to use IRBs that comply with 21 CFR Part 56. Therefore, if a sponsor and/or clinical investigator submits a study for review to an unregistered IRB, that sponsor and/or clinical investigator could be considered noncompliant with FDA regulations.

In addition, we plan to use the information accrued through the IRB registration system to distribute educational materials to IRBs that review FDA-regulated studies (see #2 above). Therefore, sponsors and/or clinical investigators who use IRBs that are not registered run the risk that the IRB may not be familiar with our current policies.

V. Availability of information

16. Will the information in the registration system be available to the public?

As discussed in the preamble to both the proposed and final rules, the information that OHRP presently has publicly available on its website will remain available. That information includes the name and location of all organizations operating an IRB and the name and location of the associated IRB(s). (This information is available at http://ohrp.cit.nih.gov/search/).

Other available information is subject to public disclosure under the Freedom of Information Act (FOIA) as well as our public information regulations in 21 CFR Part 20 and therefore can be requested. Please note, however, that certain information may be withheld from public disclosure or may require an individual's consent for public disclosure [see, e.g., 21 CFR 20.63(e)].

In addition, we will not issue reports on IRB registration nor certificates to show that an IRB is registered. As noted previously, IRB registration does not address issues regarding an IRB's competence or expertise nor does it require IRBs to meet a particular standard in order to conduct a review.

Part III

Selected FDA GCP/Clinical Trial Guidance Documents:
Drugs and Biologics

Available Therapy

Available Therapy, Guidance for Industry[1, 2]

U.S. Department of Health and Human Services
Food and Drug Administration
Center for Drug Evaluation and Research (CDER)
Center for Biologics Evaluation and Research (CBER)

July 2004

Clinical Medical

Contains Nonbinding Recommendations

This guidance is intended to provide guidance to industry on the meaning of the term "available therapy" as currently used by the Center for Drug Evaluation and Research (CDER) and the Center for Biologics Evaluation and Research (CBER) in the FDA in the specific circumstances described in the guidance.

Note: *This guidance represents the Food and Drug Administration's (FDA's) current thinking on this topic. It does not create or confer any rights for or on any person and does not operate to bind FDA or the public. You can use an alternative approach if the approach satisfies the requirements of the applicable statutes and regulations. If you want to discuss an alternative approach, contact the appropriate FDA staff. If you cannot identify the appropriate FDA staff, call the appropriate number listed on the title page of this guidance.*

This guidance represents the Food and Drug Administration's (FDA's) current thinking on this topic. It does not create or confer any rights for or on any person and does not operate to bind FDA or the public. You can use an alternative approach if the approach satisfies the requirements of the applicable statutes and regulations. If you want to discuss an alternative approach, contact the FDA staff responsible for implementing this guidance. If you cannot identify the appropriate FDA staff, call the appropriate number listed on the title page of this guidance.

I. Introduction

This document is intended to provide guidance to industry on the meaning of the term available therapy as currently used by the Center for Drug Evaluation and Research (CDER)

---

[1] Available on the FDA website at: http://www.fda.gov/RegulatoryInformation/
Guidances/ucm126586.htm
[2] This guidance has been prepared by the Center for Drug Evaluation and Research (CDER) and the Center for Biologics Evaluation and Research (CBER) at the Food and Drug Administration.

and the Center for Biologics Evaluation and Research (CBER) in the Food and Drug Administration (FDA) in the specific circumstances described in this guidance.

FDA's guidance documents, including this guidance, do not establish legally enforceable responsibilities. Instead, guidances describe the Agency's current thinking on a topic and should be viewed only as recommendations, unless specific regulatory or statutory requirements are cited. The use of the word should in Agency guidances means that something is suggested or recommended, but not required.

II. Background

Available therapy and related terms, such as existing treatments and existing therapy, appear in a number of regulations and policy statements issued by CDER and CBER, but these terms have never been formally defined by the Agency. Some confusion has arisen regarding whether available therapy refers only to products approved by FDA for the use in question, or whether the term could also refer to products used off-label or to treatments not regulated by FDA, such as surgery. This guidance is intended to inform the public of the Agency's interpretation of available therapy as used in the regulations and policy statements described in Part III.

III. Affected Regulations and Policy Statements

The regulations and policies described below incorporate the concept that the Agency can regulate a particular product in a certain manner because of a lack of available therapy or because of the product's advantage over available therapy. The language incorporating this concept is printed in bold italicized print for emphasis.

A. Treatment INDs

FDA's regulations allow the use of an investigational drug for treatment under a treatment protocol or treatment investigational new drug application (IND). According to 21 CFR 312.34(b), the investigational drug can only be used for this purpose if the following criteria are met:

- The drug is intended to treat a serious or immediately life-threatening disease.

- "There is no comparable or satisfactory alternative drug or other therapy available to treat that stage of the disease in the intended patient population" (312.34(b)(ii).

- The drug is being investigated under an IND in effect for the trial, or all clinical trials have been completed.

- The sponsor of the clinical trial is actively pursuing marketing approval of the investigational drug.

B. Subpart E Regulations

The Agency's procedures in subpart E of 21 CFR part 312, which expedite the development, evaluation, and marketing of promising therapies to treat individuals with life-threatening and severely debilitating illnesses, reflect that the Agency must make a medical risk-benefit judgment in deciding whether to approve a drug or biological product. As part of this risk-benefit analysis, the Agency will "tak[e] into consideration the severity of the disease and the absence of satisfactory alternative therapy" (21 CFR 312.84).

C. Accelerated Approval Regulations

FDA's accelerated approval procedures and restricted distribution provisions in 21 CFR subpart H are available for new drug and biological products (1) that have been studied to treat serious or life-threatening illnesses and (2) "that provide meaningful therapeutic benefit to patients over existing treatments (e.g., ability to treat patients unresponsive to, or intolerant of, available therapy, or improved patient response over available therapy)" (21 CFR 314.500 and 601.40).

D. Fast Track Drug Development Programs

FDA's fast track drug development programs are designed to facilitate the development and expedite the review of drug and biological products that are intended to treat serious or life-threatening conditions and that demonstrate the potential to address unmet medical needs (FDA guidance for industry on Fast Track Drug Development Programs: Designation, Development, and Application Review). In the guidance, the Agency defined an unmet medical need as a "medical need that is not addressed adequately by an existing therapy."

As described in the guidance, where there is no available therapy for a condition (or the only available therapy is approved under the accelerated approval regulations), a product in a drug development plan designed to evaluate the drug's potential to address the condition would meet the factors to address an unmet medical need. Where there is available therapy for the condition, the drug development program would address unmet medical needs if it evaluated any of the following:

- Improved effects on serious outcomes of the condition that are affected by alternate therapies

- Effects on serious outcomes of the condition not known to be affected by the alternatives

- Ability to provide benefits in patients who are unable to tolerate or are unresponsive to alternative agents, or ability to be used effectively in combination with other critical agents that cannot be combined with available therapy

- Ability to provide benefits similar to those of alternatives, while avoiding serious toxicity that is present in existing therapies, or avoiding less serious toxicity that is common and causes discontinuation of treatment of a serious disease

- Ability to provide benefits similar to those of alternatives but with improvement in some factor, such as compliance or convenience, that is shown to lead to improved effects on serious outcomes.

E. Priority Review Policies

CDER and CBER have established review classifications and review policies and procedures for new drug applications (NDAs), biologics license applications (BLAs), and efficacy supplements to prioritize and speed their review. Most of these Agency policies and procedures are intended to encourage the development and expedite the review of innovative drug products (i.e., subpart E regulations, accelerated approval regulations, fast track drug development programs, priority review policies), while one (treatment INDs) provides early access to investigational therapies.

A priority designation is intended to direct overall attention and resources to the evaluation of applications for products that have the potential for providing a significant treatment, preventive or diagnostic therapeutic advance, as compared to standard applications.[3]

Products regulated by CDER are eligible for priority review if they provide a significant improvement compared to marketed products in the treatment, diagnosis, or prevention of a disease. Products regulated by CBER are eligible for priority review if they provide a significant improvement in the safety or effectiveness of the treatment, diagnosis, or prevention of a serious or life-threatening disease.

IV. Policy: Definition of Available Therapy

The regulations and policies described above do not explicitly define available therapy. CDER and CBER have determined that in regulations and policy statements where the terms are not otherwise defined,

available therapy (and the terms existing treatments and existing therapy) should be interpreted as therapy that is specified in the approved labeling of regulated products, with only rare exceptions.[4]

FDA recognizes that there are cases where a safe and effective therapy for a disease or condition exists but it is not approved for that particular use by FDA. However, for purposes of the regulations and policy statements described in Section III, which are intended to permit prompt FDA approval of medically important therapies, only in exceptional cases will a treatment that is not FDA-regulated (e.g., surgery) or that is not labeled for use but is supported by compelling literature evidence (e.g., certain established oncologic treatments) be considered available therapy.

Most of the Agency programs that use the term available therapy are intended to encourage the development and expedite the review of innovative drug products. By defining available therapy to focus on approved products with labeling for use in the disease or condition at issue, FDA (1) emphasizes the importance of the approval process for establishing that a drug

---

[3] CDER Manual of Policies and Procedures (MaPP) 6020.3 explains the priority review policy and procedures. CBER Manual of Standard Operating Procedures and Policies 8405 explains the process for Complete Review and Issuance of Action Letters.

[4] Approved labeling refers to claims approved conventionally or under FDA's accelerated approval procedures.

is safe and effective for a particular use and (2) provides the greatest opportunity for development and approval of appropriately labeled drugs. For these programs, products that are used off-label for the indication at issue and products that have not had formal FDA review are rarely considered available therapy; the definition of available therapy in this guidance provides only a limited exception for particularly well-documented therapies.

Questions also have arisen concerning how the term "meaningful therapeutic benefit to patients over existing treatments" in the accelerated approval regulations (21 CFR 314.500 and 601.40) should be interpreted when the only available therapy is another treatment approved under the accelerated approval regulations. This question arises when several drugs are under investigation or application review for a specific indication based on a surrogate endpoint, or when the only product on the market has restrictions on distribution. Specifically, when one drug is approved under the accelerated approval regulations, can additional therapies be approved under the accelerated approval regulations?

We have determined that the approval of one therapy under the accelerated approval regulations (either on the basis of a surrogate endpoint or with restricted distribution) should not preclude the approval under the accelerated approval regulations of additional therapies. As a general matter, it is preferable to have more than one treatment approved under the accelerated approval provisions, because there are more bases on which an approval under these provisions may be withdrawn, and thus the availability of the therapy is less certain than it is with a conventional approval. Approval under the accelerated approval provisions may be withdrawn if, for example, post-approval studies fail to verify clinical benefit or the postmarketing restrictions are inadequate to assure safe use of the drug product (21 CFR 314.530). Such a withdrawal of approval would leave no treatment available.

Accordingly, we intend to interpret existing treatment under the accelerated approval regulations to mean, in the context of approval based on a surrogate, a treatment that has demonstrated a clinical benefit under conventional approval standards (21 CFR 314.105, 314.125, 601.2). In the context of a prior approval based on restricted distribution, existing treatment means a treatment approved for the same indication without restricted distribution.

References

FDA, CDER Manual of Policies and Procedures (MAPP) 6020.3, Priority Review

FDA, CBER Manual of Standard Operating Procedures and Policies 8405, Complete Review and Issuance of Action Letters

FDA guidance for industry on Providing Clinical Evidence of Effectiveness for Human Drug and Biological Products

FDA guidance for industry on Fast Track Drug Development Programs: Designation, Development, and Application Review

Handling and Retention of Bioavailability and Bioequivalence Testing Samples

Guidance for Industry[1]
Handling and Retention of Bioavailability and Bioequivalence Testing Samples[2]

U.S. Department of Health and Human Services
Food and Drug Administration
Center for Drug Evaluation and Research (CDER)

May 2004

OGD

Contains Nonbinding Recommendations

Inspection of clinical and analytical sites that perform bioavailability (BA) and bioequivalence (BE) studies frequently reveals the absence of reserve samples at the testing facilities where the studies are conducted. The guidance is intended to clarify how to distribute test articles and reference standards to testing facilities, how to randomly select reserve samples, and how to retain reserve samples.

Note: *This guidance represents the Food and Drug Administration's (FDA's) current thinking on this topic. It does not create or confer any rights for or on any person and does not operate to bind FDA or the public. You can use an alternative approach if the approach satisfies the requirements of the applicable statutes and regulations. If you want to discuss an alternative approach, contact the appropriate FDA staff. If you cannot identify the appropriate FDA staff, call the appropriate number listed on the title page of this guidance.*

I. Introduction

This guidance is intended to provide recommendations for study sponsors and/or drug manufacturers, contract research organizations (CROs), site management organizations (SMOs), clinical investigators, and independent third parties regarding the procedure for handling reserve samples from relevant bioavailability (BA) and bioequivalence (BE) studies, as required by 21 CFR 320.38 and 320.63. The guidance highlights (1) how the test article and reference standard for BA and BE studies should be distributed to the testing facilities, (2) how

---

[1] This guidance has been prepared by the Division of Scientific Investigations in the Center for Drug Evaluation and Research (CDER) at the Food and Drug Administration.
[2] Available on the FDA website at: http://www.fda.gov/downloads/RegulatoryInformation/Guidances/UCM126836.pdf

testing facilities should randomly select samples for testing and material to maintain as reserve samples, and (3) how the reserve samples should be retained. The guidance also clarifies and emphasizes points addressed in §§ 320.38 and 320.63.

FDA's guidance documents, including this guidance, do not establish legally enforceable responsibilities. Instead, guidances describe the Agency's current thinking on a topic and should be viewed only as recommendations, unless specific regulatory or statutory requirements are cited. The use of the word *should* in Agency guidances means that something is suggested or recommended, but not required.

II. Background

Following the generic drug scandal in the 1980s, the FDA issued an interim rule in the *Federal Register* of November 8, 1990,[3] on the retention of BA and BE testing samples. The intent of the interim rule was to deter possible bias and fraud in BA and BE testing by study sponsors and/or drug manufacturers. Following public comments, a final rule was issued in the *Federal Register* of April 28, 1993.[4] Implementing regulations are located in 21 CFR 312.57(d), 314.125(b)(17), 314.127(b), 314.150(b)(9), 320.31(d)(1), 320.38, and 320.63.

In the preamble to the final rule, the Agency stated that the study sponsor and/or drug manufacturer should not separate out the reserve samples of the test article and reference standard before sending the drug product to the testing facility.[5] This is to ensure that the reserve samples are in fact representative of the batches provided by the study sponsor and/or drug manufacturer for the testing. The study sponsor and/or drug manufacturer should send to the testing facility batches of the test article and reference standard so that the testing facility can randomly select samples for testing, and material to maintain as reserve samples. The drug product should also be maintained in the sponsor's and/or manufacturer's original container (see section III).

Also in the preamble to the final rule, the Agency noted that reserve sample retention is the responsibility of the organization that conducts the BA or BE study.[6] The intent is to eliminate the possibility of sample substitution by the study sponsor and/or drug manufacturer, or prevent the alteration of any reserve samples from a study conducted by a contractor before release of drug product samples to the FDA.

FDA's Division of Scientific Investigations (DSI) and field investigators from the Office of Regulatory Affairs (ORA) conduct inspections of clinical and analytical sites that perform BA and BE studies for study sponsors and/or drug manufacturers seeking approval of generic and new drug products. A frequent finding from these inspections is the absence of reserve samples at the testing facilities where the studies are conducted. In many cases, DSI finds that testing facilities return reserve samples to the study sponsors and/or drug manufacturers, against the direction of the regulations in 21 CFR 320.38 and 320.63. In other cases, study

---

[3] 55 FR 47034.
[4] 58 FR 25918.
[5] 58 FR 25918 at 25920.
[6]58 FR 25918 at 25921.

sponsors and/or drug manufacturers, SMOs, or contract packaging facilities designate the study test article and reference standard for each subject, and preclude the testing facilities from randomly selecting representative reserve samples from the supplies. DSI also finds that deviations from the regulations more often occur in BE studies with pharmacodynamic or clinical endpoints in which the studies are confused with clinical safety or efficacy studies. The pharmacodynamic or clinical endpoint BE studies are usually multisite, blinded studies conducted under contract (either directly with the study sponsor or drug manufacturer or through an SMO) by physicians or clinical investigators who use their own clinics or offices to conduct the studies. Moreover, some clinical investigators believe that they are not CROs and are not required to retain reserve samples. This guidance clarifies the responsibilities for retention of reserve samples.

III. Sampling Techniques

We recommend that the study sponsor and/or drug manufacturer send to the testing facility batches of the test article and reference standard packaged in such a way that the testing facility can randomly select samples for bioequivalence testing and samples to maintain as reserve samples. This will ensure that the reserve samples are in fact representative of the batches provided by the study sponsor and/or drug manufacturer and that they are retained in the study sponsor's original container. Because the study sponsor and/or drug manufacturer may provide a testing facility with a variety of container sizes and packaging, FDA is flexible in applying the representativeness requirement described in 21 CFR 320.38. For example, any of the following random sampling techniques might be used by the testing facility for the container size and packaging described[7] (bolded text is particularly relevant).

Single Container – If a single container of the test article and reference standard are provided to the testing facility, the testing facility should remove a sufficient quantity of the test article and reference standard from their respective containers to conduct the study; the remainder in each container should be retained as reserve samples in the original containers.

Multiple Containers – If multiple containers of the test article and reference standard are provided to the testing facility, the testing facility should *randomly select* enough containers of the test article and reference standard to conduct the study; the remaining containers of the test article and reference standard should be retained as the reserve sample in the original containers. Generally, multiple open bottles are discouraged. We encourage testing facilities to limit the number of open containers retained as study reserves.

Unit Dose – If the test article and reference standard are provided to the testing facility in unit dose packaging, the testing facility should *randomly select* a sufficient quantity of unit doses of the test article and reference standard to conduct the study; the remaining unit doses of the test article and of the reference standard should be retained as the reserve samples in the original unit dose packaging. *Therefore, it would be inappropriate to provide the study medications in unit dose packaging and all the reserve samples in bulk containers.*

---

[7] 58 FR 25918 at 25920.

Blinded Study – If the study is to be blinded and the test article and reference standard are provided to the testing facility in unit dose packaging with each unit dose labeled with a randomization code, *the study sponsor and/or drug manufacturer should provide the testing facility with a labeled set of the test article and reference standard sufficient to conduct the study and with additional, identically labeled sets sufficient to retain the "five times quantity" (see section V). The testing facility should randomly select a labeled set to conduct the study; the remaining labeled sets would be retained in their unit dose packaging as the reserve samples.* For a blinded study, we recommend that the study sponsor and/or drug manufacturer also provide to the testing facility a sealed code for use by FDA should it be necessary to break the code. The sealed code should be maintained at the testing facility.

IV. Retention For Multiple Studies And Shipments

If the same batches of the test articles and reference standards initially provided to the testing facility are used in performing more than one study, only one reserve sample of the test article and reference standard in sufficient quantity need to be retained. The reserve samples should be identified as having come from the same batches as used in each study. However, if additional supplies of the test article and reference standard will be used by a testing facility to perform the same study or additional studies, the testing facility should retain a sufficient quantity of reserve samples from the subsequent shipment. If a CRO with multiple testing facilities conducts more than one BE study (e.g., fed and fasted studies) for the same drug product, and the study test article and reference standard are sent to the testing facilities in different shipments, we recommend that sufficient quantity of reserve samples be kept for each study at each testing facility. These approaches are to ensure that the reserve samples are in fact representative of the batch provided by the study sponsor and/or drug manufacturer to the testing facility.

V. Quantity of Reserve Samples

The quantity of reserve samples should be sufficient to permit the Agency to perform five times all of the release tests required in the application or supplemental application. The rationale for requiring the *five times quantity* is provided in the final rule. The clinical investigator can obtain the amount that constitutes the five times quantity from the sponsor and/or drug manufacturer. For solid oral dosage forms (e.g., tablets, capsules), an upper limit of 300 units each for the test article and reference standard can be considered sufficient to meet the five times quantity. Because the Agency has limited experience with the retention and testing of non-solid oral dosage forms, the Agency is unable to recommend an upper limit for the retention of non-solid oral dosage forms at this time. In the case of a reference standard that is an extemporaneously compounded solution or suspension or a reconstitutable powder, we recommend that the pure active ingredient and the unconstituted powder be retained. For a multisite BA or BE study, we recommend that the total amount of reserve samples to be retained across **all** testing facilities satisfy the five times quantity requirement. Each site is asked to retain a reasonable amount of test article and reference standard to be determined by considering (1) the total number of testing facilities participating in the study, (2) the number of subjects expected to be enrolled at each testing facility, and (3) a minimum limit (e.g., 5 dose units) for each of the test articles and reference standards. If the reserve samples from more than one testing facility are transferred to an independent third party for storage, we recommend that the independent third party segregate the reserve samples from the various testing facilities so that any given reserve sample can be unambiguously associated with the testing facility from which it came.

VI. Responsibilities in Various Study Settings

Because of the variety of study settings potentially involved in conducting BA and BE studies, several examples are provided here. These examples are not the only possible study settings. However, in *all* instances, the chain of custody of the reserve samples used in the study should be preserved. The sponsor and/or manufacturer and any storage facility should document and maintain the transfer records for Agency verification.

A. Studies Conducted at CROs, Universities, Hospitals, or Physicians' Offices Contains Nonbinding Recommendations

CROs are the most common study sites. Many BA/BE studies of oral dosage forms are conducted at CROs to support approval of abbreviated new drug applications (ANDAs), new drug applications (NDAs), and NDA supplements. CROs typically conduct single-site, openlabel, crossover design studies with healthy volunteers as participants.

Study sponsors and drug manufacturers sometimes conduct BA and BE studies through a CRO, university faculty, hospitals, or clinical investigators in private practice. The testing facilities are usually clinical study units in universities, hospitals, or clinics run by physicians.

The responsibilities of the study sponsor and/or drug manufacturer include:

- Packaging, distributing, and shipping the test article and reference standard to the testing facility

- Monitoring the study if it is conducted under an investigational new drug application (IND) (rarely needed for most ANDA studies)

The responsibilities of the testing facility are as follows:

- The clinical investigator or designee (such as the study coordinator or research pharmacist of the testing facility) should randomly select sufficient test article and reference standard to conduct the study from the supplies received from the sponsor and/or drug manufacturer, and retain the remaining study samples as study reserves.

- The testing facility or the pharmacy of the testing facility should retain the reserve samples.

- If the testing facility does not have adequate storage, or goes out of business, the reserve samples can be transferred to an independent third party with an adequate facility for storage under conditions consistent with product labeling.

Note: When studies are conducted at universities, hospitals, or physicians' offices, the clinical investigator or physician conducting the study should not send the reserve samples back to the study sponsor and/or drug manufacturer. The goal is to eliminate the possibility for sample substitution by the study sponsor and/or drug manufacturer, and to preclude the alteration of a reserve sample from a study conducted by another entity before the release of the reserve sample to the FDA.

B. Studies Involving SMOs

When BA or BE studies are conducted by an SMO, they are frequently multisite, open-label studies of oral dosage forms in patients, or multisite, open-label studies of nonoral dosage forms with pharmacodynamic or clinical endpoints. Often, the study sponsor and/or drug manufacturer contracts with an SMO to recruit clinical investigators and to monitor a study. The SMO is involved directly or indirectly (i.e., by subcontracting to another party) in packaging and shipping of study test articles and reference standards to the testing facilities. The testing facilities are usually the clinical study units of CROs, universities, hospitals, or clinics run by physicians.

The responsibility of the study sponsor or drug manufacturer is to ship the test article and reference standard to the SMO under contract, or to the packaging facility under subcontract to the SMO.

The responsibilities of the SMO include:

- Packaging, distributing, and shipping the test article and reference standard to all testing facilities (or subcontracting a packaging facility to perform this function)

- Monitoring the study at different sites if it is conducted under an IND (rarely needed for most ANDA studies)

The SMO should **not** randomly select and retain reserve study samples. As explained in the preamble to the final rule, the Agency intended that sufficient test article and reference standard to conduct the study should be randomly selected at each testing facility, and that each testing facility should retain the remaining study samples as reserves.[8]

The responsibilities of the testing facilities are as follows:

- The clinical investigator or designee (such as the study coordinator or the research pharmacist of each testing facility) should randomly select sufficient test article and reference standard to conduct the study from the supplies received from the SMO under contract, or from the packaging facility under subcontract with the SMO, and retain the remaining study samples as study reserves.

- Each testing facility or the pharmacy of each testing facility should retain the reserve samples.

- Following the completion of the study, if one or more of the testing facilities do not have adequate storage, reserve samples can be transferred to an independent third party with an adequate facility for storage under conditions consistent with product labeling. The reserve samples should not be transferred back to an SMO or any other organization that deals

---

[8] 58 FR 25918 at 25920.

with packaging the test articles and reference standard for storage. This is to eliminate the possibility of commingling reserve samples from packaging activities (21 CFR 211.84 and 211.170) and bioequivalence studies (21 CFR 320.38 and 320.63). As stated in subsection VI.A above, the reserve samples should *not* be shipped back to the sponsor or manufacturer.

C. Blinded Studies With Pharmacodynamic or Clinical Endpoints Involving an SMO

Blinded BE studies are often conducted at multiple sites and involve nonoral dosage forms with pharmacodynamic or clinical endpoints. Often, the study sponsor and/or drug manufacturer contracts with an SMO to recruit clinical investigators and monitor the study. The SMO is involved directly or indirectly (i.e., by subcontracting to another party) in packaging and shipping study test articles and reference standards to the testing facilities. The testing facilities are usually the clinical study units of CROs, universities, hospitals, or clinics run by physicians.

In multisite, blinded BE studies, the sponsor and/or drug manufacturer needs to consider whether the study design will allow for selection and retention of reserve samples in accordance with 21 CFR 320.38 and 320.63 and the final rule. If the study design is too complex to meet the regulatory requirements for reserve samples, the study design may need to be reconsidered.

The responsibility of the study sponsor and/or drug manufacturer is to ship the test article and reference standard to the SMO under contract, or to the packaging facility under subcontract to the SMO.

The responsibilities of the SMO include:

- Packaging, distributing, and shipping test article and reference standard to all testing facilities (or subcontracting a packaging facility to perform this function). We recommend that the SMO provide the testing facilities with enough code-labeled sets to conduct the study and to retain the five times quantity. Based on inspection experience, DSI does not recommend that test article and reference standard be prenumbered for subjects, because assigning unit doses to a designated subject number precludes the random selection of drug used for dosing and drug used for reserve samples (see example below for illustration).

- Monitoring the study at different sites if it is conducted under an IND (rarely needed for most ANDA studies)

Note: *The SMO should not select reserve samples. In addition, the reserve samples should not be transferred by the testing facility back to an SMO or any other organization that deals with packaging the test articles and reference standard for storage.*

The responsibilities of the testing facilities are as follows:

- The clinical investigator or designee (such as the study coordinator or the research pharmacist) of each testing facility should randomly select sufficient test article and reference standard to conduct the study from the supplies received from the SMO under contract, or from the packaging facility under subcontract with the SMO, and retain the remaining study samples as study reserves. The clinical investigator should be aware of the sampling techniques used for blinded studies as described in section III.

- Each testing facility or the pharmacy of each testing facility should retain the reserve samples. Please note that if a placebo is used in blinded BE studies, reserve samples for the placebo should be retained along with the test article and reference standard reserves. The sealed treatment code of the study should be kept at the testing facility. This is applicable even if the reserve samples are forwarded to an independent third party (see paragraph below).

- If one or more of the testing facilities do not have adequate storage, or go out of business, the reserve samples can be forwarded to an independent third party with an adequate facility for storage under conditions consistent with product labeling.

Below is a suggested packaging and random selection plan for a blinded, multisite study of a dermatological cream product involving a SMO:

The study enrolls 300 subjects with approximately 60 subjects at five testing facilities. The five times quantity for the test article and reference standard is 50 tubes for each product. In preparation for conducting the study, the SMO prepares 200 boxes that contain one code-labeled tube of test article and one code-labeled tube of reference standard in each box. The SMO randomly distributes 40 boxes to each clinical testing facility. The clinical facility randomly selects 30 of the boxes to dose 60 subjects. The remaining 10 boxes serve as the reserve samples. In this example, staff (e.g., a pharmacist) not involved with the study may be recommended to ensure the study remains blinded. This packaging system ensures that an equal number of test article and reference standard are administered to the subjects at each site, and that an equal number of test article and reference standard will be maintained as reserve samples. Since 10 boxes are kept at each of 5 testing facilities, 50 tubes each of test article and reference standard are retained and the five times quantity reserve sample requirement is met. In addition, the requirement of random selection by each testing facility is also met.

D. In-House Studies Conducted by a Study Sponsor and/or Drug Manufacturer

Only about 7 percent of all sites inspected by DSI from 1997 to 2002 conducted in-house BA and BE studies. If a study sponsor and/or drug manufacturer conducts such a study, manufacturing reserve samples (21 CFR 211.170) and BE study reserve samples (21 CFR 320.38 and 320.63) should be separated. The in-house clinical research unit should operate as an independent unit for the purposes of sample retention. All matters (e.g., manufacturing, purchasing, packaging, transfer records) concerning the test article and reference standard should be clearly documented and available to FDA investigators during an inspection. Standard procedures concerning security and accountability of the test article and reference standard for each study should be established to eliminate the possibility of sample substitution. Sponsors conducting in-house studies can engage an independent third party to store reserve samples. If an independent third party is not used, there should be (1) a totally segregated and fully compliant in-house storage area; (2) procedures and policies in place to show that adequate test article and reference standard are retained; (3) controlled access to the reserve samples; (4) a rigorous and unbroken chain of custody for the reserve samples.

The study sponsor and/or drug manufacturer (clinical research department) should be responsible for packaging and transferring the test article and reference standard to the in-house clinical study unit.

The testing facility (in-house clinical study unit) should be responsible for:

- Documentation of all matters concerning the transfer and receipt of the test article and reference standard

- Random selection of sufficient test article and reference standard to conduct the study, and retention of the remaining study samples as reserves. The selection is generally made by the clinical investigator, study coordinator, or research pharmacist (if available) in the clinical study unit. We recommend that a staff member (e.g., a study nurse) witness the random selection process and dosing.

- Retention of reserve samples in a secure area. To ensure the authenticity of the reserve samples, access to this area should be limited. We encourage maintenance of an entry log to the storage area.

- Preparation for adequate storage of reserve samples. If the in-house testing facilities do not have adequate storage, or go out of business, the reserve samples can be forwarded to an independent third party with an adequate facility for secure storage under conditions consistent with product labeling.

E. In Vitro BE Studies

21 CFR 320.63 states:

> The applicant of an abbreviated application or a supplemental application submitted under section 505 of the Federal Food, Drug, and Cosmetic Act, or, if bioequivalence testing was performed under contract, the contract research organization shall retain reserve samples of any test article and reference standard used in conducting an in vivo or in vitro bioequivalence study required for approval of the abbreviated application or supplemental application.

Thus, the regulations for reserve samples apply to in vitro BE studies. The in vitro BE studies required for approval of nasal aerosols and nasal sprays for local action are an example of this.

Note that in vitro studies conducted to compare dissolution rates for different strengths of the same formulation are not subject to the reserve sample regulations. For an in vitro BE study, the roles of the study sponsor and/or drug manufacturer and the testing facility are similar to those described for in vivo BE studies conducted by CROs and in the examples of in vivo BE studies conducted in-house by a study sponsor and/or drug manufacturer.

VII. Exception for Inhalant Products

As stated in 21 CFR 320.38(c), each reserve sample shall consist of a sufficient quantity of samples to permit FDA to perform five times all of the release tests required in the application or supplemental application. Dose content uniformity or spray content uniformity release tests alone usually take 30 units (canisters or bottles) per batch. Performance of other release tests can suggest a need for additional units. The number of reserve sample units to be retained for three batches of test article and reference standard could exceed 1000 units (up to 250 units for each batch of the test article and reference standard) based on the five times quantity requirement. The Agency has determined that in lieu of the "five times quantity" requirement, the quantity of inhalant (nasal aerosol or nasal spray) test article and reference standard retained for testing and analyses should be at least 50 units for each batch (see the preamble to the final rule).[9]

For NDAs, at least 50 units of each of the three batches of nasal aerosol or nasal spray needed for BA studies should be retained. However, where the reference standard is another nasal aerosol or nasal spray, at least 50 units of that batch should also be retained. For ANDAs, at least 50 units of each of three batches should be retained for each of the test articles and reference standards used for in vivo or in vitro BE studies. If multiple testing facilities are used in a BA or BE study, the total amount of reserves for each product across all testing facilities would be at least 50 units, and each testing facility should retain a reasonable amount of test articles and reference standards (see section V for more details). For NDAs and ANDAs, if the in vivo or in vitro studies include placebo aerosols or sprays, at least 50 units of each placebo batch should also be retained. These recommendations apply only to nasal aerosol and nasal sprays for local action that are to be marketed as multiple dose products, typically labeled to deliver 30 or more actuations per canister or bottle.

Glossary

Clinical Investigator – An individual who actually conducts a clinical investigation (i.e., under whose immediate direction the drug is administered or dispensed to a subject) (21 CFR 312.3(b)). In this guidance, when a clinical investigation involves BA or BE studies, the clinical investigator has the responsibility of retaining the reserve samples at the testing facility or through an independent third party.

Contract Research Organization (CRO) – An independent contractor of the sponsor or manufacturer that assumes one or more of the obligations of a sponsor (e.g., design of a protocol, selection or monitoring of investigations, evaluation of reports, and preparation of materials to be submitted to the FDA) (21 CFR 312.3(b)). This guidance addresses BA and BE studies submitted to support approvals of new and generic drugs. These studies are usually conducted by CROs under contract to study sponsors and/or drug manufacturers. Many CROs have their own testing facilities, with physicians (to serve as clinical investigators) and clinical support staff (e.g., nurses, medical technologists) to conduct the BA and BE studies.

---

[9] 58 FR 25918 at 25924.

Independent Third Party – In this guidance, independent third party indicates a person that has no affiliation with the study sponsor and/or drug manufacturer.

Reference Standard – In this guidance, reference standard refers to the reference product used in a BE study. It is usually the innovator's product or a marketed product of the drug under investigation. For BA studies, the reference standard can be an oral solution of the drug under investigation.

Reserve Samples – In this guidance, reserve samples and retention samples are used interchangeably.

Site Management Organization (SMO) – In this guidance, site management organization refers to an organization that manages clinical study sites on behalf of the sponsor and/or drug manufacturer.

Sponsor-Investigator – An individual who both initiates and conducts an investigation, and under whose immediate direction the investigational drug is administered or dispensed. The term does not include any person other than an individual (21 CFR 312.3(b)).

Study Sponsor – A person who takes responsibility for and initiates a clinical investigation. The sponsor may be an individual or pharmaceutical company, governmental agency, academic institution, private organization, or other organization. The sponsor does not actually conduct the investigation unless the sponsor is a sponsor-investigator (21 CFR 312.3 (b)).

In this guidance, the term *study sponsor and/or drug manufacturer* is used in recognition of the fact that most study sponsors are pharmaceutical companies that manufacture the drugs under investigation.

Testing Facility – The entity performing the BA or BE (in vivo or in vitro) study. The testing facility can be a CRO, university, hospital, clinic of a clinical investigator, or in-house clinical study unit of a study sponsor and/or drug manufacturer, where dosing and sampling (i.e., blood, urine, or clinical endpoints) are performed. In issuing the final rule, the Agency intended that reserve samples should be kept at the testing facility.

Clinical Holds Following Clinical Investigator Misconduct

Guidance for Industry and Clinical Investigators on the Use of Clinical Holds Following Clinical Investigator Misconduct[1, 2]

U.S. Department of Health and Human Services
Food and Drug Administration
Center for Biologics Evaluation and Research (CBER)
Center for Drug Evaluation and Research (CDER)

September 2004

Contains Nonbinding Recommendations

This guidance for industry and clinical investigators provides information on one use by FDA of its authority to impose a clinical hold on a study or study site if FDA finds that human subjects are or would be exposed to unreasonable and significant risk of illness or injury. Specifically, this guidance describes circumstances in which FDA may impose a clinical hold based on credible evidence that a clinical investigator conducting the study has committed serious violations of FDA regulations on clinical trials of human drugs and biologics.

Note: *This guidance represents the Food and Drug Administration's (FDA's) current thinking on this topic. It does not create or confer any rights for or on any person and does not operate to bind FDA or the public. You can use an alternative approach if the approach satisfies the requirements of the applicable statutes and regulations. If you want to discuss an alternative approach, contact the appropriate FDA staff. If you cannot identify the appropriate FDA staff, call the appropriate number listed on the title page of this guidance.*

I. Purpose

This guidance provides information on one use by the Food and Drug Administration (FDA) of its authority to impose a clinical hold on a study or study site if FDA finds that human subjects are or would be exposed to an unreasonable and significant risk of illness or injury. Specifically, this guidance describes circumstances in which FDA may impose a clinical hold based on credible evidence that a clinical investigator conducting the study has committed serious violations of FDA regulations on clinical trials of human drugs and biologics, including 21 CFR Parts 312, 50, and 56, or has submitted false information to FDA or the sponsor in any required report. FDA may consider imposing a clinical hold in these situations where

---

[1] Available on the FDA website at: http://www.fda.gov/downloads/RegulatoryInformation/ Guidances/UCM126997.pdf

[2] This guidance was developed by the Center for Drug Development and Research (CDER) and the Center for Biologics Evaluation and Research (CBER) in consultation with the Office of the Commissioner (OC).

necessary to protect human subjects in the study from an unreasonable and significant risk of illness or injury. Such a clinical hold may be imposed on the study in which the misconduct occurred or on other studies of drugs or biological products in which the clinical investigator is directly involved or proposed to be involved. Although FDA has authority to take various enforcement actions against a clinical investigator who commits serious violations of FDA regulations, these actions may not be completed swiftly enough to protect human subjects who may be at risk in ongoing studies conducted by the investigator. Where the investigator's misconduct appears to pose an ongoing threat to the safety and welfare of such subjects, imposition of a full or partial clinical hold on ongoing or proposed studies of human drugs or biological products may be appropriate. See 21 CFR 312.42(b)(1)(i), 312.42(b)(2)(i), 312.42(b)(3)(iii), and 312.42(b)(4)(i). This guidance does not address other circumstances in which FDA may impose a clinical hold if FDA finds that human subjects are or would be exposed to an unreasonable and significant risk of illness or injury, including when there is no evidence of clinical investigator misconduct or a serious regulatory violation. This guidance finalizes the draft guidance of the same title dated April 2002.

FDA's guidance documents, including this guidance, do not establish legally enforceable responsibilities. Instead, guidances describe the FDA's current thinking on a topic and should be viewed only as recommendations, unless specific regulatory or statutory requirements are cited.

The use of the word *should* in FDA's guidances means that something is suggested or recommended, but not required.

II. Background

This section describes the responsibilities of clinical investigators, the process for bringing an enforcement action for serious clinical investigator misconduct, and the need for a more rapid means of protecting human subjects after serious misconduct has been discovered.

A. What Are a Clinical Investigator's Responsibilities?

The regulations governing the conduct of clinical trials by clinical investigators are intended to assure adequate protection of the rights, safety, and welfare of subjects involved in those trials, as well as the quality and integrity of the resulting data, while at the same time providing sufficient flexibility for clinical research.[3] A brief description of the specific responsibilities of investigators follows.

---

[3] In addition to FDA's regulations, there are a number of guidance documents available that describe FDA's current thinking on good clinical practice. These include the FDA's Guidance for Industry: E6 Good Clinical Practice Guidance; FDA's Information Sheets for Institutional Review Boards and Clinical Investigators; FDA's Guidance for Industry: Computerized Systems Used in Clinical Trials. A more comprehensive listing of useful guidances on good clinical practice can be found at the following web sites: http://www.fda.gov/oc/gcp, http://www..fda.gov/cder/guidances, and http://www.fda.gov/cber/guidelines.htm.

Clinical investigators are responsible for protecting the rights, safety and welfare of human subjects in the studies they conduct (21 CFR 312.60, 21 CFR Parts 50 and 56). Among other things, investigators must assure that an Institutional Review Board (IRB) that complies with FDA regulations conducts initial and continuing ethical review of the study (21 CFR Part 56 and § 312.66). An investigator must notify the IRB of changes in the research activity or unanticipated problems involving risks to human subjects or others, and must not make any changes in the protocol without IRB and sponsor approval, unless necessary to eliminate apparent immediate hazards to human subjects (21 CFR 312.66). An investigator must also obtain informed consent from each subject who participates in the study (21 CFR 312.60 and 21 CFR Part 50).

Clinical investigators are responsible for following the signed investigator statement (Form FDA 1572) (21 CFR 312.60). The investigator's signed statement includes a commitment to: (1) follow the study protocol, and to make changes only after notifying the sponsor, unless necessary to protect the rights, safety or welfare of the subjects; (2) personally conduct or supervise the research; and (3) inform subinvestigators and others assisting in the conduct of the investigation of their obligations in meeting these commitments (21 CFR 312.53(c)). Clinical investigators are also responsible for following the investigational plan (21 CFR 312.60).

Clinical investigators must ensure that the investigational drug is administered only to study subjects under the supervision of the investigator or a subinvestigator responsible to the investigator (21 CFR 312.61). Finally, clinical investigators mus t keep required records of the study and make required reports to the sponsor of the investigation (21 CFR 312.62 and 312.64). An investigator must prepare and maintain adequate case histories, including documentation of informed consent, and keep records of disposition of the drug (21 CFR 312.62). These records must be maintained for 2 years from the date FDA approves a marketing application for the drug under study or if FDA does not approve the drug or no application is filed for the drug, from the date the study is discontinued and FDA is informed (Id.). The investigator must make several types of reports to the sponsor, including progress reports, safety reports (prompt reports of adverse events, and immediate reports of alarming effects), and a final report (21 CFR 312.64). The investigator must also report his or her financial interests to the sponsor to permit assessment of conflicts of interest (Id.).

B. What Actions Can FDA Take to Address Clinical Investigator Misconduct?

If an inspection conducted by FDA reveals that a clinical investigator has committed violations of FDA's regulations, FDA generally will notify the investigator of the violations and take appropriate follow-up action. This notification may consist of the Form FDA 483 (Inspectional Observations) issued at the close of this inspection, or it may be in the form of a Warning Letter. In some cases, an investigator's agreement to correct the violations may be sufficient to resolve the matter. Where FDA finds that there have been serious violations of the investigator's obligations, and corrective action by the investigator cannot resolve the matter, FDA may conclude that it is appropriate to initiate an enforcement action against the investigator. First, if the inspectional findings indicate that the investigator has repeatedly or deliberately violated FDA regulations or repeatedly or deliberately submitted false information, FDA may move to disqualify the investigator from conducting future studies regulated by

FDA. Second, FDA may initiate a civil or criminal enforcement action in federal court. Such actions can take several months and frequently years to complete.

To disqualify a clinical investigator, FDA must go through an administrative process involving an opportunity for hearing (21 CFR 312.70). When a Center (i.e., CBER or CDER) has reviewed the inspectional findings and determined that there is evidence of repeated or deliberate violations or repeated or deliberate submission of false information, and that the pattern or severity of the misconduct warrants agency action, the Center issues a Notice of Initiation of Disqualification Proceedings and Opportunity to Explain (NIDPOE) letter, which furnishes the investigator with written notice of the matter and offers the investigator an opportunity to explain the matter in writing, or, at the option of the investigator, in an informal conference. If an informal conference is held, the investigator may bring an attorney. If, after hearing the investigator's explanation, the Center still believes that the investigator's actions meet the threshold for disqualification, the Center must offer the investigator an opportunity for a regulatory hearing, whose procedures are governed by 21 CFR Part 16 (21 CFR 312.70). The investigator may enter into a consent agreement or may request a hearing. At a regulatory hearing, the investigator may offer the testimony of witnesses, documentary evidence, and supporting briefs. After the hearing, the presiding officer issues a report or decision on whether the investigator has repeatedly or deliberately violated the regulations and should be disqualified.

The report is forwarded to the Commissioner, who then issues a Commissioner's decision on disqualification (21 CFR Part 16). The investigator may appeal the Commissioner's decision in federal court. A disqualification proceeding generally takes many months or years to complete.

C. How Can FDA Protect Human Subjects Following the Discovery of Clinical Investigator Misconduct?

Initiation of an enforcement action in federal court or a disqualification proceeding does not by itself halt an investigator's participation in clinical trials. Until an investigator is disqualified by FDA, the investigator remains free to participate in ongoing and new clinical investigations. There are, however, instances in which the investigator's misconduct appears to pose an ongoing risk to the safety and welfare of the human subjects under the care of that investigator. For example, where an investigator is found to have failed to monitor subjects for signs of serious toxicity associated with the experimental therapy, or falsified eligibility data, FDA may conclude that subjects under that investigator's care are at risk. Under such circumstances, protection of subjects may demand a more rapid intervention than would be offered by an enforcement action or a disqualification proceeding. As discussed above, an effective means of acting promptly to protect human subjects after the discovery of serious investigator misconduct is to impose a clinical hold on those studies or study sites involving the investigator.

III. Use of Clinical Holds to Protect Human Subjects

A. What Is a Clinical Hold?

A clinical hold is an order by FDA that immediately suspends or imposes restrictions on an ongoing or proposed clinical study. FDA has promulgated regulations authorizing clinical holds for studies involving drugs and biological products (21 CFR 312.42). Section 312.42(a) provides the scope and effect of a clinical hold order:

> A clinical hold is an order issued by FDA to the sponsor to delay a proposed clinical investigation or to suspend an ongoing investigation. The clinical hold order may apply to one or more of the investigations covered by an IND. When a proposed study is placed on clinical hold, subjects may not be given the investigational drug. When an ongoing study is placed on clinical hold, no new subjects may be recruited to the study and placed on the investigational drug; patients already in the study should be taken off therapy involving the investigational drug unless specifically permitted by FDA in the interest of patient safety.

A clinical hold may be complete or partial.[4] Delay or suspension of all clinical work under an IND is considered a complete clinical hold. Delay or suspension of only part of the clinical work under an IND is considered a partial clinical hold. A partial clinical hold could, for example, be imposed to delay or suspend one of several protocols in an IND, a part of a protocol, or a specific study site in a multi-site investigation.

FDA's regulation authorizing clinical holds on studies of drugs and biological products sets forth grounds for imposing a hold. Those grounds vary depending on the nature of the study.[5] For all types of studies, however, FDA may impose a clinical hold if it finds that "[h]uman subjects are or would be exposed to an unreasonable and significant risk of illness or injury" (21 CFR 312.42(b)(1)(i), (b)(2)(i), (b)(3)(i)(A), (b)(3)(ii)(E)(2), (b)(4)(i), (b)(5)(i), (b)(6)(i)).

---

[4] The terms complete clinical hold and partial clinical hold can be found and are further described in FDA's Guidance for Industry: Submitting and Reviewing Complete Responses to Clinical Holds October 2000. (http://www.fda.gov/cder/guidance/index.htm). Complete clinical hold means a delay or suspension of all clinical work requested under an IND. Partial clinical hold means a delay or suspension of only part of the clinical work requested under the IND (e.g., a specific protocol or part of a protocol is not allowed to proceed; however, other protocols or part of the protocol are allowed to proceed under the IND.

[5] The types of studies covered by § 312.42 include: Phase 1 studies, § 312.42(b)(1), Phase 2 and 3 studies, § 312.42(b)(2), proposed and ongoing treatment use, § 312.42(b)(3)(i)(iii), studies that are not designed to be adequate and well-controlled, § 312.42(b)(4), studies involving an exception from informed consent under § 50.24, § 312.42(b)(5), and studies involving an exception from informed consent under § 50.23, § 312.42(b)(6).

B. Under What Circumstances Would FDA Consider Imposing A Clinical Hold Following Discovery Of Clinical Investigator Misconduct?

FDA believes that, in some situations, clinical investigator misconduct may be sufficiently serious to conclude that human subjects under that investigator's care are or would be exposed to an unreasonable and significant risk of illness or injury. FDA anticipates that the use of clinical holds in instances of misconduct will be infrequent. In this section, FDA provides guidance on the circumstances in which the agency could reach such a conclusion and impose a clinical hold on the study or study sites in which an investigator is involved. Still, FDA may impose a clinical hold on a study or study site whenever it finds that human subjects are or would be exposed to an unreasonable and significant risk of illness or injury. The grounds for imposition of a clinical hold need not include a finding of misconduct or a violation of a regulation.

1. Before an enforcement action is initiated

After FDA obtains evidence about investigator misconduct, but before a decision to bring an enforcement action in federal court or to issue a NIDPOE letter has been made, there may or may not be reason to believe that human subjects under the care of the investigator are or would be exposed to an unreasonable and significant risk of illness or injury. At this stage in an inquiry into investigator misconduct, FDA would consider two factors in deciding whether to issue a clinical hold.

First, FDA would look at the nature of the violation and its significance for the rights, safety and welfare of human subjects. Certain types of violations may pose such a significant threat to subjects in the trial that suspending that part of the trial under the investigator is justified, even where the investigation into the violations is at an early stage. For example, FDA may conclude that suspending the trial is necessary to protect subjects from a significant and unreasonable risk of illness or injury, if FDA finds evidence of one or more of the following.

- Failure to report serious or life-threatening adverse events;

- Serious protocol violations, such as enrolling subjects who do not meet the entrance criteria because they have conditions that put them at increased risk from the investigational drug, or failing to carry out critical safety evaluations;

- Repeated or deliberate failure to obtain adequate informed consent, including:

  - Falsification of consent forms;

  - Repeated or deliberate failure to disclose serious risks of the investigational drug in the informed consent process;

- Falsification of study safety data;

- Failure to obtain IRB review and approval for significant protocol changes; and

- Failure to adequately supervise the clinical trial such that human subjects are or would be exposed to an unreasonable and significant risk of illness or injury.

Conversely, some types of violations would be less likely to justify a clinical hold at an early stage in FDA's investigation. For example, certain kinds of record-keeping violations would be unlikely to suggest such a significant risk of illness or injury to subjects in the trial that a clinical hold would be justified.

Second, FDA would consider the degree of certainty that there has been investigator misconduct that poses a significant risk to subjects. Nonetheless, protecting the safety of subjects is of great importance, and even preliminary (e.g., pre- inspectional information provided to FDA by the IRB, sponsor or other parties), but credible evidence raising concerns that subjects may be placed at substantial risk may warrant a hold while further information is being obtained.

2. After an enforcement action is initiated

In general, when FDA concludes that there is sufficient evidence of repeated or deliberate violations of the regulations or of repeated of deliberate submission of false information to take an enforcement action, it typically will issue a NIDPOE letter and begin a disqualification proceeding. In this case there will be a strong presumption that human subjects are or would be exposed to an unreasonable and significant risk of illness or injury. The types of violations that warrant the issuance of NIDPOE letters are always significant and, with rare exceptions, jeopardize the rights, safety and welfare of the subjects involved. Those exceptions involve violations that compromise data integrity alone without jeopardizing subjects. Minor violations of an investigator's responsibilities do not alone give rise to a NIDPOE letter. One or more of the following types of violations may give rise to NIDPOE letters, and may also give rise to clinical holds if the circumstances show that the violations pose a significant risk to subjects:[6]

- Repeated or deliberate failure to obtain or document informed consent from human subjects, which may include:

  - Repeated or deliberate omission of a description of serious risks of the experimental therapy when obtaining informed consent;

  - Repeated or deliberate failure to provide informed consent in a language understandable to the subject;

- Repeated or deliberate failure to limit administration of the investigational article to those subjects under the investigator's supervision;

- Repeated or deliberate failure to comply with conditions placed on the study by the IRB, sponsor, or FDA;

- Repeated or deliberate failure to obtain review of a study plan by an IRB, the body responsible for overseeing the rights, safety and welfare of human subjects;

- Repeated or deliberate failure to follow the signed investigator statement or protocol, e.g., by enrolling subjects who should have been excluded because of concomitant illnesses that put those subjects at greater risk;

[6] This list is not intended to be all-inclusive.

- Repeated or deliberate failure to maintain accurate study records or submit required adverse event reports to the sponsor;

- Repeated or deliberate falsification or concealment of study records, e.g., by substituting in study records the results of biological samples from subjects who met the inclusion criteria for samples of subjects who did not meet the inclusion criteria, or by fabricating subjects; and

- Repeated or deliberate failure to adequately supervise the clinical trial such that human subjects are or would be exposed to an unreasonable and significant risk or injury.

C. What Steps Will FDA Take Before Imposing A Clinical Hold to Protect Subjects from Investigator Misconduct?

The general regulations governing clinical holds require that, where FDA concludes that there may be grounds for imposing a clinical hold, "FDA will, unless patients are exposed to immediate and serious risk, attempt to discuss and satisfactorily resolve the matter with the sponsor before issuing the clinical hold order" (21 CFR 312.42(c)). If possible, as in all cases where a clinical hold is considered, FDA will contact the sponsor and attempt to resolve the matter in a way that adequately protects study subjects before imposing a clinical hold, following the timeframes described in companion guidances and regulations (e.g., Guidance with Industry:

Formal Meetings with Sponsors and Applicants for PDUFA Products and 21 CFR 312.42(e) respectively). In those cases where an inspection appears necessary to resolve issues, FDA will make every effort to ensure that the inspections are completed in a timely manner.

D. When Will FDA Lift a Clinical Hold that Was Imposed to Protect Subjects from Investigator Misconduct?

FDA will lift a clinical hold imposed to protect subjects from investigator misconduct when the grounds for the hold no longer apply. The sponsor of the affected study may, while the clinical hold is in place, present evidence to FDA to show that it has taken steps to protect study subjects, e.g., by replacing the investigator who is charged with the misconduct or, for example, in the case of a sponsor- investigator, by submitting a monitoring plan. If FDA concludes, based on this evidence, that the study subjects are no longer exposed to an unreasonable and significant risk of illness or injury, the hold will be lifted. In all instances, if a sponsor of a study that has been placed on clinical hold requests in writing that the clinical hold be removed and responds to the issues identified in the clinical hold order, FDA will respond in writing to the sponsor within 30 calendar days of receipt of the request and response (21 CFR 312.42(e)). FDA will either remove or maintain the clinical hold and will state the reasons for its decision (Id.).

Exploratory IND Studies

Exploratory IND Studies, Guidance for Industry[1, T]

U.S. Department of Health and Human Services
Food and Drug Administration
Center for Drug Evaluation and Research (CDER)

January 2006

Pharmacology/Toxicology

Contains Nonbinding Recommendations

This guidance describes the preclinical and clinical issues as well as chemistry, manufacturing and controls information that should be considered when planning exploratory studies including studies of related drugs or biologics under an investigational new drug (IND) application.

Note: *This guidance represents the Food and Drug Administration's (FDA's) current thinking on this topic. It does not create or confer any rights for or on any person and does not operate to bind FDA or the public. You can use an alternative approach if the approach satisfies the requirements of the applicable statutes and regulations. If you want to discuss an alternative approach, contact the appropriate FDA staff. If you cannot identify the appropriate FDA staff, call the appropriate number listed on the title page of this guidance.*

I. Introduction

This guidance is intended to clarify what preclinical and clinical approaches, as well as chemistry, manufacturing, and controls information, should be considered when planning exploratory studies in humans, including studies of closely related drugs or therapeutic biological products, under an investigational new drug (IND) application (21 CFR 312). Existing regulations allow a great deal of flexibility in the amount of data that needs to be submitted with an IND application, depending on the goals of the proposed investigation, the specific human testing proposed, and the expected risks. The Agency believes that sponsors have not taken full advantage of that flexibility and often provide more supporting information

---

[1] This guidance was developed by the Office of New Drugs in the Center for Drug Evaluation and Research (CDER).

This guidance contains information collection provisions that are subject to review by the Office of Management and Budget under the Paperwork Reduction Act of 1995 (44 U.S.C. 3501-3520). The collection of information in this guidance has been approved under OMB Control No. 0910-0014.

[T] Available on the FDA website at: http://www.fda.gov/downloads/Drugs/ GuidanceComplianceRegulatoryInformation/Guidances/UCM078933.pdf

in INDs than is required by regulations. This guidance is intended to clarify what manufacturing controls, preclinical testing, and clinical approaches can be considered when planning limited, early exploratory IND studies in humans.

For the purposes of this guidance the phrase *exploratory IND study* is intended to describe a clinical trial that

- is conducted early in phase 1,

- involves very limited human exposure, and

- has no therapeutic or diagnostic intent (e.g., screening studies, microdose studies).

Such exploratory IND studies are conducted prior to the traditional dose escalation, safety, and tolerance studies that ordinarily initiate a clinical drug development program. The duration of dosing in an exploratory IND study is expected to be limited (e.g., 7 days). This guidance applies to early phase 1 clinical studies of investigational new drug and biological products that assess feasibility for further development of the drug or biological product.[2]

FDA's guidance documents, including this guidance, do not establish legally enforceable responsibilities. Instead, guidances describe the Agency's current thinking on a topic and should be viewed only as recommendations, unless specific regulatory or statutory requirements are cited. The use of the word should in Agency guidances means that something is suggested or recommended, but not required.

II. Background

In its March 2004 Critical Path Report,[3] the Agency explained that to reduce the time and resources expended on candidate products that are unlikely to succeed,[4] new tools are needed to distinguish earlier in the process those candidates that hold promise from those that do not. This guidance describes some early phase 1 exploratory approaches that are consistent with regulatory requirements while maintaining needed human subject protection, but that involve fewer resources than is customary, enabling sponsors to move ahead more efficiently with the development of promising candidates.

A. Traditional Phase 1 Approach

Typically, during pharmaceutical development, large numbers of molecules are generated with the goal of identifying the most promising candidates for further development. These molecules are generally structurally related, but can differ in important ways. Promising

---

[2] Specifically, this guidance is limited to drug and certain well-characterized therapeutic biological products (e.g., recombinant therapeutic proteins and monoclonal antibodies) regulated by CDER. The guidance does not apply to human cell or tissue products, blood and blood proteins, vaccines, or to products regulated as devices.

[3] *Innovation or Stagnation, Challenge and Opportunity on the critical Path to New Medical Products* (March 2004).

[4] "A new medical compound entering phase 1 testing, often representing the culmination of upwards of a decade of preclinical screening and evaluation, is estimated to have only an 8 percent chance of reaching the market," Critical Path Report, March 2004.

candidates are often selected using in vitro testing models that examine binding to receptors, effects on enzyme activities, toxic effects, or other in vitro pharmacologic parameters; these tests usually require only small amounts of the drug. Candidates that are not rejected during these early tests are prepared in greater quantities for in vivo animal testing for efficacy and safety. Commonly, a single candidate is selected for an IND application and introduction into human subjects, initially healthy volunteers in most cases.

Before the human studies can begin, an IND must be submitted to the Agency containing, among other things, information on any risks anticipated based on the results of pharmacologic and toxicological data collected during studies of the drug in animals (21 CFR 312.23(a)(8)). These basic safety tests are most often performed in rats and dogs. The studies are designed to permit the selection of a safe starting dose for humans, to gain an understanding of which organs may be the targets of toxicity, to estimate the margin of safety between a clinical and a toxic dose, and to predict pharmacokinetic and pharmacodynamic parameters. These early tests are usually resource intensive, requiring significant investment in product synthesis, animal use, laboratory analyses, and time. Many resources are invested in, and thus wasted on, candidate products that subsequently are found to have unacceptable profiles when evaluated in humans — less than 10 percent of INDs for new molecular entities (NME) progress beyond the investigational stage to submission of a marketing application (NDA).3 In addition, animal testing does not always predict performance in humans, and potentially effective candidates may not be developed because of resource constraints.

Existing regulations allow a great deal of flexibility in terms of the amount of data that need to be submitted with any IND application, depending on the goals of the proposed investigation, the specific human testing proposed, and the expected risks. The Agency believes that sponsors have not taken full advantage of that flexibility. As a result, limited, early phase 1 studies, such as those described in this guidance, are often supported by a more extensive preclinical database than is required by the regulations.

This guidance describes preclinical and clinical approaches, and the chemistry, manufacturing, and controls information that should be considered when planning exploratory IND studies in humans, including studies of closely related drugs or therapeutic biological products, under a single IND application (21 CFR 312).

B. Exploratory IND Approach

Exploratory IND studies usually involve very limited human exposure and have no therapeutic or diagnostic intent. Such studies can serve a number of useful goals. For example, an exploratory IND study can help sponsors

- Determine whether a mechanism of action defined in experimental systems can also be observed in humans (e.g., a binding property or inhibition of an enzyme)

- Provide important information on pharmacokinetics (PK)

- Select the most promising lead product from a group of candidates[5] designed to interact with a particular therapeutic target in humans, based on PK or pharmacodynamic (PD) properties

- Explore a product's biodistribution characteristics using various imaging technologies

Whatever the goal of the study, exploratory IND studies can help identify, early in the process, promising candidates for continued development and eliminate those lacking promise. As a result, exploratory IND studies may help reduce the number of human subjects and resources, including the amount of candidate product, needed to identify promising drugs. The studies discussed in this guidance involve dosing a limited number of subjects with a limited range of doses for a limited period of time.

Existing regulations provide more flexibility with regard to the preclinical testing requirements for exploratory IND studies than for traditional IND studies. However, sponsors submitting the kinds of studies described in this guidance have not always taken full advantage of that flexibility. Sponsors often provide more supporting information in their INDs than is required by the regulations. Because exploratory IND studies involve administering either sub-pharmacologic doses of a product, or doses expected to produce a pharmacologic, but not a toxic, effect, the potential risk to human subjects is less than for a traditional phase 1 study that, for example, seeks to establish a maximally tolerated dose. *Because exploratory IND studies present fewer potential risks than do traditional phase 1 studies that look for dose-limiting toxicities, such limited exploratory IND investigations in humans can be initiated with less, or different, preclinical support than is required for traditional IND studies.*[6]

The Agency expects that this early phase 1 exploratory IND approach will apply to a number of different study paradigms. Although his guidance explores several potential applications, many others can be proposed. The Agency believes that, consistent with its Critical Path Initiative, clarifying Agency thinking about how much and what kind of testing is needed to support early studies in humans will facilitate the entry of new products into clinical testing and speed product development.

Although exploratory IND studies may be used during development of products intended for any indication, it is particularly important for manufacturers to consider this approach when developing products to treat serious diseases. Because the approach can help identify

---

[5] For the purposes of this guidance, the term *candidate*, or *candidate product*, is used to describe a drug or biologic that is being tested in early exploratory studies under an IND. This guidance **does not** distinguish between a *drug product* and a *drug substance* as some other Agency guidances do.
(Most guidances use the term drug product to refer to a finished dosage form (e.g., tablet, capsule, solution) that contains an active drug ingredient generally, but not necessarily, in association with inactive ingredients, or a finished dosage form that does not contain an active ingredient but is intended to be used as a placebo. *Drug substance* usually refers to any component that is intended to furnish pharmacological activity or other direct effect in the diagnosis, cure, mitigation, treatment, or prevention of disease, or to affect the structure or any function of the body.)
[6] Generally, these types of studies would not be carried out in pediatric patients or in pregnant or lactating women.

promising candidates more quickly and precisely, exploratory IND studies could become an important part of the armamentarium when developing drug and biological products to treat a serious or life-threatening illness. The Agency has previously articulated its commitment to ensuring that appropriate flexibility is applied when patients with a serious disease and no satisfactory alternative therapies are enrolled in a trial with therapeutic intent.[7]

III. *Content of IND Submissions*

To begin any kind of testing in humans, applicants must submit an IND application to the Agency with certain types of information (see 21 CFR 312.23 IND Content and Format). The primary purpose of the IND submission is to ensure that subjects will not face undue risk of harm. The major information that must be submitted includes:

- Information on a clinical development plan

- Chemistry, manufacturing, and controls information

- Pharmacology and toxicology information

- Previous human experience with the investigational candidate or related compounds, if there is any

The following sections discuss the first three in more detail. Because the exploratory IND studies addressed by this guidance will be first in human studies, previous human experience is not pertinent and will not be discussed. The common theme throughout is that, depending on the study, the informational requirements for exploratory IND studies are more flexible than for traditional IND studies.

A. Clinical Information

1. Introductory statement and general investigational plan

A traditional IND application describes the rationale for the proposed clinical trial program and discusses the potential outcome of the clinical investigation. The exploratory IND studies discussed here focus on a circumscribed study or group of studies, and plans for further development cannot be formulated without the results of these studies. Therefore, an exploratory IND application should articulate the rationale for selecting a compound (or compounds) and for studying them in a single trial or related trials, as this represents all that is known about the overall development plan at this stage. This section should also make it clear that the IND is intended to be withdrawn[8] after completion of the outlined study or studies.

---

[7] Subpart H Accelerated Approval of New Drugs for Serious or Life-Threatening Illnesses. See also FDA guidance for industry *Fast Track Drug Development Programs — Designation, Development, and Application Review.*
[8] The withdrawn, or inactive, IND can be referenced in any subsequent traditional IND.

2. Types of studies

Potentially useful study designs include both single- and multiple-dose studies. In single-dose studies, a sub-pharmacologic[9] or pharmacologic dose is administered to a limited number of subjects (healthy volunteers or patients). For example, microdose studies usually involve the single administration of a small dose with the goal of collecting pharmacokinetic information or performing imaging studies, or both.

endpoints. In exploratory IND studies, the duration of dosing should be limited (e.g., 7 days). For escalating dose studies done under an exploratory IND, dosing should be designed to investigate a pharmacodynamic endpoint, not to determine the limits of tolerability.

B. Chemistry, Manufacturing, and Controls Information

The regulations at 21 CFR 312.23(a)(7)(i) emphasize the graded nature of chemistry, manufacturing, and controls (CMC) information needed as development under an IND application progresses. Although in each phase of a clinical investigational program sufficient information should be submitted to ensure the proper identification, strength, quality, purity, and potency of the investigational candidate, the amount of information that will provide that assurance will vary with the phase of the investigation, the proposed duration of the investigation, the dosage form, and the amount of information already available. For the purpose of an exploratory IND application, the CMC information indicated below can be provided in a summary report to enable the Agency to make the necessary safety assessment.

The sponsor must state in the beginning of the exploratory IND application whether it believes the chemistry or manufacturing of the candidate product presents any potential for human risk (e.g., specific findings in preclinical studies associated with known risks of related compounds) (§ 312.23). If so, these potential risks should be discussed, and the steps proposed to monitor for such risks should be described.

The Agency is in the process of developing guidance explaining the stepwise approach to meeting current good manufacturing practice (CGMP) regulations. Once finalized, that guidance will be useful to persons seeking to manufacture, or prepare, products intended for use in an exploratory IND study.

1. General information for the candidate product

Except as noted below, the extent and type of chemistry and manufacturing information to be submitted in an exploratory IND application is similar to that described in current guidance for

---

[9] A radiolabeled candidate compound can be administered at doses that are known to have no pharmacologic effect in humans without an IND application in basic research studies when the compound has previously been studied in humans and the results published in the literature. These basic research investigations are conducted under the oversight of an institutional review board (IRB) and a radioactive drug research committee (21 CFR 361.1).

use of investigational products.[10] Information on each candidate product (i.e., the active ingredient) can be submitted in a summary report containing the following items.

- Description of the candidate product, including physical, chemical, and/or biological characteristics, as well as its source (e.g., synthetic, animal source, plant extract, or biotechnology-derived) and therapeutic class (e.g., radiopharmaceutic, immunosuppressant, agonist, antagonist) (see sections below for exceptions).

- Description of the dosage form and information related to the dosage form

- Description of the formulation or routes of administration intended to be used in the human trial. For oral administration, sponsors can consider using suspensions or solutions in addition to the more usual tablets, powders, and capsules. For products intended for ophthalmic, inhalational (aqueous base), or parenteral administration, sterility and apyrogenicity must be ensured. For biological candidate products, freedom from contaminants associated with their manufacture, such as viruses, mycoplasma, and foreign DNA, also should be ensured. All excipients should be generally recognized as safe[11] or part of a formulation that is approved or licensed in the United States for the same route of administration and amount,[12] or adequately qualified through appropriate animal studies.

- The grade and quality (e.g., USP, NF, ACS) of excipients used in the manufacture of the investigational candidate product, including both those components intended to appear in the product and those that may not appear, but that are used in the manufacturing process

- Name and address of the manufacturer(s) (if different from the sponsor)

- The method of preparation of the candidate product lots used in preclinical studies and intended for the proposed human study, including a brief description of the method of manufacture and the packaging procedure, as appropriate, with a description of the container and closure system. For the active substance, include a list of the starting materials, reagents, solvents, catalysts used, and purification steps employed to prepare the candidate product. For sterile products, describe the sterilization process and controls for ensuring sterility. For biological/biotechnology-derived products, also identify the source material (e.g., Master Cell Bank), describe the expression system (e.g., fermentation methods) and harvest methods, as well as methods for removal/inactivation of potential viral contaminants. We recommend the use of a detailed flow diagram that includes all materials used as the usual, most effective, presentation of this information.

- Quantitative composition of the product

---

[10] See guidance for industry *Content and Format of Investigational New Drug Applications for Phase 1 Studies of Drugs, Including Well-Characterized, Therapeutic, Biotechnology-Derived Products.*

[11] Excipients considered to be generally recognized as safe (GRAS) are included in a list that is maintained on the Internet at http://www.accessdata.fda.gov/scripts/cder/iig/index.cfm. See also 21 CFR 330.1, which explains the GRAS concept.

[12] Novel excipients should be appropriately qualified for their intended use. FDA has issued guidance on *Nonclinical Studies for Development of Pharmaceutical Excipients.*

- A brief description of adequate test methods used to ensure the identity, strength, quality, purity, and potency accompanied by the test results, or a certificate of analysis, of the candidate product lots used in toxicological studies and intended for the proposed human study. For biotechnology products produced in mammalian cells or animals, this will include tests and studies to ensure the removal and/or inactivation of potential viral contaminants.

- Information that demonstrates the stability of the product during toxicology studies and an explanation of how stability will be evaluated during the clinical studies

- For ophthalmic, inhalational (aqueous base), or parenteral dosage forms, results from sterility and pyrogenicity tests

2. Analytical characterization of candidate product

There are two scenarios under which CMC information can be provided to an IND application. In the first scenario, the *same batch* of candidate product is used in both the toxicology studies and clinical trials. This material will be qualified for human use based on the CMC information (see III.B.1, above) and results of the toxicology studies described elsewhere in this guidance. Although we recommend establishing the impurity profile to the extent possible for future reference and/or comparison, not all impurities of the candidate product may need characterization at this stage of product development. If an issue arises during the toxicology qualification of the product, the appropriate parameters can be studied further, on an as-needed basis. Impurities (e.g., chemical and microbiological) should be characterized in accordance with recommendations in Agency guidance,[13] if, and when, the sponsor files a traditional IND for further clinical investigation.

In the second scenario, the batch of candidate drug product to be used in the clinical studies may not be the same as that used in the nonclinical toxicology studies. In such a case, the sponsor should demonstrate by analytical testing that the batch to be used is *representative* of batches used in the nonclinical toxicology studies. To achieve this, relevant analytical quality test results should be sufficient to enable comparison of different batches of the product. Tests to accomplish this include:

- Identity

- Structure (e.g., optical rotation (for chiral compounds), reducing/non-reducing electrophoresis (for proteins))

- Assay for purity

- Impurity profile (e.g., product- and process-related impurities, residual solvents, heavy metals)

- Assay for potency (biologic)

- Physical characteristics (as appropriate)

---

[13] See footnote 10 and guidance for industry, *INDs for Phase 2 and Phase 3 Studies, Chemistry, Manufacturing, and Controls Information.*

• Microbiological characteristics (as appropriate)

C. Safety Program Designs — Examples

Pharmacology and toxicology information is derived from preclinical safety testing performed in animals and in vitro. Preclinical studies for small molecules are described in ICH M3 while those for biologics follow guidance described in ICH S6. Some of the toxicology tests described in this guidance may not be appropriate for biologics. The toxicology evaluation recommended for an exploratory IND application is more limited than for a traditional IND application.[14] The basis for the reduced preclinical package is the reduced scope of an exploratory IND clinical study. Although exploratory IND studies in some cases are expected to induce pharmacologic effects, they are not designed to establish maximally tolerated doses. Furthermore, the duration of drug exposure in exploratory IND studies is limited. The level of preclinical testing performed to ensure safety will depend on the scope and intended goals of the clinical trials.

There are a number of study objectives for which the preclinical safety programs may be tailored to the study design. Examples include: confirming that an expected mechanism of action can be observed in humans; measuring binding affinity or localization of drug; assessing PK and metabolism; comparing the effect on a potential therapeutic target with other therapies. Three examples are discussed in detail in the following paragraphs.

1. Clinical studies of pharmacokinetics or imaging

Microdose studies are designed to evaluate pharmacokinetics or imaging of specific targets and are designed not to induce pharmacologic effects. Because of this, the risk to human subjects is very limited, and information adequate to support the initiation of such limited human studies can be derived from limited nonclinical safety studies. A microdose is defined as less than 1/100th of the dose of a test substance calculated (based on animal data) to yield a pharmacologic effect of the test substance with a maximum dose of <100 micrograms (for imaging agents, the latter criterion applies).[15] Due to differences in molecular weights as compared to synthetic drugs, the maximum dose for protein products is ≤30 nanomoles.

FDA currently accepts the use of extended single-dose toxicity studies in animals to support single-dose studies in humans. For microdose studies, a single mammalian species (both sexes) can be used if justified by in vitro metabolism data and by comparative data on in vitro pharmacodynamic effects. The route of exposure in animals should be by the intended clinical route. In these studies, animals should be observed for 14 days post-dosing with an interim necropsy, typically on day 2, and endpoints evaluated should include body weights, clinical signs, clinical chemistries, hematology, and histopathology (high dose and control only if no pathology is seen at the high dose). The study should be designed to establish a dose inducing a

---

[14] International Conference on Harmonisation (ICH) guidance for industry *M3 Nonclinical Safety Studies for the Conduct of Human Clinical Trials for Pharmaceuticals* describes what is expected for a traditional IND.

[15] See European Medicines Agency (EMEA), Evaluation of Medicines for Human Use, "Position Paper on Non-Clinical Safety Studies to Support Clinical Trials with a Single Microdose," CPMP/SWP/2599/02Rev 1, 23 June 2004.

minimal toxic effect, or alternatively, establishing a margin of safety. To establish a margin of safety, the sponsor should demonstrate that a large multiple (e.g., 100X) of the proposed human dose does not induce adverse effects in the experimental animals. Scaling from animals to humans based on body surface area can be used to select the dose for use in the clinical trial. Scaling based on pharmacokinetic/pharmacodynamic modeling would also be appropriate if such data are available.

Because microdose studies involve only single exposures to microgram quantities of test materials and because such exposures are comparable to routine environmental exposures, routine genetic toxicology testing is not needed. For similar reasons, safety pharmacology studies are also not recommended.

2. Clinical trials to study pharmacologically relevant doses

A second example involves clinical trials designed to study pharmacologic effects of candidate products. More extensive preclinical safety data would be needed to support the safety of such studies. However, since the goal would not include defining a maximally tolerated dose, the evaluation can still be less extensive than typically needed to support a traditional IND application. See the flow chart in the Attachment to this document.

Repeat dose clinical trials lasting up to 7 days can be supported by a 2-week repeat dose toxicology study in a sensitive species accompanied by toxicokinetic evaluations. The goal of such a study would be to select safe starting and maximum doses for the clinical trial. The rat is the usual species chosen for this purpose, but other species might be selected. In addition to studies in a rodent species, additional studies in nonrodents, most often dogs, can be used to confirm that the rodent is an appropriately sensitive species. If it is known that a particular species is most appropriate for a class of compounds, studies can be limited to that species. This confirmation can be approached in a number of ways. A lack of gender difference in the rodent study can serve as a basis for testing only a single sex in the second species if only a single sex will be studied in the clinical trial.

The numbers of animals used in the confirmatory study can be fewer than normally used to attain statistically meaningful comparisons, but of sufficient number to rule out any toxicologically significant difference in sensitivity compared with rodent (e.g. four non-rodents per treatment group). The confirmatory study could be a dedicated study involving repeat administrations of a single dose level approximating the rat NOAEL[16] calculated on the basis of body surface area. Alternatively, the test in the second species could be incorporated as part of an exploratory, dose escalating study culminating in repeated doses equivalent to the rat NOAEL. The number of repeat administrations at the rat NOAEL should, at a minimum, be equal to the number of administrations, given with the same schedule, intended clinically. The route of administration should be the same as the expected clinical route, and toxicokinetic measurements should be used to assess exposure. The same endpoints assessed in the rodent study should be evaluated in the second species. If the data from the confirmatory study suggest that the rodent is not the more sensitive species, a 2-week repeated dose toxicity study should be performed in the second species to select doses for human trials. This study should

---

[16] No-observed-adverse-effect level (NOAEL).

include measurements of body weight, clinical signs, clinical chemistries, hematology, and histopathology.

In contrast to microdose studies, for clinical trials designed to evaluate higher or repeated doses, each candidate product to be tested should be evaluated for safety pharmacology.[17] Evaluation of the central nervous and respiratory systems can be performed as part the rodent toxicology studies while safety pharmacology for the cardiovascular system can be assessed in the nonrodent species, generally the dog, and can be conducted as part of the confirmatory or dose-escalation study.

In general, each product in this type of exploratory IND should be tested for potential genotoxicity unless such testing is not appropriate for the population (e.g. terminally ill patients) or product to be studied. The genetic toxicology tests should include a bacterial mutation assay using all five tester strains with and without metabolic activation[18] as well as a test for chromosomal damage either in vitro (cytogenetics assay or mouse lymphoma thymidine kinase gene mutation assay) or in vivo. The in vivo test can be a micronucleus assay performed in conjunction with the repeated dose toxicity study in the rodent species. The high dose in this case should be a maximally tolerated or limit dose.

The results from the preclinical program can be used to select starting and maximum doses for the clinical trials. The starting dose is anticipated to be no greater than 1/50 of the NOAEL from the 2-week toxicology study in the sensitive species on a mg/m2 basis. The maximum clinical dose would be the lowest of the following:

- ¼ of the 2-week rodent NOAEL on a mg/m2 basis

- Up to ½ of the AUC at the NOAEL in the 2-week rodent study, or the AUC in the dog at the rat NOAEL, whichever is lower

- The dose that produces a pharmacologic and/or pharmacodynamic response or at which target modulation is observed in the clinical trial

- Observation of an adverse clinical response

Escalation from the proposed maximal clinical dose should only be performed after consultation with and concurrence of the FDA.

It is recognized that the studies described above are most appropriate for chemical drugs. Other animal models (e.g. nonhuman primates) may be more appropriate for biologics, and some tests may be inappropriate (e.g. genetic toxicology testing) for proteins.

3. Clinical studies of MOAs related to efficacy

A third example involves clinical studies intended to evaluate mechanisms of action (MOAs). To support this approach, the FDA will accept alternative, or modified, pharmacologic and toxicological studies to select clinical starting doses and dose escalation schemes. For example,

---

[17] For details see the guidance for industry *S7A Safety Pharmacology Studies for Human Pharmaceuticals.*
[18] For details see guidance for industry *S2A: Guidance on Specific Aspects of Regulatory Genotoxicity Tests for Pharmaceuticals* and *S2B: Genotoxicity: A Standard Battery for Genotoxicity Testing for Pharmaceuticals.*

short-term, modified toxicity or safety studies in two animal species based on a dosing strategy to achieve a clinical pharmacodynamic endpoint can in some instances serve as the basis for selecting the safe clinical starting dose for a new candidate drug. These animal studies would incorporate endpoints that are mechanistically based on the pharmacology of the new chemical entity and thought to be important to clinical effectiveness. For example, if the degree of saturation of a receptor or the inhibition of an enzyme were considered possibly related to effectiveness, this parameter would be characterized and determined in the animal study and then used as an endpoint in a subsequent clinical investigation. The dose and dosing regimen determined in the animal study would be extrapolated for use in the clinical investigation. In some cases, a single species could be used if it were established as the most relevant species based on scientific evidence using the specific candidate intended for the clinical investigation.

Although the production of frank toxicity is not the primary intended goal of the nonclinical study, relevant informative endpoints (e.g., hematology and histopathology) selected as important for clinical safety evaluation should be investigated. For example, an antibody that binds with a high degree of selectivity to a tumor-associated antigen could be studied in accordance with this third category. The mechanism of action of antibody-based products is generally associated with their binding properties and the effect on functions associated with immunoglobulins. Pharmacology and toxicology studies provide information about the selection of doses used in clinical studies through evidence of both a safe upper and potentially efficacious lower limit of exposure. These doses might be consistent with target plasma levels of the drug based on animal models of disease. The upper safe levels could be established in animal studies that show a lack of toxicity at these levels.

D. GLP Compliance

It is expected that all preclinical safety studies supporting the safety of an exploratory IND application will be performed in a manner consistent with good laboratory practices (GLP) (21 CFR Part 58). The GLP provisions apply to a broad variety of studies, test articles, and test systems. Sponsors are encouraged to discuss any need for an exemption from GLP provisions with the FDA prior to conducting safety related studies, for example, during a pre-IND meeting. Sponsors must justify any nonconformance with GLP provisions (21 CFR 312.23(a)(8)(iii)).

IV. Conclusion

Existing regulations allow a great deal of flexibility in the amount of data that needs to be submitted with any IND application, depending on the goals of an investigation, the specific human testing being proposed, and the expected risks. Sponsors have not taken full advantage of that flexibility, and limited, early phase 1 studies, such as those described in this guidance, are often supported by a more extensive preclinical database than is needed for those studies alone.

The common theme throughout this guidance is that, depending on the study, the preclinical testing programs for exploratory IND studies can be less extensive than for traditional IND studies. This is because for the approaches discussed in this guidance, which involve administering sub-pharmacologic doses of a candidate product or products, the potential risks to human subjects are less than for a traditional phase 1 study.

The Agency is undertaking a number of efforts to reduce the time spent in early drug development on products that are unlikely to succeed. This guidance describes some exploratory approaches that are consistent with regulatory requirements, but that will enable sponsors to move ahead more efficiently with the development of promising candidate products while maintaining needed human subject protections.

Attachment

A Preclinical Toxicology Testing Strategy for Exploratory INDs Designed To Administer Pharmacologically Active Doses

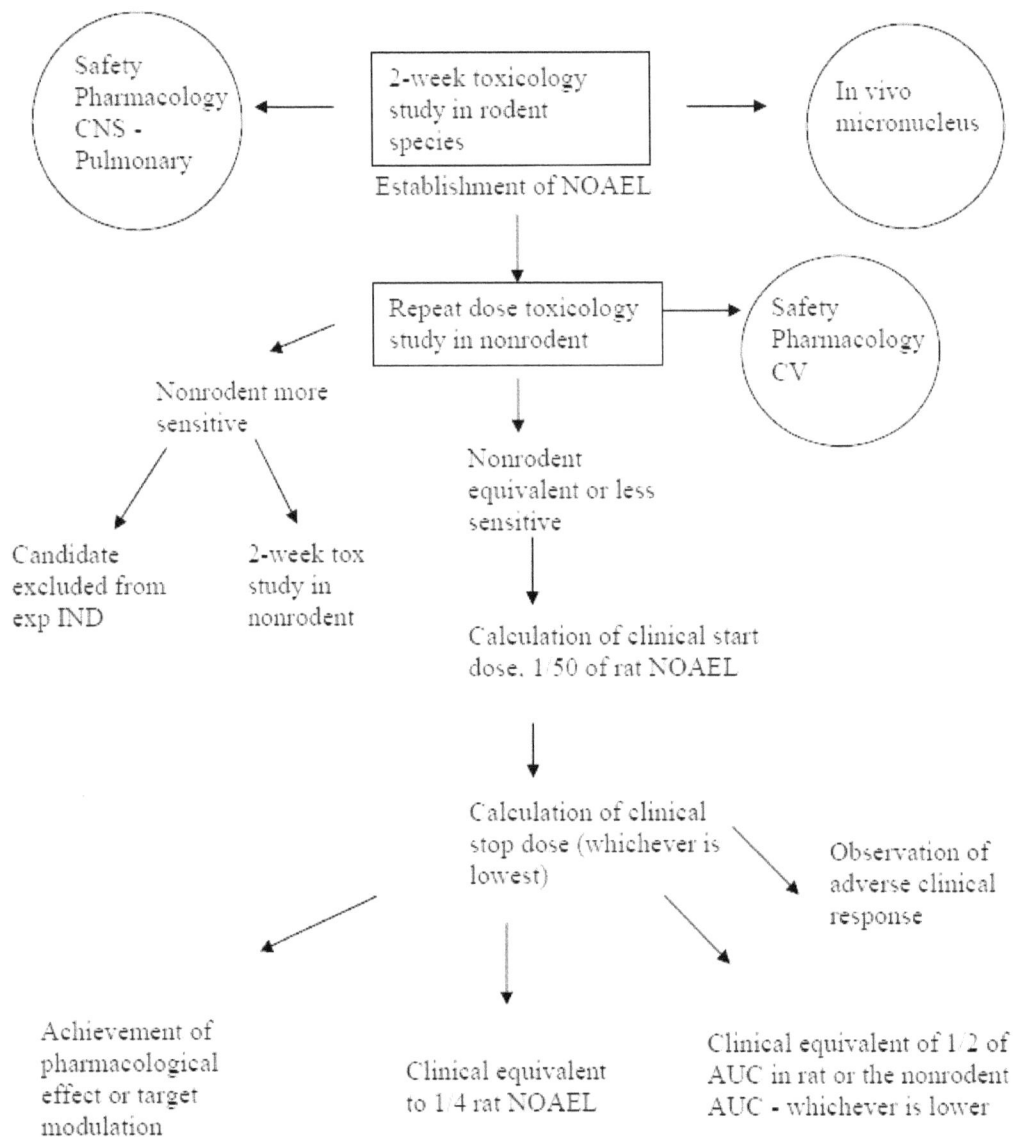

Food-Effect Bioavailability and Fed Bioequivalence Studies

Guidance for Industry,
Food-Effect Bioavailability and Fed Bioequivalence Studies [†, 1]

U.S. Department of Health and Human Services
Food and Drug Administration
Center for Drug Evaluation and Research (CDER)

December 2002

BP

Contains Nonbinding Recommendations

This guidance provides recommendations to sponsors and/or applicants planning to conduct food-effect bioavailability (BA) and fed bioequivalence (BE) studies for orally administered drug products as part of investigational new drug applications (INDs), new drug applications (NDAs) and abbreviated new drug applications (ANDAs), and supplemental applications.

Note: *This guidance represents the Food and Drug Administration's (FDA's) current thinking on this topic. It does not create or confer any rights for or on any person and does not operate to bind FDA or the public. You can use an alternative approach if the approach satisfies the requirements of the applicable statutes and regulations. If you want to discuss an alternative approach, contact the appropriate FDA staff. If you cannot identify the appropriate FDA staff, call the appropriate number listed on the title page of this guidance.*

I. Introduction

This guidance provides recommendations to sponsors and/or applicants planning to conduct food-effect bioavailability (BA) and fed bioequivalence (BE) studies for orally administered drug products as part of investigational new drug applications (INDs), new drug applications (NDAs), abbreviated new drug applications (ANDAs), and supplements to these applications. This guidance applies to both immediate-release and modified-release drug products. The guidance addresses how to meet the BA and BE requirements in 21 CFR 320, 314.50 (d) (3), and 314.94 (a) (7) as they apply to oral dosage forms. This guidance provides recommendations

---

[†] Available on the FDA website at: http://www.fda.gov/downloads/RegulatoryInformation/ Guidances/UCM126833.pdf

[1] This guidance has been prepared by the Food Effect Working Group of the Biopharmaceutics Coordinating Committee in the Office of Pharmaceutical Science, Center for Drug Evaluation and Research (CDER) at the Food and Drug Administration (FDA).

for food-effect BA and fed BE study designs, data analysis, and product labeling. It also provides information on when food-effect BA and fed BE studies should be performed.[2]

II. Background

Food effect BA studies are usually conducted for new drugs and drug products during the IND period to assess the effects of food on the rate and extent of absorption of a drug when the drug product is administered shortly after a meal (fed conditions), as compared to administration under fasting conditions. Fed BE studies, on the other hand, are conducted for ANDAs to demonstrate their bioequivalence to the reference listed drug (RLD) under fed conditions.

A. Potential Mechanisms of Food Effects on BA

Food can change the BA of a drug and can influence the BE between test and reference products.

Food effects on BA can have clinically significant consequences. Food can alter BA by various means, including

- Delay gastric emptying
- Stimulate bile flow
- Change gastrointestinal (GI) pH
- Increase splanchnic blood flow
- Change luminal metabolism of a drug substance
- Physically or chemically interact with a dosage form or a drug substance

Food effects on BA are generally greatest when the drug product is administered shortly after a meal is ingested. The nutrient and caloric contents of the meal, the meal volume, and the meal temperature can cause physiological changes in the GI tract in a way that affects drug product transit time, luminal dissolution, drug permeability, and systemic availability. In general, meals that are high in total calories and fat content are more likely to affect the GI physiology and thereby result in a larger effect on the BA of a drug substance or drug product. We recommend use of high-calorie and high-fat meals during food-effect BA and fed BE studies.

B. Food Effects on Drug Products

Administration of a drug product with food may change the BA by affecting either the drug substance or the drug product. In practice, it is difficult to determine the exact mechanism by which food changes the BA of a drug product without performing specific mechanistic studies. Important food effects on BA are least likely to occur with many rapidly dissolving, immediate-release drug products containing highly soluble and highly permeable drug substances (BCS

---

[2] See also the guidance for industry on *Bioavailablity and Bioequivalence Studies for Orally Administered Drug Products: General Considerations.*

Class I) because absorption of the drug substances in Class I is usually pH- and site-independent and thus insensitive to differences in dissolution.[3] However, for some drugs in this class, food can influence BA when there is a high first-pass effect, extensive adsorption, complexation, or instability of the drug substance in the GI tract. In some cases, excipients or interactions between excipients and the food-induced changes in gut physiology can contribute to these food effects and influence the demonstration of BE. For rapidly dissolving formulations of BCS Class I drug substances, food can affect Cmax and the time at which this occurs (Tmax) by delaying gastric emptying and prolonging intestinal transit time. However, we expect the food effect on these measures to be similar for test and reference products in fed BE studies.

For other immediate-release drug products (BCS Class II, III, and IV) and for all modified release drug products, food effects are most likely to result from a more complex combination of factors that influence the in vivo dissolution of the drug product and/or the absorption of the drug substance. In these cases, the relative direction and magnitude of food effects on formulation BA and the effects on the demonstration of BE are difficult, if not impossible, to predict without conducting a fed BE study.

III. Recommendations for Food-Effect BA and Fed BE Studies

This section of the guidance provides recommendations on when food-effect BA studies should be conducted as part of INDs and NDAs and when fed BE studies should be conducted as part of ANDAs. For postapproval changes in an approved immediate- or modified-release drug product that requires in vivo redocumentation of BE under fasting conditions, fed BE studies are generally unnecessary.

A. Immediate-Release Drug Products

1. INDs/NDAs

We recommend that a food-effect BA study be conducted for all new chemical entities (NCEs) during the IND period.

Food-effect BA studies should be conducted early in the drug development process to guide and select formulations for further development. Food-effect BA information should be available to design clinical safety and efficacy studies and to provide information for the CLINICAL PHARMACOLOGY and/or DOSAGE AND ADMINISTRATION sections of product labels. If a sponsor makes changes in components, composition, and/or method of manufacture in the clinical trial formulation prior to approval, BE should be demonstrated between the to-be-marketed formulation and the clinical trial formulation.

Sponsors may wish to use relevant principles described in the guidance for industry on SUPAC-IR: Immediate Release Solid Oral Dosage Forms: Scale-Up and Post-Approval Changes: Chemistry, Manufacturing, and Controls, In Vitro Dissolution Testing, and In Vivo

---

[3] See the guidance for industry on *Waiver of In Vivo Bioavailability and Bioequivalence Studies for Immediate Release Solid Oral Dosage Forms Based on a Biopharmaceutics Classification System.*

Bioequivalence Documentation (SUPAC-IR guidance) to determine if in vivo BE studies are recommended. These BE studies, if indicated, should generally be conducted under fasting conditions.

2. ANDAs

In addition to a BE study under fasting conditions, we recommend a BE study under fed conditions for all orally administered immediate-release drug products, with the following exceptions:

- When both test product and RLD are rapidly dissolving, have similar dissolution profiles, and contain a drug substance with high solubility and high permeability (BCS Class I) (see footnote 3), or

- When the DOSAGE AND ADMINISTRATION section of the RLD label states that the product should be taken only on an empty stomach, or

- When the RLD label does not make any statements about the effect of food on absorption or administration.

B. Modified-Release Drug Products

We recommend that food-effect BA and fed BE studies be performed for all modifiedrelease dosage forms.

1. INDs/NDAs

We recommend a study comparing the BA under fasting and fed conditions for all orally administered modified-release drug products.

When changes occur in components, composition, and/or method of manufacture between the to-be-marketed formulation and the primary clinical trial material, the sponsor may wish to use relevant principles described in the guidance for industry on *SUPAC-MR: Modified Release Solid Oral Dosage Forms: Scale-Up and Post-Approval Changes: Chemistry, Manufacturing, and Controls: In Vitro Dissolution Testing and In Vivo Bioequivalence Documentation* (SUPAC-MR guidance) to determine if documentation of in vivo BE is recommended. These BE studies, if indicated, should generally be conducted under fasting conditions.

2. ANDAs

In addition to a BE study under fasting conditions, a BE study under fed conditions should be conducted for all orally administered modified-release drug products.

IV. Study Considerations

This section provides general considerations for designing food effect BA and fed BE studies. A sponsor may propose alternative study designs and data analyses. The scientific rationale and justification for these study designs and analyses should be provided in the study protocol. Sponsors may choose to conduct additional studies for a better understanding of the drug product and to provide optimal labeling statements for dosage and administration (e.g. different meals and different times of drug intake in relation to meals). In studying modified-release dosage forms, consideration should be given to the possibility that co-administration with food can result in dose dumping, in which the complete dose may be more rapidly released from the dosage form than intended, creating a potential safety risk for the study subjects.

A. General Design

We recommend a randomized, balanced, single-dose, two-treatment (fed vs. fasting), two-period, two-sequence crossover design for studying the effects of food on the BA of either an immediate-release or a modified-release drug product. The formulation to be tested should be administered on an empty stomach (fasting condition) in one period and following a test meal (fed condition) in the other period. We recommend a similar, two-treatment, two-period, twosequence crossover design for a fed BE study except that the treatments should consist of both test and reference formulations administered following a test meal (fed condition). An adequate washout period should separate the two treatments in food-effect BA and fed BE studies.

B. Subject Selection

Both food-effect BA and fed BE studies can be carried out in healthy volunteers drawn from the general population. Studies in the patient population are also appropriate if safety concerns preclude the enrollment of healthy subjects. A sufficient number of subjects should complete the study to achieve adequate power for a statistical assessment of food effects on BA to claim an absence of food effects, or to claim BE in a fed BE study (see DATA ANALYSIS AND LABELING section). A minimum of 12 subjects should complete the food-effect BA and fed BE studies.

C. Dosage Strength

In general, the highest strength of a drug product intended to be marketed should be tested in food-effect BA and fed BE studies. In some cases, clinical safety concerns can prevent the use of the highest strength and warrant the use of lower strengths of the dosage form. For ANDAs, the same lot and strength used in the fasting BE study should be tested in the fed BE study. For products with multiple strengths in ANDAs, if a fed BE study has been performed on the highest strength, BE determination of one or more lower strengths can be waived based on dissolution profile comparisons (for details see the guidance on *Bioavailablity and Bioequivalence Studies for Orally Administered Drug Products - General Considerations*).

D. Test Meal

We recommend that food-effect BA and fed BE studies be conducted using meal conditions that are expected to provide the greatest effects on GI physiology so that systemic drug availability is maximally affected. A high-fat (approximately 50 percent of total caloric content of the meal) and high-calorie (approximately 800 to 1000 calories) meal is recommended as a test meal for food-effect BA and fed BE studies. This test meal should derive approximately

150, 250, and 500-600 calories from protein, carbohydrate, and fat, respectively.[4] The caloric breakdown of the test meal should be provided in the study report. If the caloric breakdown of the meal is significantly different from the one described above, the sponsor should provide a scientific rationale for this difference. In NDAs, it is recognized that a sponsor can choose to conduct food-effect BA studies using meals with different combinations of fats, carbohydrates, and proteins for exploratory or label purposes. However, one of the meals for the food-effect BA studies should be the high-fat, high-calorie test meal described above.

---

[4] An example test meal would be two eggs fried in butter, two strips of bacon, two slices of toast with butter, four ounces of hash brown potatoes and eight ounces of whole milk. Substitutions in this test meal can be made as long as the meal provides a similar amount of calories from protein, carbohydrate, and fat and has comparable meal volume and viscosity.

Study and Evaluation of Gender Differences in the Clinical Evaluation of Drugs

Guideline for the Study and Evaluation of Gender Differences in the Clinical Evaluation of Drugs, Guidance for Industry [Ŧ, 1]

Department of Health and Human Services

Food and Drug Administration

[Docket No. 93D—236]

Contains Nonbinding Recommendations

This guideline presents guidance on FDA's expectations regarding inclusion of both genders in drug development.

Note: *This guidance represents the Food and Drug Administration's (FDA's) current thinking on this topic. It does not create or confer any rights for or on any person and does not operate to bind FDA or the public. You can use an alternative approach if the approach satisfies the requirements of the applicable statutes and regulations. If you want to discuss an alternative approach, contact the appropriate FDA staff. If you cannot identify the appropriate FDA staff, call the appropriate number listed on the title page of this guidance.*

Summary

The Food and Drug Administration (FDA) is publishing a guideline entitled "Guideline for the Study and Evaluation of Gender Differences in the Clinical Evaluation of Drugs." This guideline provides new guidance on FDA's expectations regarding inclusion of both genders in drug development and revises the section "Women of Childbearing Potential" in the 1977 guideline entitled, "General Considerations for the Clinical Evaluation of Drugs" (HEW Publication No. (FDA) 77-3040).

[Ŧ] Available on the FDA website at: http://www.fda.gov/downloads/RegulatoryInformation/Guidances/UCM126835.pdf
[1] Federal Register / Vol. 58, No. 139 / Thursday, July 22, 1993 / Notices

Supplementary Information

I. Introduction

In this document, FDA is publishing a new guideline on FDA's expectations regarding inclusion of patients of both genders in drug development, analyses of clinical data by gender, assessment of potential pharmacokinetic differences between genders, and conduct of specific additional studies in women, where indicated. This guideline revises the section of the 1977 guideline, entitled "General Considerations for the Clinical Evaluation of Drugs," that excluded women of childbearing potential from participation in early studies of drugs. For the purpose of this document, the agency will refer to the "General Considerations for the Clinical Evaluation of Drugs" as the "1977 guideline."

Although the new guideline outlines in some detail the specific considerations related to the evaluation of gender differences during evaluation of drug products, the agency views the principles of inclusion of women in product development programs and analysis of subgroup differences as being broader standards which apply equally to the clinical development of biological products and medical devices.

The new guideline reflects good drug development practice implicit in the law and regulations. Certain requirements, such as inclusion of adequate numbers of women and by-gender analyses, have been emphasized in the past. However, as with any new guideline, where sponsors have developed drugs in good faith relying on existing guidelines, they will have an opportunity to satisfy newly appreciated data needs after approval where this is compatible with the public health and the law. This new guideline does not change FDA's commitment to safe development of drugs but gives more flexibility to institutional review boards (IRB's), investigators, and patients in determining how best to ensure safety.

II. Background

A. Participation of Women in Clinical Studies

Over the past decade there has been growing concern that the drug development process does not produce adequate information about the effects of drugs in women. This concern arises from a number of sources.

Analyses of published clinical trials in certain therapeutic areas (notably cardiovascular disease) have indicated that there had been little or no participation of women in many of the studies. Certain major studies of the role of aspirin in cardiovascular and cerebrovascular disease, for example, did not include women, and this omission left the scientific community with doubts about whether aspirin was, in fact, effective in women for these indications. Similarly, published studies of anti-anginal drugs often had few or no women in them. It has been suggested that a similar situation might exist for the studies intended to support marketing approval of new drugs.

In addition, FDA notes that there has been little study of the effects of such aspects of female physiology as the menstrual cycle and menopause. or of the effects of drugs widely used in women such as oral contraceptives and systemic progestins and estrogens, on drug action and pharmacokinetics.

Concern has also been expressed that the 1977 policy excluding women of childbearing potential from early drug studies may have led to a more general lack of participation of women in drug development studies, and thus to a paucity of information about the effects of drugs in women. In addition to concerns about whether the policy interfered with development of adequate data on drug therapy in women, the 1977 guideline, seen from the viewpoint of the 1990's, has appeared rigid and paternalistic, leaving virtually no room for the exercise of judgment by responsible female research subjects, physician investigators, and IRB's.

Concerns about the adequacy of data on the effects of drugs in women have arisen at a time when FDA, drug developers, and the scientific community have focused increasingly on the need to individualize treatment in the face of the wide variety of demographic, disease-related, and individual patient-related factors that can lead to different responses to drugs in subsets of the population. Optimal use of drugs requires identification of these factors so that appropriate adjustments in dose, concomitant therapy, or monitoring can be made.

Subgroup-specific differences in response can arise because of variation in a drug's pharmacokinetics (i.e., the drug's concentration in plasma or elsewhere as a function of time) or pharmacodynamics (the body's response to a given concentration of the drug).

B. Pharmacokinetic and Pharmacodynamic Differences Among Patients

Important variations in pharmacokinetics can arise from many factors:

1. A number of demographic characteristics may affect pharmacokinetics: Older people are more likely to have decreased renal function, which may cause drugs excreted by the kidney to accumulate; younger people metabolize theophylline more rapidly; ethnic groups differ in the prevalence of metabolic abnormalities such as slow acetylation and G6PD deficiency; women metabolize certain substances at rates different from men (for example. they metabolize alcohol and ondansetron more slowly).

2. Diseases other than the one being studied may alter the pharmacokinetics of many drugs: Kidney disease may decrease the ability to excrete drugs in the urine; liver disease can interfere with the metabolism of drugs or with their excretion into the bile.

3. The presence of other drugs may lead to pharmacokinetic interactions: Quinidine and fluoxetine inhibit the metabolism of imipramine and desipramine, as well as that of many other drugs metabolized by cytochrome P450 2D6 (debrisoquin hydroxylase); ketoconazole and erythromycin inhibit the metabolism of terfenadine. In such cases, toxic blood concentrations of the drug whose metabolism is inhibited can occur even while a constant dose of the drug is maintained.

4. In addition, other differences between individual subjects may affect pharmacokinetics. For example, small body size or muscle mass may lead to higher blood concentrations after a given dose.

Documented subgroup pharmacodynamic differences are fewer, but have been observed, including increased sensitivity to beta-blockers in Asians, decreased sensitivity to betablockers in the elderly, decreased responsiveness to the blood pressurelowering effects of adrenocortical extract (ACE) inhibitors and betablockers in African-Americans, and increased sensitivity to the central nervous system effects of midazolam in older people.

Despite the many examples of documented pharmacokinetic and pharmacodynamic differences in population subsets, there has often been insufficient attention in the course of drug development to looking for such differencas among individuals in responses to drugs, including differences related to gender. In the case of gender, some have suggested the lack of information may have resulted from the exclusion of women from clinical trials. A number of studies have evaluated this possibility.

In 1983 and 1989, FDA examined the relative numbers of individuals from two important demographic groups, women and the elderly, in the data bases of new drug applications (NDA's). FDA found, in general, that the proportions of women and men included in the clinical trials were similar to the respective proportions of women and men who had the diseases for which the drugs were being studied, taking into account the age range of the population studied. The General Accounting Office (GAO) conducted a larger study of drugs approved during the period 1988 through 1991, with generally similar findings. Thus, women typically represent a majority of patients in NDA data bases of drugs used to treat conditions more common (or more commonly treated) in women (e.g., arthritis and depression) and a minority, although usually a sizable one of about 30 percent or more, in conditions that occur predominantly in males in the age ranges usually included in clinical trials (e.g., angina pectoris). Appendix I of the guideline includes additional details of these surveys.

Although women have been included in the later phases of clinical trials, inclusion alone is not sufficient for adequate assessment of potential gender differences. There must be an effort to use the data to discover such differences. An FDA guideline issued in 1988 ("Guideline for the Format and Content of the Clinical and Statistical Sections of New Drug Applications") called for analyses of gender-related differences in response. FDA and GAO examined NDA's to see whether analyses of this kind were being conducted and submitted. Both examinations found that in many cases (about half) the data bases were not being analyzed to determine whether there were gender, age, or race differences in response to drugs.

A further reason for the lack of information about potential gender differences in drug response is the lack of specific studies of pharmacokinetics in women, even where gender-related differences in pharmacokinetics might be expected or important. There are a variety of potential differences of this type, including differences due to menopause or the menstrual cycle, or to concomitant oral contraceptive or estrogen use, as well as differences based on different body fat proportion, and differences in weight or muscle mass.

C. FDA Guidance on Individualization of Treatment

Since 1988, FDA has taken several major steps to encourage development of data that support informed individualization of treatment:

1. The agency's 1988 guideline entitled, "Guideline for the Format and Content of the Clinical and Statistical Sections of New Drug Applications," calls for analyses of NDA data to identify variations among population subsets in favorable responses (effectiveness) and unfavorable responses (adverse reactions) to drugs. The population subsets that should be evaluated routinely include demographic subsets, such as different genders, age groups and races, people receiving other drug therapy, and people with concomitant illness.

2. The agency has addressed specifically the need to develop information on a particular demographic subset, the elderly, in the 1989 guideline entitled. "Guideline for the Study of Drugs Likely to be Used in the Elderly."

3. In the Federal Register of November 1, 1990 (55 FR 46134). The agency proposed to amend the labeling regulation (21 CFR 201.57) to require a "Geriatric Use" section that would contain available information on experience with the drug in the elderly and describe any needed modifications in the use of the drug in that population. In the Federal Register of October 16, 1992 (57 FR 47423). the agency proposed to amend the same regulation to facilitate inclusion of information on the use of drugs in children.

D. Changes in the Guideline

The new guideline discusses FDA's expectations regarding inclusion of patients of both genders in drug development. analyses of clinical data by gender. assessment of potential pharmacokinetic differences between genders, and, where appropriate, assessment of pharmacodynamic differences and the conduct of specific additional studies in women. The policy applies to all drug or disease specific clinical guidelines based on the 1977 guideline. that exclude women of childbearing potential from participation in early studies of drugs.

III. Revised Policy on Inclusion of Women of Childbearing Potential in Clinical Trials

A. The 1977 Guideline — "General Considerations for the Clinical Evaluation of Drugs"

The 1977 guideline set forth a policy on. among other things. the inclusion of women of childbearing potential in clinical trials. The policy stated that. In general, women of childbearing potential should be excluded from the earliest studies of a new drug. that is, phase 1 and early phase 2 studies. Phase 1 refers to the first introduction of a new drug into humans. who are often. But not always. healthy volunteers. to study the basic tolerability of the drug. Its metabolism. and its short-term pharmacokinetics. With the exception of some early studies in life-threatening diseases. phase 1 studies usually do not have therapeutic intent. Phase 2 refers to the initial controlled trials of a drug to study its effectiveness. Before the first such study. there is generally no evidence that the drug is of therapeutic value in humans.

If adequate information on effectiveness and relative safety were amassed during phase 1 and early phase 2. the guideline stated that women of childbearing potential could be included in subsequent studies of effectiveness, that is, later phase 2 and phase 3 studies, so long as animal teratogenicity and the female part of animal fertility studies had been completed. The policy did not specifically address the manner in which the early human evidence of safety and effectiveness and the results of animal reproduction studies should be used to make decisions about participation of women in later trials, leaving these considerations to the usual risk-benefit assessment made by the patient, physician, and IRB, with subsequent FDA review.

In tlie 1977 guideline, the term "women of childbearing potential" was defined very strictly, essentially referring to all premenopausal women physiologically capable of becoming pregnant, including women on oral, injectable, or mechanical contraceptives, single women, celibate women, and women whose partners had been sterilized by vasectomy. There was no provision for the use of pregnancy testing to identify women who could participate in studies without a risk of fetal exposure. The 1977 guideline also noted, however, that women of childbearing potential could receive investigational drugs in the earliest phases of testing, even in the absence of adequate reproduction studies in animals, when the drugs were intended for life-saving or life-prolonging treatment.

The effect of the 1977 guideline has been that women generally have not been included in phase 1 nontherapeutic studies or in the earliest controlled effectiveness studies (i.e., early phase 2), except for studies of lifethreatening illnesses, such as acquired immune deficiency syndrome (AIDS) and cancer.

B. Reasons for Revising the 1977 Policy

The policy set forth in the 1977 guideline has been under discussion for several years within and outside the agency, and there has been increasing sentiment that it should be revised. For example, in October 1992, FDA and the Food and Drug Law Institute cosponsored a meeting on women in clinical trials of FDA-regulated products at which many speakers described the current restrictions as paternalistic and overprotective, denying young women the opportunity available to men and older women to participate in early drug development research.

Although the 1977 guideline has not resulted in a failure to include adequate numbers of women in the later phases of clinical trials, it has restricted the early accumulation of information about response to drugs in women that could be utilized in designing phase 2 and 3 trials, and has perhaps delayed appreciation of gender-related variation in drug effects. The early exclusion also may have perpetuated, in a subtle way, a view of the male as the primary focus of medicine and drug development, with women considered secondarily. There is reason to believe that earlier participation of women in studies would increase the likelihood that gender-specific data might be used to make appropriate adjustments in larger clinical studies (e.g., different doses in women or weight adjusted (milligram per kilogram) dosing instead of fixed doses).

The agency believes that removal of the prohibition on participation of women of childbearing potential in phase 1 and early phase 2 trials is consistent with congressional efforts to prevent unwarranted discrimination against such women. For example, in the employment context, the Pregnancy Discrimination Act, as interpreted by the U.S. Supreme Court in the landmark case of *International Union, United Automobile, Aerospace and Agricultural Implement Workers, UAW v. Johnson Controls, Inc.*, 111 S.Ct. 1196 (1991), prohibits the blanket exclusion of pregnant women from jobs they are qualified to perform solely because the working conditions of those jobs pose potential risks to exposed fetuses. The Court emphasized that "decisions about the welfare of future children must be left to the parents who conceive, bear, support, and raise them, rather than to the employers who hire those parents."

While the purposes of clinical trials to develop safe and effective drugs are manifestly different from the purposes of private employment, FDA takes serious note of the Court's position on a woman's right to participate in decisions about fetal risk and believes it is appropriate to consider the Court's opinion in developing policy on the inclusion of women in clinical trials.

C. Current FDA Position on Participation of Women of Childbearing Potential in Early Clinical Studies

The agency has reconsidered the 1977 guideline and has concluded that it should be revised. This does not reflect a lack of concern for potential fetal exposure or indifference to potential fetal damage, but rather the agency's opinion that (1) exclusion of women from early trials is not medically necessary because the risk of fetal exposure can be minimized by patient behavior and laboratory testing, and (2) initial determinations about whether that risk is adequately addressed are properly left to patients, physicians, local IRB's, and sponsors, with appropriate review and guidance by FDA. as are all other aspects of the safety of proposed investigations.

The agency is, therefore, withdrawing the restriction on the participation of women of childbearing potential in early clinical trials. including clinical pharmacology studies (e.g. dose tolerance, bioavailability, and mechanism of action studies), and early therapeutic studies. It is expected that, in accordance with good medical practice, appropriate precautions against becoming pregnant and exposing a fetus to a potentially dangerous agent during the course of study will be taken by women participating in clinical trials. It is also expected that women will receive adequate counseling about the importance of such precautions, that efforts will be made to be sure that a woman entering a trial is not pregnant at the time the trial begins (i.e., a pregnancy test detecting the beta subunit of the hCG molecule is negative), and that the woman participant is fully informed about the current state of the animal reproduction studies and any other information about the teratogenic potential of the drug. As is the case for all studies carried out under an investigational new drug application (IND), the adequacy of the precautions taken will be considered by FDA in its review of protocols. In situations where enrollment continues over a prolonged period (unlikely for early clinical studies) and significant new information about teratogenicity becomes available. the sponsor has the responsibility to transmit this information quickly to the investigator and to current as well as potential study participants in the informed consent process.

The agency recognizes that this change in FDA's policy will not, by itself, cause drug companies or IRB's to alter restrictions they might impose on the participation of women of childbearing potential. We do not at this time perceive a regulatory basis for requiring routinely that women in general or women of childbearing potential be included in particular trials, such as phase 1 studies, However, as this guideline delineates, careful characterization of drug effects by gender is expected by the agency, and FDA is determined to remove the unnecessary Federal impediment to inclusion of women in the earliest stages of drug development. The agency is confident that the interplay of ethical, social, medical, legal and political forces will allow greater participation of women in the early stages of clinical trials.

In some cases, there may be a basis for requiring participation of women in early studies. When the disease under study is serious and affects women, and especially when a promising drug for the disease is being developed and made available rapidly under FDA's accelerated approval or early access procedures, a case can be made for requiring that women participate in clinical studies at an early stage. When such a drug becomes available under expanded access mechanisms (for example, treatment IND or parallel track) or is marketed rapidly under subpart E procedures (because an effect on survival or irreversible morbidity has been shown in the earliest controlled trials), it is medically important that a representative sample of the entire population likely to receive the drug has been studied, including representatives of both genders. Under these circumstances, clinical protocols should not place unwarranted restrictions on the participation of women.

The agency advises that this guideline represents its current position on the clinical evaluation of drugs in humans. This guideline does not bind the agency, and it does not create or confer any rights, privileges, or benefits for or on any person.

IV. Comments

Interested persons may, on or before November 19, 1993, submit to the Dockets Management Branch (address above) written comments regarding this guideline. Two copies of any comments should be submitted, except that individuals may submit one copy. Comments are to be identified with the docket number found in brackets in the heading of this document. Received comments may be seen in the office above between 9 a.m. and 4 p.m., Monday through Friday. These comments will be considered in determining whether further amendments to, or revisions of, the guideline are warranted.

The new guideline replaces that portion of the 1977 guideline that dealt with women of childbearing potential. The text of the new guideline on gender differences follows:

Guideline for the Study and Evaluation of Gender Differences in the Clinical Evaluation of Drugs

1. Introduction

The Food and Drug Administration (FDA) advises that this guideline represents its current position on the clinical evaluation of drugs in humans. This guideline does not bind the agency, and it does not create or confer any rights, privileges. or benefits for or on any person.

The principles of inclusion of women in product development programs and analysis of subgroup differences outlined in this guideline also apply to the clinical development of biological products and medical devices.

A. Abstract

In general, drugs should be studied prior to approval in subjects representing the full range of patients likely to receive the drug once it is marketed. Although in most cases, drugs behave qualitatively similarly in demographic (age, gender, race) and other (concomitant illness, concomitant drugs) subsets of the population, there are many quantitative differences, for example, in dose-response, maximum size of effect, or in the risk of an adverse

effect. Recognition of these differences can allow safer and more effective use of drugs. Rarely, there may be qualitative differences as well. It is very difficult to evaluate subsets of the overall population as thoroughly as the entire population, but sponsors are expected to include a full range of patients in their studies, carry out appropriate analyses to evaluate potential subset differences in the patients they have studied, study possible pharmacokinetic differences in patient subsets. and carry out targeted studies to look for subset pharmacodynamic differences that are especially probable. are suggested by existing data. or that would be particularly important if present. Study protocols are also expected to provide appropriate precautions against exposure of fetuses to potentially dangerous agents. Where animal data suggest possible effects on fertility, such as decreased sperm production, special studies in humans may be needed to evaluate this potential toxicity.

B. Underlying Observations

The following general observations and conclusions underlie the recommendations set forth in this guideline:

1. Variations in response to drugs, including gender-related differences, can arise from pharmacokinetic differences (that is. differences in the way a drug is absorbed, excreted, metabolized, or distributed) or pharmacodynamic differences (i.e., differences in the pharmacologic or clinical response to a given concentration of the drug in blood or other tissue).

2. Gender-related variations in drug effects may arise from a variety of sources. Some of these are specifically associated with gender, e.g., effects of endogenous and exogenous hormones. Gender-related differences could also arise. however. not because of gender itself. but because the frequency of a particular characteristic (for example, small size, concomitant hepatic disease or concomitant drug treatment, or habits such as smoking or alcohol use) is different in one gender, even if the characteristic could occur in either gender. Proper management of patients of both genders thus requires that physicians know all the factors that can influence the pharmacokinetics of a drug. An approach is needed that will identify, better than is done at present, all such factors. Understanding how various factors may influence pharmacokinetics will greatly enhance our ability to treat people of both genders appropriately.

3. For a number of practical and theoretical reasons. the evaluation of possible gender-related differences in response should focus initially on the evaluation of potential pharmacokinetic differences. Such differences are known to occur and have, at least to date, been documented much more commonly than documented pharmacodynamic differences. Moreover, pharmacokinetic differences are relatively easy to discover. Once reliable assays are developed for a drug and its metabolites (such assays are now almost always available early in the development of the drug). techniques exist for readily assessing aender-related or other subgroup-related pharmacokinetic differences.

Formal pharmacokinetic studies are one means of answering questions about specific subgroups. Another approach is use of a screening procedure, a "pharmacokinetic screen" (see "Guideline for the Study of Drugs Likely To Be Used in the Elderly"). Carried out in phase 2 and 3 study populations, the pharmacokinetic screen can greatly increase the ability to detect pharmacokinetic differences in subpopulations and individuals, even when these differences are not anticipated. By obtaining a small number of blood concentration determinations in most or all phase 2 and 3 patients, it is possible to detect markedly atypical pharmacokinetic behavior in individuals, such as that seen in slow metabolizers of debrisoquin. and pharmacokinetic differences in population subsets, such as patient populations of different gender, age, or race, or patients with particular underlying diseases or concomitant therapy. The screen may also detect interactions of two factors, e.g. gender and age. The relative ease with which pharmacokinetic differences among population subsets can be assessed contrasts with the difficulty of developing precise relationships of most clinical responses to drug dose or to the drug concentration in blood, which usually would be necessary when attempting to observe pharmacodynamic differences between two subgroups.

A final reason to emphasize pharmacokinetic evaluation is that it must be carried out to allow relevant assessment of pharmacodynamic differences or relationships. Assessing pharmacodynamic differences between groups or establishing blood concentration-response relationships is possible only when groups are reasonably well matched for blood concentrations. Enough pharmacokinetic data must therefore be available to permit the investigator to administer doses that will produce comparable blood concentrations in the subsets to be compared or, alternatively, to compare subsets that have been titrated to similar blood concentrations.

4. The number of documented gender-related pharmacodynamic differences of clinical consequence is at this time small, and conducting formal pharmacodynamic/effectiveness studies to detect them may be difficult, depending on the clinical endpoint. Such studies are therefore not routinely necessary. The by-gender analyses of clinical trials that include both men and women, however, which are specified in the 1988 guideline entitled "Guideline for the Format and Content of the Clinical and Statistical Sections of New Drug Applications" are not difficult to carry out. Particularly if these analyses are accompanied by blood concentration data for each patient, they can detect important pharmacodynamic/effectiveness differences related to gender.

C. Inclusion of Both Genders in Clinical Studies

The patients included in clinical studies should. in general, reflect the population that will receive the drug when it is marketed. For most drugs, therefore. representatives of both genders should be included in clinical trials in numbers adequate to allow detection of clinically significant gender-related differences in drug response. Although it may be reasonable to exclude certain patients at early stages because of characteristics that might make evaluation of therapy more difficult (e.g., patients on concomitant therapy) such exclusions should usually be abandoned as soon as possible in later development so that possible drug-drug and drug-disease interactions can be detected. Thus, for example, there is ordinarily no good reason to exclude women using oral contraceptives or estrogen replacement from trials. Rather, they should be included and differences in responses between them and patients not on such therapy examined. Pharmacokinetic interaction studies (or screening approaches) to look at the interactions resulting from concomitant treatment are also useful.

Ordinarily, patients of both genders should be included in the same trials. This permits direct comparisons of genders within the studies. In some cases, however, it may be appropriate to conduct studies in a single gender, e.g., to evaluate the effects of phases of the menstrual cycle on drug response.

Although clinical or pharmacokinetic data collected during phase 3 may provide evidence of gender-related differences, these data may become available too late to affect the design and dose-selection of the pivotal controlled trials. Inclusion of women in the earliest phases of clinical development, particularly in early pharmacokinetic studies, is, therefore, encouraged so that information on gender differences may be used to refine the design of later trials. Note that the strict limitation on the participation of women of childbearing potential in phase 1 and early phase 2 trials that was imposed by the 1977 guideline entitled, "General Considerations for the Clinical Evaluation of Drugs," has been eliminated.

There is no regulatory or scientific basis for routine exclusion of women from bioequivalence trials. For certain drugs, however, it is possible that changes during the menstrual cycle may lead to increases in intra-subject variability. Such variability could be related to hormonally-mediated differences in metabolism or changes in fluid balance. Sponsors of bioequivalence trials are encouraged to examine available information on the pharmacokinetics and metabolism of the test drugs and related drugs to determine whether there is a basis for concern about variability in pharmacokinetics during the menstrual cycle. Where the available information does raise such concern, measures could be taken to reduce or adjust for variability, e.g., administration of each drug at the same phase of the menstrual cycle, or inclusion of larger numbers of subjects. Sponsors are encouraged to collect data that will contribute to the understanding of the relationship between hormonal variations and pharmacokinetics.

D. Analysis of Effectiveness and Adverse Effects by Gender

FDA's guideline on the clinical and statistical sections of NDA's calls for analyses of effectiveness, adverse effects. dose-response, and, if available, blood concentration-response, to look for the influence of: (1) Demographic features, such as age. gender. and race; and (2) other patient characteristics, such as body size (body weight, lean body mass, fat mass), renal. cardiac, and hepatic status, the presence of concomitant illness, and concomitant use of drugs. including ethanol and nicotine. Analyses to detect the influence of gender should be carried out both for individual studies and in the overall integrated analyses of effectiveness and safety. Such analyses of subsets with particular characteristics can be expected to detect only relatively large gender-related differences, but in general, small differences are not likely to be clinically important. The results of these analyses may suggest the need for more formal dose-response or blood concentration-response studies in men or women or in other patient subsets. Depending on the magnitude of the findings. or their potential importance (e.g., they would be more important for drugs with low therapeutic indices), these additional studies might be carried out before or after marketing.

E. Defining the Pharmacokinetics of the Drug in Both Genders

The factors most commonly having a major influence on pharmacokinetics are renal function. for drugs excreted by the kidney, and hepatic function, for drugs that are metabolized or excreted by the liver; these should be assessed directly as part of the ordinary development of drugs. The pharmacokinetic effects of other subgroup characteristics such as gender can be assessed either by a pharmacokinetic screening approach, described in the 1989 guideline entitled, "Guideline for the Study of Drugs Likely to Be Used in the Elderly," or by formal pharmacokinetic studies in specific gender or age groups.

Using either a specific pharmacokinetic study or a pharmacokinetic screen, the pharmacokinetics of a drug should be defined for both genders. In general, it is prudent to at least carry out pilot studies to look for major pharmacokinetic differences before conducting definiti ve controlled trials, so that differences that might lead to the need for different dosing regimens can be detected. Such studies are particularly important for drugs with low therapeutic indices, where the smaller average size of women alone might be sufficient to require modified dosing, and for drugs with nonlinear kinetics, where the somewhat higher milligram per kilogram dose caused by a woman's smaller size could lead to much larger differences in blood concentrations of drug. Gender may interact with other factors, such as age. The potential for such interactions should be explored.

Three pharmacokinetic issues related specifically to women that should be considered during drug development are: (1) The influence of menstrual status on the drug's pharmacokinetics, including both comparisons of premenopausal and postmenopausal patients and examination of within-cycle changes; (2) the influence of concomitant supplementary estrogen treatment or systemic contraceptives (oral contraceptives, long-acting progesterone) on the drug's pharmacokinetics; and (3) the influence of the drug on the pharmacokinetics of oral contraceptives. Which of these influences should be studied in a given case would depend on the drug's excretion, metabolism, and other pharmacokinetic properties, and on the steepness of the dose-response curve.

Hormonal status during the menstrual cycle may affect plasma volume and the volume of distribution (and thus clearance) of drugs. The activity of certain cytochrome P450 enzymes may be influenced by estrogen levels and, in addition, microsomal oxidation by these enzymes may decline in the elderly more in men than women. Oral contraceptives can cause decreased clearance of drugs (e.g., imipramine, diazepam, chlordiazepoxide, phenytoin, caffeine, and cyclosporine), apparently by inhibiting hepatic metabolism. They can also increase clearance by inducing drug metabolism (e.g., of acetaminophen, salicylic acid, morphine, lorazepam, temazepam, oxazepam. and clofibrate). Certain anticonvulsants (carbamazepine, phenytoin) and antibiotics (rifampin) can reduce the effectiveness of oral contraceptives. Many of the potential interactions of gender and genderrelated characteristics (e.g., use of oral contraceptives) can be evaluated with the pharmacokinetic screen. In some cases. specific studies will be needed.

F. Gender-Specific Pharmacodynamic Studies

Because documented demographic differences in pharmacodynamics appear to be relatively uncommon. it is not necessary to carry out separate pharmacodynamic/effectiveness studies in each gender routinely. Evidence of such differences should be sought, however. in the data from clinical trials by carrying out the by-gender analyses suggested in the guideline on the clinical and statistical sections of NDA's. These analyses of controlled trials involving both genders are probably more likely to detect differences than studies carried out entirely in one gender. Experience has shown that gender differences can be detected with such approaches.

If the by-gender analyses suggest gender-related differences, or if such differences would be particularly important, e.g., because of a low therapeutic index, additional formal studies to seek such differences between the blood level-response curves of men and women should be conducted. Even in the absence of a particular concern based on the by-gender analyses, if there is a readily measured pharmacodynamic endpoint, such as blood pressure or rate of ventricular premature beats, and if there are good dose-response data for the overall population. it should be feasible to develop dose response data from population subsets (e.g., both genders) in the critical clinical trials.

G. Precautions in Clinical Trials Including Women of Childbearing Potential

Appropriate precautions should be taken in clinical studies to guard against inadvertent exposure of fetuses to potentially toxic agents and to inform subjects and patients of potential risk and the need for precautions. In all cases, the informed consent document and investigator's brochure should include all available information regarding the potential risk of fetal toxicity. If animal reproductive toxicity studies are complete, the results should be presented, with some explanation of their significance in humans. If these studies have not been completed, other pertinent information should be provided. such as a general assessment of fetal toxicity in drugs with related structures or pharmacologic effects. If no relevant information is available, the informed consent should explicitly note the potential for fetal risk.

In general, it is expected that reproductive toxicity studies will be completed before there is large-scale exposure of women of childbearing potential, i.e., usually by the end of phase 2 and before any expanded access program is implemented.

Except in the case of trials in tended for the study of drug effects during pregnancy. clinical protocols should also include measures that will minimize the possibility of fetal exposure to the investigational drug. These would ordinarily include providing for the use of a reliable method of contraception (or abstinence) for the duration of drug exposur'e (which may exceed the length of the study), use of pregnancy testing (beta HCG) to detect unsuspected pregnancy prior to initiation of study treatment, and timing of studies (easier with studies of short duration) to coincide with, or immediately follow, menstruation. Female subjects should be referred to a study physician or other counselor knowledgeable in the selection and use of contraceptive approaches.

H. Potential Effects on Fertility

Where abnormalities of repro uctive organs or their function (spermatogenesis or ovulation) have been observed in experimental animals, the decision to include patients of reproductive age in a clinical study should be based on a careful risk-benefit evaluation, taking into account the nature of the abnormalities, the dosage needed to induce them, the consistency of findings in different species, the severity of the illness being treated, the potential importance of the drug, the availability of alternative treatment, and the duration of therapy. Where patients of reproductive potential are included in studies of drugs showing reproductive toxicity in animals, the clinical studies should include appropriate monitoring and/or laboratory studies to allow detection of these effects. Long-term followup will usually be needed to evaluate the effects of such drugs in humans.

Appendix I

1. Surveys of Participation of Women in Clinical Trials in New Drug Applications (NDA 's)

The extent of participation of women in the data bases of NDA's has been examined several times in recent years, by FDA in 1983 and 1989, and by the General Accounting Office (GAO) in 1992. In general, the genders were represented tc approximately the extent one would predict from the gender prevalence of the condition treated by the drug in the age group studied. The relative disease prevalence in men and women can vary with age. Consider, for example, the participation of women in studies of anti-anginal drugs. Almost all patients in angina studies, which require vigorous treadmill exercise tests, are under 75 years old and the large majority are under 65. Although eventually women develop symptomatic coronary artery disease in their 60's, 70's, and 80's, and become similar to men in the prevalence of this condition, they are much less likely than men to be affected in their 40's, 50's, and early 60's. The overall NDA data base for an anti-anginal drug, made up primarily of people 50 to 65, will therefore include a significantly greater proportion of men than women. Efforts to include more very old patients in trials, i.e., patients in their 70's and 80's, should lead to a greater proportion of women in trials of anti-anginal drugs.

Results of the FDA and GAO surveys are described below. Also included is an analysis of gender distribution in recently approved or submitted NDA's for antidepressant drugs. This analysis was conducted to evaluate the frequently heard claim that this class of drugs is studied predominantly (or even exclusively) in males despite the wide use of antidepressants in women.

A. The 1983 Survey

Primarily carried out to assess the inclusion of the elderly in NDA's, the 1983 survey looked at the age and gender prevalence of patients included in 11 pending NDA's. The NDA's were chosen because they were readily available and did not need to be retrieved from storage; figures ware taken by FDA staff from the pending applications. In one case (ranitidine), the values represent only domestic patients for only one claim, leading to a small number of patients; many more patients (those included in foreign studies, or in siddies of other claims) were available for safety evaluation.

Table 1 shows the results of the survey. As expected, the non-steroidal anti-inflammatory drugs (NSAID's) were studied predominantly in women, because arthritis, especially rheumatoid arthritis, is more common in women. This predominance was slightly less prominent in the case of zomepirac, which was studied extensively for pain (gender-neutral), in addition to arthritis. The hypnotic drug (triazolam) and the antibiotics (cefoperazone andnetilmycin) were studied in approximately equal proportions of men and women. The patient populations included in the NDA's for verapamil, for angina, and bumetanide, for heart failure, were about two-thirds male, and about two-thirds of the patients were less than 60 years old, an age group in which angina and heart failure are more prevalent in men than in women. In the patients over age 70, representing 10 percent of the bumetanide patients and 7 percent of verapamil patients, the gender distribution was about equal (49 percent women in the verapamil studies and 45 percent women in the bumetanide studies). Studies of ranitidine for duodenal ulcer, a predominantly male disease, included about 75 percent males. Other indications for this drug, such as gastric ulcer, would be expected to have a different gender distribution. The two anti-cancer drugs in this survey were studied principally for exclusively male conditions, cancer of the prostate and testis.

B. The 1989 Survey

In an effort to avoid possible selection bias, all drugs approved in 1988 were surveyed; this time the sponsors provided the data. FDA asked them to provide data reflecting "the principal data base used for safety review" in the latest safety update and asked that phase 1 subjects/patients be excluded. Sponsors gave either data on all patients or only patients given the test drug; the estimates of gender exposure should not be greatly affected by this difference.

Table 2 shows the results of the 1989 survey for 12 of the 20 drugs approved in 1988. Because sponsors had little control over gender distributions in the small populations available for study, four orphan drugs were omitted from the survey (tiopronin for prevention of cystine stones; ethanolamine oleate for esophageal varices; ifosfamide, thirdline therapy for testicular cancer; and mesna, a prophylactic agent for ifosfamide-induced hemorrhagic cystitis). Also omitted were three contrast agents for single dose uses (but these agents are in the 1992 GAO survey), and a topical product (oxiconazole cream) for which gender distribution was not available.

Again, the anti-inflammatory drug (diclofenac) was studied predominantly in women (more than two-thirds of the patients), as was nimodipine, for prevention of vascular spasm after subarachnoid hemorrhage, also a female-predominant condition. Pergolide, an anti-Parkinson's disease drug; astemizole, an antihistamine; and octreotide, a drug for symptoms of carcinoid tumor, were studied in about equal numbers of men and women. The studies of the cardiovascular drugs nicardipine (angina and hypertension) and carteolol (hypertension) included 59 and 67 percent men, respectively, reflecting the male gender predominance of angina, and perhaps hypertension. in the relatively young (two-thirds of the patients were under the age of 60) populations studied. Nizatidine and misoprostol were studied extensively in duodenal ulcer, a predominantly male disease, with about 70 percent of patients being male, although approval of misoprostol was for a different claim. Cefotiam, an intravenous antibiotic, was studied mainly in elderly patients (65 percent over 60; 36 percent over 70);

about two-thirds were male, for unclear reasons. The topicals were studied in a predominantly young population (about 90 percent under the age of 60), more often in males. Certain tinea infections (tinea cruris and tinea pedis) are more common in males, accounting for the high proportion (72 percent) of males in studies of naftifine. Why photoplex was studied somewhat more in males (63 percent) is not clear.

C. The GAO Survey

In 1992, the GAO analyzed the gender, age, and race distribution of all NDA's approved from January 1988 through June 1991. Data were collected by means of a questionnaire sent to the sponsor of each drug. The number of patients receiving the test drug during drug development, domestic studies only, was requested, and patients were broken down by gender, age (<15, 15 to 49, 50 to 64, >65), and race. The age distribution data allow a separate analysis of women of childbearing potential (taken here as women age 15 to 49). Data are available for 53 drugs (of 63 drugs approved during the 3 1/2-year period, 4 drugs intended for single gender use and 6 whose sponsors provided no, or no usable, questionnaire were omitted).

The results of the GAO survey are given in Tables 3A and 3B for phase 2 and 3 patients. The tables show gender distribution overall for the whole data base and for the 15 to 49 age group as well. For anti-inflammatory, antiinfective, central nervous system/ anesthetic, topical, antihistamine, and cancer drugs, women constituted 40 percent or more of the patients studied, with occasional exceptions. The most striking exception is mefloquine, where only 11 percent of patients were women. This occurred because the primary studies of mefloquine for treatment of malaria were conducted in Thai military personnel. Women fairly consistently represented less than 40 percent of the patients for anti-ulcer drugs (duodenal ulcer, a malepredominant condition, was a principal disease studied for nizatidine, omeprazole, and misoprostol) but accounted for 55 percent of the patients in studies of dipentum, a drug for ulcerative colitis (ulcerative colitis is more common in women). Women consistently made up less than 40 percent of the populations studied for cardiovascular disease, including D. Antidepressants populations used to evaluate agents used to diagnose or evaluate coronary artery disease, except for nimodipine (for spasm after subarachnoid bleed) and adenosine (for supraventricular tachycardia). For drugs to treat ventricular arrhythmias and angina, both commonly the result of coronary disease, the fraction of women ranged from 15 percent (bepridil, for unresponsive angina) to 20 to 30 percent (propafenone, moricizine, and indecainide), reflecting the lower rate of coronary artery disease in younger women and the fact that most patients in studies are under 60 years old.

Studies of drugs for hypertension (carteolol, doxazosin, nicardipine, isradipine, ramapril, pinacidil) included 27 to 42 percent women. In some cases, these drugs were being evaluated for other claims, such as angina or heart failure, which are male predominant in the age groups studied. For all of the antihypertensives, there were at least 290 women in the domestic data base, enough to detect significant gender differences in response.

D. Antidepressants

By chance, none of the surveys included any antidepressant drugs, a class of drug frequently cited as needing study in women, both because women are frequently given antidepressants and because of suspected interactions of the drugs with the menstrual cycle.

Of interest is the observation that there was no tendency for women to represent a lower percentage of patients in the 15 to 49 age group than in the overall population. There is thus no suggestion in these data that the restriction on participation of women of childbearing potential in early trials carries over to later phase 2 or 3 trials.

Table 4 shows gender participation for sertraline and paroxetine, the two most recently approved antidepressants, as well as two agents likely to be approved within the next year. Women, as expected based on past experience, represented 58 to 65 percent of the patients.

II. Tables

TABLE 1

Drug	n	Percent of total	
		Female	Male
Anti-inflammatory:			
Benoxaprofen (Oraflex)	3,446	64	36
Ketoprofen (Orudis)	1,579	68	32
Zomepirac (Zomax)	3,479	60	40
Cardiovascular:			
Verapamil (Isoptin)	1,810	36	64
Bumetanide (Bumex)	838	27	72
Hypnotic:			
Triazolam (Halcion)	4,254	49	51
Antibiotic:			
Cefoperazone (Cefobid)	1,958	52	48
Netilmycin (Netromycin)	3,376	43	57
Anti-ulcer:			
Ranitidine (Zantac)	193	23	77
Anti-cancer (prostate, testes):			
Leuprolide (Lupron)	387	17	83
Etoposide (Vepesid)	259	16	84

TABLE 2

Drug	n	Percent of total	
		Female	Male
Anti-inflammatory:			
Diclofenac (Voltaren)	8,175	69	31
Cardiovascular/cerebrovascular:			
Nicardipine (Cardene)	2,962	41	59
Carteolol (Cartrol)	1,536	33	67
Nimodipine (Nimotop)	1,301	64	36
Anti-ulcer:			
Nizatidine (Axid)	2,063	31	69
Misoprostol (Cytotec)	8,687	28	72
Antibiotic:			
Cefotiam (Ceradon)	844	33	67
Anti-Parkinson:			
Pergolide (Permax)	1,836	45	55
Antihistamine:			
Astemizole (Hismanal)	1,356	48	52
Anti-carcinoid symptoms:			
Octreotide (Sandostatin)	455	49	51
Topical (tinea, sunscreen):			
Naftifine (Naftin)	452	28	72
Photoplex	227	37	63

TABLE 3A.— ALL AGES

Drug	n	Percent of total	
		Female	Male
Anti-inflammatory/Analgesic:			
Dezocine (Dalgan)	1,417	60	40
Diclofenac (Voltaren)	1,714	64	36
Etodolac (Lodine)	5,395	65	35
Ketorolac (Toradol)	1,248	64	36
Anti-infectives:			
Ofloxacin (Floxin)	3,585	56	44
Cefmetazole (Zefazone)	2,769	67	33
Cefixime (Suprox)	1,859	60	40
Fluconazole (Diflucan)	983	36	64
Naftifine (Naftin)	222	38	62
Cefpiramide	1,325	39	61
Mefloquine (Lariam)	1,319	11	89
Oxiconazole (Oxistat)	886	35	65
Central Nervous System/Anesthetic:			
Clomipramine (Anaframil)	3,826	54	46
Propofol (Dipravan)	696	48	52
Clozapine (Clozaril)	581	37	63
Estazolam (Prosan)	1,243	50	50
Pipecuronium (Arduan)	580	52	48
Doxacurium (Nuromax)	987	39	61
Pergolide (Permax)	1,667	43	57
Cardiovascular:			
Nimodipine (Nimotop)	343	69	31
Adenosine (Adenocard)	109	48	52
Doxazosin (Cardura)	698	42	58
Pinacidil (Pindac)	1,774	36	64
Nicardipine (Cardene)	1,915	37	63
Benazepril (Lotensin)	2,130	32	68
Isradipine (Dynacirc)	1,842	27	73
Propafenone (Rhythmol)	3,328	30	70
Ramapril (Altace)	1,723	33	67
Carteolol (Cartrol)	1,253	28	72
Moricizine (Ethmozine)	1,017	21	79
Indecainide (Decabid)	761	23	77
Bepridil (Vascor)	884	15	85
Cancer:			
Octreotide (Sandostatin)	569	38	62
Carboplatin (Paraplatin)	2,214	77	23
Levamisole (Ergamisol)	1,038	48	52
Ondansetron (Zofran)	939	29	71
Diagnostics:			
Technescan Mag 3	160	43	57
Ioversol (Optiray)	1,101	45	55
Gadopentetate (Magnevist)	410	41	59
TC-99M Sestamibi (Cardolyte)	1,102	29	71
TC-99M Exametazime (Ceretec)	202	28	72
Iotralan (Osmovist)	545	31	69
Topicals:			
Photoplex	371	40	60
Fluticasone (Cutivate)	730	42	58
Halobetasol (Ultravate)	662	46	54
Metipranolol (Optipranolol)	465	53	47
Cefotiam (Ceradon)	715	34	66
Rev-Eyes	646	47	53
Gastrointestina 1:			
Olsalazine (Dipentum)	98	55	45
Nizatidine (Axid)	3,854	35	65
Misoprostol (Cytotec)	1,917	37	63
Omeprazole (Losec)	2,189	26	74
Antihistamine:			
Astemizole (Hismanal)	979	41	59

TABLE 3B.—AGES 15 TO 49

Drug	n	Percent of total	
		Female	Male
Anti-inflammatory/Analgesic:			
Dezocine (Dalgan)	1,142	61	39
Diclofenac (Voltaren)	577	55	45
Etodolac (Lodine)	3,155	65	35
Ketorolac (Toradol)	NA	NA	NA
Anti-infectives:			
Ofloxacin (Floxin)	2,890	60	40
Cefmetazole (Zefazone)	1,621	72	28
Cefixime (Suprox)	879	70	30
Fluconazole (Diflucan)	759	64	36
Naftifine (Naftin)	151	36	64
Cefpiramide	362	44	56
Mefloquine (Lariam)	1,189	9	91
Oxiconazole (Oxistat)	NA	NA	NA
Central Nervous System/Anesthetic:			
Clomipramine (Anafranil)	3,277	55	45
Propofol (Dipravan)	514	58	42
Clozapine (Clozaril)	510	35	65
Estazolam (Prosan)	784	42	58
Pipecuronium (Arduan)	263	57	43
Doxacurium (Nuromax)	623	37	63
Pergolide (Permax)	357	63	37
Cardiovascular:			
Nimodipine (Nimotop)	195	63	37
Adenosine (Adenocard)	62	43	57
Doxazosin (Cardura)	62	43	57
Pinacidil (Pindac)	682	37	63
Nicardipine (Cardene)	596	39	61
Benazepril (Lotensin)	602	27	73
Isradipine (Dynacirc)	692	27	73
Propafenone (Rhythmol)	604	46	54
Ramapril (Altace)	622	23	77
Carteolol (Cartrol)	410	24	76
Moricizine (Ethmozine)	193	31	69
Indecainide (Decabid)	94	44	56
Bepridil (Vascor)	93	13	8
Cancer:			
Octreotide (Sandostatin)	391	34	66
Carboplatin (Paraplatin)	563	70	30
Levamisole (Ergamisol)	195	50	50
Ondansetron (Zofran)	288	19	81
Diagnostics:			
Technescan Mag 3	101	47	53
Ioversol (Optiray)	370	51	49
Gadopentetate (Magnevist)	183	29	71
TC–99M Sestamibi (Cardolyte)	402	34	66
TC–99M Exametazime (Ceretec)	26	50	50
Iotralan (Osmovist)	327	34	66
Topicals:			
Photoplex	296	34	66
Fluticasone (Cutivate)	405	45	55
Halobetasol (Ultravate)	360	45	55
Metipranolol (Optipranolol)	70	41	59
Cefotiam (Ceradon)	NA	NA	NA
Rev-Eyes	531	47	53
Gastrointestinal:			
Olsalazine (Dipentum)	72	60	40
Nizatidine (Axid)	2,302	32	68
Misoprostol (Cytotec)	945	33	67
Omeprazole (Losec)	NA	NA	NA
Antihistamine:			
Astemizole (Hismanal)	NA	NA	NA

TABLE 4.—ALL AGES

Drug	Date	n	Percent of total	
			Female	Male
Sertaline (Zoloft)	1991	2,979	58	42

TABLE 4.—ALL AGES—Continued

Drug	Date	n	Percent of total	
			Female	Male
Paroxetine (Paxil)	1992	4,126	65	35
Pending No. 1	NA	2,181	62	38
Pending No. 2	NA	2,256	62	38

Good Pharmacovigilance Practices and Pharmacoepidemiologic Assessment

Good Pharmacovigilance Practices and Pharmacoepidemiologic Assessment, Guidance for Industry [†, 1]

U.S. Department of Health and Human Services
Food and Drug Administration
Center for Drug Evaluation and Research (CDER)
Center for Biologics Evaluation and Research (CBER)

March 2005

Clinical Medical

Contains Nonbinding Recommendations

This guidance is intended to assist industry with risk management activities for drug products, including biological drug products, in the Center for Drug Evaluation and Research (CDER) and the Center for Biologics Evaluation and Research (CBER).

Note: *This guidance represents the Food and Drug Administration's (FDA's) current thinking on this topic. It does not create or confer any rights for or on any person and does not operate to bind FDA or the public. You can use an alternative approach if the approach satisfies the requirements of the applicable statutes and regulations. If you want to discuss an alternative approach, contact the appropriate FDA staff. If you cannot identify the appropriate FDA staff, call the appropriate number listed on the title page of this guidance.*

I. Introduction

This document provides guidance to industry on good pharmacovigilance practices and pharmacoepidemiologic assessment of observational data regarding drugs, including biological drug products (excluding blood and blood components).[2] Specifically, this document provides guidance on (1) safety signal identification, (2) pharmacoepidemiologic assessment and safety signal interpretation, and (3) pharmacovigilance plan development.

[†] Available on the FDA website at: http://www.fda.gov/downloads/RegulatoryInformation/ Guidances/UCM126834.pdf

[1] This guidance has been prepared by the PDUFA III Pharmacovigilance Working Group, which includes members from the Center for Drug Evaluation and Research (CDER) and the Center for Biologics Evaluation and Research (CBER) at the Food and Drug Administration.

[2] For ease of reference, this guidance uses the term *product* or *drug* to refer to all products (excluding blood and blood components) regulated by CDER and CBER. Similarly, for ease of reference, this guidance uses the term *approval* to refer to both drug approval and biologic licensure.

FDA's guidance documents, including this guidance, do not establish legally enforceable responsibilities. Instead, guidances describe the Agency's current thinking on a topic and should be viewed only as recommendations, unless specific regulatory or statutory requirements are cited. The use of the word *should* in Agency guidances means that something is suggested or recommended, but not required.

II. Background

A. PDUFA III's Risk Management Guidance Goal

On June 12, 2002, Congress reauthorized, for the second time, the Prescription Drug User Fee Act (PDUFA III). In the context of PDUFA III, FDA agreed to satisfy certain performance goals. One of those goals was to produce guidance for industry on risk management activities for drug and biological products. As an initial step towards satisfying that goal, FDA sought public comment on risk management. Specifically, FDA issued three concept papers. Each paper focused on one aspect of risk management, including (1) conducting premarketing risk assessment, (2) developing and implementing risk minimization tools, and (3) performing postmarketing pharmacovigilance and pharmacoepidemiologic assessments. In addition to receiving numerous written comments regarding the three concept papers, FDA held a public workshop on April 9 – 11, 2003, to discuss the concept papers. FDA considered all of the comments received in developing three draft guidance documents on risk management activities.

The draft guidance documents were published on May 5, 2004, and the public was provided with an opportunity to comment on them until July 6, 2004. FDA considered all of the comments received in producing the final guidance documents.

1. *Premarketing Risk Assessment (Premarketing Guidance)*

2. *Development and Use of Risk Minimization Action Plans (RiskMAP Guidance)*

3. *Good Pharmacovigilance Practices and Pharmacoepidemiologic Assessment (Pharmacovigilance Guidance)*

B. Overview of the Risk Management Guidances

Like the concept papers and draft guidances that preceded them, each of the three final guidance documents focuses on one aspect of risk management. The *Premarketing Guidance* and the *Pharmacovigilance Guidance* focus on premarketing and postmarketing risk assessment, respectively. The *RiskMAP Guidance* focuses on risk minimization. Together, risk assessment and risk minimization form what FDA calls *risk management*. Specifically, risk management is an iterative process of (1) assessing a product's benefit-risk balance, (2) developing and implementing tools to minimize its risks while preserving its benefits, (3) evaluating tool effectiveness and reassessing the benefit-risk balance, and (4) making adjustments, as appropriate, to the risk minimization tools to further improve the benefit-risk balance. This four-part process should be continuous throughout a product's lifecycle, with the results of risk assessment informing the sponsor's decisions regarding risk minimization.

When reviewing the recommendations provided in this guidance, sponsors and applicants should keep the following points in mind:

- Many recommendations in this guidance are not intended to be generally applicable to all products.

  Industry already performs risk assessment and risk minimization activities for products during development and marketing. The Federal Food, Drug, and Cosmetic Act (FDCA) and FDA implementing regulations establish requirements for *routine* risk assessment and risk minimization (see e.g., FDA requirements for professional labeling, and adverse event monitoring and reporting). As a result, many of the recommendations presented here focus on situations when a product may pose a clinically important and unusual type or level of risk. To the extent possible, we have specified in the text whether a recommendation is intended for all products or only this subset of products.

- It is of critical importance to protect patients and their privacy during the generation of safety data and the development of risk minimization action plans.

  During all risk assessment and risk minimization activities, sponsors must comply with applicable regulatory requirements involving human subjects research and patient privacy.[3]

- To the extent possible, this guidance conforms with FDA's commitment to harmonize international definitions and standards as appropriate.

  The topics covered in this guidance are being discussed in a variety of international forums. We are participating in these discussions and believe that, to the extent possible, the recommendations in this guidance reflect current thinking on related issues.

- When planning risk assessment and risk minimization activities, sponsors should consider input from health care participants likely to be affected by these activities (e.g., from consumers, pharmacists and pharmacies, physicians, nurses, and third party payers).

- There are points of overlap among the three guidances.

  We have tried to note in the text of each guidance when areas of overlap occur and when referencing one of the other guidances might be useful.

III. The Role of Pharmacovigilance and Pharmacoepidemiology in Risk Management

Risk assessment during product development should be conducted in a thorough and rigorous manner; however, it is impossible to identify all safety concerns during clinical trials. Once a

---

[3] See 45 CFR part 46 and 21 CFR parts 50 and 56. See also the Health Insurance Portability and Accountability Act of 1996 (HIPAA) (Public Law 104-191) and the Standards for Privacy of Individually Identifiable Health Information (the Privacy Rule) (45 CFR part 160 and subparts A and E of part 164). The Privacy Rule specifically permits covered entities to report adverse events and other information related to the quality, effectiveness, and safety of FDA-regulated products both to manufacturers and directly to FDA (45 CFR 164.512(b)(1)(i) and (iii), and 45 CFR 164.512(a)(1)). For additional guidance on patient privacy protection, see http://www.hhs.gov/ocr/hipaa.

product is marketed, there is generally a large increase in the number of patients exposed, including those with co-morbid conditions and those being treated with concomitant medical products. Therefore, postmarketing safety data collection and risk assessment based on observational data are critical for evaluating and characterizing a product's risk profile and for making informed decisions on risk minimization.

This guidance document focuses on pharmacovigilance activities in the post-approval period. This guidance uses the term *pharmacovigilance* to mean all scientific and data gathering activities relating to the detection, assessment, and understanding of adverse events. This includes the use of pharmacoepidemiologic studies. These activities are undertaken with the goal of identifying adverse events and understanding, to the extent possible, their nature, frequency, and potential risk factors.

Pharmacovigilance principally involves the identification and evaluation of safety signals. In this guidance document, *safety signal* refers to a concern about an excess of adverse events compared to what would be expected to be associated with a product's use. Signals can arise from postmarketing data and other sources, such as preclinical data and events associated with other products in the same pharmacologic class. It is possible that even a single well-documented case report can be viewed as a signal, particularly if the report describes a positive rechallenge or if the event is extremely rare in the absence of drug use. Signals generally indicate the need for further investigation, which may or may not lead to the conclusion that the product caused the event. After a signal is identified, it should be further assessed to determine whether it represents a potential safety risk and whether other action should be taken.

IV. Identifying and Describing Safety Signals: from Case Reports to Case Series

Good pharmacovigilance practice is generally based on acquiring complete data from spontaneous adverse event reports, also known as case reports. The reports are used to develop case series for interpretation.

A. Good Reporting Practice

Spontaneous case reports of adverse events submitted to the sponsor and FDA, and reports from other sources, such as the medical literature or clinical studies, may generate signals of adverse effects of drugs. The quality of the reports is critical for appropriate evaluation of the relationship between the product and adverse events. FDA recommends that sponsors make a reasonable attempt to obtain complete information for case assessment during initial contacts and subsequent follow-up, especially for serious events,[4] and encourages sponsors to use

---

[4] Good reporting practices are extensively addressed in a proposed FDA regulation and guidance documents. See (1) Safety Reporting Requirements for Human Drug and Biological Products, Proposed Rule, 68 FR 12406 (March 14, 2003), (2) FDA guidance for industry on *Postmarketing Reporting of Adverse Experiences*, (3) FDA guidance for industry on *E2C Clinical Safety Data Management: Periodic Safety Update Report (PSUR)*, (4)

trained health care practitioners to query reporters. Computer-assisted interview technology, targeted questionnaires, or other methods developed to target specific events can help focus the line of questioning. When the report is from a consumer, it is often important to obtain permission to contact the health care practitioner familiar with the patient's adverse event to obtain further medical information and to retrieve relevant medical records, as needed.

FDA suggests that the intensity and method of case follow-up be driven by the seriousness of the event reported, the report's origin (e.g., health care practitioner, patient, literature), and other factors. FDA recommends that the most aggressive follow-up efforts be directed towards serious adverse event reports, especially of adverse events not known to occur with the drug.

B. Characteristics of a Good Case Report

Good case reports include the following elements:

1. Description of the adverse events or disease experience, including time to onset of signs or symptoms;

2. Suspected and concomitant product therapy details (i.e., dose, lot number, schedule, dates, duration), including over-the-counter medications, dietary supplements, and recently discontinued medications;

3. Patient characteristics, including demographic information (e.g., age, race, sex), baseline medical condition prior to product therapy, co-morbid conditions, use of concomitant medications, relevant family history of disease, and presence of other risk factors;

4. Documentation of the diagnosis of the events, including methods used to make the diagnosis;

5. Clinical course of the event and patient outcomes (e.g., hospitalization or death);[5]

6. Relevant therapeutic measures and laboratory data at baseline, during therapy, and subsequent to therapy, including blood levels, as appropriate;

7. Information about response to dechallenge and rechallenge; and

8. Any other relevant information (e.g., other details relating to the event or information on benefits received by the patient, if important to the assessment of the event).

FDA guidance for industry on *Postmarketing Adverse Experience Reporting for Human Drug and Licensed Biological Products: Clarification of What to Report.*

[5] Patient outcomes may not be available at the time of initial reporting. In these cases, follow-up reports can convey important information about the course of the event and serious outcomes, such as hospitalization or death.

For reports of medication errors, good case reports also include full descriptions of the following, when such information is available:

1. Products involved (including the trade (proprietary) and established (proper) name, manufacturer, dosage form, strength, concentration, and type and size of container);

2. Sequence of events leading up to the error;

3. Work environment in which the error occurred; and

4. Types of personnel involved with the error, type(s) of error, and contributing factors.

FDA recommends that sponsors capture in the case narrative section of a medication error report all appropriate information outlined in the National Coordinating Council for Medication Error Reporting and Prevention (NCC MERP) Taxonomy.[6] Although sponsors are not required to use the taxonomy, FDA has found the taxonomy to be a useful tool to categorize and analyze reports of medication errors. It provides a standard language and structure for medication error-related data collected through reports.

C. Developing a Case Series

FDA suggests that sponsors initially evaluate a signal generated from postmarketing spontaneous reports through a careful review of the cases and a search for additional cases. Additional cases could be identified from the sponsor's global adverse event databases, the published literature, and other available databases, such as FDA's Adverse Event Reporting System (AERS) or Vaccine Adverse Events Reporting System (VAERS), using thorough database search strategies based on updated coding terminology (e.g., the Medical Dictionary for Regulatory Activities (MedDRA)). When available, FDA recommends that standardized case definitions (i.e., formal criteria for including or excluding a case) be used to assess potential cases for inclusion in a case series.[7] In general, FDA suggests that case-level review occur before other investigations or analyses. FDA recommends that emphasis usually be placed on review of serious, unlabeled adverse events, although other events may warrant further investigation (see section IV.F. for more details).

As part of the case-level review, FDA suggests that sponsors evaluate individual case reports for clinical content and completeness, and follow up with reporters, as necessary. It is important to remove any duplicate reports. In assessing case reports, FDA recommends that sponsors look for features that may suggest a causal relationship between the use of a product and the adverse event, including:

1. Occurrence of the adverse event in the expected time (e.g., type 1 allergic reactions occurring within days of therapy, cancers developing after years of therapy);

2. Absence of symptoms related to the event prior to exposure;

---

[6] See http://www.nccmerp.org for the definition of a medication error and taxonomy of medication errors.
[7] See, for example, Institute of Medicine (IOM) Immunization Safety Review on Vaccines and Autism, 2004.

3.   Evidence of positive dechallenge or positive rechallenge;

4.   Consistency of the event with the established pharmacological/toxicological effects of the product, or for vaccines, consistency with established infectious or immunologic mechanisms of injury;

5.   Consistency of the event with the known effects of other products in the class;

6.   Existence of other supporting evidence from preclinical studies, clinical trials, and/or pharmacoepidemiologic studies; and

7.   Absence of alternative explanations for the event (e.g., no concomitant medications that could contribute to the event; no co- or pre-morbid medical conditions).

Confounded cases are common, especially among patients with complicated medical conditions. Confounded cases (i.e., cases with adverse events that have possible etiologies other than the product of concern) could still represent adverse effects of the product under review. FDA recommends that sponsors carefully evaluate these cases and not routinely exclude them. Separate analyses of unconfounded cases may be useful.

For any individual case report, it is rarely possible to know with a high level of certainty whether the event was caused by the product. To date, there are no internationally agreed upon standards or criteria for assessing causality in individual cases, especially for events that often occur spontaneously (e.g. stroke, pulmonary embolism). Rigorous pharmacoepidemiologic studies, such as case-control studies and cohort studies with appropriate follow-up, are usually employed to further examine the potential association between a product and an adverse event.

FDA does not recommend any specific categorization of causality, but the categories *probable*, *possible*, or *unlikely* have been used previously.[8] If a causality assessment is undertaken, FDA suggests that the causal categories be specified and described in sufficient detail to understand the underlying logic in the classification.

If the safety signal relates to a medication error, FDA recommends that sponsors report all known contributing factors that led to the event. A number of references are available to assist sponsors in capturing a complete account of the event.[9] FDA recommends that sponsors follow up to the extent possible with reporters to capture a complete account of the event, focusing on the *medication use systems* (e.g., prescribing/order process, dispensing process, administration process). This data may be informative in developing strategies to minimize future errors.

---

[8] See World Health Organization, the Uppsala Monitoring Center, 2000, *Safety Monitoring of Medicinal Product*, for additional categorizations of causality.

[9] See Cohen MR (ed), 1999, *Medication Errors*, American Pharmaceutical Association, Washington DC; Cousins DD (ed), 1998, *Medication Use: A Systems Approach to Reducing Errors*, Joint Commission on Accreditation of Healthcare Organizations, Oakbrook Terrace, IL.

D. Summary Descriptive Analysis of a Case Series

In the event that one or more cases suggest a safety signal warranting additional investigation, FDA recommends that a case series be assembled and descriptive clinical information be summarized to characterize the potential safety risk and, if possible, to identify risk factors. A case series commonly includes an analysis of the following:

1. The clinical and laboratory manifestations and course of the event;

2. Demographic characteristics of patients with events (e.g., age, gender, race);

3. Exposure duration;

4. Time from initiation of product exposure to the adverse event;

5. Doses used in cases, including labeled doses, greater than labeled doses, and overdoses;

6. Use of concomitant medications;

7. The presence of co-morbid conditions, particularly those known to cause the adverse event, such as underlying hepatic or renal impairment;

8. The route of administration (e.g., oral vs. parenteral);

9. Lot numbers, if available, for products used in patients with events; and

10. Changes in event reporting rate over calendar time or product life cycle.

E. Use of Data Mining to Identify Product-Event Combinations

At various stages of risk identification and assessment, systematic examination of the reported adverse events by using statistical or mathematical tools, or so-called data mining, can provide additional information about the existence of an excess of adverse events reported for a product.

By applying data mining techniques to large adverse event databases, such as FDA's AERS or VAERS, it may be possible to identify unusual or unexpected product-event combinations warranting further investigation. Data mining can be used to augment existing signal detection strategies and is especially useful for assessing patterns, time trends, and events associated with drug-drug interactions. Data mining is not a tool for establishing causal attributions between products and adverse events.

The methods of data mining currently in use usually generate a score comparing (1) the fraction of all reports for a particular event (e.g., liver failure) for a specific drug (i.e., the "observed reporting fraction") with (2) the fraction of reports for the same particular event for all drugs (i.e.,"the expected reporting fraction").[10] This analysis can be refined by adjusting for aspects of reporting (e.g., the reporting year) or characteristics of the patient (e.g., age or gender) that might influence the amount of reporting. In addition, it may be possible to limit data mining to an analysis for drugs of a specific class or for drugs that are used to treat a particular disease.

The score (or statistic) generated by data mining quantifies the disproportionality between the observed and expected values for a given product-event combination. This score is compared to a threshold that is chosen by the analyst. A potential excess of adverse events is operationally defined as any product-event combination with a score exceeding the specified threshold. When applying data mining to large databases (such as AERS), it is not unusual for a product to have several product-event combinations with scores above a specified threshold. The lower the threshold, the greater the likelihood that more combinations will exceed the threshold and will warrant further investigation.

Several data mining methods have been described and may be worth considering, such as the Multi-Item Gamma Poisson Shrinker (MGPS) algorithm[11, 12], the Proportional Reporting Ratio (PRR) method[13, 14] and the Neural Network approach.[15] Except when the observed number of cases with the drug event combination is small (e.g., less than 20) or the expected number of cases with the drug event combination is < 1, the MGPS and PRR methods will generally identify similar drug event combinations for further investigation.[16]

Although all of these approaches are inherently exploratory or hypothesis generating, they may provide insights into the patterns of adverse events reported for a given product relative to other products in the same class or to all other products. FDA exercises caution when making such comparisons, because voluntary adverse event reporting systems such as AERS or VAERS are subject to a variety of reporting biases (e.g., some observations could reflect concomitant treatment, not the product itself, and other factors, including the disease being

---

[10] Evans SJ, 2000, Pharmacovigilance: A science or fielding emergencies? *Statistics in Medicine* 19(23):3199-209; Evans SJW, Waller PC, and Davis S, 2001, Use of proportional reporting ratios (PRRs) for signal generation from spontaneous adverse drug reaction reports, *Pharmacoepidemiology and Drug Safety* 10:483-6.

[11] DuMouchel W and Pregibon D, 2001, Empirical Bayes screening for multi-item associations, *Seventh ACM SigKDD International Conference on Knowledge Discovery and Data Mining.*

[12] Szarfman A, Machado SG, and O'Neill RT, 2002, Use of screening algorithms and computer systems to efficiently signal higher-than-expected combinations of drugs and events in the US FDA's spontaneous reports database, *Drug Safety* 25(6): 381-92.

[13] Evans SJW, Waller P, and Davis S, 1998, Proportional reporting ratios: the uses of epidemiological methods for signal generation [abstract], *Pharmacoepidemiology and Drug Safety* 7:S102.

[14] Evans SJ, 2000, Pharmacovigilance: A science or fielding emergencies? *Statistics in Medicine* 19(23):3199-209; Evans SJW, Waller PC, and Davis S, 2001, Use of proportional reporting ratios (PRRs) for signal generation from spontaneous adverse drug reaction reports, *Pharmacoepidemiology and Drug Safety* 10:483-6.

[15] Bate A et al., 1998, A Bayesian neural network method for adverse drug reaction signal generation, *European Journal of Clinical Pharmacology* 54:315-21.

[16] This conclusion is based on the experience of FDA and of William DuMouchel, Ph.D., Chief Scientist, Lincoln Technologies, Wellsley, MA, as summarized in an email communication from Dr. DuMouchel to Ana Szarfman, M.D., Ph.D., Medical Officer, OPaSS, CDER, on October 13, 2004.

---

treated, other comorbidities or unrecorded confounders, may cause the events to be reported). In addition, AERS or VAERS data may be affected by the submission of incomplete or duplicate reports, underreporting, or reporting stimulated by publicity or litigation. As reporting biases may differ by product and change over time, and could change differently for different events, it is not possible to predict their impact on data mining scores.

Use of data mining techniques is not a required part of signal identification or evaluation. If data mining results are submitted to FDA, they should be presented in the larger appropriate clinical epidemiological context. This should include (1) a description of the database used, (2) a description of the data mining tool used (e.g., statistical algorithm, and the drugs, events and stratifications selected for the analyses) or an appropriate reference, and (3) a careful assessment of individual case reports and any other relevant safety information related to the particular drugevent combination of interest (e.g., results from preclinical, clinical, pharmacoepidemiologic, or other available studies).

F. Safety Signals That May Warrant Further Investigation

FDA believes that the methods described above will permit a sponsor to identify and preliminarily characterize a safety signal. The actual risk to patients cannot be known from these data because it is not possible to characterize all events definitively and because there is invariably under-reporting of some extent and incomplete information about duration of therapy, numbers treated, etc. Safety signals that may warrant further investigation may include, but are not limited to, the following:

1. New unlabeled adverse events, especially if serious;

2. An apparent increase in the severity of a labeled event;

3. Occurrence of serious events thought to be extremely rare in the general population;

4. New product-product, product-device, product-food, or product-dietary supplement interactions;

5. Identification of a previously unrecognized at-risk population (e.g., populations with specific racial or genetic predispositions or co-morbidities);

6. Confusion about a product's name, labeling, packaging, or use;

7. Concerns arising from the way a product is used (e.g., adverse events seen at higher than labeled doses or in populations not recommended for treatment);

8. Concerns arising from potential inadequacies of a currently implemented risk minimization action plan (e.g., reports of serious adverse events that appear to reflect failure of a RiskMAP goal);[17] and

9. Other concerns identified by the sponsor or FDA.

G. Putting the Signal into Context: Calculating Reporting Rates vs. Incidence Rates

If a sponsor determines that a concern about an excess of adverse events or safety signal warrants further investigation and analysis, it is important to put the signal into context. For this reason, calculations of the rate at which new cases of adverse events occur in the product-exposed population (i.e., the incidence rate) are the hallmark of pharmacoepidemiologic risk assessment.

In pharmacoepidemiologic studies (see section V.A), the numerator (number of new cases) and denominator (number of exposed patients and time of exposure or, if known, time at risk) may be readily ascertainable. In contrast, for spontaneously reported events, it is not possible to identify all cases because of under-reporting, and the size of the population at risk is at best an estimate.

Limitations in national denominator estimates arise because:

1. Accurate national estimates of the number of patients exposed to a medical product and their duration of exposure may not be available;

2. It may be difficult to exclude patients who are not at risk for an event, for example, because their exposure is too brief or their dose is too low;[18] and

3. A product may be used in different populations for different indications, but use estimates are not available for the specific population of interest.

Although we recognize these limitations, we recommend that sponsors calculate crude adverse event reporting rates as a valuable step in the investigation and assessment of adverse events.

FDA suggests that sponsors calculate reporting rates by using the total number of spontaneously reported cases in the United States in the numerator and estimates of national patient exposure to product in the denominator.[19, 20] FDA recommends that whenever possible, the number of patients or person time exposed to the product nationwide be the estimated denominator for a reporting rate. FDA suggests that other surrogates for exposure, such as numbers of prescriptions or kilograms of product sold, only be used when patient-level

---

[17] For a detailed discussion of risk minimization action plan evaluation, please consult the *RiskMAP Guidance*.
[18] See *Current Challenges in Pharmacovigilance: Pragmatic Approaches*, Report of the Council for International Organizations of Medical Sciences (CIOMS) Working Group V, Geneva, 2001.
[19] See Rodriguez EM, Staffa JA, Graham DJ, 2001, *The role of databases in drug postmarketing surveillance*, Pharmacoepidemiology and Drug Safety, 10:407-10.
[20] In addition to U.S. reporting rates, sponsors can provide global reporting rates, when relevant.

estimates are unavailable. FDA recommends that sponsors submit a detailed explanation of the rationale for selection of a denominator and a method of estimation.

Comparisons of reporting rates and their temporal trends can be valuable, particularly across similar products or across different product classes prescribed for the same indication. However, such comparisons are subject to substantial limitations in interpretation because of the inherent uncertainties in the numerator and denominator used. As a result, FDA suggests that a comparison of two or more reporting rates be viewed with extreme caution and generally considered exploratory or hypothesis-generating. Reporting rates can by no means be considered incidence rates, for either absolute or comparative purposes.

To provide further context for incidence rates or reporting rates, it is helpful to have an estimate of the background rate of occurrence for the event being evaluated in the general population or, ideally, in a subpopulation with characteristics similar to that of the exposed population (e.g., premenopausal women, diabetics). These background rates can be derived from: (1) national health statistics, (2) published medical literature, or (3) ad hoc studies, particularly of subpopulations, using large automated databases or ongoing epidemiologic investigations with primary data collection. FDA suggests that comparisons of incidence rates or reporting rates to background rate estimates take into account potential differences in the data sources, diagnostic criteria, and duration of time at risk.

While the extent of under-reporting is unknown, it is usually assumed to be substantial and may vary according to the type of product, seriousness of the event, population using the product, and other factors. As a result, a reporting rate higher than the background rate may, in some cases, be a strong indicator that the true incidence rate is sufficiently high to be of concern. However, many other factors affect the reporting of product-related adverse events (e.g., publicity, newness of product to the market) and these factors should be considered when interpreting a high reporting rate. Also, because of under-reporting, the fact that a reporting rate is less than the background rate does not necessarily show that the product is not associated with an increased risk of an adverse event.

V. Beyond Case Review: Investigating a Signal Through Observational Studies

FDA recognizes that there are a variety of methods for investigating a safety signal. Signals warranting additional investigation can be further evaluated through carefully designed nonrandomized observational studies of the product's use in the "real world" and randomized trials. The *Premarketing Guidance* discusses a number of types of randomized trials, including the large simple safety study, which is a risk assessment method that could be used either pre- or post-approval.

This document focuses on three types of non-randomized observational studies: (1) pharmacoepidemiologic studies, (2) registries, and (3) surveys. By focusing this guidance on certain risk assessment methods, we do not intend to advocate the use of these approaches over others. FDA encourages sponsors to consider all methods to evaluate a particular safety signal. FDA recommends that sponsors choose the method best suited to the particular signal and research question of interest. Sponsors planning to evaluate a safety signal are encouraged to communicate with FDA as their plans progress.

A. Pharmacoepidemiologic Studies

Pharmacoepidemiologic studies can be of various designs, including cohort (prospective or retrospective), case-control, nested case-control, case-crossover, or other models.[21] The results of such studies may be used to characterize one or more safety signals associated with a product, or may examine the natural history of a disease or drug utilization patterns. Unlike a case series, a pharmacoepidemiologic study which is designed to assess the risk attributed to a drug exposure has a protocol and control group and tests prespecified hypotheses. Pharmacoepidemiologic studies can allow for the estimation of the relative risk of an outcome associated with a product, and some (e.g., cohort studies) can also provide estimates of risk (incidence rate) for an adverse event. Sponsors can initiate pharmacoepidemiologic studies at any time. They are sometimes started at the time of initial marketing, based on questions that remain after review of the premarketing data. More often, however, they are initiated when a safety signal has been identified after approval. Finally, there may also be occasions when a pharmacoepidemiologic study is initiated prior to marketing (e.g., to study the natural history of disease or patterns of product use, or to estimate background rates for adverse events).

For uncommon or delayed adverse events, pharmacoepidemiologic studies may be the only practical choice for evaluation, even though they can be limited by low statistical power. Clinical trials are impractical in almost all cases when the event rates of concern are less common than 1:2000-3000 (an exception may be larger trials conducted for some vaccines, which could move the threshold to 1:10,000). It may also be difficult to use clinical trials: (1) to evaluate a safety signal associated with chronic exposure to a product, exposure in populations with co-morbid conditions, or taking multiple concomitant medications, or (2) to identify certain risk factors for a particular adverse event. On the other hand, for evaluation of more common events, which are seen relatively often in untreated patients, clinical trials may be preferable to observational studies.

Because pharmacoepidemiologic studies are observational in nature, they may be subject to confounding, effect modification, and other bias, which may make results of these types of studies more difficult to interpret than the results of clinical trials. Some of these problems can be surmounted when the relative risk to exposed patients is high.

Because different products pose different benefit-risk considerations (e.g., seriousness of the disease being treated, nature and frequency of the safety signal under evaluation), it is impossible to delineate a universal set of criteria for the point at which a pharmacoepidemiologic study should be initiated, and the decision should be made on a case-by-case basis. When an important adverse event–product association leads to questions on the product's benefit-risk balance, FDA recommends that sponsors consider whether the particular signal should be addressed with one or more pharmacoepidemiologic studies. If a sponsor determines that a pharmacoepidemiologic study is the best method for evaluating a

---

[21] *Guidelines for Good Pharmacoepidemiology*, International Society for Pharmacoepidemiology, 2004 (http://www.pharmacoepi.org/resources/guidelines_08027.cfm)

particular signal, the design and size of the proposed study would depend on the objectives of the study and the expected frequency of the events of interest.

When performing a pharmacoepidemiologic study, FDA suggests that investigators seek to minimize bias and to account for possible confounding. Confounding by indication is one example of an important concern in performing a pharmacoepidemiologic study.[22] Because of the effects of bias, confounding, or effect modification, pharmacoepidemiologic studies evaluating the same hypothesis may provide different or even conflicting results. It is almost always prudent to conduct more than one study, in more than one environment and even use different designs. Agreement of the results from more than one study helps to provide reassurance that the observed results are robust.

There are a number of references describing methodologies for pharmacoepidemiologic studies, discussing their strengths and limitations,[23] and providing guidelines to facilitate the conduct, interpretation, and documentation of such studies.[24] Consequently, this guidance document does not comprehensively address these topics. However, a protocol for a pharmacoepidemiologic study generally includes:

1. Clearly specified study objectives;

2. A critical review of the literature; and

3. A detailed description of the research methods, including:

    - the population to be studied;

    - the case definitions to be used;

    - the data sources to be used (including a rationale for data sources if from outside the U.S.);

    - the projected study size and statistical power calculations; and

    - the methods for data collection, management, and analysis.

Depending on the type of pharmacoepidemiologic study planned, there are a variety of data sources that may be used, ranging from the prospective collection of data to the use of existing data, such as data from previously conducted clinical trials or large databases. In recent years, a number of pharmacoepidemiologic studies have been conducted in automated claims databases (e.g., HMO, Medicaid) that allow retrieval of records on product exposure and patient outcomes. In addition, recently, comprehensive electronic medical record databases

---

[22] See, for example, Strom BL (ed), 2000, *Pharmacoepidemiology*, 3rd edition, Chichester: John Wiley and Sons, Ltd; Hartzema AG, Porta M, and Tilson HH (eds), 1998, *Pharmacoepidemiology: An Introduction*, 3rd edition, Cincinnati, OH: Harvey Whitney Books.
[23] Ibid.
[24] *Guidelines for Good Pharmacoepidemiology*, International Society for Pharmacoepidemiology, 2004 (http://www.pharmacoepi.org/resources/guidelines_08027.cfm).

have also been used for studying drug safety issues. Depending on study objectives, factors that may affect the choice of databases include the following:

1. Demographic characteristics of patients enrolled in the health plans (e.g., age, geographic location);

2. Turnover rate of patients in the health plans;

3. Plan coverage of the medications of interest;

4. Size and characteristics of the exposed population available for study;

5. Availability of the outcomes of interest;

6. Ability to identify conditions of interest using standard medical coding systems (e.g., International Classification of Diseases (ICD-9)), procedure codes or prescriptions that could be used as markers;

7. Access to medical records; and

8. Access to patients for data not captured electronically.

For most pharmacoepidemiologic studies, FDA recommends that sponsors validate diagnostic findings through a detailed review of at least a sample of medical records. If the validation of the specific outcome or exposure of interest using the proposed database has been previously reported, FDA recommends that the literature supporting the validity of the proposed study be submitted for review.

FDA encourages sponsors to communicate with the Agency when pharmacoepidemiologic studies are being developed.

B. Registries

The term *registry* as used in pharmacovigilance and pharmacoepidemiology can have varied meanings. In this guidance document, a registry is "an organized system for the collection, storage, retrieval, analysis, and dissemination of information on individual persons exposed to a specific medical intervention who have either a particular disease, a condition (e.g., a risk factor) that predisposes [them] to the occurrence of a health-related event, or prior exposure to substances (or circumstances) known or suspected to cause adverse health effects."[25] Whenever possible, a control or comparison group should be included, (i.e., individuals with a

---

[25] See Frequently Asked Questions About Medical and Public Health Registries, The National Committee on Vital and Health Statistics, at http://www.ncvhs.hhs.gov.

disease or risk factor who are not treated or are exposed to medical interventions other than the intervention of interest).[26]

Through the creation of registries, a sponsor can evaluate safety signals identified from spontaneous case reports, literature reports, or other sources, and evaluate factors that affect the risk of adverse outcomes, such as dose, timing of exposure, or patient characteristics.[27] Registries can be particularly useful for:

1. Collecting outcome information not available in large automated databases; and

2. Collecting information from multiple sources (e.g., physician records, hospital summaries, pathology reports, vital statistics), particularly when patients receive care from multiple providers over time.

A sponsor can initiate a registry at any time. It may be appropriate to initiate the registry at or before initial marketing, when a new indication is approved, or when there is a need to evaluate safety signals identified from spontaneous case reports. In deciding whether to establish a registry, FDA recommends that a sponsor consider the following factors:

1. The types of additional risk information desired;

2. The attainability of that information through other methods; and

3. The feasibility of establishing the registry.

Sponsors electing to initiate a registry should develop written protocols that provide: (1) objectives for the registry, (2) a review of the literature, and (3) a summary of relevant animal and human data. FDA suggests that protocols also contain detailed descriptions of: (1) plans for systematic patient recruitment and follow-up, (2) methods for data collection, management, and analysis, and (3) conditions under which the registry will be terminated. A registry-based monitoring system should include carefully designed data collection forms to ensure data quality, integrity, and validation of registry findings against a sample of medical records or through interviews with health care providers. FDA recommends that the size of the registry and the period during which data will be collected be consistent with the safety questions under study and we encourage sponsors to discuss their registry development plans with FDA.

C. Surveys

Patient or health care provider surveys can gather information to assess, for example:

1. A safety signal;

2. Knowledge about labeled adverse events;

---

[26] See for example, *FDA Guidance for Industry, Establishing Pregnancy Exposure Registries*, August 2002 http://www.fda.gov/cder/guidance/3626fnl.pdf.
[27] Ibid.

3. Use of a product as labeled, particularly when the indicated use is for a restricted population or numerous contraindications exist;

4. Compliance with the elements of a RiskMAP (e.g., whether or not a Medication Guide was provided at the time of product dispensing); and [28]

5. Confusion in the practicing community over sound-alike or look-alike trade (or proprietary) names.

Like a registry, a survey can be initiated by a sponsor at any time. It can be conducted at the time of initial marketing (i.e., to fulfill a postmarketing commitment) or when there is a desire to evaluate safety signals identified from spontaneous case reports.

FDA suggests that sponsors electing to initiate a survey develop a written protocol that provides objectives for the survey and a detailed description of the research methods, including: (1) patient or provider recruitment and follow-up, (2) projected sample size, and (3) methods for data collection, management, and analysis.[29] FDA recommends that a survey-based monitoring system include carefully designed survey instruments and validation of survey findings against a sample of medical or pharmacy records or through interviews with health care providers, whenever possible. FDA recommends that survey instruments be validated or piloted before implementation. FDA suggests that sponsors consider whether survey translation and cultural validation would be important.

Sponsors are encouraged to discuss their survey development plans with FDA.

VI. Interpreting Safety Signals: From Signal to Potential Safety Risk

After identifying a safety signal, FDA recommends that a sponsor conduct a careful case level review and summarize the resulting case series descriptively. To help further characterize a safety signal, a sponsor can also: (1) employ data mining techniques, and (2) calculate reporting rates for comparison to background rates. Based on these findings and other available data (e.g., from preclinical or other sources), FDA suggests that a sponsor consider further study (e.g., observational studies) to establish whether or not a potential safety risk exists.

When evaluation of a safety signal suggests that it may represent a potential safety risk, FDA recommends that a sponsor submit a synthesis of all available safety information and analyses performed, ranging from preclinical findings to current observations. This submission should include the following:

1. Spontaneously reported and published case reports, with denominator or exposure information to aid interpretation;

2. Background rate for the event in general and specific patient populations, if available;

---

[28] For a detailed discussion of RiskMAP evaluation, please consult the *RiskMAP Guidance*.
[29] See 21 CFR parts 50 and 56 for FDA's regulations governing the protection of human subjects.

3. Relative risks, odds ratios, or other measures of association derived from pharmacoepidemiologic studies;

4. Biologic effects observed in preclinical studies and pharmacokinetic or pharmacodynamic effects;

5. Safety findings from controlled clinical trials; and

6. General marketing experience with similar products in the class.

After the available safety information is presented and interpreted, it may be possible to assess the degree of causality between use of a product and an adverse event. FDA suggests that the sponsor's submission provide an assessment of the benefit-risk balance of the product for the population of users as a whole and for identified at-risk patient populations, and, if appropriate, (1) propose steps to further investigate the signal through additional studies, and (2) propose risk minimization actions.[30] FDA will make its own assessment of the potential safety risk posed by the signal in question, taking into account the information provided by the sponsor and any additional relevant information known to FDA (e.g., information on other products in the same class) and will communicate its conclusions to the sponsor whenever possible. Factors that are typically considered include:

1. Strength of the association (e.g., relative risk of the adverse event associated with the product);

2. Temporal relationship of product use and the event;

3. Consistency of findings across available data sources;

4. Evidence of a dose-response for the effect;

5. Biologic plausibility;

6. Seriousness of the event relative to the disease being treated;

7. Potential to mitigate the risk in the population;

8. Feasibility of further study using observational or controlled clinical study designs; and

9. Degree of benefit the product provides, including availability of other therapies.

---

[30] In the vast majority of cases, risk communication that incorporates appropriate language into the product's labeling will be adequate for risk minimization. In rare instances, however, a sponsor may consider implementing a RiskMAP. Please refer to the *RiskMAP Guidance* for a complete discussion of RiskMAP development.

As noted in section II, risk management is an iterative process and steps to further investigate a potential safety risk, assess the product's benefit-risk balance, and implement risk minimization tools would best occur in a logical sequence, not simultaneously. Not all steps may be recommended, depending on the results of earlier steps.[31] FDA recommends that assessment of causality and of strategies to minimize product risk occur on an ongoing basis, taking into account the findings from newly completed studies.

VII. Beyond Routine Pharmacovigilance: Developing a Pharmacovigilance Plan

For most products, routine pharmacovigilance (i.e., compliance with applicable postmarket requirements under the FDCA and FDA implementing regulations) is sufficient for postmarketing risk assessment. However, in certain limited instances, unusual safety risks may become evident before approval or after a product is marketed that could suggest that consideration by the sponsor of a pharmacovigilance plan may be appropriate. A pharmacovigilance plan is a plan developed by a sponsor that is focused on detecting new safety risks and/or evaluating already identified safety risks. Specifically, a pharmacovigilance plan describes pharmacovigilance efforts above and beyond routine postmarketing spontaneous reporting, and is designed to enhance and expedite the sponsor's acquisition of safety information.[32] The development of pharmacovigilance plans may be useful at the time of product launch or when a safety risk is identified during product marketing. FDA recommends that a sponsor's decision to develop a pharmacovigilance plan be based on scientific and logistical factors, including the following:

1. The likelihood that the adverse event represents a potential safety risk;

2. The frequency with which the event occurs (e.g., incidence rate, reporting rate, or other measures available);

3. The severity of the event;

4. The nature of the population(s) at risk;

5. The range of patients for which the product is indicated (broad range or selected populations only); and

---

[31] For additional discussion of the relationship between risk assessment and risk minimization, please consult the *RiskMAP Guidance*.

[32] As used in this document, the term "pharmacovigilance plan" is defined differently than in the ICH draft E2E document (version 4.1). As used in the ICH document, a "pharmacovigilance plan" would be routinely developed (i.e., even when a sponsor does not anticipate that enhanced pharmacovigilance efforts are necessary). In contrast, as discussed above, FDA is only recommending that pharmacovigilance plans be developed when warranted by unusual safety risks. This ICH guidance is available on the Internet at http://www.fda.gov/cder/guidance/index.htm under the topic ICH Efficacy. The draft E2E guidance was made available on March 30, 2004 (69 FR 16579). ICH agreed on the final version of the E2E guidance in November, 2004.

6. The method by which the product is dispensed (through pharmacies or performance linked systems only).[33]

A pharmacovigilance plan may be developed by itself or as part of a Risk Minimization Action Plan (RiskMAP), as described in the RiskMAP Guidance. Sponsors may meet with representatives from the appropriate Office of New Drugs review division and the Office of Drug Safety in CDER, or the appropriate Product Office and the Division of Epidemiology, Office of Biostatistics and Epidemiology in CBER regarding the specifics of a given product's pharmacovigilance plan.

FDA believes that for a product without safety risks identified pre- or post-approval and for which at-risk populations are thought to have been adequately studied, routine spontaneous reporting will be sufficient for postmarketing surveillance. On the other hand, pharmacovigilance plans may be appropriate for products for which: (1) serious safety risks have been identified pre- or post-approval, or (2) at-risk populations have not been adequately studied.

Sponsors may discuss with the Agency the nature of the safety concerns posed by such a product and the determination whether a pharmacovigilance plan is appropriate.

A pharmacovigilance plan could include one or more of the following elements:

1. Submission of specific serious adverse event reports in an expedited manner beyond routine required reporting (i.e., as 15-day reports);

2. Submission of adverse event report summaries at more frequent, prespecified intervals (e.g., quarterly rather than annually);

3. Active surveillance to identify adverse events that may or may not be reported through passive surveillance. Active surveillance can be (1) drug based: identifying adverse events in patients taking certain products, (2) setting based: identifying adverse events in certain health care settings where they are likely to present for treatment (e.g., emergency departments, etc.), or (3) event based: identifying adverse events that are likely to be associated with medical products (e.g., acute liver failure);

4. Additional pharmacoepidemiologic studies (for example, in automated claims databases or other databases) using cohort, case-control, or other appropriate study designs (see section V);

5. Creation of registries or implementation of patient or health care provider surveys (see section V); and

6. Additional controlled clinical trials.[34]

---

[33] For a detailed discussion of controlled access systems, please consult the *RiskMAP Guidance*.

[34] For a discussion of risk assessment in controlled clinical trials, please consult the *Premarketing Guidance*.

As data emerges, FDA recommends that a sponsor re-evaluate the safety risk and the effectiveness of its pharmacovigilance plan. Such re-evaluation may result in revisions to the pharmacovigilance plan for a product. In some circumstances, FDA may decide to bring questions on potential safety risks and pharmacovigilance plans before its Drug Safety and Risk Management Advisory Committee or the FDA Advisory Committee dealing with the specific product in question. Such committees may be convened when FDA seeks: (1) general advice on the design of pharmacoepidemiologic studies, (2) comment on specific pharmacoepidemiology studies developed by sponsors or FDA for a specific product and safety question, or (3) advice on the interpretation of early signals from a case series and on the need for further investigation in pharmacoepidemiologic studies. While additional information is being developed, sponsors working with FDA can take interim actions to communicate information about potential safety risks (e.g., through labeling) to minimize the risk to users of the product.

IND Exemptions for Studies of Lawfully Marketed Drug or Biologicial Products for the Treatment of Cancer

IND Exemptions for Studies of Lawfully Marketed Drug or Biological Products for the Treatment of Cancer, Guidance for Industry [T, 1]

U.S. Department of Health and Human Services
Food and Drug Administration
Center for Drug Evaluation and Research (CDER)
Center for Biologics Evaluation and Research (CBER)

January 2004

Clinical Medical
Revision 1

Contains Nonbinding Recommendations

This guidance is intended to assist sponsors in deciding whether a study of marketed drugs or biological products for treating cancer falls within the exemption under § 312.2(b)(1) (21 CFR 312.2(b)(1)) from the general requirement to submit an investigational new drug application (IND).

Note: This guidance represents the Food and Drug Administration's (FDA's) current thinking on this topic. It does not create or confer any rights for or on any person and does not operate to bind FDA or the public. You can use an alternative approach if the approach satisfies the requirements of the applicable statutes and regulations. If you want to discuss an alternative approach, contact the appropriate FDA staff. If you cannot identify the appropriate FDA staff, call the appropriate number listed on the title page of this guidance.

I. Introduction

This guidance is intended to assist sponsors in deciding whether a study of marketed drugs or biological products for treating cancer falls within the exemption under § 312.2(b)(1) (21 CFR 312.2(b)(1)) from the general requirement to submit an investigational new drug application (IND). The guidance discusses the Agency's current thinking on when studies of marketed

---

[T] Available on the FDA website at: http://www.fda.gov/downloads/RegulatoryInformation/ Guidances/UCM126837.pdf

[1] This guidance has been prepared by the Division of Oncology Drug Products in the Center for Drug Evaluation and Research (CDER) and by the Center for Biologics Evaluation and Research (CBER) at the Food and Drug Administration.

cancer products are exempt from IND regulation based on a risk assessment. The Agency hopes that clarifying its policy will help sponsors identify which studies are exempt, thus saving them from submitting unnecessary IND applications.

This guidance revises the guidance of the same title published in September 2003. In the September 2003 version, the Agency's final statement was that it believed that most randomized studies of a size that could support a labeling supplement would likely not be exempt from IND regulation under § 312.2(b)(1)(i), (ii). This is because they would be intended to support approval of a new indication, a significant change in the product labeling, or a significant change in advertising. Experience has shown that this interpretation was formulated too broadly and inappropriately referred to size alone. The Agency has decided to revise this guidance by removing that statement (the last sentence in section V.B). Whether a study could support a change in labeling is a complex determination, based on study design, size, and other factors.

FDA's guidance documents, including this guidance, do not establish legally enforceable responsibilities. Instead, guidances describe the Agency's current thinking on a topic and should be viewed only as recommendations, unless specific regulatory or statutory requirements are cited. The use of the word *should* in Agency guidances means that something is suggested or recommended, but not required.

II. Background

Generally, regulations in part 312 (21 CFR part 312) require sponsors who wish to study a drug or biological product in humans to submit an IND to the Agency.[2] However, these regulations also provide for the exemption of some studies from the requirement to submit an IND if they meet certain criteria. Each year, many INDs for cancer drugs are submitted that contain studies that the Agency determines are exempt. This guidance is intended to help applicants identify which studies may be exempt.

A. Regulations

Regulations in § 312.2(b)(1) provide for the exemption of some studies for some drugs from IND regulations if the studies meet the following five criteria:

1. The study is not intended to support FDA approval of a new indication or a significant change in the product labeling.

2. The study is not intended to support a significant change in the advertising for the product.

---

[2] Part 312 applies to all clinical investigations of products that are subject to section 505 of the Federal Food, Drug, and Cosmetic Act (21 U.S.C. 355) or to the licensing provisions of the Public Health Service Act (58 Stat. 632, as amended (42 U.S.C. 201 et seq.)).

3. The investigation does not involve a route of administration or dosage level or use in a patient population or other factor that significantly increases the risks (or decreases the acceptability of the risks) associated with the use of the drug product.

4. The study is conducted in compliance with institutional review board (IRB) and informed consent regulations set forth in parts 56 and 50 (21 CFR parts 56 and 50).

5. The study is conducted in compliance with § 312.7 (promotion and charging for investigational drugs).

Requirements 1, 2, 4, and 5 are not directly related to the specific protocol submitted, and their interpretation is similar for oncologic and nononcologic therapies. Requirement 3 is protocol related and has special meaning in the oncology therapy setting, particularly with respect to doses above the labeled dose, use with other treatments, and use in different populations.

In the preamble to the IND regulations, which published in the *Federal Register* on March 19, 1987, the Agency explained that the exemption was not necessarily intended to tie the investigator to the doses and routes of administration and patient population described in the approved labeling, but to permit deviations from the approved labeling to the extent that such changes are supported by the scientific literature and generally known clinical experience. The Agency recognizes that a considerable amount of professional judgment is exercised in determining whether the planned investigation significantly increases the risk associated with the use of the drug. FDA maintains that "because the assessment of risks involved in a therapeutic procedure is an everyday part of the practice of medicine, the individual investigator should usually be able to determine the applicability of the exemption."[3]

B. 1996 Agency Cancer Initiative

In 1996, as part of the President's National Performance Review, the Agency launched its *Reinventing the Regulation of Cancer Drugs* initiative with the goal of accelerating the approval of and expanding patient access to cancer drugs.[4] As part of this initiative, the Agency explained that many sponsor-investigators were submitting INDs for exploratory studies for so-called offlabel indications for two reasons: (1) IRBs incorrectly believe an IND is required, or (2) the pharmaceutical manufacturer agrees to provide a drug free of charge, but mistakenly concludes that the FDA will view this as promotional activity. With the intent of clarifying the Agency's policy and decreasing the number of unnecessary submissions, the Agency emphasized that it would no longer accept INDs considered exempt under § 312.2(b)(1). (See § 312.2(b)(4).) Furthermore, FDA stated that providing a drug for study would not, in and of itself, be viewed as a promotional activity if the manufacturer or distributor provides the product for a

---

[3] New Drug, Antibiotic, and Biologic Drug Product Regulations, Federal Register, March 19, 1987, Vol. 52, Nr. 53, p. 8802.

[4] Reinventing the Regulation of Cancer Drugs – Accelerating Approval and Expanding Access (March 1996), CBER, Office of Communication, Training, and Manufacturer Assistance, Voice Information System at 1-800-835-4709 or 301-827-1800, document ID number 0281. Available on the Internet at http://www.fda.gov/cber/genadmin/reincanc.htm

---

physicianinitiated, bona fide clinical investigation. The Agency explained that it is the responsibility of the investigator to determine whether an IND is necessary.

Despite the Agency's attempts to clarify its policy on IND exemptions, many cancer drug IND applications that the Agency determines are exempt from IND regulation are still being submitted unnecessarily. From 1997 to 1999, a majority of investigator IND submissions for marketed cancer drugs were considered exempt (204, 205, and 140 applications in 1997, 1998, and 1999, respectively).

III. Risk/Benefit Analysis in the Practice of Oncology

As noted above, a critical question in determining whether a study is exempt involves criterion 3 in the exemption regulations (§ 312.2(b)(1)(iii)): The investigation may not *significantly increase the risk* associated with use of a drug product. The question of increased risk is determined by assessing the deviation in the planned investigation from the use described in the approved label. In oncology, modifications of labeled dosing recommendations are common and occur as part of oncologists' clinical practice. As outlined below, oncologists are familiar with evaluating the risk of off-label dosing regimens for cancer drug and biological products.

- Treatment with cancer drugs may be associated with significant risk from known toxicity. Because effectiveness is often related to dose, a dose close to the *maximal tolerated dose* is often selected for studies of cancer drugs. This same dose usually becomes the recommended dose in labeling when the new cancer drug is approved with the knowledge that the dose may be altered if it is not tolerated by a patient. Because it is not generally possible to have maximal efficacy in a population without inducing toxicity in some patients, it is not uncommon to observe severe or even lethal side effects from cancer drugs in some patients. In general, these circumstances mean that the toxicity, even potentially lethal toxicity, of cancer drugs is described in approved labeling.

- Off-label therapy with cancer drugs is common in practice. When there is no established therapy for a cancer, or stage of cancer, it is common for oncologists to try different regimens or combinations of established drugs. A 1996 GAO report (*Prescription Drugs, Implications of Drug Labeling and Off-Label Use*) showed that there was substantial off-label use in situations where satisfactory treatment was not available, and lower rates of off-label use when there was an effective therapy. In their daily practice, many oncologists treat cancer patients with regimens that include off-label use of drugs. They evaluate the published data and past clinical experience to assess the risk of such treatments. Such treatment of individual patients with approved drugs within their clinical practice does not require an IND (§ 312.2(d)).

- In many cases, as discussed in the examples in section V below, drug administration to patients with similar off-label regimens in the context of an investigation seems to involve no increased risk to patients, and an investigator could conclude that such a study would not *significantly increase the risk* associated with the labeled use of a drug product and the study could be conducted without an IND. Oversight by an IRB and informed consent in compliance with parts 56 and 50, respectively, would be required as usual (§

312.2(b)(1)(iv)). On request, FDA will advise on the applicability of the IND exemption to a planned clinical investigation (§ 312.2(e)).

IV. Determining Application Status

A. Agency Determination

As explained in FDA's 1996 cancer initiative and the IND exemption regulation, FDA will not accept applications for clinical studies that it determines to be exempt from the requirement for an IND (§ 312.2(b)(4)). Although § 312.2(b)(1) does not require a submission for a determination of exempt status, whenever an IND application is submitted, FDA staff perform an initial limited review of the application to determine whether the study is exempt. The protocolrelated criterion FDA considers in assessing exemption is: The investigation may not involve a route of administration or dosage level or use in a patient population or other factor that significantly increases the risks (or decreases the acceptability of the risks) associated with the use of the drug product (§ 312.2(b)(1)(iii)). Thus, when determining if the risk is significantly increased, FDA staff examine the parts of the protocol that concern dose, schedule, route of administration, and patient population. If the Agency's initial limited review determines that a study protocol is exempt from the requirement for an IND, the Agency performs no further review of the application. A letter is sent to the sponsor giving notice of the exemption.

B. Investigator Determination

When determining if an IND needs to be submitted to study marketed drugs for treating cancer, investigators must apply the exemption criteria listed in § 312.2(b)(1)(i-v) in light of the discussion in this guidance. Planned studies may be considered exempt from the requirements of an IND if the studies involve a new use, dosage, schedule, route of administration, or new combination of marketed cancer products in a patient population with cancer and the following conditions apply:

- The studies are not intended to support FDA approval of a new indication or a significant change in the product labeling.

- The studies are not intended to support a significant change in the advertising for the product.

- Investigators and their IRBs determine that based on the scientific literature and generally known clinical experience, there is no *significant increase in the risk associated with the use of the drug product.*

- The studies are to be conducted in compliance with IRB and informed consent regulations, pursuant to parts 50 and 56.

- The studies will not be used to promote unapproved indications, in compliance with § 312.7.

V. Examples of Studies

The following examples of studies are being provided to illustrate the Agency's current thinking on the types of studies that the Agency considers to be exempt from IND regulation based on a risk assessment.

A. Studies That Generally Are Exempt

As noted above, of the five criteria in § 312.2(b)(1), four are not protocol related and one is protocol related. The following are examples of general categories of studies of marketed cancer drugs that would likely be exempt from IND regulation based on protocol-related issues.

1. Single-arm, phase 2 trials using marketed drugs to treat a cancer different from that indicated in the approved labeling and using doses and schedules similar to those in the marketed drug labeling are usually exempt. An exception may exist when standard therapy in the population to be studied is very effective (e.g., is associated with a survival benefit); in that case, use of another regimen may expose patients to the risk of receiving an ineffective therapy and an IND would be necessary.

2. Phase 1 oncology trials of marketed drugs may be considered exempt if such therapy is appropriate for the patient population (i.e., if patients have residual cancer) and if there is no effective therapy (i.e., therapy producing cure or a documented increase in survival) that the patients have not yet received. It remains the investigator's responsibility to use starting doses that appear safe based on approved labeling or detailed literature reports, use incremental changes in dose or schedule, and carefully evaluate toxicity prior to dose escalation.

3. The study of new combinations of drugs would not ordinarily constitute a significant risk if these combinations have been described in the professional medical literature. Even when the regimen described in the literature does not use exactly the doses planned for study, incremental differences in doses from those described in the literature would not normally pose a significant risk and would not require an IND.

   Because of the danger of synergistic toxicity (i.e., enhanced effects from the combination) occurring with a new drug combination, if there are no data from the literature on its safety, the initial study of a new drug combination should ordinarily be performed under an IND. Synergistic toxicity may be anticipated when one agent interferes with the metabolism or elimination of the other agent; when both agents target the same metabolic pathway or cellular function; or when one agent targets signaling pathways that are reasonably expected to modulate sensitivity to the other agent. If it is determined that synergistic toxicity is likely, animal studies should be considered for determining a safe starting dose for the drug combination in humans.

4. Studies of new routes or schedules of administration not described in the approved labeling are generally exempt if there is sufficient clinical experience described in the

literature documenting safety to determine that treatment is safe. On the other hand, initial experience with a new route of administration should be based on studies in animals, and an IND should be submitted.

5. Studies of high-dose therapy in cancer patients are likely to be considered exempt if the studies use adequately evaluated regimens that appear to have an acceptable therapeutic ratio for the population being studied. Similarly, phase 1 studies involving incremental changes from such well-described regimens are generally exempt.

B. Studies That Generally Are Not Exempt

As noted above, of the five criteria in § 312.2(b)(1), four are not protocol related and one is protocol related. The following are examples of general categories of studies of marketed cancer drugs that would likely *not* be exempt from IND regulation because of protocol-related issues.

1. Studies of cytotoxic drugs are normally not exempt in patients for whom cytotoxic therapy would not be considered standard therapy and would require special justification. Any use of cytotoxic agents in nonmalignant disease (e.g., rheumatoid arthritis, multiple sclerosis) would, most likely, be considered to alter the acceptability of the risk of the agent.

2. Studies of adjuvant chemotherapy (chemotherapy given after surgery to remove cancer) are likely not exempt for the following reasons:

    • If the population studied has a low risk of cancer recurring after surgery, treatment with any toxic therapy may indicate a significantly increased risk.

    • If standard adjuvant therapy is available and produces a survival benefit, substitution of new therapy for standard therapy poses a significant risk that the new therapy will not produce the same survival benefit.

    • If adjuvant trials are properly designed, they usually will be able to demonstrate whether the new therapy is safe and effective, and such results may lead to a marketing application. As discussed earlier, under regulations at § 312.2(b)(1), all investigations intended to support marketing of a new product indication, significant change in product labeling, or a significant change in the advertising for a product require an IND. During FDA review of INDs intended to support marketing applications, the Agency will provide feedback about the acceptability of trial design for this purpose.

3. Studies involving substitution of a new agent of unproven activity are generally not exempt in settings where standard therapy provides a cure or increase in survival. For instance, in the first-line treatment of testicular cancer, ovarian cancer, breast cancer, leukemia, and lymphoma, studies of new agents without proven efficacy would likely not be exempt. In this case, the critical judgment is whether it is ethical to withhold standard therapy while testing a new agent.

4. Studies are generally not exempt in settings where animal studies should be conducted to determine a safe starting dose or schedule.

   For example:

   • Initial studies of a marketed drug given by a new route of administration are likely not exempt.

   • Unless adequately described in the literature, initial studies of new drug combinations should usually be performed under an IND because of the possible occurrence of synergistic toxicity. As noted earlier, synergistic toxicity may be anticipated when one agent interferes with the metabolism or elimination of the other agent; when both agents target the same metabolic pathway or cellular function; or when one agent targets signaling pathways that are reasonably expected to modulate sensitivity to the other agent.

   • Initial studies in humans of changes in the schedule of drug administration should generally be submitted in an IND. Some drugs have demonstrated significantly greater toxicity when given by an alternative schedule (e.g., methotrexate demonstrates much more hematologic toxicity when given by prolonged administration compared to intermittent administration).

   • Initial studies of drugs intended to be chemosensitizers, radiosensitizers, or resistance modulators should generally be submitted in an IND. Animal studies should be used to estimate the effect of the modulator on toxicity and to allow estimation of a safe starting dose in humans.

5. Studies intended to support approval of a new indication, a significant change in the product labeling, or a significant change in advertising are not exempt (§ 312.2(b)(1)(i), (ii)).

Premarketing Risk Assessment

Premarketing Risk Assessment, Guidance for Industry [T, 1]

U.S. Department of Health and Human Services
Food and Drug Administration
Center for Drug Evaluation and Research (CDER)
Center for Biologics Evaluation and Research (CBER)

March 2005

Clinical Medical

Contains Nonbinding Recommendations

This guidance is intended to assist industry with risk management activities for drug products, including biological drug products, in the Center for Drug Evaluation and Research (CDER) and the Center for Biologics Evaluation and Research (CBER).

Note: *This guidance represents the Food and Drug Administration's (FDA's) current thinking on this topic. It does not create or confer any rights for or on any person and does not operate to bind FDA or the public. You can use an alternative approach if the approach satisfies the requirements of the applicable statutes and regulations. If you want to discuss an alternative approach, contact the appropriate FDA staff. If you cannot identify the appropriate FDA staff, call the appropriate number listed on the title page of this guidance.*

I. Introduction

This document provides guidance to industry on good risk assessment practices during the development of prescription drug products, including biological drug products.[2] This is one of three guidances that were developed to address risk management activities. Specifically, this document discusses the generation, acquisition, analysis, and presentation of premarketing safety data.

FDA's guidance documents, including this guidance, do not establish legally enforceable responsibilities. Instead, guidances describe the Agency's current thinking on a topic and should be viewed only as recommendations, unless specific regulatory or statutory

---

[T] Available on the FDA website at: http://www.fda.gov/downloads/RegulatoryInformation/ Guidances/UCM126958.pdf

[1] This guidance has been prepared by the Center for Drug Evaluation and Research (CDER) and the Center for Biologics Evaluation and Research (CBER) at the Food and Drug Administration.

[2] For ease of reference, this guidance uses the terms *product* and *drug* to refer to all products (excluding blood and blood components) regulated by CDER or CBER, including vaccines. Similarly, for ease of reference, this draft guidance uses the term *approval* to refer to both drug approval and biologic licensure.

requirements are cited. The use of the word *should* in Agency guidances means that something is suggested or recommended, but not required.

II. Background

A. PDUFA III's Risk Management Guidance Goal

On June 12, 2002, Congress reauthorized, for the second time, the Prescription Drug User Fee Act (PDUFA III). In the context of PDUFA III, FDA agreed to satisfy certain performance goals. One of those goals was to produce guidance for industry on risk management activities for drug and biological products. As an initial step towards satisfying that goal, FDA sought public comment on risk management. Specifically, FDA issued three concept papers. Each paper focused on one aspect of risk management, including (1) conducting premarketing risk assessment, (2) developing and implementing risk minimization tools, and (3) performing postmarketing pharmacovigilance and pharmacoepidemiologic assessments. In addition to receiving numerous written comments regarding the three concept papers, FDA held a public workshop on April 9-11, 2003, to discuss the concept papers. FDA considered all of the comments received in developing three draft guidance documents on risk management activities.

The draft guidance documents were published on May 5, 2004, and the public was provided with an opportunity to comment on them until July 6, 2004. FDA considered all of the comments received in producing the final guidance documents.

- *Premarketing Risk Assessment (Premarketing Guidance)*

- *Development and Use of Risk Minimization Action Plans (RiskMAP Guidance)*

- *Good Pharmacovigilance Practices and Pharmacoepidemiologic Assessment (Pharmacovigilance Guidance).*

B. Overview of the Risk Management Guidances

Like the concept papers and draft guidances that preceded them, each of the three final guidance documents focuses on one aspect of risk management. The *Premarketing Guidance* and the *Pharmacovigilance Guidance* focus on premarketing and postmarketing risk assessment, respectively. The *RiskMAP Guidance* focuses on risk minimization. Together, risk assessment and risk minimization form what FDA calls *risk management*. Specifically, risk management is an iterative process of (1) assessing a product's benefit-risk balance, (2) developing and implementing tools to minimize its risks while preserving its benefits, (3) evaluating tool effectiveness and reassessing the benefit-risk balance, and (4) making adjustments, as appropriate, to the risk minimization tools to further improve the benefit-risk balance. This four-part process should be continuous throughout a product's lifecycle, with the results of risk assessment informing the sponsor's decisions regarding risk minimization.

When reviewing the recommendations provided in this guidance, sponsors and applicants should keep the following points in mind:

- Many recommendations in this guidance are not intended to be generally applicable to all products.

Industry already performs risk assessment and risk minimization activities for products during development and marketing. The Federal Food, Drug, and Cosmetic Act (FDCA) and FDA implementing regulations establish requirements for *routine* risk assessment and risk minimization (see e.g., FDA requirements for professional labeling and adverse event monitoring and reporting). As a result, many of the recommendations presented here focus on situations in which a product may pose a clinically important and unusual type or level of risk. To the extent possible, we have specified in the text whether a recommendation is intended for all products or only this subset of products.

• It is of critical importance to protect patients and their privacy during the generation of safety data and the development of risk minimization action plans.

During all risk assessment and risk minimization activities, sponsors must comply with applicable regulatory requirements involving human subjects research and patient privacy.[3]

• To the extent possible, this guidance reflects FDA's commitment to harmonization of international definitions and standards.

• When planning risk assessment and risk minimization activities, sponsors should consider input from healthcare participants likely to be affected by these activities (e.g., from consumers, pharmacists and pharmacies, physicians, nurses, and third party payers).

• There are points of overlap among the three guidances.

We have tried to note in the text of each guidance when areas of overlap occur and when referencing one of the other guidances might be useful.

III. The Role of Risk Assessment in Risk Management

Risk management is an iterative process designed to optimize the benefit-risk balance for regulated products. Risk assessment consists of identifying and characterizing the nature, frequency, and severity of the risks associated with the use of a product. Risk assessment occurs throughout a product's lifecycle, from the early identification of a potential product, through the premarketing development process, and after approval during marketing. Premarketing risk assessment represents the first step in this process, and this guidance focuses on risk assessment prior to marketing.

It is critical to FDA's decision on product approval that a product's underlying risks and benefits be adequately assessed during the premarketing period. Sponsors seeking approval

---

[3] See 45 CFR part 46 and 21 CFR parts 50 and 56. See also the Health Insurance Portability and Accountability Act of 1996 (HIPAA) (Public Law 104-191) and the Standards for Privacy of Individually Identifiable Health Information (the Privacy Rule) (45 CFR part 160 and subparts A and E of part 164). The Privacy Rule specifically permits covered entities to report adverse events and other information related to the quality, effectiveness, and safety of FDA-regulated products both to manufacturers and directly to FDA (45 CFR 164.512(b)(1)(i) and (iii), and 45 CFR 164.512(a)(1)). For additional guidance on patient privacy protection, see http://www.hhs.gov/ocr/hipaa.

must provide from the clinical trials a body of evidence that adequately characterizes the product's safety profile.[4]

This guidance provides general recommendations for assessing risk. The adequacy of the assessment of risk is a matter of both quantity (ensuring that enough patients are studied) and quality (the appropriateness of the assessments performed, the appropriateness and breadth of the patient populations studied, and how results are analyzed). Quantity is, in part, considered in other Agency guidances,[5] but it is discussed further here. This guidance also addresses the qualitative aspects of risk assessment.

Although risk assessment continues through all stages of product development, this guidance focuses on risk assessment during the later stages of clinical development, particularly during phase 3 studies. The guidance is not intended to cover basic aspects of preclinical safety assessments (i.e., animal toxicity testing) or routine clinical pharmacology programs. Good clinical risk assessment in the later stages of drug development should be guided by the results of comprehensive preclinical safety assessments and a rigorous, thoughtful clinical pharmacology program (including elucidation of metabolic pathways, identification of possible drug-drug interactions, and determination of any effects from hepatic and/or renal impairment). These issues are addressed in other FDA guidances and guidances developed under the auspices of the International Conference for Harmonisation of Technical Requirements for Registration of Pharmaceuticals for Human Use (ICH).

IV. Generating Risk Information During Clinical Trials

Providing detailed guidance on what constitutes an adequate safety database for all products is impossible. The nature and extent of safety data that would provide sufficient information about risk for purposes of approving a product are individualized decisions based on a number of factors (several of which are discussed below). In reaching a final decision on approvability, both existing risk information and any outstanding questions regarding safety are considered in a product's risk assessment and weighed against the product's demonstrated benefits. The fewer a product's demonstrated benefits, the less acceptable may be higher levels of demonstrated risks. Likewise, the fewer the benefits, generally, the less uncertainty may be accepted about a product's risks.

To maximize the information gained from clinical trials, FDA recommends that from the outset of development, sponsors pay careful attention to the overall design of the safety evaluation.

---

[4] Section 505(d)(1) of the Federal Food, Drug, and Cosmetic Act (21 U.S.C. 355(d)(1)) requires the conduct of "adequate tests by all methods reasonably applicable to show whether or not . . . [a] drug is safe for use under the [labeled] conditions. . . ." See also 21 CFR 314.50(d)(5)(vi). Section 351 of the Public Health Service Act (42 U.S.C. 262) requires a demonstration that a biologic is "safe, pure, and potent." See also 21 CFR 601.2.

[5] See the guidance for industry *E1A The Extent of Population Exposure to Assess Clinical Safety: For Drugs Intended for Long-Term Treatment of Non-Life-Threatening Conditions*, International Conference on Harmonisation (ICH).

Potential problems that may be suspected because of preclinical data or because of effects of related drugs should be targeted for evaluation. And, because it is impossible to predict every important risk, as experience accrues, sponsors should refine or modify their safety evaluations.

A. Size of the Premarketing Safety Database

Even large clinical development programs cannot reasonably be expected to identify all risks associated with a product. Therefore, it is expected that, even for a product that is rigorously tested preapproval, some risks will become apparent only after approval, when the product is used in tens of thousands or even millions of patients in the general population. Although no preapproval database can possibly be sized to detect all safety issues that might occur with the product once marketed in the full population, the larger and more comprehensive the preapproval database, the more likely it is that serious adverse events will be detected during drug development.

The appropriate size of a safety database supporting a new product will depend on a number of factors specific to that product, including:

- Its novelty (i.e., whether it represents a new treatment or is similar to available treatment)

- The availability of alternative therapies and the relative safety of those alternatives as compared to the new product

- The intended population and condition being treated

- The intended duration of use

Safety databases for products intended to treat life-threatening diseases, especially in circumstances where there are no alternative satisfactory treatments, are usually smaller than for products intended to treat diseases that are neither life-threatening nor associated with major, irreversible morbidity. A larger safety database may be appropriate if a product's preclinical assessment or human clinical pharmacology studies identify signals of risk that warrant additional clinical data to properly define the risk. The appropriate size of the preapproval safety database may warrant specific discussion with the relevant review division. For instance, 21 CFR 312.82(b) (subpart E) provides that for drugs intended to treat life-threatening and seriously debilitating illnesses, end-of-phase 1 meetings can be used to agree on the design of phase 2 trials "with the goal that such testing will be adequate to provide sufficient data on the drug's safety and effectiveness to support a decision on its approvability for marketing."

For products intended for short-term or acute use (e.g., treatments that continue for, or are cumulatively administered for, less than 6 months), FDA believes it is difficult to offer general guidance on the appropriate target size of clinical safety databases. This is because of the wide range of indications and diseases (e.g., acute strokes to mild headaches) that may be targeted by such therapies. Sponsors are therefore encouraged to discuss with the relevant review division the appropriate size of the safety database for such products. Because products intended for lifethreatening and severely debilitating diseases are often approved with relatively small safety databases, relatively greater uncertainty remains regarding their adverse effects. Similarly, when products offer a unique, clinically important benefit to a population or patient group, less certainty in characterizing risk prior to approval may be acceptable.

For products intended for long-term treatment of non-life-threatening conditions, (e.g., continuous treatment for 6 months or more or recurrent intermittent treatment where cumulative treatment equals or exceeds 6 months), the ICH and FDA have generally recommended that 1500 subjects be exposed to the investigational product (with 300 to 600 exposed for 6 months, and 100 exposed for 1 year).[6] For those products characterized as chronic use products in the ICH guidance E1A, FDA recommends that the 1500 subjects include only those who have been exposed to the product in multiple dose studies, because many adverse events of concern (e.g., hepatotoxicity, hematologic events) do not appear with single doses or very short-term exposure. Also, the 300 to 600 subjects exposed for 6 months and 100 subjects exposed for 1 year should have been exposed to relevant doses (i.e., doses generally in the therapeutic range)

We note that it is common for well-conducted clinical development programs to explore doses higher than those ultimately proposed for marketing. For example, a dose tested in clinical trials may offer no efficacy advantage and show some dose-related toxicities; therefore, the sponsor does not propose the dose for marketing when the application is submitted. In such cases, data from subjects exposed to doses in excess of those ultimately proposed are highly informative for the safety evaluation and should be counted as contributing to the relevant safety database.

The E1A guidance describes a number of circumstances in which a safety database larger than 1500 patients may be appropriate, including the following:

1. There is concern that the drug would cause late developing adverse events, or cause adverse events that increase in severity or frequency over time. The concern could arise from:

   - Data from animal studies

   - Clinical information from other agents with related chemical structures or from a related pharmacologic class

   - Pharmacokinetic or pharmacodynamic properties known to be associated with such adverse events

2. There is a need to quantitate the occurrence rate of an expected specific low-frequency adverse event. Examples would include situations where a specific serious adverse event has been identified in similar products or where a serious event that could represent an alert event is observed in early clinical trials.

3. A larger database would help make risk-benefit decisions in situations when the benefit from the product:

---

[6] See the guidance for industry *E1A The Extent of Population Exposure to Assess Clinical Safety: For Drugs Intended for Long-term Treatment of Non-Life-Threatening Conditions*, ICH.

- Is small (e.g., symptomatic improvement in less serious medical conditions)

- Will be experienced by only a fraction of the treated patients (e.g., certain preventive therapies administered to healthy populations)

- Is of uncertain magnitude (e.g., efficacy determination on a surrogate endpoint)

4. Concern exists that a product may add to an already significant background rate of morbidity or mortality, and clinical trials should be designed with a sufficient number of patients to provide adequate statistical power to detect prespecified increases over the baseline morbidity or mortality.

The determination of whether the above provisions of the ICH E1A guidance are appropriate for a particular product development program and how these considerations would best be addressed by that program calls for evaluation on a case-by-case basis. Therefore, FDA recommends that this issue be discussed with the relevant review division at the end-of-phase 2 meeting, if not earlier.

In addition to the considerations provided in E1A, there are other circumstances in which a larger database may be appropriate.

1. The proposed treatment is for a healthy population (e.g., the product under development is for chemoprevention or is a preventive vaccine).

2. An effective alternative to the investigational product is already available and has been shown to be safe.

FDA is not suggesting that development of a database larger than that described in E1A is required or should be the norm. Rather, the appropriate database size would depend on the circumstances affecting a particular product, including the considerations outlined above. Therefore, FDA recommends that sponsors communicate with the review division responsible for their product early in the development program (e.g., at the pre-IND meeting) on the appropriate size of the safety database. FDA also recommends that sponsors revisit the issue at appropriate regulatory milestones (e.g., end-of-phase 2 and pre-NDA meetings).

B. Considerations for Developing a Premarketing Safety Database

Although the characteristics of an appropriate safety database are product-specific, some general principles can be applied. In general, efforts to ensure the quality and completeness of a safety database should be comparable to those made to support efficacy. Because data from multiple trials are often examined when assessing safety, it is particularly critical to examine terminology, assessment methods, and use of standard terms (e.g., use of the Medical Dictionary for Regulatory Activities (MedDRA)) to be sure that information is not obscured or distorted. Ascertainment and evaluation of the reasons for leaving assigned therapy during study (deaths and dropouts for any reason) are particularly important for a full understanding of a product's safety profile.

The following elements should be considered by sponsors when developing proposals for their clinical programs as these programs pertain to risk assessment.

1. Long-Term Controlled Safety Studies

It is common in many clinical programs for much of subject exposure data and almost all of long-term exposure data to come from single-arm or uncontrolled studies. Although these data can be informative, it may be preferable in some circumstances to develop controlled, long-term safety data. Such data allow for comparisons of event rates and facilitate accurate attribution of adverse events. Control groups may be given an active comparator or a placebo, depending on the disease being treated (i.e., the ethical and medical feasibility of using a placebo versus an active comparator will depend on the disease being treated).

The usefulness of active comparators in long-term safety studies depends on the adverse events of interest.

- Generally, serious events that rarely occur spontaneously (e.g., severe hepatocellular injury or aplastic anemia) would be considered significant and interpretable whenever (1) they are clearly documented and (2) there is no likely alternative explanation, since the expected rate is essentially zero in populations of any feasible size. As a result, the events can usually be appropriately interpreted and regarded as a signal of concern whether or not there is a control group.

- On the other hand, control groups are needed to detect increases in rates of events that are relatively common in the treated population (e.g., sudden death in patients with ischemic cardiac disease). Control groups are particularly important when an adverse event could be considered part of the disease being treated (e.g., asthma exacerbations occurring with inhalation treatments for asthma).

Therefore, FDA decisions as to when long-term comparative safety studies should be conducted for a product should be based on the intended use of the product, the nature of the labeled patient population (e.g., more useful if there is a high rate of serious adverse events), and earlier clinical and preclinical safety assessments. Although it is clear that long-term controlled studies will not usually be conducted, such studies may be particularly useful when a safety issue is identified during earlier development of the drug. In these cases, safety studies designed to test specific safety hypotheses may be appropriate. This would be especially true in situations where the safety issue of concern is more common with cumulative exposure. (See section IV.D below for further discussion of comparative trials.)

2. A Diverse Safety Database

Premarketing safety databases should include, to the extent possible, a population sufficiently diverse to adequately represent the expected target population, particularly in phase 3 studies. FDA has previously addressed this issue in a memorandum,[7] and the recommendations provided here are intended to supplement that document. To the extent feasible, only patients with obvious contraindications or other clinical considerations that clearly dictate exclusion

---

[7] The memorandum from Janet Woodcock, M.D., to Michael Friedman, M.D., dated July 20, 1998, and titled *FDAMA – Women and Minority Guidance Requirements* (with its attached report) discusses the regulations related to diversity. The memorandum can be found on the CDER guidance page under Modernization Act guidance (http://www.fda.gov/cder/guidance/women.pdf).

should be excluded from study entry. Inclusion of a diverse population allows for the development of safety data in a broad population that includes patients sometimes excluded from clinical trials, such as the elderly (particularly the very old), patients with concomitant diseases, and patients taking concomitant medications. Broadening inclusion criteria in phase 3 enhances the generalizability of the safety (and efficacy) findings. Although some phase 3 efficacy studies may target certain demographic or disease characteristics (and have narrower inclusion and exclusion criteria), overall, the phase 3 studies should include a substantial amount of data from less restricted populations.

3. Exploring Dose Effects Throughout the Clinical Program

Currently, it is common for only one dose, or perhaps a few doses, to be studied during drug development beyond phase 2. Yet, a number of characteristics common to many phase 2 studies limit the ability of these trials to provide definitive data on exposure-response or adequate data for definitive phase 3 dose selection. These characteristics of phase 2 studies (in comparison to phase 3 studies) include the following:

- Shorter durations of exposure

- Common use of pharmacodynamic (PD) endpoints, rather than clinical outcomes

- Smaller numbers of patients exposed

- Narrowly restrictive entry criteria

Although phase 3 trials do not necessarily need to examine a range of doses, such an examination is highly desirable, particularly when phase 2 studies cannot reasonably be considered to have established a single most appropriate dose. When a dose is not established in phase 2, more than one dose level should be examined in phase 3 trials of fixed dose products to better characterize the relationship between product exposure and resulting clinical benefit and risk. Dose-response data from phase 3 trials with multiple dose levels will help to better define the relationship of clinical response to dose for both safety and effectiveness. Furthermore, inadequate exploration of a product's dose-response relationship in clinical trials can raise safety concerns, since recommending doses in labeling that exceed the amount needed for effectiveness may increase risk to patients through dose-related toxicities with no potential for gain. Exposure-response data from phase 3 trials can also provide critical information on whether dose adjustments should be made for special populations. Finally, demonstrating a dose-response relationship in late phase clinical trials with meaningful clinical endpoints may aid the assessment of efficacy, since showing a dose ordering to efficacy can be compelling evidence of effectiveness.[8] When multiple dose levels are examined in phase 3 trials, the appropriate choice of doses to be included in these studies would be based on prior efficacy and safety information, including prior dose-ranging studies. In these circumstances, an end-of-phase 2 meeting with the appropriate review division would be particularly useful.

---

[8] See FDA's guidance for industry *Exposure-Response Relationships — Study Design, Data Analysis, and Regulatory Applications*.

C. Detecting Unanticipated Interactions as Part of a Safety Assessment

Even a well-conducted and reasonably complete general clinical pharmacology program does not guarantee a full understanding of all possible risks related to product interactions. Therefore, risk assessment programs should examine a number of interactions during controlled safety and effectiveness trials and, where appropriate, in specific, targeted safety trials. This examination for unanticipated interactions should include the potential for the following:

- Drug-drug interactions in addition to those resulting from known metabolic pathways (e.g., the effect of azole antibiotics on a CYP 3A4 dependent drug)

  We recommend that these examinations target a limited number of specific drugs, such as likely concomitant medications (e.g., for a new cholesterol lowering treatment, examining the consequences of concomitant use of HMG CoA reductase inhibitors and/or binding resins). The interactions of interest could be based, for example, on known or expected patterns of use, indications sought, or populations that are likely users of the drug.

- Product-demographic relationships — by ensuring sufficient diversity of the population (including gender, age, and race) to permit some assessments of safety concerns in demographic population subsets of the intended population

- Product-disease interactions — by ensuring sufficient variability in disease state and concomitant diseases

- Product-dietary supplement interactions for commonly used supplements that are likely to be co-administered or for which reasonable concerns exist (e.g., examination of the interactions between a new drug for the treatment of depression and St. John's Wort).

Again, FDA recommends that any such examinations target likely concomitant use based, for example, on indications sought, intended patterns of use, or the population of intended users of the drug and based on a history of drug and dietary supplement use elicited from subjects.

Generally, a sponsor determines its product's intended use and intended population(s) during product development. Decisions as to which interactions to either explore or specifically test in clinical trials could be based on these determinations and/or surveys and epidemiologic analyses.

One important way to detect unexpected relationships is by systematic incorporation of pharmacokinetic (PK) assessments (e.g., universal steady state sampling or population PK analyses) into some or all of the later phase clinical trials, including any specific safety trials. PK assessments can aid in the detection of unexpected PK interactions and, in some cases, could suggest exposure-response relationships for both safety and efficacy. Such data would allow for better assessment of whether pharmacokinetics contribute to any adverse events seen in the clinical trials, particularly rare, serious, and unanticipated events.

When a product has one or more well-established, valid biomarkers pertinent to a known safety concern, the marker should be studied during the PK studies and clinical development (e.g., creatine phosphokinase assessments used in the evaluation of new HMG CoA reductase inhibitors as a marker for rhabdomyolysis, or assessments of QT/QTc effects for new antihistamines).

D. Developing Comparative Safety Data

Depending on the drug and its indication, much of the safety data in an application may be derived from placebo-controlled trials and single-arm safety studies, with little or no comparative safety data. Although comparative safety data from controlled trials comparing the drug to an active control (these could also include placebo group) generally are not necessary, situations in which such data would be desirable include the following:

- The background rate of adverse events is high.

  The new drug may seem to have a high rate of adverse events in a single-arm study when, in fact, the rate is typical of that for other drugs. The additional use of a placebo would help to show whether either drug actually caused the adverse events.

- There is a well-established treatment with an effect on survival or irreversible morbidity.

  In such cases, not only are comparative data important scientifically, but the use of the comparator would likely be required ethically, as a placebo control could not be used and a single-arm trial would generally be uninformative.

- The sponsor hopes to claim superiority for safety or effectiveness.

  If a comparative effectiveness claim were sought, it would be expected that the studies would also address comparative safety, since a gain in effectiveness could be outweighed by or negated by an accompanying safety disadvantage.

In situations where there is a well-established related therapy, a comparative study of the new agent against that well-established therapy would be desirable (e.g., a new NSAID-like drug could be compared to a market-leading NSAID). Such a study could show whether the toxicity profile for the established therapy is generally similar to that for the novel therapy or whether important differences exist.

V. Special Considerations for Risk Assessment

Although many of the previous comments and recommendations are intended to apply to new product development programs generally, some risk assessment issues would apply only in certain circumstances or to certain types of products.[9]

A. Risk Assessment During Product Development

The following are examples of how risk assessment strategies could be tailored to suit special situations, where appropriate.

---

[9] The *Pharmacovigilance Guidance* discusses additional risk assessment strategies that may be initiated either pre or postapproval. In particular, the *Pharmacovigilance Guidance* includes a detailed discussion of pharmacoepidemiologic safety studies. Although such studies would principally be initiated after marketing, the *Pharmacovigilance Guidance* discusses certain situations when they could be initiated preapproval.

- If a product is intended to be chronically used (particularly when it has a very long halflife) and/or has dose-related toxicities, it can be useful to examine whether a lower or less frequent maintenance dose would be appropriate.

- If a product's proposed dosing includes a proposed titration scheme, the scheme could be based on specific studies to define how titration is best performed and the effects of titration on safety and efficacy.

- Certain kinds of adverse effects are not likely to be detected or readily reported by patients without special attention. When a drug has the potential for such effects, additional testing or specific assessments within existing trials may be appropriate.

  For example, for a new drug with recognized CNS effects (especially sedating effects), sponsors should conduct an assessment of cognitive function, motor skills, and mood. Similarly, since many antidepressants have significant effects on sexual function, new antidepressants should be assessed for these effects. The use of targeted safety questionnaires or specific psychometric or other validated instruments is often important for such assessments, since routine adverse event monitoring and safety assessments tend to underestimate or even entirely miss such effects.

- If a product is to be studied in pediatric patients, special safety issues should be considered (e.g., effects on growth and neurocognitive development if the drug is to be given to very young children/infants; safety of excipients for the very young; universal immunization recommendations and school entry requirements for immunization).

- A sponsor may consider reserving blood samples (or any other bodily fluids/tissues collected during clinical trials) from some or all patients in phase 3 studies for possible assessments at a later time, particularly in circumstances when earlier safety data signal an unusual or important concern. Such later assessments could include pharmacogenomic markers, assessments for immunogenicity, or measurements of other biomarkers that might prove helpful clinically. Having samples available for retrospective analysis of pharmacogenomic markers could help to link the occurrence of serious adverse event to particular genetic markers (e.g., haplotypes).

In unusual circumstances, a large, simple, safety study (LSSS) may be appropriate. An LSSS is usually a randomized clinical study designed to assess limited, specific outcomes in a large number of patients. These outcomes — generally important safety endpoints or safety concerns suggested by earlier studies — should be defined a priori with the study specifically designed to assess them. Although the large simple study model arose in the context of effectiveness assessment, and thus always involved randomized, controlled trials, an LSSS could in some cases be useful even without a control group — for example, to assess the rate of rare events (i.e., events so uncommon that usual safety studies would not be expected to provide good estimates of risk). Although an LSSS would most commonly be performed postapproval, either as a phase 4 commitment to address a lingering safety issue that does not preclude approval or outside of a formal phase 4 commitment in response to a new safety concern that arises after marketing, there are instances where an LSSS may be appropriate prior to approval. This would be the case when, for instance, there is a significant safety signal of concern (e.g., hepatotoxicity, myotoxicity) arising out of the developing clinical trial database that is not sufficiently resolved by the available data or is unlikely to be sufficiently addressed by the remaining ongoing studies. In these circumstances, an LSSS may be appropriate if the safety signal cannot otherwise be better delineated and the safety signal would have an impact on approvability.[10]

In addition, a sponsor seeking to develop a product for preventive use in at-risk, but otherwise healthy, individuals could conduct a large trial to investigate the product's safety. The use of a large trial may increase the chance of showing the product to have an acceptable benefit-risk profile in such cases, because the potential for benefit in the exposed population would generally be small. Such large trials, though not always LSSSs in a strict sense, may in some cases appropriately employ limited, targeted evaluations of both efficacy and safety endpoints, similar to an LSSS.

B. Assessing and Minimizing the Potential for Medication Errors

Sponsors can help minimize the occurrence of medication errors by assessing, prior to marketing, common sources of medication errors. Such errors may arise because of the product's inherent properties or because of the inadvertent contribution of the proposed proprietary name, the established name, the proposed labeling (e.g., container, carton, patient/consumer labeling, or professional package insert), and the proposed packaging.

Some medication errors, especially those involving parenteral products, have been detected in clinical trials prior to marketing. When occurring in clinical trials, events such as improper dilution or improper administration techniques, which may result in non-optimal dosing, should be carefully examined as warning signs that the product could be subject to dosing errors that warrant changes in labeling, packaging, or design. Even if errors are not observed in trials, careful consideration should be given during development to the implications of the design of the product, its packaging, and any device used to administer or deliver the product. For example, when a concentrated product that requires further dilution prior to intravenous administration is being developed, packaging is important. Packaging such a product in a syringe would make it possible to inject the product as a bolus without proper dilution, increasing risks to patients. Similarly, when developing a product that is administered or delivered by a device, the implications of mechanical failure of the device should be examined. Any such occurrences seen or considered during product development should be documented, reported, and analyzed for potential remedial actions (e.g., redesign of the device or modification of instructions for use).

Medication errors arising from confusion because of the similarity of the drug name, when written and spoken, to the name of another drug are less likely to be detected prior to marketing due to the controlled environment of clinical trials. However, the many well-documented cases of medication errors associated with similar proprietary names, confusing labels and labeling, and product packaging suggest it is important that sponsors carefully consider these issues before marketing a product.

Premarketing assessments should focus on:

---

[10] As mentioned in the RiskMAP Guidance, an LSSS could also be a method of evaluating the effectiveness of RiskMAP tools in actual practice prior to approval.

- Identifying all medication errors that occur during product development

- Identifying the reasons or causes for each identified error (e.g., dosage form, packaging, labeling, or confusion due to trade names when written or spoken)

- Assessing the resultant risk in the context of how and in whom the product will be used

- Identifying the means to minimize, reduce or eliminate the medication errors by ensuring the proper naming, labeling, design, and packaging of the product

Depending on the nature of the product, the indication, how it is administered, who will be receiving it, and the context in which it will be used, one or more of the following techniques may be helpful in assessing and preventing medication errors:

- Conducting a Failure Mode and Effects Analysis[11, 12]

- Use of expert panels

- Use of computer-assisted analysis

- Use of direct observation during clinical trials

- Directed interviews of consumers and medical and pharmacy personnel to better understand comprehension

- Use of focus groups

- Use of simulated prescription and over-the-counter (OTC) use studies

Additional information on the application of these assessment techniques will be published in a future guidance document.

C. Addressing Safety Aspects During Product Development

FDA recommends addressing the potential for the following serious adverse effects as a part of the new drug application (NDA) for all new small molecule drugs:

- Drug-related QTc prolongation

- Drug-related liver toxicity

- Drug-related nephrotoxicity

- Drug-related bone marrow toxicity

- Drug-drug interactions

- Polymorphic metabolism

---

[11] Stamatis, D.H., *Failure Mode and Effect Analysis: FMEA From Theory to Execution*,. Milwaukee: American Society for Quality, Quality Press, 2003.
[12] Cohen, Michael R. ed., *Medication Errors: Causes, Prevention, and Risk Management*, Washington D.C.: American Pharmaceutical Association, 1999.

Prior experience has shown that these effects can often be identified when properly assessed in clinical development programs. Although FDA believes it is important to address these potential effects in all NDAs, adequately addressing all of these considerations would not necessarily involve the generation of additional data or the conduct of specific trials. (For some issues, such as QTc, specifically conducted preclinical and clinical studies are generally recommended.) For example, a drug that is intended to be topically applied may be shown to have no systemic bioavailability; therefore, systemic toxicities would be of no practical concern.

Some of the above-listed potential effects are relevant to biological products; some are not. In addition, for biological products such as cytokines, antibodies, other recombinant proteins, and cell-, gene-, and tissue-based therapeutics, it may be appropriate to assess other issues. The issues listed here are dependent on the specific nature of the biological product under development.

- Potentially important issues for biological products include assessments of immunogenicity, both the incidence and consequences of neutralizing antibody formation and the potential for adverse events related to binding antibody formation.

- For gene-based biological products, transfection of nontarget cells and transmissibility of infection to close contacts, and the genetic stability of products intended for longpersistence transfections constitute important safety issues.

- For cell-based products, assessments of adverse events related to distribution, migration, and growth beyond the initial intended administration are important, as are adverse events related to cell survival and demise. Such events may not appear for a long time after product administration.

A complete discussion of assessment of safety issues unique to biological products is beyond the scope of this guidance. We recommend that sponsors address the unique safety concerns pertaining to the development of any particular biological product with the relevant product office.

VI. Data Analysis And Presentation

Many aspects of data analysis and presentation have been previously addressed in guidance, most notably in FDA's Guideline for the *Format and Content of the Clinical and Statistical Sections of an Application* and the ICH guidances *E3 Structure and Content of Clinical Study Reports* and *M4 Common Technical Document for the Registration of Pharmaceuticals for Human Use.* We do not repeat that guidance here, but offer new guidance on selected issues.

With regard to the guidance offered in this section of the document, it is important to emphasize that the regulatory approach to the evaluation of the safety of a product usually differs substantially from the evaluation of effectiveness. Most studies in the later phases of drug or biologic development are directed toward establishing effectiveness. In such studies, critical efficacy endpoints are identified in advance, and statistical planning is conducted based on being able to make definitive statistical inferences about efficacy. In contrast, these later phase trials are not generally designed to test specified hypotheses about safety or to measure or identify adverse events with any prespecified level of sensitivity. Therefore, the premarket safety evaluation is often, by its nature, exploratory and is intended to identify common adverse events related to the therapy, as well as to help identify signals for serious and/or less common adverse events.

A. Describing Adverse Events to Identify Safety Signals

Because individual investigators may use different terms to describe a particular adverse event, FDA recommends that sponsors ensure that each investigator's verbatim terms are coded to standardized, preferred terms specified in a coding convention or dictionary. Proper coding allows similar events that were reported using different verbatim language to be appropriately grouped. Consistent and accurate coding of adverse events allows large amounts of data regarding these events to be analyzed and summarized and maximizes the likelihood that safety signals will be detected. Inaccurate coding, inconsistent coding of similar verbatim terms, and inappropriate "lumping" of unrelated verbatim terms or "splitting" of related verbatim terms can obscure safety signals.

In general, FDA suggests that sponsors use one coding convention or dictionary (e.g., MedDRA) throughout a clinical program with the understanding that, due to the duration of product development, the coding convention used may undergo revisions. Use of more than one coding convention or dictionary can result in coding differences that prevent adverse event data from being appropriately grouped and analyzed. To the extent possible, sponsors should use a single version of the selected convention or dictionary without revisions. However, if this is not possible, it is important to appropriately group and analyze adverse events taking into account the revisions in subsequent versions. It is not advisable to analyze adverse event data using one version and then base proposed labeling on a different version.

1. Accuracy of Coding

Sponsors should explore the accuracy of the coding process with respect to both investigators and the persons who code adverse events.

- Investigators may sometimes choos verbatim terms that do not accurately communicate the adverse event that occurred.

  — The severity or magnitude of an event may be inappropriately exaggerated (e.g., if an investigator terms a case of isolated elevated transaminases *acute liver failure* despite the absence of evidence of associated hyperbilirubinemia, coagulopathy, or encephalopathy, which are components of the standard definition of acute liver failure).

  — Conversely, the significance or existence of an event may be masked (e.g., if an investigator uses a term that is nonspecific and possibly unimportant to describe a subject's discontinuation from a study when the discontinuation is due to a serious adverse event).

  If an adverse event is mischaracterized, sponsors could consider, in consultation with FDA, recharacterizing the event to make it consistent with accepted case definitions. We recommend that recharacterization be the exception rather than the rule and, when one, be well documented with an audit trail.

- We recommend that in addition to ensuring that investigators have accurately characterized adverse events, sponsors confirm that verbatim terms used by investigators have been appropriately coded.

  Sponsors should strive to identify obvious coding mistakes as well as any instances when a potentially serious verbatim term may have been inappropriately mapped to a more benign coding term, thus minimizing the potential severity of an adverse event. One example is coding the verbatim term *facial edema* (suggesting an allergic reaction) as the nonspecific term *edema*; another is coding the verbatim term *suicidal ideation* as the more benign term *emotional lability*.

- Prior to analyzing a product's safety database, sponsors should ensure that adverse events were coded with minimal variability across studies and individual coders.

  Consistency is important because adverse event coding may be performed over time, as studies are completed, and by many different individuals. Both of these factors are potential sources of variability in the coding process. FDA recommends that to examine the extent of variability in the coding process, sponsors focus on a subset of preferred terms, particularly terms that are vague and commonly coded differently by different people. For example, a sponsor might evaluate the consistency of coding verbatim terms such as weakness and asthenia or dizziness and vertigo. NOS (not otherwise specified)-type codes, such as ECG abnormality NOS, are also coding terms to which a variety of verbatim terms may often be mapped. These should be examined for consistency as well. Sponsors should pay special attention to terms that could represent serious or otherwise important adverse reactions.

In addition to considering an adverse event independently and as it is initially coded, sponsors should also consider a coded event in conjunction with other coded events in some circumstances. Certain adverse events or toxicities (particularly those with a constellation of symptoms, signs, or laboratory findings) may be defined as an amalgamation of multiple preferred coding terms. Sponsors should identify these events (e.g., acute liver failure) based on recognized definitions.

2. Coding Considerations During Adverse Event Analysis

When analyzing an adverse event, sponsors should consider the following:

- Combining related coding terms can either amplify weak safety signals or obscure important toxicities.

  For example, the combination of dyspnea, cough, wheezing, or pleuritis might provide a more sensitive, although less specific, appraisal of pulmonary toxicity than any single term. Conversely, by combining terms for serious, unusual events with terms for more common, less serious events (e.g., constipation might include cases of toxic megacolon), the more important events could be obscured.

- Coding methods can divide the same event into many terms. Dividing adverse event terms can decrease the apparent incidence of an adverse event (e.g., including pedal edema, generalized edema, and peripheral edema as separate terms could obscure the overall finding of fluid retention).

Although potentially important safety events cannot always be anticipated in a clinical development program, sponsors, in consultation with the Agency, should prospectively group adverse event terms and develop case definitions or use accepted standardized definitions whenever possible.

- A prospective grouping approach is particularly important for syndromes such as serotonin syndrome, Parkinsonism, and drug withdrawal, which are not well characterized by a single term.

- Some groupings can be constructed only after safety data are obtained, at which time consultation with FDA might be considered.

- Sponsors should explain such groupings explicitly in their applications so that FDA reviewers have a clear understanding of what terms were grouped and the rationale for the groupings.

- For safety signals that are identified toward the end of a development program, the pre-NDA meeting would be a reasonable time to confer with FDA regarding such groupings or case definitions.

B. Analyzing Temporal or Other Associations

For individual safety reports, the temporal relationship between product exposure and adverseevent is a critical consideration in the assessment of potential causality. However, temporal factors, including the duration of the event itself, are often overlooked during the assessment of aggregate safety data. Simple comparisons of adverse event frequencies between (or among) treatment groups, which are commonly included in product applications and reproduced in tabular format in labeling, generally do not take into account the time dependency of adverse events. Temporal associations can help further understand causality, adaptation, and tolerance, but may be obscured when only frequencies of adverse events are compared.

Temporal analyses may be warranted for important adverse events whether they arise from controlled clinical trial data or treatment cohorts. In both cases, analyzing changes over time may be important for assessing risk and potential causality. Analyses of temporal associations are particularly worth conducting in situations where prior experience (e.g., experience from similar products) has shown that a temporal relationship between product exposure and ensuing adverse events is likely to exist. In addition, in the context of controlled clinical trials, temporal analyses may provide insight into the relative importance of differences in adverse event frequencies between study groups.

Descriptions of risk as a function of subjects' duration of exposure to a product, or as a function of time since initial exposure, can contribute to the understanding of the product's safety profile. Assessments of risk within discrete time intervals over the observation period (i.e., a hazard rate curve) can be used to illustrate changes in risk over time (e.g., flu-like symptoms with interferons that tend to occur at the initiation of treatment but diminish in frequency over time). It may be useful for sponsors to consider event rates (events per unit of time) in reconciling apparent differences in the frequencies of events between studies when there are disparities in subjects' time of exposure or time at risk.

For important events that do not occur at a constant rate with respect to time and for events in studies where the size of the population at risk (denominator) changes over time, a life-table or Kaplan-Meier approach may be of value for evaluating risks of adverse events. Clinically important events (e.g., those events for which the occurrence of even a few cases in a database may be significant) are of particular interest. Examples of such events include the development of restenosis following coronary angioplasty, cardiac toxicity, and seizures.

Temporal associations identified in previous experience with related products can help focus sponsor analyses of potential temporal associations for a product under study, but sponsors should balance this approach with an attempt to detect unanticipated events and associations as well. Knowledge of a product's pharmacokinetic and pharmacodynamic profiles, as well as an appreciation of physiologic, metabolic, and host immune responses, may be important in understanding the possible timing of treatment-related adverse events.

It is important to consider study and concomitant treatment regimens (i.e., single treatment; short course of treatment; continuous, intermittent, titrated, or symptom-based treatment) in temporal analyses. Other important factors to consider in planning and interpreting temporal analyses are (1) the initiation or withdrawal of therapies and (2) changes in the severity or frequency of subjects' preexisting conditions over time.

For events that decrease in frequency over time and are found to be associated with the initiation of treatment, supplemental analyses may be of value to discriminate the relative contributions of adaptation, tolerance, dose reduction, symptomatic treatment, decreases in reporting, depletion of susceptibles, and subject dropout.

C. Analyzing Dose Effect as a Contribution to Risk Assessment

Sponsors should analyze event rates by dose for clinically important adverse events that may be product related and events that might be expected based on a product's pharmacologic class or preclinical data.

For studies involving the evaluation of a range of doses, dose response is most commonly assessed by analyzing adverse event frequencies by administered dose. In such studies, it may also be useful to consider event frequencies by weight-adjusted or body surface area-adjusted dose, especially if most patients are given the same dose regardless of body weight or size. It should be recognized, however, that when doses are adjusted by a subject's weight or body surface area, women are commonly overrepresented on the upper end of the range of adjusted doses, and men are commonly overrepresented on the lower end of this range. For products administered over prolonged periods, it may be useful to analyze event rates based on cumulative

dose. In addition, when specific demographic or baseline disease-related subgroups may be at particular risk of incurring adverse events, exploration of dose-response relationships by subgroup is important. Subgroup analyses have the potential to provide a more reliable and relevant estimate of risk for important subgroups of the target population. Alternatively, multiplicity issues could result in an apparent signal that does not represent a real finding (i.e., a false positive).

Although the most reliable information on dose response comes from randomized fixed dose studies, potentially useful information may emerge from titration studies and from associations between adverse events and plasma drug concentrations.

For dose titration or flexible dose studies, it would generally be useful to assess the relationship between adverse event frequencies and the actual doses subjects received preceding the adverse events or the cumulative dose they received at the onset of the events. The choice is a function of the mode of action, pharmacokinetics, and pharmacodynamics of the product.

For products with a stepped dosing algorithm (i.e., incremental dosing based on age or weight), the actual cut points of the paradigm are often selected relatively early in product development. Although the cut points may be based on the best knowledge available at the time, it is useful in such cases to make a specific effort to explore safety (and efficacy) just above and below these points. For example, if the dose of a product is to be 100 mg for patients weighing less than 80 kg and 150 mg for patients weighing 80 kg or more, an assessment of the comparative safety profiles of patients weighing from 75 to 79.9 kg versus patients weighing from 80 to 84.9 kg would be valuable.

As is typical of most safety evaluations, the likelihood of observing false positive signals increases with the number of analyses conducted. Positive associations between adverse events and dose, as well as signals that emerge from subgroup analysis, should be considered with this in mind. Such associations should be examined for consistency across studies, if possible.

D. Role of Data Pooling in Risk Assessment

Data pooling is the integration of patient-level outcome data from several clinical studies to assess a safety outcome of interest. Generally, data pooling is performed to achieve larger sample sizes and data sets because individual clinical studies are not designed with sufficient sample size to estimate the frequency of low incidence events or to compare differences in rates or relative rates between the test drug (exposed group) and the control (unexposed group). Use of pooled data does not imply that individual study results should not be examined and considered. When pooling data, sponsors should consider the possibility that various sources of systematic differences can interfere with interpretation of a pooled result. To ensure that pooling is appropriate, sponsors should confirm that study designs, as well as ascertainment and measurement strategies employed in the studies that are pooled, are reasonably similar.

Used appropriately, pooled analyses can enhance the power to detect an association between product use and an event and provide more reliable estimates of the magnitude of risk over time. Pooled analyses can also provide insight into a positive signal observed in a single study by allowing a broader comparison. This can protect against undue weight being given to chance findings in individual studies. However, a finding from a single study should not be automatically dismissed because of the results of a pooled analysis, especially if it is detected in a study of superior design or in a different population. Any pooled analysis resulting in a reduced statistical association between a product and an observed risk or magnitude of risk, as compared to the original safety signal obtained from one or more of the contributing studies, should be carefully examined.

Some issues for consideration in deciding whether pooling is appropriate include possible differences in the duration of studies, heterogeneity of patient populations, and case ascertainment differences across studies (i.e., different methods for detecting the safety outcomes of interest, such as differences in the intensities of patient follow-up). When there is clinical heterogeneity among trials with regard to the safety outcome of interest (e.g., major disparity in findings for particular safety endpoints), sponsors should present risk information that details the range of results observed in the individual studies, rather than producing a summary value from a pooled analysis.

E. Using Pooled Data for Risk Assessment

All placebo-controlled studies in a clinical development program should be considered and evaluated for appropriateness for inclusion in a pooled analysis. Decisions to exclude certain placebo-controlled studies from, or to add other types of studies (such as active-controlled studies or open-label studies) to, a pooled analysis would depend on the objectives of the analysis. Such analyses should be conducted in a manner that is consistent with the following guiding principles:

- Generally, phase 1 pharmacokinetic and pharmacodynamic studies should be excluded.

  These are usually single- or multiple-dose trials of a short duration conducted in healthy subjects or in patients with refractory or incurable end-stage disease who have confounding symptoms. Unless a risk were limited to a short period immediately after the first dose, inclusion of these studies in a pooled analysis would not increase the statistical power or contribute to the precision of the risk estimates. However, inclusion of these studies could (1) diminish the magnitude of apparent risk by including a population with little or no possibility of having had the adverse reaction or (2) increase the apparent magnitude of risk because of significant baseline symptoms unrelated to the drug.

- The risk of the safety outcome of interest should be expressed in reference to total person-time (exposure time) or be evaluated using a time-to-event analysis.

  When the duration of drug exposure for the individual subjects included in a pooled analysis varies, sponsors should not express the risk merely in terms of event frequency (that is, using persons as the denominator). Use of the person-time approach relies on the assumption that the risk is constant over the period of the studies. Whenever there is concern regarding a non-constant nature of a risk, a time-to-event log-rank type analysis may be helpful, as it is a robust approach even when risk is not constant over time.

- The patient population in the pooled analysis should be relatively homogeneous with respect to factors that may affect the safety outcome of interest (e.g., dose received, duration of therapy).

The pooled analysis should be of a size sufficient to allow analyses of demographic subgroups (gender, age, race, geographic locations).

- The studies included in a pooled analysis should have used similar methods of adverse event ascertainment, including ascertainment of the cause of dropouts.

Study-specific incidence rate should be calculated and compared for any signs of case ascertainment differences. Since study-to-study variation is to be expected, it is a challenge to distinguish between possible case ascertainment differences and study-tostudy variation.

There are some situations in which pooling may be relatively straightforward. For example, a pooled analysis of similarly designed phase 3 studies could readily be used to create a table of common adverse events. This type of analysis is typically less subject to the problems discussed above because (1) the studies are similar in study design and patient population and (2) the intent of such an analysis is often more descriptive than quantitative. However, if a specific safety concern is raised during the clinical development program, the guiding principles discussed above should be closely followed when conducting a prespecified pooled analysis.

F. Rigorous Ascertainment of Reasons for Withdrawals from Studies

Subjects may drop out or withdraw from clinical trials for many reasons, including perceived lack of efficacy, side effects, serious adverse events, or an unwillingness to expend the effort necessary to continue. The reasons for dropout are not always clear. This lack of information may be largely irrelevant (e.g., discontinuation due to moving from the area) or indicative of an important safety problem (e.g., stroke). Therefore, regardless of the reason for withdrawal, sponsors should attempt to account for all dropouts.

- Sponsors should try to ascertain what precipitated dropout or withdrawal in all cases, particularly if a safety issue was a part of the reason for withdrawal.

It is not helpful to simply record vague explanations such as "withdrew consent," "failed to return," "administratively withdrawn," or "lost to follow-up."

- Participants who leave a study because of serious or significant safety issues should be followed closely until the adverse events are fully and permanently resolved or stabilized (if complete resolution is not anticipated), with follow-up data recorded in the case report forms.

- Follow-up information should be pursued on patients withdrawn from the study (for reasons other than withdrawing consent in the absence of an adverse event).

If this information is not obtainable, FDA recommends that the measures taken to obtain follow-up information be reflected on the case report forms and the resultant failure to obtain the information should be discussed in the clinical discussion of safety.

- Patients considering withdrawing consent should be encouraged to provide the reason, and patients who withdraw should be encouraged to provide information as to whether the withdrawal of consent resulted from a serious or significant safety issue.

- Some patients withdraw due to abnormal laboratory values, vital signs, or ECG findings that are not characterized as adverse events. Sponsors should include information on these types of discontinuations in addition to information on discontinuations due to adverse events.

G. Long-term Follow-up

In some cases, it is recommended that all subjects be followed to the end of the study or even after the formal end of the study (e.g., where the drug has a very long half-life, is deposited in an organ such as bone or brain, or has the potential for causing irreversible effects, such as cancer).

The concern over adequate follow-up for ascertaining important safety events in such cases is particularly critical in long-term treatment and clinical outcome studies. In such cases, FDA recommends the follow-up for late safety events, even for subjects off therapy, include those subjects who drop out of the trial or who finish the study early due to meeting a primary outcome of interest. The duration of follow-up, however, would be dependent on the circumstances of the product development and therefore should be discussed with the appropriate review division (e.g., during end-of-phase 2 meetings).

H. Important Aspects of Data Presentation

We recommend that once a product's safety data have been analyzed, comprehensive risk assessment information be presented succinctly. FDA and ICH have provided extensive guidance regarding the presentation of safety data,[13],[14],[15] and we offer these additional recommendations, which have not been addressed previously.

- For selected adverse events, adverse event rates using a range of more restrictive to less restrictive definitions (e.g., myocardial infarction versus myocardial ischemia) should be summarized.

  The events chosen for such a summary might be limited to more serious events and events that are recognized to be associated with the relevant class of drugs;

- For a drug that is a new member of an established class of drugs, the adverse events that are important for the class of drug should be fully characterized in the NDA's integrated summary of safety.

  That characterization should include an analysis of the incidence of the pertinent adverse events, as well as any associated laboratory, vital sign, or ECG data. For example, the characterization of a drug joining a class that is associated with orthostatic hypotension would include analyses of orthostatic blood pressure changes as well as the incidence of syncope, dizziness, falls, or other events. We recommend that when sponsors are establishing case definitions for particular adverse events, they consider definitions previously used for the other drugs in the class or, if available, standard definitions.

---

[13] See *Guideline for the Format and Content of the Clinical and Statistical Section of an Application.*
[14] See the guidance for industry *E3 Structure and Content of Clinical Study Reports*, ICH.
[15] See the guidance for industry *M4 Common Technical Document for the Registration of Pharmaceuticals for Human Use*, ICH.

- The distribution of important variables across the pooled data, such as gender, age, extent of exposure, concomitant medical conditions, and concomitant medications (especially those that are used commonly to treat the indication being studied), should be included in the integrated summary of safety.

- The effect of differential discontinuation rates by treatment on adverse event occurrence should be characterized (e.g., when placebo-treated patients drop out of a trial earlier than patients being treated with an active drug). This differential discontinuation can lead to misleading adverse event incidences unless patient exposure is used as the denominator for risk calculations.

- Case report forms (CRFs) submitted for patients who died or discontinued a study prematurely due to an adverse event should include copies of relevant hospital records, autopsy reports, biopsy reports, and radiological reports, when feasible. The possibility that such information may be reported to FDA should be stated in the informed consent document with a notation that the patient would not be identified in such reports.

These source documents should become a formal part of the official CRF and be properly referenced.

- Narrative summaries (as previously described in guidance[16]) of important adverse events (e.g., deaths, events leading to discontinuation, other serious adverse events) should provide the detail necessary to permit an adequate understanding of the nature of the adverse event experienced by the study subject. (This level of detail may be unnecessary for events expected in the population (e.g., late deaths in a cancer trial). This issue should be discussed with the appropriate review division.)

Narrative summaries should not merely provide, in text format, the data that are already presented in the case report tabulation, as this adds little value. A valuable narrative summary would provide a complete synthesis of all available clinical data and an informed discussion of the case, allowing a better understanding of what the patient experienced. The following is a list of components that would be found in a useful narrative summary:

- Patient age and gender

- Signs and symptoms related to the adverse event being discussed

- An assessment of the relationship of exposure duration to the development of the adverse event

- Pertinent medical history

- Concomitant medications with start dates relative to the adverse event

- Pertinent physical exam findings

- Pertinent test results (e.g., lab data, ECG data, biopsy data)

- Discussion of the diagnosis as supported by available clinical data

---

[16] See the guidance for industry *E3 Structure and Content of Clinical Study Reports*, ICH.

– For events without a definitive diagnosis, a list of the differential diagnoses

– Treatment provided

– Re-challenge results (if performed)

– Outcomes and follow-up information

Providing Regulatory Submissions in Electronic Format – Human Pharmaceutical Product Applications and Related Submissions Using the eCTD Specifications

Providing Regulatory Submissions in Electronic Format – Human Pharmaceutical Product Applications and Related Submissions Using the eCTD Specifications, Guidance for Industry [Ŧ, 1]

U.S. Department of Health and Human Services
Food and Drug Administration
Center for Drug Evaluation and Research (CDER)
Center for Biologics Evaluation and Research (CBER)

October 2005

Electronic Submissions

Contains Nonbinding Recommendations

This guidance discusses issues related to the electronic submission of new drug applications (NDAs), abbreviated new drug applications (ANDAs), biologics license applications (BLAs), investigational new drug applications (INDs), master files, advertising material, and promotional labeling using the electronic common technical document (eCTD) specifications.

Note: *This guidance represents the Food and Drug Administration's (FDA's) current thinking on this topic. It does not create or confer any rights for or on any person and does not operate to bind FDA or the public. You can use an alternative approach if the approach satisfies the requirements of the applicable statutes and regulations. If you want to discuss an alternative approach, contact the appropriate FDA staff. If you cannot identify the appropriate FDA staff, call the appropriate number listed on the title page of this guidance.*

Technical specifications associated with this guidance will be provided as stand alone documents. They will be updated periodically. To ensure that you have the most recent versions, check the appropriate center's guidance Web page. For CBER, this Web site is

---

[Ŧ] Available on the FDA website at: http://www.fda.gov/RegulatoryInformation/ Guidances/ucm126959.htm
[1] This guidance has been developed by the Center for Drug Evaluation and Research (CDER) and the Center for Biologics Evaluation and Research (CBER).

http://www.fda.gov/cber/esub/esub.htm. For CDER, this Web site is
http://www.fda.gov/cder/regulatory/ersr/ectd.htm.

I. Introduction

This is one in a series of guidance documents intended to assist applicants making regulatory submissions to the FDA in electronic format using the electronic common technical document (eCTD) specifications. This guidance discusses issues related to the electronic submission of applications for human pharmaceutical products[2] and related submissions, including abbreviated new drug applications (ANDAs), biologics license applications (BLAs), investigational new drug applications (INDs), new drug application (NDAs), master files (e.g., drug master files), advertising material, and promotional labeling.[3] At this time, this does not include applications supporting combination products.

The goals of the guidance are to enhance the receipt, processing, and review of electronic submissions to the FDA. Specifically, this guidance makes recommendations regarding the use of the eCTD backbone files developed through the International Conference on Harmonisation (ICH) to facilitate efficient submission handling. In addition, the guidance provides more specificity than in previous guidances for electronic submissions with regard to the organization of individual submissions. Finally, the guidance harmonizes the organization and formatting of electronic submissions for multiple submission types.

This guidance refers to a series of technical specifications associated with the guidance. They are being provided as stand alone documents to make them more accessible to the user. **The associated specifications will be updated periodically. To ensure that you have the most recent versions, check the appropriate center's guidance Web page.**

FDA's guidance documents, including this guidance, do not establish legally enforceable responsibilities. Instead, guidances describe the Agency's current thinking on a topic and should be viewed only as recommendations, unless specific regulatory or statutory requirements are cited. The use of the word should in Agency guidances means that something is suggested or recommended, but not required.

---

[2] Human pharmaceutical products include those products that meet the definition of drug under the Food, Drug and Cosmetic Act, including those that are chemically synthesized and those derived from living sources (biologic products).

[3] Agency guidance documents on electronic submissions will be updated regularly to reflect the evolving nature of the technology and the experience of those using this technology.

Paperwork Reduction Act of 1995: This guidance contains information collection provisions that are subject to review by the Office of Management and Budget (OMB) under the Paperwork Reduction Act of 1995 (44 U.S.C. 3501-3520). The collections of information in this guidance have been approved under OMB Control Nos. 0910-0014, 0910-0001, and 0910-0338.

II. General Issues

This portion of the guidance makes recommendations on general organizational issues related to the electronic submission of applications for human pharmaceutical products using the eCTD specifications. The requirements for *the content* of such applications are described in our regulations in chapter 21 of the Code of Federal Regulations (CFR). Additional recommendations on the contents of applications are provided in Agency guidances, which are available on the Agency Web page.

A. Scope

This guidance applies to marketing applications (ANDAs, BLAs, NDAs), investigational applications (INDs), and related submissions (master files, advertising material, and promotional labeling). The guidance applies equally to original submissions, supplements, annual reports, and amendments to these applications and related submissions, including correspondence. This guidance does not apply to electronic submission of prelicense or preapproval inspection materials.

B. Guidance on the Content of Applications and Related Submissions

This document provides general guidance on how to organize application information for electronic submission to the Agency using the eCTD specifications. Guidance on the information to be included in the technical sections of applications and submissions is described in a series of guidance documents based on the International Conference on Harmonisation of Technical Requirements for Registration of Pharmaceuticals for Human Use (ICH) common technical document (CTD): *M4: Organization of the CTD, M4Q: The CTD – Quality; M4S – The CTD Safety;* and *M4E: The CTD – Efficacy.*

C. ICH eCTD Specification

The recommendations made here on how to organize application information are based on the ICH CTD and the electronic CTD (eCTD), which was developed by the ICH M2 expert working group. Although the CTD and the eCTD were designed for marketing applications, they could apply equally to other submission types, including INDs, master files, advertising material, and promotional labeling.[4] Details on the specification for the ICH eCTD can be found in the guidance document *M2 eCTD: Electronic Common Technical Document Specification.*

D. Document Granularity and Table of Contents Headings

Submissions are a collection of documents. A document is a collection of information that includes forms, reports, and datasets. When making an electronic submission, *each document should be provided as a separate file*.[5] The documents, whether for a marketing application, an investigational application, or a related submission, should be organized based

---

[4] Advertising and promotional labeling provided with marketing applications.
[5] Some documents are provided in more than one file because a file containing everything would be too large. See specifications for the size limitations for a file.

on the five modules in the CTD: module 1 includes administrative information and prescribing information, module 2 includes CTD summary documents, module 3 includes information on quality, module 4 includes the nonclinical study reports, and module 5 includes the clinical study reports.

A table of contents is defined by headings arranged in a hierarchical fashion. See the associated specification, Comprehensive Table of Contents Headings and Hierarchy for the comprehensive listing of headings and hierarchy Because this is a comprehensive listing, not all headings are applicable to all submissions or submission types. All of the information you need to submit is covered by these headings. If you think other headings are needed, you should contact our electronic submission coordinators prior to using any other headings (see section II.S of this guidance).   Reviewers will not be able to access documents associated with headings not listed in the "Comprehensive Table of Contents Headings and Hierarchy."

Unless otherwise specified, documents should be organized so that the subject matter of the document is specifically associated with the lowest heading in the table of contents hierarchy. For example, if you look at the associated document "Comprehensive Table of Contents Headings and Hierarchy," the headings "Meeting request" and "Meeting background material" are the lowest headings in the "Meeting" hierarchy. Therefore, the meeting request and meeting background material would be in two separate documents — the meeting request in one document and the meeting background material in another document.

A document can be associated with more than one heading. However, the actual electronic file would only be provided once. The eCTD specifications provide details on how to refer to an electronic file.

E. Electronic Submissions

Under our regulations (21 CFR 11.2(b)(2)), applicants and sponsors are expected to contact us for details on how to proceed with electronic submissions. These details are usually provided in guidance documents. For example, we are already receiving marketing application submissions for human pharmaceutical products in electronic format based on details provided in the guidances for industry *Providing Regulatory Submissions in Electronic Format – NDAs, Providing Regulatory Submissions in Electronic Format – ANDAs, Providing Regulatory Submissions to the Center for Biologics Evaluation and Research (CBER) in Electronic Format – Biologics Marketing Applications, and Providing Regulatory Submissions in Electronic Format – General Considerations.*[6] However, we recommend that you begin submitting eCTD backbone files as described in this guidance because we believe that having the information in the eCTD backbone files will result in greater efficiency in the future. In time, the other guidances may be withdrawn because they may no longer be needed.

When we are ready to receive a particular submission type in electronic format only, we usually identify it in public docket 92S-0251. Under 21 CFR part 11, you then have the option of providing that submission type in electronic format according to FDA guidance so that the Agency may adequately process, archive, and review the files.

---

[6] This includes mixed electronic and paper submissions.

Once you begin to submit a specific application in electronic format based on this guidance, subsequent submissions to the application, including amendments and supplements, should include eCTD backbone files. Without the eCTD backbone files, we will not be able to adequately manage, process, archive, or review the submissions. If you choose to submit an original application using the eCTD backbone files, you should obtain an application number in advance by contacting the appropriate center. You may obtain the number at any time and the numbers will not be reused.

We believe it is most beneficial to begin your eCTD-based submissions with the initial submission of an application. Contact the appropriate center first if you wish to make eCTD-based submissions to pending applications. You should avoid the submission of any paper documents when you follow the recommendations in this document. The maximum benefit will be derived once an application is in electronic format. This is particularly true for the IND, where submissions are provided over a long period of time. You should submit the electronic information for all files in the eCTD backbone files following the specifications associated with this guidance.

F. Document Information for Previous Submissions

If you decide to submit a specific application in electronic format based on this guidance, you do not have to provide eCTD backbone files for the previous submissions to the application. For example, if you submitted an original application in 2001 and now submit an amendment to the application using the eCTD backbone files, you do not have to go back and submit the document information for the files submitted in 2001.

G. Referencing Previously Submitted Documents[7]

If a document was submitted in electronic format with the eCTD backbone files, you should not submit additional copies when referencing the previously submitted document. Instead, you should include the information by reference by providing in the text of the document (1) the application or master file number, (2) the date of submission (e.g., letter date), (3) the document name, and (4) the page number of the referenced document along with a hypertext link to the location of the information (see section II.Q of this guidance). If a document replaces or appends a document previously submitted with an eCTD backbone file, then you should include this information in the appropriate eCTD backbone file. The details on how to include this information in the eCTD backbone file are provided in the associated specifications for eCTD backbone files.

If a document was previously submitted in either paper or electronic format **without the proper eCTD backbone files**, you should reference the document as with any paper submission. In the text of the document, you should include (1) the application or master file number, (2) the date of submission (e.g., letter date), (3) the document name, (4) the page number, and (5) the submission identification (e.g., submission serial number, volume number,

---

[7] Previously submitted documents include previously submitted information by reference for master files, market applications, and investigational applications discussed under 21 CFR 312.23(a)(11)(b), 314.50(g)(1), 314.420(b), and 601.51(a).

electronic folder, and file name) of the referenced document. In such cases, providing an electronic copy of the previously submitted documents can increase the utility of the submission. These documents, like all documents in the submission, should be appropriately described in the eCTD backbone files. These files are considered new in the eCTD backbone files.

When referring to documents that are part of other applications, please remember to include the appropriate letters of authorization with the submission (e.g., 21 CFR 314.420(d)).

H. Refuse to File

We may refuse to file an application or supplement under our regulations (e.g., 21 CFR 314.101 and 601.2) if the submission is illegible, uninterpretable, or otherwise clearly inadequate, including having incompatible formats or inadequate organization. These regulations apply to both paper and electronic submissions. The absence of electronic datasets in an acceptable format to permit review and analysis may be considered inadequate, resulting in a refuse-to-file decision.[8] Following the recommendations in this guidance document will help ensure that your electronic application meets the requirements of FDA regulations and can be archived, processed, and reviewed within specified time frames using our tools.

I. Submission of Paper Copies

When providing applications in electronic format using the eCTD backbone files, paper copies of the application, including review copies and desk copies, are not required and should not be sent.

J. Scanned Documents

Scanned documents submitted electronically as images are not as useful for review as documents that are text based. Image-based documents are more difficult to read and cannot be electronically searched. It takes longer to print image-based documents, and they occupy more storage space than text-based documents. For these reasons, we strongly urge that you provide text-based documents, rather than image files, whenever possible. We understand that certain documents may only be available as image files. Handwritten documents and documents that were generated independent from the company, such as journal publications, may be available only in paper. Documents that may only be available in paper can be scanned and submitted in electronic format as image-based files. However, we expect documents such as study reports recently generated by the company or recently generated as the result of the company's request to be available as text-based documents. We understand that legacy study reports, those generated years ago, may only be available in paper. For these reports, especially those for pivotal studies, you may want to consider converting these documents from image files to text-based files. Optical Character Recognition that has been validated is an option.

---

[8] See more on this in CBER's SOPP 8404.

K. The FDA District Office Copy

FDA District offices have access to documents submitted in electronic format. Therefore, when sending submissions in electronic format, you need not provide any documentation to the FDA Office of Regulatory Affairs District Office.

L. Electronic Signatures

Documents required by regulations to be submitted with an original signature (e.g., FDA form 356h, FDA form 1571) should be submitted with electronic signatures that follow the controls described under 21 CFR part 11.

M. Number of Copies of Electronic Files

You should send a single copy of the electronic portions of a submission to the appropriate central document room facility. Copies should not be sent directly to the reviewer or review division. Electronic documents that bypass the controls for electronic files described in 21 CFR 11 are not considered official documents for review.

N. Naming Electronic Files

To function properly, the eCTD backbone files must have specific names (e.g., index.xml, us-regional.xml). For other files without a specified name, you should provide a name that is indicative of the contents (e.g., protocol-101). The file name should allow a reviewer to infer some concept of the file's contents relative to other files. The file name should be less than or equal to 64 characters including the appropriate file extension. You should use only letters (lower case), numbers, or hyphens in the name. You should not use blank spaces. When naming files, it is important to remember that — to avoid truncation — the length of the entire path of the file should not exceed 230 characters.

O. Naming Folders

The terms *folder* and *subfolder* are used in this guidance and are intended to be synonymous with *directory* and *subdirectory*. The main submission, regional administrative folders, and certain subfolders should have specific names for proper and efficient processing of the submission. Recommendations regarding naming the main submission folders and regional administrative folders can be found in section III, below. Other specific folder names can be found in the specifications associated with this guidance. You can use only letters (lower case), numbers, or hyphens in the name. You should not use blank spaces. The length of the folder name should not exceed 64 characters. When naming folders, it is important to remember that the length of the entire path should not exceed 230 characters. You should not include empty folders in the submission.

P. File Formats

We recommend that you send electronic documents in the file formats specified in this guidance. We will not be able to manage, process, archive, or review documents provided in other file formats.

The following file formats should be used:

- PDF for reports and forms

- SAS XPORT (version 5) transport files (XPT) for datasets

- ASCII text files (e.g., SAS program files, NONMEM control files) using txt for the file extension

- XML for documents, data, and document information files

- Stylesheets (XSL) and document type definition (DTD) for the XML document information files

- Microsoft Word for draft labeling (because Microsoft Word can change, check our Web site for the current version)

In the future, we may consider other electronic file formats for use with electronic submissions, or we may consider the use of the current formats with other electronic submissions. We intend to publish guidance to advise on the use of file formats for specific types of submissions for use in the future.

Q. PDF Bookmarks and Hypertext Links

For documents with a table of contents, provide bookmarks and hypertext links for each item listed in the table of contents including tables, figures, publications, references, and associated appendices. These bookmarks and hypertext links are essential for efficient navigation through documents. You should make the bookmark hierarchy identical to the table of contents. Navigation efficiency is also improved by providing hypertext links throughout the body of the document to supporting annotations, related sections, references, appendices, tables, or figures that are not located on the same page.

It is possible to link to other documents in a submission using relative paths when creating hypertext linking. Absolute links that reference specific drives and root directories are not functional once the submission is loaded onto the document repository. For example, the link path ../../../123456/0001/.. will work, but the link c:1234560001… will not work. However, you should keep in mind that some documents may be subsequently replaced or appended, possibly rendering the link obsolete, so linking should be used cautiously.

When creating bookmarks and hypertext links, choose the magnification setting *Inherit Zoom* so that the destination page displays at the same magnification level that the reviewer is using for the rest of the document.

R. Sending Electronic Submissions

All submissions provided in electronic format must be sent to the appropriate central document room facility for processing to maintain the integrity of the submission as required under 21 CFR part 11. Electronic documents sent directly to division document rooms or to reviewers bypass the controls established for the receipt and archiving of documents and are not considered official documents for review. See the associated specifications for more information, including electronic transmission.

S. Technical Problems or Questions

If you have any questions on technical issues related to providing electronic submissions according to the recommendations in this guidance, contact the electronic submission coordinator at esub@cder.fda.gov. Specific technical issues related to submissions to CBER should be sent to esubprep@cber.fda.gov. Specific questions pertaining to content should be directed to the appropriate review division or office.

III. Organizing the Main SUBMISSION Folder

All documents in the electronic submission should be placed in a main submission folder using a four-digit sequence number for the application with the original submission for an application designated 0000. You should assign numbers for each submission to the same application with consecutive numbers. For example, the folder for the 3rd submission to an application, whether it is an amendment, supplement, or general correspondence is numbered 0002. The 4th submission is numbered 0003. This also applies to applications where previous submissions were not based on the ICH eCTD specifications. For example, if the submission is the 25th and the previous 24 were in paper, you would number the folder 0024. You should place the eCTD backbone file for modules 2 to 5 for the submission in this folder (*index.xml*). You should place the checksum file (e.g., index-md5.txt) in the same folder. Sequence numbers are used to differentiate between submissions for the same application and do not need to correspond to the order they are received by the Agency.

We recommend that you use subfolders to organize files in a submission, including for each module *m1*, *m2*, *m3*, *m4*, and *m5*, respectively. There is a subfolder *util* to organize eCTD technical files in the submission. Place these subfolders in the sequence number folder (e.g., folder named 0000 for the initial submission to an application). Do not include empty subfolders.

The following sections provide guidance for organizing the folders and files in the *m1*, *m2*, *m3*, *m4*, *m5*, and *util* folders. In addition, you can find instructions on preparing the submission of an electronic application to CBER at http://www.fda.gov/cber/esub/esub.htm

A. Module 1 Administrative Information and Prescribing Information Folder

Module 1 contains administrative and labeling documents. The organization of the documents in module 1 is the same for all applications and related submissions. The subject matter for each document should be assigned to the lowest level of the hierarchy outlined in the associated document "Comprehensive Table of Contents Headings and Hierarchy." Note that some headings apply only to specific applications or specific submissions. You should create a folder named us and place it in the folder named m1. The documents for module 1 should be placed in the us folder including the us-regional.xml file pertaining to the eCTD backbone files for module 1. Below are some additional details on providing specific types of documents.

1. eCTD backbone document information files

The details on creating this file are in the associated document "eCTD Backbone Files Specification for Module 1."

2. Cover letter (optional)

If you decide to include a cover letter, we recommend you include the following information:

- Description of the submission including appropriate regulatory information
- Description of the submission including the approximate size of the submission (e.g., 2 gigabytes), the format used for DLT tapes, and the type and number of electronic media used (e.g., three CDROMs), if applicable
- Statement that the submission is virus free with a description of the software (name, version, and company) used to check the files for viruses
- Regulatory and technical point of contact for the submission

3. Labeling

The following section describes how to provide specific labeling documents.

a. Labeling history

    You can provide a history summarizing labeling changes as a single PDF file. The following information will help us confirm changes made to the labeling:

    - Complete list of the labeling changes being proposed in the current submission and the explanation for the changes
    - Date of the last approved labeling
    - History of all changes since the last approved labeling. With each change, you should note the submission that originally described the change and the explanation for the change.
    - List of supplements pending approval that may affect the review of the labeling in the current submission

b. Content of labeling

    See the guidance for industry on Providing Regulatory Submissions in Electronic Format — Content of Labeling for details on providing the content of labeling files.

c. Labeling samples

    Each labeling sample (e.g., carton labels, container labels, package inserts) should be provided as individual PDF files. The samples should (1) include all panels, if applicable; (2) be provided in their actual size; and (3) reflect the actual color proposed for use.

4. Advertisements and promotional material

Advertisements and promotional labeling include material submitted under 21 CFR 314.81(b)(3)(i) or 601.12(f)(4) as part of the postmarketing reporting regulations for approved applications, submitted under the requirements of 21 CFR 314.550 and 601.45 (part of the accelerated approval requirements and restricted distribution for drug and biological products), or voluntarily submitted to INDs. You should submit promotional material to the appropriate application. You should not mix submissions of advertisements and promotional labeling with submissions containing other types of information.

Each promotional piece should be provided as an individual PDF file. In cases when promotional writing or images cover more than one page (e.g., a brochure spread), the reviewer should be able to view the entire layout at one time. For three-dimensional objects, you should provide a digital image of the object in sufficient detail to allow us to review the promotional material. In addition, you should provide information adequate to determine the size of the object (e.g., point size, dimensions). A dimensional piece shown flat, such as a flattened carton, can also be submitted.

If you choose to include cover letters with your submissions of advertising and promotional material, they should be provided as individual PDF files and indicate for the reviewer any additional important information, such as which materials need priority reviews.

If references are provided, each reference should be submitted as an individual PDF file and placed in the appropriate module based on subject matter. If possible, you should highlight the sections of the full reference that you refer to in the promotional materials. When a reference is used to support a claim in proposed promotional materials voluntarily submitted for advisory opinion or Agency comment, you should provide a hypertext link to the page of the reference or labeling that contains the supporting information.

For promotional materials submitted as part of the postmarketing reporting requirements, you may choose to provide hypertext links to references or labeling. References improve the efficiency of a review.

5. Marketing annual reports

In the postmarketing study commitments files, you should include a bookmark for each study described.

6. Information amendments

You should include documents that are provided in information amendments in the appropriate module using the appropriate headings to describe the subject matter. In the unusual case when information amendments do not fit appropriately under any heading in the CTD, you should place the documents in module 1 under the heading "information amendment: Information not covered under modules 2 to 5." You should provide a separate PDF file for each subject covered. Documents that apply to more than one module should be placed under the heading "Multiple module information amendments."

B. Module 2 Summary Folder

You should place the documents for module 2 in the *m2* folder. The subject matter for each document should be specific for the lowest level of the hierarchy outlined in the associated document "Comprehensive Table of Contents Headings and Hierarchy." Each document should be provided as an individual PDF file. The subfolders described in the *M2 eCTD: Electronic Common Technical Document Specification* are not necessary for the review of the submission. If you choose to use the additional subfolder, we will maintain the subfolder structure so links will function properly.

C. Module 3 Quality Folder

The organization of the module 3 folder is the same for all applications and related submissions. You should place the documents for module 3 in the *m3* folder. The subject matter for each document should be specific for the lowest level of the hierarchy outlined in the associated document "Comprehensive Table of Contents Headings and Hierarchy." Each document should be provided as an individual PDF file. The subfolders described in the *M2 eCTD: Electronic Common Technical Document Specification* are not necessary for the review of the submission. If you choose to use the additional subfolder, we will maintain the subfolder structure used so links will function properly.

You should provide the files pertaining to Key Literature References (CTD section 3.3) as individual PDF files. The filenames should be short and meaningful.

D. Module 4 Safety Folder

The organization of the module 4 folder is the same for all applications and related submissions. You should place the documents for module 4 in the *m4* folder. The subject matter for each document should be specific for the lowest level of the hierarchy outlined in the associated document "Comprehensive Table of Contents Headings and Hierarchy." The headings for study reports should also be specific for the lowest level of the hierarchy. Each document should be provided as an individual PDF file. The subfolders described in the *M2 eCTD: Electronic Common Technical Document Specification* are not necessary for the review of the submission. If you choose to use the additional subfolder, we will maintain the subfolder structure so links will function properly.

1. Study reports

Typically, a single document should be provided for each study report included in this module. However, if you provide the study reports as multiple documents, you should confine the subject matter of each document to a single item in the following list.

- Synopsis

- Study report body

- Protocol and amendments

- Signatures of principal or coordinating investigator(s)

- Audit certificates and reports

- Documentation of statistical methods and interim analysis plans

- Documentation of interlaboratory standardization methods of quality assurance procedures if used

- Publications based on the study

- Important publications referenced in the report

- Compliance and/or drug concentration data

- Individual subject data listings

    — Data tabulations

      - Data tabulations datasets

      - Data definitions

    — Data listing

      - Data listing datasets

      - Data definitions

    — Analysis datasets

      - Analysis datasets

      - Analysis programs

      - Data definitions

    — IND safety reports

In the following examples, you should provide the study reports as separate documents

- Documents previously submitted. If you have provided a document in a previous submission (e.g., protocol), you should provide a reference to the protocol, not resubmit the protocol.

- Additional information added. If you think you will want to add information to the study report over time (e.g., audit information, publication based on the study), you should provide the study reports as separate documents and then the new information can be provided as a separate file, rather than replacing the entire study report.

- Different file formats. If you submit the individual animal data listings as datasets (e.g., SAS transport files), you should provide these as separate files from the study reports (e.g., submitted as PDF files).

When providing a study report, you should include the study tagging file (STF) described in the associated document "The eCTD Backbone File Specifcation for Study Tagging Files."

2. Literature references

You should provide each literature reference as an individual PDF file. The filenames should be short and meaningful.

3. Datasets

See the associated document "Study Data Specifications" for details on providing datasets and related files (e.g., data definition file, program files)

E. Module 5 Clinical Study Reports Folder

The organization of the module 5 folder is the same for all applications and related submissions. You should place the documents for module 5 in the *m5* folder. The subject matter for each document should be specific for the lowest level of the hierarchy outlined in the associated document "Comprehensive Table of Contents Headings and Hierarchy." One exception is that legacy study reports can be provided as a single document. Each document should be provided as an individual PDF file. The subfolders described in the guidance *M2 eCTD: Electronic Common Technical Document Specification* are not necessary for the review of the submission. If you choose to use the additional subfolder, we will maintain the subfolder structure so links will function properly.

1. Tabular listing of all clinical studies

You should provide the tabular listing of all clinical studies as a single PDF file.

2. Study reports

Typically, clinical study reports are provided as more than one document based on the ICH E3 guidance document when providing a study.[9] In addition, if you have provided a document in a previous submission (e.g., protocol), you should provide a reference to the protocol rather than resubmitting the protocol. In cases when a legacy report has already been prepared as a single electronic document, you can provide the entire study report, other than the case report forms (CRFs) and individual data listings, as a single document. The individual documents that should be included in a study report are listed below:

- Synopsis[10] (E3 2)

- Study report (E3 1, 3 to 15)

- Protocol and amendments (E3 16.1.1)

- Sample case report forms (E3 16.1.2)

- List of IECs or IRBs (E3 16.1.3) and consent forms

- List and description of investigators (E3 16.1.4) and sites

---

[9] When providing a study report, you should include the study tagging file (STF) described in the associated document "The eCTD Backbone File Specification for Study Tagging Files."

[10] The synopsis should be provided as a document separate from the study report.

- Signatures of principal or coordinating investigator(s) or sponsor's responsible medical officer (E3 16.1.5)

- Listing of patients receiving test drug(s) from specified batch (E3 16.1.6)

- Randomizations scheme (E3 16.1.7)

- Audit certificates (E3 16.1.8) and reports

- Documentation of statistical methods (E3 16.1.9) and interim analysis plans

- Documentation of interlaboratory standardization methods of quality assurance procedures if used (E3 16.1.10)

- Publications based on the study (E3 16.1.11)

- Important publications referenced in the report (E3 16.1.12)

- Discontinued patients (E3 16.2.1)

- Protocol deviations (E3 16.2.2)

- Patients excluded from the efficacy studies (E3 16.2.3)

- Demographic data (E3 16.2.4)

- Compliance and/or drug concentration data (E3 16.2.5)

- Individual efficacy response data (E3 16.2.6)

- Adverse event listings (E3 16.2.7)

- Listing of individual laboratory measurements by patient (E3 16.2.8)

- Case report forms (E3 16.3)

- Individual patient data listings (CRTs) (E3 16.4)
    — Data tabulations
      - Data tabulations datasets
      - Data definitions
      - Annotated case report form
    — Data listing
      - Data listing datasets
      - Data definitions
      - Annotated case report form
    — Analysis datasets
      - Analysis datasets
      - Analysis programs

- Data definitions

- Annotated case report form

— Subject profiles

— IND safety reports

3. Case report forms

You should provide an individual subject's complete CRF as a single PDF file. If a paper CRF was used in the clinical trial, the electronic CRF should be a scanned image of the paper CRF including all original entries with all modifications, addenda, corrections, comments, annotations, and any extemporaneous additions. If electronic data capture was used in the clinical trial, you should submit a PDF-generated form or other PDF representation of the information (e.g., subject profile).

You should use the subject's unique identifier as the title of the document and the file name. These names are used to assist reviewers in finding the CRF for an individual subject. Each CRF must have bookmarks as part of the comprehensive table of contents required under 21 CFR 314.50(b). We recommend bookmarks for each CRF domain and study visit to help the reviewer navigate the CRFs. For addenda and corrections, making a hypertext link from the amended item to the corrected page or addendum is a useful way to avoid confusion. Bookmarks for these items should be displayed at the bottom of the hierarchy.

4. Datasets

See the associated document "Study Data Specifications" for details on providing datasets and related files (e.g., data definition files, program files). For subject profiles, you should use the subject's unique identifier in the title of the document and the file name.

5. Periodic safety update reports

To facilitate electronic submissions, we have divided the postmarketing periodic adverse drug experience report into three parts: (1) individual case safety reports (ICSRs), (2) ICSR attachments, if applicable, and (3) descriptive information. The descriptive information includes the narrative summary and analysis of the information in the report (i.e., periodic ICSRs and ICSR attachments), an analysis of the 15-day alert reports submitted during the reporting interval (i.e., expedited ICSRs and ICSR attachments), and the history of actions taken since the last report because of adverse drug experiences (e.g., labeling changes, studies initiated) as described in 21 CFR 314.80(c)(2)(ii)(a) and (c) and 600.80(c)(2)(ii)(A) and (C)). You should supply the descriptive information as an individual PDF file. You should provide bookmarks for each of the sections and subsections of this report. ICSR and ICSR attachments should be provided as described in the guidance for industry Providing Regulatory Submissions in Electronic Format — Postmarketing Periodic Adverse Drug Experience Reports.

6. Literature references

You should provide each literature reference as an individual PDF file. The filenames should be short and meaningful.

IV. Utility Folder

You should create two folders, *dtd* and *style* and place them in the *util* folder.

A. Document Type Definition Folder

You should place the document type definition (DTD) that you used to create the eCTD backbone file (regional.xml), the DTD you used to create the FDA Regional eCTD backbone file (us-index.xml), and the DTD used for the STF in the folder named *dtd*. You should use the most recent DTD.[11]

B. Style Folder

You should use the most recent stylesheet. See the guidance for industry M2 eCTD: Electronic Common Technical Document Specification.

---

[11] See the FDA Web site at http://www.fda.gov/cder/regulatory/ersr/

Development and Use of Risk Minimization Action Plans

Development and Use of Risk Minimization Action Plans, Guidance for Industry [†, 1]

U.S. Department of Health and Human Services
Food and Drug Administration
Center for Drug Evaluation and Research (CDER)
Center for Biologics Evaluation and Research (CBER)

March 2005

Clinical Medical

Contains Nonbinding Recommendations

This guidance is intended to assist industry with risk management activities for drug products, including biological drug products, in the Center for Drug Evaluation and Research (CDER) and the Center for Biologics Evaluation and Research (CBER). It addresses the development, implementation, and evaluation of risk minimization action plans for drug products.

Note: This guidance represents the Food and Drug Administration's (FDA's) current thinking on this topic. It does not create or confer any rights for or on any person and does not operate to bind FDA or the public. You can use an alternative approach if the approach satisfies the requirements of the applicable statutes and regulations. If you want to discuss an alternative approach, contact the appropriate FDA staff. If you cannot identify the appropriate FDA staff, call the appropriate number listed on the title page of this guidance.

I. Introduction

This document provides guidance to industry on the development, implementation, and evaluation of risk minimization action plans for prescription drug products, including biological drug products.[2] In particular, it gives guidance on (1) initiating and designing plans called risk minimization action plans or RiskMAPs to minimize identified product risks, (2) selecting and developing tools to minimize those risks, (3) evaluating RiskMAPs and

---

[†] Available on the FDA website at: http://www.fda.gov/downloads/RegulatoryInformation/ Guidances/UCM126830.pdf

[1] This guidance has been prepared by the PDUFA III Risk Management Working Group, which includes members from the Center for Drug Evaluation and Research (CDER) and the Center for Biologics Evaluation and Research (CBER) at the Food and Drug Administration (FDA).

[2] For ease of reference, this guidance uses the term *product* or *drug* to refer to all drug products (excluding blood and blood components) regulated by CDER or CBER. Similarly, for ease of reference, this guidance uses the term *approval* to refer to both drug approval and biologic licensure.

monitoring tools, and (4) communicating with FDA about RiskMAPs, and (5) the recommended components of a RiskMAP submission to FDA.

FDA's guidance documents, including this guidance, do not establish legally enforceable responsibilities. Instead, guidances describe the Agency's current thinking on a topic and should be viewed only as recommendations, unless specific regulatory or statutory requirements are cited. The use of the word *should* in Agency guidances means that something is suggested or recommended, but not required.

II. Background

A. PDUFA III's Risk Management Guidance Goal

On June 12, 2002, Congress reauthorized, for the second time, the Prescription Drug User Fee Act (PDUFA III). In the context of PDUFA III, FDA agreed to satisfy certain performance goals. One of those goals was to produce guidance for industry on risk management activities for drug and biological products. As an initial step towards satisfying that goal, FDA sought public comment on risk management. Specifically, FDA issued three concept papers. Each paper focused on one aspect of risk management, including (1) conducting premarketing risk assessment, (2) developing and implementing risk minimization tools, and (3) performing postmarketing pharmacovigilance and pharmacoepidemiologic assessments. In addition to receiving numerous written comments regarding the three concept papers, FDA held a public workshop on April 9–11, 2003, to discuss the concept papers. FDA considered all of the comments received in developing the three draft guidance documents on risk management activities. The draft guidance documents were published on May 5, 2004, and the public was provided with an opportunity to comment on them until July 6, 2004. FDA considered all of the comments received in producing the final guidance documents:

1. *Premarketing Risk Assessment (Premarketing Guidance)*

2. *Development and Use of Risk Minimization Action Plans (RiskMAP Guidance)*

3. *Good Pharmacovigilance Practices and Pharmacoepidemiologic Assessment (Pharmacovigilance Guidance)*

B. Overview of the Risk Management Guidance Documents

Like the concept papers and draft guidances that preceded them, each of the three final guidance documents focuses on one aspect of risk management. The *Premarketing Guidance* and the *Pharmacovigilance Guidance* focus on premarketing and postmarketing risk assessment, respectively. The *RiskMAP Guidance* focuses on risk minimization. Together, risk assessment and risk minimization form what FDA calls *risk management*. Specifically, risk management is an iterative process of (1) assessing a product's benefit-risk balance, (2) developing and implementing tools to minimize its risks while preserving its benefits, (3) evaluating tool effectiveness and reassessing the benefit-risk balance, and (4) making adjustments, as appropriate, to the risk minimization tools to further improve the benefit-risk balance. This fourpart process should be continuous throughout a product's lifecycle, with the results of risk assessment informing the sponsor's decisions regarding risk minimization.

When reviewing the recommendations provided in this guidance, sponsors and applicants should keep the following points in mind:

- Many recommendations in this guidance are not intended to be generally applicable to all products.

  Industry already performs risk assessment and risk minimization activities for products during development and marketing. The Federal Food, Drug, and Cosmetic Act (FDCA) and FDA implementing regulations establish requirements for *routine* risk assessment and risk minimization (see e.g., FDA requirements for professional labeling and adverse event monitoring and reporting). As a result, many of the recommendations presented here focus on situations in which a product may pose a clinically important and unusual type or level of risk. To the extent possible, we have specified in the text whether a recommendation is intended for all products or only this subset of products.

- It is of critical importance to protect patients and their privacy during the generation of safety data and the development of risk minimization action plans.

  During all risk assessment and risk minimization activities, sponsors must comply with applicable regulatory requirements involving human subjects research and patient privacy.[3]

- To the extent possible, this guidance reflects FDA's commitment to harmonization of international definitions and standards.

- When planning risk assessment and risk minimization activities, sponsors should consider input from healthcare participants likely to be affected by these activities (e.g., from consumers, pharmacists and pharmacies, physicians, nurses, and third-party payers).

- There are points of overlap among the three guidances.

  We have tried to note in the text of each guidance when areas of overlap occur and when referencing one of the other guidances might be useful.

III. The Role of Risk Minimization and Riskmaps in Risk Management

As described in section II.B, FDA views risk management as an iterative process encompassing the assessment of risks and benefits, the minimization of risks, and the maximization of benefits.

Specifically, the premarketing guidance and the pharmacovigilance guidance discuss how sponsors should engage in evidence-based risk assessment for all products in development and

---

[3] See 45 CFR part 46 and 21 CFR parts 50 and 56. See also the Health Insurance Portability and Accountability Act of 1996 (HIPAA) (Public Law 104-191) and the Standards for Privacy of Individually Identifiable Health Information (the Privacy Rule) (45 CFR part 160 and subparts A and E of part 164). The Privacy Rule specifically permits covered entities to report adverse events and other information related to the quality, effectiveness, and safety of FDA-regulated products both to manufacturers and directly to FDA (45 CFR 164.512(b)(1)(i) and (iii) and 45 CFR 164.512(a)(1)). For additional guidance on patient privacy protection, see http://www.hhs.gov/ocr/hipaa.

on the market to define the nature and extent of a product's risks in relation to its benefits. The goal of risk minimization is to minimize a product's risks while preserving its benefits. For the majority of products, routine risk minimization measures are sufficient to minimize risks and preserve benefits. Only a few products are likely to merit consideration for additional risk minimization efforts (see section III.D). Efforts to maximize benefits to improve the overall balance of risks and benefits can be pursued in concert with risk minimization efforts and can be discussed with FDA.

A. Relationship Between a Product's Benefits and Risks

The statutory standard for FDA approval of a product is that the product is safe and effective for its labeled indications under its labeled conditions of use (see sections 201(p)(1) and 505(d) of the Federal Food, Drug, and Cosmetic Act (21 U.S.C. 321(p)(1) and 355(d)). FDA's determination that a product is safe, however, does not suggest an absence of risk. Rather, a product is considered to be safe if the clinical significance and probability of its beneficial effects outweigh the likelihood and medical importance of its harmful or undesirable effects. In other words, a product is considered safe if it has an appropriate benefit-risk balance for the intended population and use.

Benefit and risk information emerges continually throughout a product's lifecycle (i.e., during the investigational and marketing phases) and can reflect the results of both labeled and off-label uses. Benefits and risks can result in a range of corresponding positive and negative effects on patient outcomes that may (1) be cosmetic, symptomatic, or curative; (2) alter the course of the disease; or (3) affect mortality. Benefits and risks are difficult to quantify and compare because they may apply to different individuals and are usually measured and valued differently.

Examples of factors to weigh are (1) population risks and benefits, (2) individual benefits from treatment, (3) risks of nontreatment or alternative products, and (4) modest population benefits in the context of a serious adverse effect that occurs rarely or unpredictably. Benefits as well as risks are also patient-specific and are influenced by such factors as (1) the severity of the disease being treated, (2) the outcome of the disease if untreated, (3) the probability and magnitude of any treatment effect, (4) existing therapeutic options, and (5) the individual's understanding of risks and benefits and the value they attach to each of them. Thus, assessment and comparison of a product's benefits and risks is a complicated process that is influenced by a wide range of societal, healthcare, and individualized patient factors.

B. Determining an Appropriate Risk Minimization Approach

To help ensure safe and effective use of their products, sponsors have always sought to maximize benefits and minimize risks. FDA believes that, for most products, routine risk minimization measures are sufficient. Such measures involve, for example, FDA-approved professional labeling describing the conditions in which the drug can be used safely and effectively, updated from time to time to incorporate information from postmarketing surveillance or studiesk concerns. Efforts to make FDA-approved professional labeling clearer, more concise, and better focused on information of clinical relevance reflect the Agency's belief that communication of risks and benefits through product labeling is the cornerstone of risk management efforts for prescription drugs.[4] For most products, routine risk management will be sufficient and a RiskMAP need not be considered.

There are, however a small number of products for which a RiskMAP should be considered (see section III.D). FDA recommends that RiskMAPs be used judiciously to minimize risks without encumbering drug availability or otherwise interfering with the delivery of product benefits to patients.

This guidance focuses on the development, implementation, and evaluation of RiskMAPs.

C. Definition of Risk Minimization Action Plan (RiskMAP)

As used in this document, the term RiskMAP means a strategic safety program designed to meet specific *goals* and *objectives* in minimizing known risks of a product while preserving its benefits. A RiskMAP targets one or more safety-related health outcomes or goals and uses one or more *tools* to achieve those goals.[5] A RiskMAP could also be considered as a selectively used type of Safety Action Plan as defined in the International Conference on Harmonization (ICH) guidance *E2E: Pharmacovigilance Planning (E2E guidance).*[6]

FDA recommends that RiskMAP goals target the achievement of particular health outcomes related to known safety risks. FDA suggests that sponsors state goals in a way that aims to achieve maximum risk reduction. The following are examples of RiskMAP goals: "patients on X drug should not also be prescribed Y drug" or "fetal exposures to Z drug should not occur." FDA recommends that goals be stated in absolute terms. Although it might not be possible to ensure that absolutely no one on X drug receives Y drug, FDA believes that a *goal*, as the term implies, is a statement of the ideal outcome of a RiskMAP.

FDA recommends that RiskMAP goals be translated into pragmatic, specific, and measurable program *objectives* that result in processes or behaviors leading to achievement of the RiskMAP goals. Objectives can be thought of as intermediate steps to achieving the overall RiskMAP goal. A RiskMAP goal can be translated into different objectives, depending upon the frequency, type, and severity of the specific risk or risks being minimized. For example, a goal may be the elimination of dangerous concomitant prescribing. The objectives could include lowering physician co-prescribing rates and/or pharmacist co-dispensing rates. As described in

---

[4] For example, see the Proposed Rule on Requirements on Content and Format of Labeling for Human Prescription Drugs and Biologics; Requirements for P rescription Drug Product Labels that published in the Federal Registeron December 22, 2000 (65 FR 81081).

[5] Although all products with RiskMAPs would also have FDA-approved professional labeling, the term tool as used in this document means a risk minimization action in addition to routine risk minimization measures. Some tools may be incorporated into a product's FDA-approved labeling, such as Medication Guides or patient package inserts. As used in this document, the FDA-approved professional labeling refers to that portion of approved labeling that is directed to the healthcare practitioner audience. See section IV for a more detailed discussion of other non-routine risk minimization tools that focus on targeted education and outreach.

[6] This ICH guidance is available on the Internet at http://www.fda.gov/cder/guidance/index.htm under the topic ICH Efficacy. The draft E2E guidance was made available on March 30, 2004 (69 FR 16579). ICH agreed on the final version of the E2E guidance in November 2004.

---

greater detail in section IV, many processes or systems to minimize known safety risks are available or under development for use in RiskMAPs. These systems include:

- targeted education and outreach to communicate risks and appropriate safety behaviors to healthcare practitioners or patients

- reminder systems, processes, or forms to foster reduced-risk prescribing and use

- performance-linked access systems that guide prescribing, dispensing, and use of the product to target the population and conditions of use most likely to confer benefits and to minimize particular risks

For certain types of risks (e.g., teratogenicity of category X drug products), it may be possible to develop systems with similar processes and procedures that can be used industrywide.

The use of these systems can occur outside of a RiskMAP. For example, while most drugs do not need a RiskMAP, many would still benefit from a program of physician and patient education and outreach. At times, communication of potential product risks may be warranted before a sponsor agrees to do a RiskMAP or an agreed upon RiskMAP is completed.

D. Determining When a RiskMAP Should Be Considered[7]

As described in the premarketing guidance and pharmacovigilance guidance, evidence-based risk identification, assessment, and characterization are processes that continue throughout a product's lifecycle. Therefore, a risk warranting the consideration of a RiskMAP could emerge during premarketing or postmarketing risk assessment.[8] The Agency recommends that the appropriate information for consideration in making such a determination include, as applicable, (1) data from the clinical development program, postmarketing surveillance, and phase 4 studies, and (2) the product's intended population and use.

Although it is expected and hoped that sponsors will determine when a RiskMAP would be appropriate, FDA may recommend a RiskMAP based on the Agency's own interpretation of risk information.

Decisions to develop, submit, or implement a RiskMAP are always made on a case-by-case basis, but several considerations are common to most determinations of whether development of a RiskMAP may be desirable:

- Nature and rate of known risks versus benefits: Comparing the characteristics of the product's adverse effects and benefits may help clarify whether a RiskMAP could improve the product's benefit-risk balance. The characteristics to be weighed might include the (1) types, magnitude, and frequency of risks and benefits; (2) populations at greatest risk and/or those likely to derive the most benefit; (3) existence of treatment alternatives and their risks and benefits; and (4) reversibility of adverse events observed.

---

[7] This guidance is directed primarily toward sponsors of innovator products. However, a generic product may have the same benefit-risk balance as an innovator product and so may be cons idered for a similar RiskMAP.

[8] See section VII for a detailed discussion of RiskMAP submissions.

- Preventability of adverse effects: Serious adverse effects that can be minimized or avoided by preventive measures around drug prescribing are the preferred candidates for RiskMAPs.

- Probability of benefit: If factors are identified that can predict effectiveness, a RiskMAP could help encourage appropriate use to increase benefits relative to known risks.

Consider the following examples:

- Opiate drug products have important benefits in alleviating pain but are associated with significant risk of overdose, abuse, and addiction. The Agency recommends that sponsors of Schedule II controlled substances, including Schedule II extended release or high concentration opiate drug products, consider developing RiskMAPs for these products.

- Drugs that provide important benefits, but that are human teratogens would often be appropriate for a RiskMAP to minimize in utero exposure.

- Some drugs may warrant RiskMAP consideration because safe and effective use call for specialized healthcare skills, training, or facilities to manage the therapeutic or serious side effects of the drug.

Involving all stakeholders during the initial phases of considering whether a RiskMAP is appropriate allows input and buy-in by all parties who will later have roles in implementing the RiskMAP. If a RiskMAP is appropriate, stakeholders can help shape the RiskMAP to foster its success in the healthcare delivery environment. Therefore, we recommend public discussion about the appropriateness of a RiskMAP through the FDA advisory committee process. Such public advisory committee meetings can also be used to address (1) whether a RiskMAP is appropriate, (2) what the goals and objectives of the RiskMAP could be (see footnote 6), (3) the circumstances under which a RiskMAP tool might be revised or terminated, and (4) whether a RiskMAP itself is no longer appropriate. The FDA advisory committee structure and processes are well suited to foster such discussions as they arise on a case-by-case basis.

IV. Tools for Achieving Riskmap Goals and Objectives

A risk minimization tool is a process or system intended to minimize known risks. Tools can communicate particular information regarding optimal product use and can also provide guidance on prescribing, dispensing, and/or using a product in the most appropriate situations or patient populations. A number of tools are available; FDA encourages and anticipates the development of additional tools.

A. Relationship of RiskMAP Tools to Objectives and Goals

Risk minimization tools are designed to help achieve one or more RiskMAP objectives that are directed at the overall RiskMAP goal or goals. One or more tools can be chosen to achieve a particular objective. For example, a goal might be that patients with condition A should not be exposed to product B. An objective for achieving this goal might be to communicate to patients that if they have condition A, they should not take product B. Depending on the likelihood and severity of the adverse event associated with product B in a patient with condition A, a variety of tools could be applied to achieve this objective. One possible tool would be patient labeling explaining that a patient with condition A should not take product B. On the other hand, if the potential harm to a patient with condition A is severe and/or likely to occur, a more active tool may be appropriate. For example, the sponsor could choose to develop a patient agreement where, before receiving the product, the patient formally acknowledges their understanding and/or agreement not to take product B if he or she has condition A.

B. Categories of RiskMAP Tools

A variety of tools are currently used in risk minimization plans. These fall within three categories: (1) targeted education and outreach, (2) reminder systems, and (3) performancelinked access systems. A RiskMAP might include tools from one or more categories, depending on its risk minimization goals. FDA notes that the use of tools in different categories does not imply greater or lesser safety risks, but rather indicates the particular circumstances put in place to achieve the objectives and goals.

1. Targeted Education and Outreach

FDA recommends that sponsors consider tools in the targeted education and outreach category (1) when routine risk minimization is known or likely to be insufficient to minimize product risks or (2) as a component of RiskMAPs using reminder or performance-linked access systems (see sections IV.B.2 and 3 below).

Tools in this category employ specific, targeted education and outreach efforts about risks to increase appropriate knowledge and behaviors of key people or groups (e.g., healthcare practitioners and consumers) that have the capacity to prevent or mitigate the product risks of concern.

FDA acknowledges that tools in this category are occasionally used for products where the benefit/risk balance does not necessarily warrant a RiskMAP. Educational efforts by sponsors might include one or more of the tools described below without a RiskMAP being in place.

Sponsors are encouraged to continue using tools, such as education and outreach, as an extension of their routine risk minimization efforts even without a RiskMAP.

Examples of tools in this category are as follows:

- healthcare practitioner letters
- training programs for healthcare practitioners or patients
- continuing education for healthcare practitioners such as product-focused programs developed by sponsors and/or sponsor-supported accredited CE programs
- prominent professional or public notifications
- patient labeling such as Medication Guides and patient package inserts
- promotional techniques such as direct-to-consumer advertising highlighting appropriate patient use or product risks

- patient-sponsor interaction and education systems such as disease management and patient access programs

In addition to informing healthcare practitioners and patients about conditions of use contributing to product risk, educational tools can inform them of conditions of use that are important to achieve the product's benefits. For example, a patient who takes a product according to labeled instructions is more likely to achieve maximum product effectiveness. On the other hand, deviations from the labeled dose, frequency of dosing, storage conditions, or other labeled conditions of use might compromise the benefit achieved, yet still expose the patient to productrelated risks. Risks and benefits can have different dose-response relationships. Risks can persist and even exceed benefits when products are used in ways that minimize effectiveness.

Therefore, educational tools can be used to explain how to use products in ways that both maximize benefits and minimize risks.

2. Reminder Systems

We recommend that tools in the reminder systems category be used in addition to tools in the targeted education and outreach category when targeted education and outreach tools are known or likely to be insufficient to minimize identified risks.

Tools in this category include systems that prompt, remind, double-check or otherwise guide healthcare practitioners and/or patients in prescribing, dispensing, receiving, or using a product in ways that minimize risk. Examples of tools in this category are as follows:

- Patient education that includes acknowledgment of having read the material and an agreement to follow instructions. These agreements are sometimes called consent forms.

- Healthcare provider training programs that include testing or some other documentation of physicians' knowledge and understanding.

- Enrollment of physicians, pharmacies, and/or patients in special data collection systems that also reinforce appropriate product use.

- Limited number of doses in any single prescription or limitations on refills of the product.

- Specialized product packaging to enhance safe use of the product.

- Specialized systems or records that are used to attest that safety measures have been satisfied (e.g., prescription stickers, physician attestation of capabilities).

3. Performance-Linked Access Systems

Performance-linked access systems include systems that link product access to laboratory testing results or other documentation. Tools in this category, because they are very burdensome and can disrupt usual patient care, should be considered only when (1) products have significant or otherwise unique benefits in a particular patient group or condition, but unusual risks also exist, such as irreversible disability or death, and (2) routine risk minimization measures, targeted education and outreach tools, and reminder systems are known or likely to be insufficient to minimize those risks.

Examples of tools in this category include:

- the sponsor's use of compulsory reminder systems, as described in the previous section (e.g., the product is not made available unless there is an agreement or acknowledgment, documented qualifications, enrollment, and/or appropriate testing or laboratory records)

- prescription only by specially certified healthcare practitioners

- product dispensing limited to pharmacies or practitioners that elect to be specially certified

- product dispensing only to patients with evidence or other documentation of safe-use conditions (e.g., lab test results)

Performance-linked access systems should seek to avoid unnecessary or unintended restrictions or fragmentation of healthcare services that may limit access by physicians, pharmacists, or patients, or that may lead to discontinuities in medical or pharmacy care.

C. Description of RiskMAP Tools

FDA plans to develop a RiskMAP Web site that will include (1) descriptions of tools that are currently used in RiskMAPs and (2) other information relevant to RiskMAP development (see section IV.D below). The information will be made available consistent with federal law and regulations governing disclosure of information by FDA to the public. The list of tools will be intended to assist sponsors in designing a RiskMAP but will not suggest that the listed tools are FDA-approved or -validated. On the contrary, FDA does not suggest that the tools listed on the Web site are the only tools that could be useful and encourages sponsors to develop tools that may be optimal for their particular products. See also Section V.D on making information from RiskMAP evaluations available to the public.

D. Selecting and Developing the Best Tools

Given the variety of available tools, FDA recommends that a sponsor carefully consider which tool or tools are most appropriate, given the goals and objectives of its product's RiskMAP. A tool could be developed or selected based on its individual impact and/or because of its impact when used in coordination with other tools. Generally, the best tools would be those that have a high likelihood of achieving their objective based on positive performance in other RiskMAPs or in similar settings and populations. Relevant non-RiskMAP evidence and experience can be found in healthcare quality initiatives, public health education and outreach, marketing, and other outcomes-based research (see section V for a more detailed discussion of evaluating tools' effectiveness).

Although FDA suggests that the best tool or tools be selected on a case-by-case basis, the following are generally applicable considerations in designing a RiskMAP. In choosing tools for a RiskMAP, FDA recommends that sponsors:

- Maintain the widest possible access to the product with the least burden to the healthcare system that is compatible with adequate risk minimization (e.g., a reminder system tool should not be used if targeted education and outreach would likely be sufficient).

- Identify the key stakeholders who have the capacity to minimize the product's risks (such as physicians, pharmacists, pharmacies, nurses, patients, and third-party payers) and define the anticipated role of each group.

- Seek input from the key stakeholders on the feasibility of implementing and accepting the tool in usual healthcare practices, disease conditions, or lifestyles, if possible. Examples of considerations could include (but would not be limited to) patient and healthcare practitioner autonomy, time effectiveness, economic issues, and technological feasibility.

- Acknowledge the importance of using tools with the least burdensome effect on healthcare practitioner-patient, pharmacist-patient, and/or other healthcare relationships.

- Design the RiskMAP to be:

  1. compatible with current technology

  2. applicable to both outpatient and inpatient use

  3. accessible to patients in diverse locales, including non-urban settings

  4. consistent with existing tools and programs, or systems that have been shown to be effective with similar products, indications, or risks

- Select tools based on available evidence of effectiveness in achieving the specified objective (e.g., tools effectively used in pregnancy prevention).

- Consider indirect evidence of tool effectiveness in a related area that supports the rationale, design, or method of use (e.g., tools applied in modifying patient or healthcare practitioner behaviors in medical care settings).

- Consider, and seek to avoid, unintended consequences of tool implementation that obstruct risk minimization and product benefit, such as obstructing patient access or driving patients to seek alternative product sources (e.g., Internet sales, counterfeit products) or less appropriate products.

FDA recognizes that once it approves a product for marketing, healthcare practitioners are the most important managers of product risks. FDA believes that by including information in the FDA-approved professional labeling on the conditions in which medical products can be used safely and effectively by their intended population and for their intended use or uses, the Agency and the sponsor encourage healthcare practitioners to prescribe medical products in circumstances that yield a favorable benefit-risk balance. However, as the Agency has long recognized, the FDCA and FDA regulations establish requirements governing the safety and effectiveness of medical products. FDA does not have authority under these provisions to control decisions made by qualified healthcare practitioners to prescribe products for conditions other than those described in FDA-approved professional labeling, or to otherwise regulate medical or surgical practice.

E. Mechanisms Available to the FDA to Minimize Risks

This guidance focuses on the tools that industry can incorporate into RiskMAPs. As noted, FDA has a variety of risk management measures at its disposal under the FDCA and FDA regulations (see e.g., FDA requirements for professional labeling and adverse event monitoring and reporting).

FDA must occasionally invoke other mechanisms to minimize the risks from medical products that pose serious risks to the public health. These tools include:

- FDA-requested product recalls, warning and untitled letters, and import alerts

- safety alerts, guidance documents, and regulations

- judicial enforcement procedures such as seizures or injunctions

Further information on these mechanisms is available on the Internet at http://www.fda.gov.

V. RiskMAP Evaluation: Assessing the Effectiveness of Tools and the Plan

As FDA and sponsors seek additional knowledge about the design, effectiveness, burdens, and potential unintended consequences of RiskMAPs, it is important to collect as much information as possible on plan performance. RiskMAPs and their component objectives and tools should be monitored and evaluated in a timely manner to identify areas for improvement.

A. Rationale for RiskMAP Evaluation

At least two studies have documented poor or limited implementation and effectiveness of traditional risk minimization tools. In particular, the studies examined situations in which labeling changes (with or without Dear Healthcare Practitioner letters) were used to reduce safety problems.[9] The iterative process of risk assessment, risk minimization, and reevaluation previously described is intended to avoid repeating these experiences by identifying poorly performing or ineffective RiskMAPs or RiskMAP components as soon as possible. Ultimately, RiskMAP evaluation is intended to ensure that the energy and resources expended on risk minimization are actually achieving the desired goals of continued benefits with mi nimized risks.

FDA considers evaluation of the effectiveness of a RiskMAP to be important and recommends that every RiskMAP contain a plan for periodically evaluating its effectiveness after implementation (see section VII for a detailed discussion of RiskMAP submissions to FDA).[10]

---

[9] Smalley W, D Shatin, D Wysowski, J Gurwitz, S Andrade et al., 2000, *Contraindicated Use of Cisapride: Impact of Food and Drug Administration Regulatory Action*. JAMA 284(23):3036-3039; Weatherby LB, BL Nordstrom, D Fife, and AM Walker, 2002, *The Impact Of Wording in "Dear Doctor" Letters and In Black Box Labels*. Clin Pharmacol Ther 72:735-742.

[10] As noted in section III.B, sponsors should not develop a RiskMAP for a product for which routine risk minimization measures are sufficient. Similarly, formal evaluation plans and performance measures should not be developed for these products. Instead, evaluation by routine postmarketing surveillance should be sufficient, although some products may also have a Pharmacovigilance Plan as described in the *Pharmacovigilance Guidance*. If a RiskMAP is later developed for this type of product based on new risk information, then a sponsor should consider submitting a formal evaluation plan.

The evaluation of RiskMAPs can take several forms. Most critical is determining the performance of the overall RiskMAP in achieving its targeted health outcomes or goals. Separate but related assessments can be done for (1) individual tool performance, (2) acceptability of RiskMAP tools by consumers and healthcare practitioners, and (3) compliance with important RiskMAP processes or procedures.

Generally, FDA anticipates that RiskMAP evaluations would involve the analysis of observational or descriptive data. The specific types of data gathered in a RiskMAP evaluation will determine whether it would be appropriate to include a statistical analysis of evaluation results.

B. Considerations in Designing a RiskMAP Evaluation Plan

FDA recommends that RiskMAP evaluation plans be tailored to the specific product and designed to assess whether the RiskMAP's goals have been achieved through its objectives and tools. The following are generally applicable guidelines for sponsors designing RiskMAP evaluation plans.

1. Selecting Evidence-Based Performance Measures

The Agency recommends that sponsors select well-defined, evidence-based, and objective performance measures tailored to the particular RiskMAP to determine whether the RiskMAP's goals or objectives are being achieved. An appropriate measure could be a number, percentage, or rate of an outcome, event, process, knowledge, or behavior. Ideally, the chosen measure would directly measure the RiskMAP's health outcome goal. For example, for a RiskMAP with a goal of preventing a particular complication outcome from product use, a sample performance measure could be the complication rate. For evaluation purposes, a target for that measure could be established to be no more than a specified number or rate of that complication. In some cases, however, a health outcome cannot be practically or accurately measured. In those cases, other measures can be used that are closely related to the health outcome, such as the following:

- Surrogates for health outcome measures (e.g., emergency room visits for an adverse consequence, pregnancy test results for determining if pregnancy occurred). The sensitivity, specificity, and predictive value of surrogate markers should be established before their use as a performance measure.

- Process measures that reflect desirable safety behaviors (e.g., performance of recommended laboratory monitoring, signatures attesting to knowledge or discussions of risk).

- Assessments of comprehension, knowledge, attitudes, and/or desired safety behaviors about drug safety risks (e.g., provider, pharmacist, or patient surveys).

FDA recommends that the validity of a measure be judged by how closely it is related to the desired health outcome goal of the RiskMAP. Simply stated, the more closely related a measure is to the RiskMAP goal, the greater its degree of validity. For example, if the RiskMAP goal is avoidance of liver failure, then ascertainment of the rate of liver failure in the user population would be a highly valid performance measure. Hospitalization for severe liver injury would be another, but less direct, assessment of the RiskMAP goal. The frequency of liver function monitoring in users could be used to see if RiskMAP processes to prevent liver failure were being followed, but since liver function monitoring may not be tightly linked to the occurrence of liver failure, such process monitoring would have limited validity as an indicator of successful prevention of liver failure.

2. Compensating for an Evaluation Method's Limitations

Most evaluation measures have limitations. FDA suggests that, in choosing among evaluation methods and measures, sponsors consider their strengths and limitations. The following are examples of some of the limitations of evaluation methods:

- Spontaneous adverse event data are a potentially biased outcome measure because reporting of adverse events varies due to many factors and represents an unknown and variable fraction of the adverse outcomes that are actually occurring. As a result, systematic data collection or active surveillance of adverse events in populations with well-defined exposure to the product would be preferred for purposes of evaluation.

- Population-based evaluation methods can use administrative or claims-based data systems that capture service or payment claims to measure rates of events, although it is usually recommended that medical records be examined to validate the actual occurrence of coded diagnoses and procedures. Administrative data may come from various insurers, purchasing groups, or networks that are tied to employment or entitlement programs, so it is important to determine if an administrative data system is representative of the general population being treated with the product. Also, unless enrollment in an administrative claims system is large, the number of patients exposed to any single product is likely to be limited, as will be the power to detect uncommon adverse events.[11] In addition, there may be data processing time lags of several months or longer before administrative data can be retrieved and analyzed.

- Active surveillance using sentinel reporting sites may be useful for evaluating adverse events, but it is costly and may not detect rare events. Surveys of healthcare practitioners or patients using various modes (in-person, mail, telephone, electronic) can be another useful form of active surveillance of knowledge, attitudes, policies, and practices of healthcare practitioners, institutions, and patients about recommended RiskMAP tools and their associated processes. However, issues relating to response rates, representativeness, and reporting biases may limit the accuracy of survey results.[12]

These examples illustrate how using only one evaluation method could skew assessment of the performance of a RiskMAP. Therefore, FDA recommends that, whenever feasible, sponsors design evaluation plans to include at least two different quantitative, representative, and minimally biased evaluation methods for each critical RiskMAP goal. By using two methods, one method can compensate for the limitations of the other. For example, surveys of healthcare practitioners may indicate high compliance with systems for preventing product

---

[11] For further discussion of administrative claims systems, please consult the pharmacovigilance guidance.
[12] For a more detailed discussion of survey development and implementation, please consult the pharmacovigilance guidance.

complications. However, systematically collected or spontaneous reports might show that product complications are occurring, thus suggesting that prevention efforts in actual practice may be ineffective or incompletely applied. If it is not practical to use two complementary and representative methods, FDA suggests using other quantitative methods such as multiple site sampling or audits that aim for high coverage or response rates by the affected population. If RiskMAPs use multiple tools or interventions, it may be useful to consider using evaluation methods applicable to the program as a whole. For example, a systematic program evaluation model, such as Failure Modes and Effect Analysis (FMEA),[13, 14] can provide a framework for evaluating the individual RiskMAP components and the relative importance of each in achieving the overall RiskMAP goal or goals.

3. Evaluating the Effectiveness of Tools in Addition to RiskMAP Goals

FDA recommends that sponsors periodically evaluate each RiskMAP tool to ensure it is materially contributing to the achievement of RiskMAP objectives or goals. Tools that do not perform well may compromise attainment of RiskMAP goals, add unnecessary costs or burdens, or limit access to product benefits without minimizing risks. Tools that are implemented incompletely or in a substandard fashion could result in additional tools being adopted unnecessarily. For all these reasons, evaluating tools is important. Data from such evaluations may make it possible to improve a tool's effectiveness or eliminate the use of a tool that fails to contribute to achieving a RiskMAP goal. By eliminating ineffective tools, resources can be concentrated on useful tools.

Distinguishing between the evaluation of RiskMAP goals and tools is important because the achievement of goals and the performance of tools may not be linked. For example, the overall goal of a RiskMAP may be achieved despite individual tools performing poorly. The reverse situation may also occur, with component tools performing well but without appropriate progress in achieving the RiskMAP goal. This situation may occur if a surrogate objective correlates poorly to the desired health outcome. The first example (i.e., the RiskMAP goal may be achieved despite individual tools performing poorly) may afford an opportunity to discontinue a tool, whereas its converse may trigger the implementation of new or improved tools, or even a redesign of the overall RiskMAP. Two important factors that contribute to tool effectiveness are its acceptability and unintended consequences. Since tool performance will often depend upon the understanding, cooperation, efforts, and resources of healthcare providers, pharmacists, and patients, evaluation of acceptability and unintended consequences for individual tools may help to improve the use of tools and thus their performance.

4. Evaluating RiskMAP Tools Prior to Implementation

FDA recommends that, to the extent possible, sponsors evaluate tools for effectiveness before implementation. As discussed in section IV.D, FDA suggests that in selecting tools to include in a RiskMAP, a sponsor consider tools that are likely to be effective. For example, the success

---

[13] Stamatis DH, *Failure Mode and Effects Analysis: FMEA From Theory to Execution*, Milwaukee: American Society for Quality, Quality Press, 2003.

[14] Cohen Michael R ed, *Medication Errors: Causes, Prevention, and Risk Management*, Washington, DC: American Pharmaceutical Association, 1999.

of potential RiskMAP tools might be predicted to some extent by evidence in the scientific literature or from their use in other RiskMAPs. Application of computer modeling or simulation techniques may also assist in projecting potential outcomes of implementation of various combinations of RiskMAP tools.

Besides using literature evidence and past RiskMAP experience to identify tools with a known track record of effectiveness, sponsors can pretest or pilot test a tool before implementation. Such testing, ideally with a comparison group or time period, can help to assess comprehension, acceptance, feasibility, and other factors that influence how readily RiskMAP tools will fit into patient lifestyles and the everyday practices of healthcare practitioners. Pretesting can potentially avoid wasted time, expense, and escalation of RiskMAP tools by discriminating between high- and low-performing tools. For example, if a preventable risk is identified in Phase 2 trials, Phase 3 trials could provide an opportunity to pretest targeted education and outreach tools.

FDA recommends that pretesting methods be chosen on a case-by-case basis, depending on the product, tool, objective, and goal. For example, in certain preapproval situations, large simple safety studies may be a means of generating useful information about the effectiveness of RiskMAP tools in conditions close to actual practice.[15] On the other hand, for certain tools such as targeted education and outreach, published *best practices* could be used as guidelines for implementation. If time is particularly limited, multiple interviews or focus group testing can assist in determining acceptance or comprehension of a RiskMAP tool by major stakeholder groups. This action might be particularly useful in situations where risks and benefits are closely matched, and RiskMAP goals may include the making of informed therapeutic choices by patients and prescribers.

FDA recognizes that, in some cases, tools cannot be pretested for logistical reasons. Pretesting of tools may not be practical in situations in which newly recognized adverse events dictate the importance of rapid implementation of a RiskMAP after approval and marketing. In such instances, sponsors should seek to employ tools with a proven track record of effectiveness. In general, the greater the rate or severity of risks to be minimized, the more critical it becomes to have compelling evidence of effectiveness of the tool through some form of testing or prior use.

C. FDA Assessment of RiskMAP Evaluation Results

FDA recommends that if a sponsor makes a RiskMAP submission to the Agency, the submission describe when the sponsor will send periodic evaluation results to FDA. As discussed in section VII.B, the Agency recommends that sponsors analyze evaluation results and requests that sponsors provide FDA with (1) the data, (2) all analyses, (3) conclusions regarding effectiveness, and (4) any proposed modifications to the RiskMAP. FDA, in turn, generally would perform its own assessment of RiskMAP effectiveness according to the principles of this and the other risk management guidances. At a minimum, FDA and sponsors would discuss their respective RiskMAP evaluations in a meeting or teleconference. In cases where risks are frequent and/or severe, or where results are ambiguous or uncertain, or where

---

[15] For a detailed discussion of large simple safety studies, please consult the premarketing guidance.

there is disagreement between the sponsor and FDA in the interpretation of the RiskMAP or tool effectiveness, public and expert input would be sought through the FDA Advisory Committee process. This will also allow airing and discussion of important information about effective and ineffective RiskMAPs and tools.

D. Making Information From RiskMAP Evaluations Available to the Public

As discussed in section IV.C, FDA plans to maintain a RiskMAP Web site that will describe all publicly available information about implemented RiskMAPs (and their tools). On the same Web site, FDA intends to make available, in summary format, information that has been publicly discussed or is otherwise publicly available (from sponsors or other sources) about the effectiveness of particular RiskMAP tools in achieving risk minimization objectives. The summaries may derive from materials presented and discussed at FDA Advisory Committee meetings where the effectiveness of a particular RiskMAP has been discussed and potential modifications have been entertained.

VI. Communicating with FDA Regarding Riskmap Development and Design Issues

As discussed in section III.D, because risk and benefit information emerge continually throughout a product's lifecycle, a sponsor could decide, or FDA could recommend, that a RiskMAP is appropriate at several different times. These times include:

- before approval, when a risk is identified from clinical studies, nonclinical studies, or in similar class of products, and risk minimization is appropriate as the product is introduced into the marketplace

- after marketing, if pharmacovigilance efforts identify a new serious risk and minimization of the risk will contribute to a favorable benefit-risk balance

- when marketing a generic product that references an innovator drug with a RiskMAP

If a sponsor would like to initiate a dialogue with FDA to benefit from the Agency's experience in reviewing previously implemented plans, the Agency recommends that the sponsor contact the product's review division. The review division is the primary contact for a sponsor. The review division may choose to consult with other Offices in assisting the sponsor in developing a RiskMAP. These consulting offices could include CDER's Office of Drug Safety (ODS), CBER's Office of Biostatics and Epidemiology (OBE), or CDER's Office of Generic Drugs (OGD), as appropriate. In any particular case, it is helpful if the sponsor and FDA:

- share information and analyses regarding the product's risks and benefits

- discuss the choice of RiskMAP goals, objectives, and tools

- discuss the evaluation plan, including (1) times for evaluation, (2) performance measures and their targets, and (3) analyses

Sponsors may wish to discuss RiskMAP issues with FDA at pre-defined meeting times (e.g., end-of-phase-2 meetings), if appropriate, or request meetings where RiskMAPs can be specifically considered. To maximize the value of their discussions with FDA, we recommend that sponsors who seek the Agency's guidance apprise reviewers of the rationale for and data underlying RiskMAPs under consideration. FDA requests that sponsors also share relevant background information and questions for discussion before their meetings with FDA.

Both CDER and CBER will develop internal Manuals of Policies and Procedures (MaPPs) (or standard operating procedures (SOPs)) regarding the review of RiskMAPS. The procedures will define milestone points at which RiskMAP discussion is logical and will promote consistency in RiskMAP review and design. All RiskMAPs involving reminder tools or performance-linked access systems will be considered at the Center level as a secondary method of ensuring consistency across product classes and across divisions.

If the sponsor decides to submit a RiskMAP before marketing approval of the product, most times the RiskMAP will be submitted to the new drug application (NDA) or biologics license application (BLA) for the product in question. However, if a risk is identified early (e.g., the product is a teratogen), and the sponsor wishes to institute formal risk management activities during Phases 1 to 3 studies, the sponsor can submit the RiskMAP to the investigational new drug application (IND). If a RiskMAP is being considered in a product's postmarket phase, FDA recommends that it be submitted as a supplement to the relevant NDA or BLA. Additional user fees will only be applicable to a supplement if FDA determines that new clinical data are required for its approval. This would be unlikely for a RiskMAP supplement.

FDA encourages early and open discussion of safety concerns and whether such concerns may merit a RiskMAP. Early discussion of RiskMAPs could provide the opportunity to pretest risk minimization tools.

VII. Recommended Elements of a Riskmap Submission to FDA

A. Contents of a RiskMAP Submission to FDA

FDA suggests that a RiskMAP submission to FDA include the following sections, as well as a table of contents:

- Background

- Goals and Objectives

- Strategy and Tools

- Evaluation Plan

1. Background

FDA suggests that the Background section explain why a RiskMAP is being considered and created. We recommend that it describe the risks to be minimized and the benefits that would be preserved by implementation of a RiskMAP. Further, we suggest that this section describe, to the extent possible, the type, severity, frequency, and duration of the product's risks, with particular attention to the risk or risks addressed by the RiskMAP.

The following are sample questions regarding risk characterization that we recommend be addressed in the Background section:

- What is the rationale for the RiskMAP?

- What is the risk the RiskMAP addresses? Is there more than one risk to be minimized? If there is, how do they relate to each other with regard to the following bulleted items?

- What is the magnitude and severity of the risk?

- Who is at highest risk?

- Are particular populations at risk (e.g., children, pregnant women, the elderly)?

- Is the risk predictable?

- Is the risk preventable?

- Is the risk reversible?

- Is the risk time-limited, continuous, or cumulative?

These questions are similar in intent to what the ICH calls a Safety Specification in its E2E guidance.[16]

FDA recommends that this section include a discussion that considers the product's risks in the context of its benefits. The following are sample questions that address benefit characterization.

- What is the overall nature or extent of benefit and what are the expected benefits over time (i.e., long-term benefits)?

- How do the populations most likely to benefit from this product compare to those that may be at highest risk?

- How would implementation of a RiskMAP affect individual and population benefits? Will it increase the likelihood that benefits will exceed risks in patients using the product? Will the RiskMAP affect access to the product by patients who benefit from it?

- Could certain individuals and/or populations likely to benefit from the product potentially have less access to the product because of the tools in the RiskMAP?

We suggest that the Background section include a discussion, if pertinent, about the successes and failures of other regulatory authorities, systems of healthcare, or sponsor actions in minimizing the risks of concern for this product. Information provided by the sponsor regarding relevant past experiences, domestically or in other countries, will assist in harmonizing plans as well as avoiding the cost of implementing RiskMAP tools already

---

[16] Available on the Internet at http://www.fda.gov/cder/guidance/index.htm under the topic ICH Efficacy.

deemed unsuccessful. We encourage sponsors to provide applicable information or evaluations from past experiences with products or programs that are similar to the proposed RiskMAP.

2. Goals and Objectives

FDA suggests that the Goals and Objectives section describe the goals and objectives of the RiskMAP.[17] In addition, we recommend that this section describe how the stated objectives will individually and collectively contribute to achieving the goal or goals.

3. Strategy and Tools

FDA suggests that the Strategy and Tools section define the overall strategy and tools to be used to minimize the risk or risks targeted by the RiskMAP. We recommend that the sponsor provide a rationale for choosing the overall strategy. We suggest that the sponsor describe how each tool fits into the overall RiskMAP and its relationship to the other tools. FDA suggests that the sponsor also provide the rationale for choosing each tool (see section IV.D for a discussion of considerations in choosing tools). In particular, we recommend that the sponsor describe the available evidence regarding the tool's effectiveness and, where applicable, provide results from pretesting. In addition, we suggest that the sponsor state whether it sought input from patient or healthcare interests, and if it did, we suggest that the sponsor describe the feedback that was received regarding the feasibility of its RiskMAP. FDA plans to maintain a Web site that will describe publicly available summary information about effectiveness of RiskMAP tools (see section V.D).

We recommend this section also include an implementation scheme that describes how and when each RiskMAP tool would be implemented and coordinated. FDA suggests that sponsors specify overall timelines and milestones. For example, this section could address whether targeted education and outreach tools would be implemented before, or concurrently with, other tools.

4. Evaluation Plan

FDA suggests that the Evaluation Plan section describe the evaluation measurements or measures that will be used to periodically assess the effectiveness of the RiskMAP's goals, objectives, and tools. For a detailed discussion of RiskMAP evaluation, see section V.

We recommend that this section include:

- The proposed evaluation methods for assessing RiskMAP effectiveness (e.g., claimsbased data systems, surveys, registries) and the rationales for the sponsor's chosen measures.

- Targeted values for each measure and the time frame for achieving them. FDA recommends the sponsor include interpretations of expected results under best- and worst-case scenarios. In addition, we suggest the sponsor specify what values of measures at specific time points will trigger consideration of RiskMAP modification.

---

[17] See section IV for a discussion of goals and objectives.

- The nature and timing of data collection, analyses, and audits or monitoring that will be used to assess the performance of each individual tool in achieving the RiskMAP's objectives and goals. Again, we suggest specifying target values for measures.

- A schedule for submitting progress reports to FDA regarding the evaluation results for the RiskMAP's individual tools, objectives, and goals (see section VII.B for a discussion of progress reports). We recommend that the timing and frequency of progress reports be based primarily on the nature of the risk, tools used, and outcomes under consideration. FDA recommends that progress reports be included in periodic safety update reports or traditional periodic reports.

Where applicable and possible, we recommend that the Evaluation Plan section discuss potential unintended and untoward consequences of the RiskMAP. Such a discussion would be particularly valuable if there are therapeutic alternatives with similar benefits and risks. We suggest that sponsors discuss how unintended consequences would be assessed after RiskMAP implementation. The goal of the assessment would be to ensure that overall population risks are minimized and specific product benefits, including access, are preserved.

B. Contents of a RiskMAP Progress Report

FDA recommends that a RiskMAP progress report contain the following sections, accompanied by a table of contents:

- Summary of the RiskMAP
- Methodology
- Data
- Results
- Discussion and Conclusions

1. Summary

We suggest that the Summary section briefly provide background on and an overview of the RiskMAP, and describe the overall RiskMAP goals and objectives, as well as its strategy and tools. We recommend that this section also summarize (1) the evaluation methods used and (2) the relevant measures and time frames for achieving targeted values.

2. Methodology

We recommend that the Methodology section provide a brief overview of the evaluation methods used (e.g., ascertainment of outcomes, comprehension testing, patient surveys, process audits). FDA suggests that it describe the evaluation plan, sources of potential measurement error or bias for the outcome of interest, and any analytical methods used to account for them.

Since RiskMAP evaluations will often rely upon observational data, we recommend that the analytical plan address issues such as measurement errors, sensitivity, and specificity of the measures, as well as power for detecting differences where appropriate.

3. Data

To the extent possible, we recommend that the Data section of a RiskMAP progress report contain data that would allow FDA to analyze the information and make conclusions independently.

4. Results

To the extent possible, we recommend that the Results section of a RiskMAP progress report contain the primary data from each evaluation method and analyses of the evaluation data, statistical estimation if appropriate, and the sponsor's comparison of tool, objective, and/or goal achievement relative to targeted performance measures.

5. Discussion and Conclusions

FDA recommends that this section describe whether the RiskMAP has met or is making progress in meeting the stated measures for each tool, objective, and goal. We suggest that this discussion take all available data, evaluations, and analyses into consideration.

Progress towards achieving RiskMAP goals or performance measures should be reported. Where appropriate, sponsors are encouraged to propose modifications to the RiskMAP and discuss them with FDA.

Information Program on Clinical Trials for Serious or Life-Threatening Diseases and Conditions

Information Program on Clinical Trials for Serious or Life-Threatening Diseases and Conditions, Guidance for Industry [Ŧ, 1]

U.S. Department of Health and Human Services
Food and Drug Administration
Center for Drug Evaluation and Research (CDER)
Center for Biologics Evaluation and Research (CBER)

March 2002

Procedural

Contains Nonbinding Recommendations

This guidance is intended to assist sponsors who will be submitting information to the Clinical Trials Data Bank. It addresses statutory and procedural issues for submitting information to the data bank.

Note: *This guidance represents the Food and Drug Administration's (FDA's) current thinking on this topic. It does not create or confer any rights for or on any person and does not operate to bind FDA or the public. You can use an alternative approach if the approach satisfies the requirements of the applicable statutes and regulations. If you want to discuss an alternative approach, contact the appropriate FDA staff. If you cannot identify the appropriate FDA staff, call the appropriate number listed on the title page of this guidance.*

I. Introduction

This guidance is intended to assist sponsors who will be submitting information to the Clinical Trials Data Bank. The data bank was established as required under section 113 of the Food and Drug Administration Modernization Act of 1997 (Modernization Act). This guidance combines the statutory and procedural issues discussed in two previously published draft

---

[Ŧ] Available on the FDA website at: http://www.fda.gov/downloads/RegulatoryInformation/ Guidances/UCM126838.pdf

[1] This guidance has been prepared by the Implementation Team for section 113 of the Food and Drug Administration Modernization Act of 1997, including individuals from the Office of the Commissioner, the Center for Drug Evaluation and Research (CDER), the Center for Biologics Evaluation and Research (CBER), and the Center for Devices and Radiological Health (CDRH), at the Food and Drug Administration.

guidances on this topic. It was finalized after considering comments received on the two draft guidances.

II. Background

Section 113 of the Modernization Act creates a public resource for information on studies of drugs, including biological drug products, to treat serious or life-threatening diseases and conditions conducted under FDA's investigational new drug (IND) regulations (21 CFR part 312). Section 113 of the Modernization Act, enacted November 21, 1997, amends section 402 of the Public Health Service Act (42 U.S.C. 282). It directs the Secretary of Health and Human Services, acting through the Director of NIH, to establish, maintain, and operate a data bank of information on clinical trials for drugs to treat serious or life-threatening diseases and conditions.

The Clinical Trials Data Bank is intended to be a central resource, providing current information on clinical trials to individuals with serious or life-threatening diseases or conditions, to other members of the public, and to health care providers and researchers. Specifically, section 113 of the Modernization Act requires that the Clinical Trials Data Bank contain (1) information about Federally and privately funded clinical trials for experimental treatments (drug and biological products) for patients with serious or life-threatening diseases or conditions, (2) a description of the purpose of each experimental drug, (3) patient eligibility criteria, (4) a description of the location of clinical trial sites, and (5) a point of contact for patients wanting to enroll in the trial. Section 113 of the Modernization Act requires that information provided through the Clinical Trials Data Bank be in a form that can be readily understood by the public. 42 U.S.C. 282(j)(3)(A).

The National Institutes of Health (NIH), through its National Library of Medicine (NLM) and with input from the FDA and others, developed the Clinical Trials Data Bank. The first version of the Clinical Trials Data Bank was made available to the public on February 29, 2000, on the Internet.[2] At that time, the data bank included primarily NIH-sponsored trials.

On March 29, 2000, FDA made available in the Federal Register a draft guidance entitled *Information Program on Clinical Trials for Serious or Life-Threatening Diseases: Establishment of a Data Bank*.[3] The draft guidance provided recommendations for industry on the submission of protocol information to the Clinical Trials Data Bank. It included information about the types of clinical trials for which submissions are required under section 113 of the Modernization Act, as well as the content of those submissions.

FDA made available a second draft guidance entitled *Information Program on Clinical Trials for Serious or Life-Threatening Diseases: Implementation Plan*, in the Federal Register on July 9, 2001.[4] The second draft guidance addressed procedural issues, including how to submit required and voluntary protocol information to the Clinical Trials Data Bank, as well as issues related to submitting certification to the Secretary that disclosure of information for a particular protocol

---

[2] See http://clinicaltrials.gov
[3] See 65 FR 16620 and http://www.fda.gov/cder/guidance/3585dft.htm
[4] See 66 FR 35798 and http://www.fda.gov/cder/guidance/4602dft.htm

would substantially interfere with the timely enrollment of subjects in the clinical investigation. The second draft guidance also proposed a time frame for submitting the information. This final guidance combines the two draft guidances into a single guidance.

III. Requirements Under Section 113 of the Modernization Act for IND Sponsors

A. What information must I submit to the Clinical Trials Data Bank?

Section 113 of the Modernization Act requires you to submit information to the data bank about a clinical trial conducted under an investigational new drug (IND) application if it is for a drug to treat a serious or life-threatening disease or condition and it is a trial to test effectiveness (42 U.S.C. 282(j)(3)(A)). If you wish, you can also provide information on non-effectiveness trials or for drugs to treat conditions not considered serious or life-threatening.

Section 113 of the Modernization Act requires that you submit a description of the purpose of each experimental drug, patient eligibility criteria for participation in the trial, a description of the location of clinical trial sites, and a point of contact for those wanting to enroll in the trial. Section 113 requires that the data bank provide this information in a form that can be readily understood by members of the public (42 U.S.C. 282(j)(3)(A)).

To ensure that information available through the Clinical Trial Data Bank is in a form that is readily understood, we have established four data elements, which are listed below. The data elements are made up of the following data fields: (1) descriptive information, (2) recruitment information, (3) location and contact information, and (4) administrative data. We have established the Protocol Registration System (PRS), a Web-based data processing program, to facilitate collection of this information for the data bank. The four data elements, which are listed below, as well as definitions applicable to the PRS, can be viewed at http://prsinfo.clinicaltrials.gov/.

1. Descriptive Information

Brief Title (in lay language)

Brief Summary (in lay language)

Study Design/Study Phase/Study Type

Condition or Disease

Intervention

2. Recruitment Information

Study Status Information including

• Overall Study Status (e.g., recruiting, no longer recruiting)

• Individual Site Status

Eligibility Criteria/Gender/Age

3. Location and Contact Information

Location of Trial

Contact information (includes an option to list a central contact person for all trial sites)

4. Administrative Data

Unique Protocol ID Number

Study Sponsor

Verification date

To verify the existence of an IND and to assist in administrative tracking, we ask that you also include in your submission the IND number and serial number and designate whether the IND is located in the Center for Drug Evaluation and Research (CDER) or the Center for Biologics Evaluation and Research (CBER). This administrative information is in a separate data field and will not be made public.

B. When should I begin submitting clinical trial information?

Section 113 of the Modernization Act requires that sponsors submit information no later than 21 days after the trial is opened for enrollment[5] (42 U.S.C. 282(j)(3)). Section 113 does not specify when sponsors must submit information about clinical trials that are existing and ongoing. To provide a transitional period for sponsors of clinical trials that are currently ongoing and expected to continue enrolling patients for more than 45 days, we ask that you submit information within 45 days after this guidance is made available through the *Federal Register*. We encourage you to submit information through the PRS for inclusion in the data bank as soon as possible.[6]

C. Can I submit my information at specified intervals rather than on a rolling basis?

As discussed above, you must submit information about new protocols open for enrollment within 21 days after the trial is open for enrollment (42 U.S.C. 282(j)(3)), and we request that you submit information about existing ongoing trials within 45 days after this guidance is published. Supplemental information can be submitted at 30-day intervals. Such information includes amendments to the protocol with respect to one of the data elements, or interruptions, continuations, or completion of enrollment for a study. Protocol changes related to eligibility or status information, such as routine opening and closing of trial sites, can be made at 30-day intervals. FDA strongly encourages you to update information about trials that

---

[5] Section 113 says "not later than 21 days after the approval of the protocol." Because the Agency does not approve protocols, we have interpreted this to mean within 21 days after the trial is open for enrollment.
[6] See http://prsinfo.clinicaltrials.gov

are unexpectedly closed (e.g., clinical hold) within 10 days after the closing or sooner if possible.

To ensure that the information available through the data bank is timely and accurate, FDA also encourages you to review, verify, and update all active protocol records on a semi-annual basis, at a minimum.

D. What is a trial for a serious or life-threatening disease or condition?

FDA has defined serious and life-threatening diseases and conditions in previous documents.

Most recently, FDA discussed issues related to products intended to treat serious or lifethreatening diseases and conditions in the guidance for industry on *Fast Track Drug Development Programs — Designation, Development, and Application Review* (November 1998).[7] In that guidance, we stated that all conditions meeting the definition of life-threatening, as set forth at 21 CFR 312.81(a), would also be serious conditions. The term *life-threatening* is defined as (1) diseases or conditions where the likelihood of death is high unless the course of the disease is interrupted and (2) diseases or conditions with potentially fatal outcomes, where the endpoint of clinical trial analysis is survival (21 CFR 312.81(a)). All references in this document to serious diseases or conditions include life-threatening diseases and conditions.

As FDA reiterated in the *Fast Track Guidance*, the seriousness of a disease is a matter of judgment, but generally is based on such factors as survival, day-to-day functioning, and the likelihood that the disease, if left untreated, will progress from a less severe condition to a more serious one. For example, acquired immunodeficiency syndrome (AIDS), all other stages of human immunodeficiency virus (HIV) infection, Alzheimer's disease, angina pectoris, heart failure, cancer, and many other diseases are clearly serious in their full manifestations. Furthermore, many chronic illnesses that are generally well managed by available therapy can have serious outcomes. For example, inflammatory bowel disease, asthma, rheumatoid arthritis, diabetes mellitus, systemic lupus erythematosus, depression, psychoses, and many other diseases can be serious in some or all of their phases or for certain populations.

Any investigational drug that has received fast track designation would be considered a drug to treat a serious disease or condition.[8] Information on effectiveness trials for drugs that have received fast track designation would qualify for submission to the Clinical Trials Data Bank.

E. What is a trial to test effectiveness?

Not all trials carried out under 21 CFR part 312 are trials to test effectiveness. FDA considers all phase 2, phase 3, and phase 4 trials with efficacy endpoints as trials to test effectiveness.[9]

---

[7] CDER guidances are available at http://www.fda.gov/cder/guidance/index.htm

[8] That a drug is intended to treat a serious or life-threatening disease or condition, however, does not mean that it fills an unmet medical need and qualifies for fast track designation under section 506 of the Food Drug and Cosmetic Act (21 U.S.C. 356).

[9] Listing a trial in the Clinical Trials Data Bank is not a guarantee that the trial design is considered adequate to support approval of a drug, nor does it reflect any judgment on the conduct, analysis, or outcome of the study.

F. Which trials are provided to the public through the Clinical Trials Data Bank?

Section 113 of the Modernization Act requires sponsors to submit information about clinical trials of experimental treatments for serious diseases and conditions when conducted under the IND regulations. 42 U.S.C. 282(j)(3)(A). Such information can be submitted at any time with the consent of the protocol sponsor, and must be submitted within 21 days after a trial to test effectiveness begins. In addition, section 113 of the Modernization Act states that information on all treatment IND protocols and all Group C protocols[10] must be included in the Clinical Trials Data Bank.

Although it is not specifically discussed in section 113 of the Modernization Act, there are situations in which there may be a significant number of patients with the disease or condition for which the drug is being developed who are not adequately treated by existing therapy, who do not meet the eligibility criteria for enrollment, or who are otherwise unable to participate in a controlled clinical study. In these situations, sponsors may have initiated one or more expanded access protocols that include such patients. In such cases, FDA strongly recommends that sponsors also consider submitting information to the Clinical Trials Data Bank about the availability of any expanded access protocol for treatment use in addition to required submissions.

For protocols not specifically mentioned above, sponsors should review each protocol submitted to an IND to determine if the protocol is for a serious disease or condition and if it is a trial to test effectiveness. If the protocol meets these criteria, the sponsor must submit information about the trial to the Clinical Trials Data Bank, unless the sponsor provides detailed certification to FDA that such a disclosure would substantially interfere with the timely enrollment of subjects in the investigation (42 U.S.C. 282(j)(3) and (j)(4)). Sponsors with questions on whether protocols meet the criteria for submission to the Clinical Trials Data Bank are encouraged to contact the appropriate review division for additional guidance.

G. Must I include information about foreign trial sites?

Yes, you must include information about foreign trials when those trials are conducted under an IND submitted to FDA and the trial meets the criteria for submission to the Clinical Trials Data Bank. Section 113 of the Modernization Act requires sponsors to submit information about specified clinical trials that are "under regulations promulgated pursuant to section 505(i) of the Federal Food, Drug, and Cosmetic Act," which are FDA's IND regulations (42 U.S.C. 282(j)(3)). Sponsors may voluntarily conduct a foreign trial under the IND regulations. Sponsors are not required to submit information to the Clinical Trials Data Bank when a foreign trial is not conducted under an IND.

---

[10] "Group C protocols" refers to investigational drugs designated by FDA for the treatment of specific cancers. These drugs have reproducible efficacy in one or more specific tumor types. Such a drug has altered or is likely to alter the pattern of treatment of disease and can be safely administered by properly trained physicians without specialized supportive care facilities. See National Cancer Institute Handbook for Investigators, Appendix XV, "Policy for Group C Drug Distribution," http://ctep.info.nih.gov/HandbookText/Appendix_XV.htm#Proc_Mgmt_GrpC_Prot.

IV. Implementation Issues

A. How do I submit information to the Clinical Trials Data Bank?

To facilitate the submission process, we have established the Web-based PRS at *ClinicalTrials.gov*. The system allows for entry of required and voluntary information about clinical trials. You or your designee can initiate submission of clinical trial information to *ClinicalTrials.gov* by completing a registration form at http://prsinfo.clinicaltrials.gov/.

After you have entered the data, the PRS generates a receipt for use by sponsors. An electronic copy of the receipt will be sent to the FDA.

B. What information about trial sites must be included?

Section 113 of the Modernization Act requires sponsors to submit a description of the location of trial sites and a point of contact. To ensure an adequate description, we recommend that you provide for each individual trial site the full name of the organization, city, state, postal code, and country where the protocol is being conducted; and a central contact name and phone number.

You can also provide the names and phone numbers of individual site contacts.

C. How long does it take for information to be made available on ClinicalTrials.gov?

Studies will be made available to the public through *ClinicalTrials.gov* within two to five days after submission by the sponsor.

D. How long will information about studies remain available through ClinicalTrials.gov?

NLM intends to maintain the Data Bank as a long-term registry of clinical trials. Therefore, in addition to information about open trials, information about closed trials will also be available through *ClinicalTrials.gov*, even after accrual and analysis are completed and the product is approved.

E. Can information be transferred from a sponsor computer to the PRS?

Yes. Information can be transferred according to the format specified by the PRS. The PRS has a mechanism for uploading and downloading XML-formatted protocol records. Instructions for transferring information are provided at http://prsinfo.clinicaltrials.gov/

F. Can intermediaries acting on behalf of a sponsor submit data?

Yes. For example, in some cases a sponsor might want to contract with an information management company to serve as an intermediary in preparing data for inclusion in *ClinicalTrials.gov*. The information management company, when authorized by the sponsor, could act on behalf of the sponsor for this purpose.

G. Can sponsors designate multiple individuals to be data providers?

Yes. When sponsors register to become a PRS data provider, they will be given information, including instructions, for creating additional users for their accounts. A sponsor can control access to the account by designating users and administrators for the account.

H. What happens to the information submitted to the Clinical Trials Data Bank?

Except for the IND number, serial number, and FDA center designation, all information submitted through the PRS is made available to the public at http://clinicaltrials.gov.

I. Can I submit other information to the Clinical Trials Data Bank?

Yes. PRS is designed to permit you to submit more detailed information about a protocol. Additional data fields (e.g., projected enrollment) and their definitions are included in the PRS. You also can submit protocol information about other clinical trials under IND, including trials for a disease or condition that is not serious or any trial that is not designed to test effectiveness.

Finally, you can submit information about results of a trial. This information, which, according to the structure of the Clinical Trials Data Bank, must come from the published literature, should be linked by including the unique MEDLINE identifier for citations of publications.

You can use the *link* section provided to allow pointers to Web pages directly relevant to the protocol. If you link to other Web pages from your entries, you should ensure that the links do not misbrand your products, for example, by promoting the products before the product or an indication is approved. (See 21 U.S.C. 321(n), 331(a)(b)(c)(d), 352(a)(n) http://www.fda.gov/opacom/laws/fdcact/fdcact1.htm.) When inputting links to other web pages, the database will instruct you that the links should be directly relevant to the protocol, and that you should not link to sites whose primary goal is to advertise or sell commercial products or services.

J. Should I continue submitting information to the ACTIS and PDQ databases?

No. All information for AIDS and cancer protocols that meet the requirements of section 113 of the Modernization Act must now be submitted to *ClinicalTrials.gov* through the PRS. Data from the current AIDS Clinical Trials Information System (ACTIS) and Physician's Data Query (PDQ) databases are included in *ClinicalTrials.gov*. Information from the Rare Diseases and National Institute of Aging Databases is also included in *ClinicalTrials.gov*.

K. Are there exemptions for submitting clinical trials information?

Information about an investigation will not be included in the data bank if you provide a detailed certification to the Secretary of Health and Human Services that disclosure of such information would substantially interfere with timely enrollment of subjects in the clinical trial and the Secretary does not disagree. If there is disagreement, the Secretary will provide a detailed written determination that such disclosure would not substantially interfere with such enrollment (42 U.S.C. 282(j)(4)).

FDA has not identified specific instances when disclosure of information would substantially interfere with enrollment of subjects in a clinical investigation. We solicited comments on this topic for the purpose of including a listing of acceptable reasons for certification in the final guidance. We received no comments. Therefore, if you identify a specific instance when disclosure of information would interfere with enrollment of subjects in a clinical investigation, FDA will consider your request on a case-by-case-basis.

All requests for exemption should be forwarded to Director, Office of Special Health Issues, Office of Communications and Constituent Relations, Office of the Commissioner, HF-12, 5600 Fishers Lane Rockville, MD 20857, or by email at *113trials@oc.fda.gov*, or by fax at 301-443- 4555.

L. Is Institutional Review Board preapproval of the protocol listing required?

No. Section 113 of the Modernization Act does not require prior IRB approval when submitting this information to the Clinical Trials Data Bank. Current FDA guidance recommends that IRB review of listings need not occur when, as here, the system format limits the information provided to basic information, such as title, purpose of the study, protocol summary, basic eligibility criteria, study site locations, and how to contact the site for further information.[11]

M. Will FDA monitor compliance?

A copy of the protocol listing in ClinicalTrials.gov will be sent to the FDA. FDA's Office of Special Health Issues intends to initiate a one-year pilot educational program in 2002 that will include a component to evaluate compliance. The primary objective of the pilot program is to educate sponsors about the existence of the guidance document and the availability of the online PRS data entry tool. The secondary objective of the pilot program is to evaluate the success of the educational initiative. The pilot, which will measure the number of protocols (voluntary and required) made available through the ClinicalTrials.gov database, will provide FDA with compliance information.

---

[11] The 1998 update of Information Sheets: Guidance for Institutional Review Boards and Clinical Investigators provides guidance on IRB review and approval of listings of clinical trials on the Internet. See http://www.fda.gov/oc/ohrt/irbs/toc4.html#recruiting.

Part IV

Selected FDA GCP/Clinical Trial Guidance Documents:
Medical Devices

Analyte Specific Reagents (ASRs): Frequently Asked Questions

Analyte Specific Reagents (ASRs): Frequently Asked Questions, Guidance for Industry and FDA Staff – Commercially Distributed [Ŧ, 1]

U.S. Department of Health and Human Services
Food and Drug Administration
Center for Devices and Radiological Health
Office of In Vitro Diagnostic Device Evaluation and Safety
Center for Biologic Evaluation and Research

Contains Nonbinding Recommendations

This guidance document is intended to clarify the regulations regarding commercially distributed analyte specific reagents (ASRs) (21 CFR 809.10(e), 809.30, and 864.4020), and the role and responsibilities of ASR manufacturers. This document is not intended to provide guidance on the role of clinical laboratories in the development of laboratory developed tests (LDTs). The guidance follows the substance, spirit, and meaning of the ASR regulations already in place.

Note: *This guidance represents the Food and Drug Administration's (FDA's) current thinking on this topic. It does not create or confer any rights for or on any person and does not operate to bind FDA or the public. You can use an alternative approach if the approach satisfies the requirements of the applicable statutes and regulations. If you want to discuss an alternative approach, contact the appropriate FDA staff. If you cannot identify the appropriate FDA staff, call the appropriate number listed on the title page of this guidance.*

Introduction

This guidance document is intended to clarify the regulations regarding commercially distributed analyte specific reagents (ASRs) (21 CFR 809.10(e), 809.30, and 864.4020), and the role and responsibilities of ASR manufacturers. This document is not intended to provide guidance on the role of clinical laboratories in the development of laboratory developed tests (LDTs). The guidance follows the substance, spirit, and meaning of the ASR regulations already in place.

The guidance addresses some frequently asked questions about how ASRs may be marketed, and provides FDA's Office of In Vitro Diagnostic Device Evaluation and Safety's (OIVD's) and the Center for Biologics Evaluation and Research's (CBER's) responses to those questions. Except where otherwise indicated, the use of the term "ASR" in this guidance

---

[Ŧ] Available on the FDA website at: http://www.fda.gov/downloads/MedicalDevices/
DeviceRegulationandGuidance/GuidanceDocuments/UCM071269.pdf

document refers to commercially distributed ASRs and the term "manufacturer" refers to manufacturers of commercially distributed ASRs.

FDA is providing this guidance in order to eliminate confusion regarding particular marketing practices among ASR manufacturers. As noted in this guidance document, ASRs are building blocks of LDTs. (See section II of this guidance.) ASRs are defined and classified in a rule codified at 21 CFR 864.4020. With this guidance document, FDA seeks to advise ASR manufacturers that it views the following practices as being inconsistent with the marketing of an ASR, as defined under 21 CFR 864.4020:

- Combining, or promoting for use, a single ASR with another product such as other ASRs, general purpose reagents, controls, designated laboratory instrument(s), software, etc.

- Promoting an ASR with specific analytical or clinical performance claims, instructions for use in a particular test, or instructions for validation of a specific test using the ASR.

Some manufacturers have believed that when they combine a Class I ASR, which is exempt from premarket notification requirements under section 510(l) of the Federal Food, Drug, and Cosmetic Act (the Act), 21 U.S.C. 360(l), with other products, or with instructions for use in a specific test, the product remains exempt because of the presence of an ASR. However, as explained in this guidance, when an ASR is marketed in the ways described above, FDA views the product as no longer being an ASR within the meaning of 21 CFR 860.4020 and instead views it as another type of in vitro diagnostic device (IVD) or device component not covered by the ASR regulations and, therefore, not necessarily exempt from premarket notification.

FDA's guidance documents, including this guidance, do not establish legally enforceable responsibilities. Instead, guidances describe the Agency's current thinking on a topic and should be viewed only as recommendations, unless specific regulatory or statutory requirements are cited. The use of the word should in Agency guidances means that something is suggested or recommended, but not required.

The Least Burdensome Approach

The issues identified in this guidance document represent those that we believe need to be addressed before your device can be marketed. In developing the guidance, we carefully considered the relevant statutory criteria for Agency decision-making. We also considered the burden that may be incurred in your attempt to follow the statutory and regulatory criteria in the manner suggested by the guidance and in your attempt to address the issues we have identified. We believe that we have considered the least burdensome approach to resolving the issues presented in the guidance document. If, however, you believe that there is a less burdensome way to address the issues, you should follow the procedures outlined in the document, "A Suggested Approach to Resolving Least Burdensome Issues." It is available on our Center web page at: http://www.fda.gov/cdrh/modact/leastburdensome.html.

Frequently Asked Questions

I. The ASR Rule

1. What is the definition of an ASR?

ASRs are defined as "antibodies, both polyclonal and monoclonal, specific receptor proteins, ligands, nucleic acid sequences, and similar reagents which, through specific binding or chemical reactions with substances in a specimen, are intended for use in a diagnostic application for identification and quantification of an individual chemical substance or ligand in biological specimens." 21 CFR 864.4020(a). ASRs are medical devices that are regulated by FDA. They are subject to general controls, including current Good Manufacturing Practices (cGMPs), 21 CFR Part 820, as well as the specific provisions of the ASR regulations (21 CFR 809.10(e), 809.30, 864.4020).

2. What is the ASR rule?

This guidance document refers to three regulations as "the ASR rule." Published in 1997, the regulations define and classify ASRs (21 CFR 864.4020), impose restrictions on the sale, distribution, and use of ASRs (21 CFR 809.30), and establish requirements for ASR labeling (21 CFR 809.10(e)).

3. What was the objective of the ASR rule?

The ASR rule was designed to accomplish several policy objectives. One of the primary goals of the rule was to ensure the quality of the primary, active reagents of finished IVDs or LDTs. Another focus of the rule is the requirement for appropriate labeling to be appended to test results when ASRs are used by clinical laboratories in LDTs, so that healthcare users can understand when tests are being developed and validated by the laboratory and have not undergone FDA clearance or approval. 62 FR 62244. FDA adopted the approach of regulating most ASRs by means of general controls and exempting them from premarket notification requirements as the least burdensome approach. This approach relies primarily on cGMPs, medical device reporting, and labeling requirements to adequately control the risks associated with these devices. In addition, laboratories that develop tests using ASRs must be in compliance with the Clinical Laboratory Improvement Amendments (CLIA), 42 U.S.C. 263a 62 FR 62252.

4. What does the ASR rule require?

The rule classifies most ASRs as Class I devices subject to general controls under section 513(a)(1)(A) of the Act, but exempt from premarket notification. The general controls require ASR manufacturers to register and list their devices, 21 CFR 807.20(a), submit medical device reports (21 CFR Part 803), follow labeling requirements, 21 CFR 809.10(e), and follow cGMPs, 21 CFR 809.20(b). The rule also restricts the sale, use, distribution, labeling, advertising and promotion of ASRs. 21 CFR 809.30. One of these restrictions allows only physicians and other persons authorized by applicable State law to order LDTs that are developed using ASRs. 21 CFR 809.30(f). Another restriction requires the laboratory that develops an LDT using an ASR to add a statement disclosing that the laboratory developed the test and it has not been cleared or approved by FDA when reporting the test result to the practitioner. 21 CFR 809.30(e).

The restrictions also prohibit advertising and promotional materials for ASRs from making any claims for clinical or analytical performance. 21 CFR 809.30(d)(4). Consistent with this restriction, the labeling for Class I, exempt ASRs must bear the statement, "Analyte Specific Reagent. Analytical and performance characteristics are not established." 21 CFR 809.10(e)(1)(x). Manufacturers who wish to make analytical and/or clinical performance claims for a product should submit an application to FDA for premarket review rather than marketing the product as an ASR. For example, performance claims might include statements such as, "This ASR can be used for quantification of [an analyte] to determine [a diagnosis]".

5. Are some ASRs Class II or Class III, requiring a premarket submission?

Yes. Although most ASRs are Class I, there are some ASRs that are Class II and Class III and that must be cleared or approved by FDA before they can be marketed in the United States. 21 CFR 864.4020. FDA classifies medical devices, including diagnostic devices such as ASRs, into Class I, II, or III according to the level of regulatory control that is necessary to provide a reasonable assurance of safety and effectiveness. These classifications include consideration of the level of risk associated with the device. 21 U.S.C. 360c. The classification of an ASR determines the appropriate premarket process.

An ASR is a Class II device if the reagent is used as a component in a blood banking test of a type that has been classified as a Class II device (e.g., certain cytomegalovirus serological and treponema pallidum nontreponemal test reagents). 21 CFR 864.4020(b)(2).

An ASR is a Class III device if the reagent is intended as a component in tests intended either:

- to diagnose a contagious condition that is highly likely to result in a fatal outcome and prompt, accurate diagnosis offers the opportunity to mitigate the public health impact of the condition (e.g., human immunodeficiency virus (HIV/AIDS) or tuberculosis (TB)); or

- for use in donor screening for conditions for which FDA has recommended or required testing in order to safeguard the blood supply or establish the safe use of blood and blood products (e.g., tests for hepatitis or for identifying blood groups). 21 CFR 864.4020(b)(3).

FDA considers ASRs intended to be used as a component in tests for diagnosis of HIV (including monitoring for viral load or HIV drug resistance mutations) to be Class III ASRs.

6. How does a manufacturer know whether its device is an ASR?

We recommend that ASR manufacturers consult this document for guidance on whether their product is or is not within the scope of the ASR rule. Manufacturers should contact FDA if they are unsure about the classification of their device to discuss any applicable regulatory requirements. Manufacturers who wish to obtain FDA advice on this matter in advance of marketing may consult with OIVD, or, with CBER for questions about HIV ASRs or ASRs for blood or cellular and tissue products.

II. What Meets the ASR Definition?

There has been some confusion about which products fall within and outside the definition of an ASR. Some of this confusion relates to a misunderstanding that a product is an ASR if it is labeled as one, even if the product contains analytical or clinical performance claims and does not meet the definition of an ASR.

In the preamble to the ASR rule, FDA stated that ASRs are the "active ingredients" of tests that are used to identify one specific disease or condition. ASRs are purchased by manufacturers who use them as components of tests that are cleared or approved by FDA and also by clinical laboratories that use the ASRs to develop LDTs used exclusively by that laboratory. 62 FR 62243, 62244. This is in contrast to what the preamble referred to as a "kit or system for 'in vitro diagnostic use'" that has a proposed intended use, indications for use, instructions for use, and performance characteristics. 62 FR 62243, 62250.

The ASR rule was intended to require that ASR manufacturers take certain actions, such as following cGMPs, to help ensure the quality of these reagents so that IVD manufacturers and laboratories who purchase them can produce tests that are safe and effective. A premise underlying the rule, however, is that laboratories and IVD manufacturers, rather than ASR manufacturers, design and develop the test in which the ASR is used, using their judgment and knowledge, and provide all necessary verification and validation.

Based upon this description, together with the ASR definition, FDA views an ASR as having the following characteristics:

- used to detect a single ligand or target (e.g., protein, single nucleotide change, epitope);

- not labeled with instructions for use or performance claims; and

- not promoted for use on specific designated instruments or in specific tests.

7. What are some examples of entities that FDA considers to be ASRs?

Examples of entities that are ASRs include:

- a single antibody (e.g., an anti-troponin I polyclonal or monoclonal antibody, whether untagged or tagged, e.g., conjugated to horseradish peroxidase),

- a single forward/reverse oligonucleotide primer[2] pair (e.g., a primer pair for amplification of a single amplicon, such as for amplification of the ΔF508 locus of the gene encoding the cystic fibrosis transmembrane regulator (CFTR)), or single forward or reverse primer individually,

- a nucleic acid probe[3] (whether untagged or tagged, e.g., conjugated to biotin or Cy™3) intended to bind a single complementary amplified or unamplified nucleic acid sequence,

---

[2] For the purposes of this guidance, a primer is defined as a nucleic acid sequence that is intended to initiate amplification by binding selectively to a complementary sequence in a large nucleic acid polymer.
[3] For the purposes of this guidance, a probe is defined as a molecule that is intended to isolate, bind, or identify a specific target or ligand.

- a single purified protein or peptide (e.g., purified B-type natriuretic peptide).

The above-listed examples would not be considered ASRs if they are marketed with clinical or analytical performance claims (e.g., quantification of an infectious agent, assessment of cardiac risk).

In addition to the examples listed here, there may be other products that can be appropriately marketed as ASRs provided they meet the criteria listed above, i.e., the ASR is used to detect a single ligand or target (e.g., protein, single nucleotide change, epitope), is not labeled with instructions for use or performance claims, and is not promoted for use on specific designated instruments or in specific tests. In the future, with the development of novel technologies, there may be products that meet the definition of an ASR (21 CFR 864.4020) that are dissimilar to the examples listed above. In those cases, ASR manufacturers should contact FDA with questions about specific products.

8. What are some examples of entities that FDA does not consider to be ASRs?

- Multiple individual ASRs (e.g., antibodies, probes, primer pairs) bundled together in a single pre-configured or optimized mixture so that they must be used together in the resulting LDT. For example, a set of 5 primer pairs combined in a single tube that are used to detect 5 different viral genotypes requires that all of these pairs be used together, and that they work together to accurately detect all five genotypes. This is an analytical claim for the product, and FDA does not consider this type of product to be an ASR.

- Products that include or require more than a single ASR (i.e., the product includes some or all of the products needed to conduct a particular test such as more than one ASR, general reagents, controls, equipment, software, etc.) and/or has instructions for use. FDA does not consider such a product to be an ASR but rather an IVD or IVD component not covered by the ASR rule.

- Reagents that are designed to require use in a specific assay or on a designated instrument (e.g., arrayed on beads). The requirement for reagents and designated instruments to be used together constitutes a performance claim that they will work properly when used in combination, since those specific reagents are intended for use with that specific instrument. FDA does not consider such a product to be an ASR but rather an IVD or IVD component not covered by the ASR rule.

When manufacturers have assembled ingredients towards the development of a test, such as in the examples listed above, the product is no longer an ASR. A laboratory cannot validate that the way those individual ASRs and other components are combined by the manufacturer (e.g., the identity, concentration, purity) is appropriate for meeting the intended use and specifications of their in-house test. However, laboratories themselves are not precluded by the ASR rule from selecting and combining individual ASRs and other components in the development of their own in-house tests.

Other types of devices that do not meet the definition of an ASR include:

- Control material or calibrators.

- Products that have specific performance claims, or procedural instructions, or interpretations for use.

- Reagents offered with software for interpretation of results.

- Software for interpretation of assay results.

- Microarrays.

Manufacturers who wish to market a product in a fashion that is similar to the examples listed above should discuss the classification of their product with FDA prior to marketing.

9. How do General Purpose Reagents compare to ASRs?

A General Purpose Reagent (GPR) is "a chemical reagent that has general laboratory application, that is used to collect, prepare, and examine specimens from the human body for diagnostic purposes, and that is not labeled or otherwise intended for a specific diagnostic application." 21 CFR 864.4010(a). Like ASRs, GPRs are not labeled for a specific clinical or diagnostic use. Because GPRs are not analyte-specific, they should have the potential to be combined with, or used in conjunction with more than one type of ASR by the laboratory or IVD manufacturer that develops the finished test. In contrast, as stated above, an ASR is a specific chemical component, probe, or antibody that by its design determines which individual chemical substance or ligand can be detected.

III. Manufacturer Marketing Practices

10. To whom can manufacturers sell ASRs?

ASRs may only be sold to:

- in vitro diagnostic manufacturers;

- clinical laboratories regulated under the Clinical Laboratory Improvement Amendments of 1988 (CLIA), as qualified to perform high complexity testing, or clinical laboratories regulated under VHA Directive 1106; and

- organizations that use the reagents to make tests for purposes other than providing diagnostic information to patients and practitioners, e.g., forensic, academic, research and other nonclinical laboratories. 21 CFR 809.30 (b).

11. Can a manufacturer or distributor promote specific ASRs and GPRs for use together in developing a test?

No. As explained above, ASRs are considered specific individual "building blocks" of LDTs and finished IVDs tests, while GPRs are not intended for any specific diagnostic application. 21 CFR 864.4010, 864.4020. A product that is promoted for use with an ASR is intended for a specific diagnostic use with that ASR and therefore would not meet the GPR definition of being for "general laboratory application." Similarly, a product that is promoted for use with a specific GPR is intended for a particular intended use rather than as an ASR, which is a building block of LDTs. Therefore, a manufacturer who wishes to market its products as

GPRs or ASRs should not promote or sell them together, including in finished or partial IVD test configurations.

We recommend that manufacturers who wish to market multiple different products as ASRs, e.g. multiple ASR primer sets to amplify different mutation loci in a single gene, rather than as IVDs or IVD components, avoid marketing the ASRs in a manner that suggests that use of particular ASRs together will provide a particular effect, or that these devices should be used together for a specific purpose.

12. Can the manufacturer include instructions with an ASR?

ASR manufacturers should not provide instructions for developing or performing an assay with an ASR. As explained above, FDA does not view reagents that are sold with instructions for developing or performing a test as ASRs because ASRs are intended to be building blocks for tests and are intended to be used to develop a test for which the developer must establish instructions for use and performance claims. 21 CFR 809.10(e)(1)(x), 809.30(d)(4). Instructions for use of an ASR in a particular test would constitute a claim that, when used as directed, the ASR will perform to detect a particular chemical substance or ligand.

On the other hand, instructions for proper storage and handling of an ASR must be provided. 21 CFR 809.10(e)(1)(vi). In addition, scientific information may be included on chemical/molecular composition, concentration or mass, nucleic acid sequence, binding affinity, cross-reactivities, known mutations associated with the sequence, and interaction with substances of known clinical significance. 21 CFR 809.10(e)(1)(iv).

13. Can a manufacturer or distributor tell a laboratory which ASRs are useful for a particular application, for example, which monoclonal antibodies or probes are useful for leukemia or lymphoma testing?

ASR manufacturers and distributors should not make claims to physicians or laboratories regarding analytical or clinical performance for ASRs. The laboratories, not the ASR manufacturers or distributors, are responsible for the design and performance of the test. 21 CFR 809.30(d)(4).

ASR labeling may indicate the affinity of the reagent, such as "anti-estrogen receptor antibody" or "ΔF508 CFTR nucleic acid probe." Other similar information, such as the affinity, target, or sequence of a DNA probe or a protein sequence, may also be provided because it describes the ligand to which the ASR is specific but does not claim to produce a particular clinical or analytical result. But a name such as "Cardiac Risk ASR" describes a specific clinical use for the product and FDA, therefore, would not consider such a product to be an ASR.

14. Can an ASR manufacturer supply quality control materials/reagents that can be used with an ASR?

Yes, but these materials should be promoted independently of specific ASRs. Marketing such materials for use with an ASR would indicate that the ASR manufacturer is actually marketing a product which could trigger premarket review requirements. Quality control materials should be promoted and sold using existing FDA classifications for quality control material (e.g., 21 CFR 862.1660, 862.3280, 864.8625).

15. Can a manufacturer or distributor market software for use with an ASR?

If an ASR manufacturer chooses to market software for use with its product, then the products together would not be considered an ASR. Software does not meet the definition of an ASR. FDA views marketing practices that directly suggest or state that particular software is needed to achieve a function of an ASR to cause the ASR part of the combination to fall outside of the ASR definition because the ASR would now be intended for use with the software. As a result, FDA's view is that ASR manufacturers should not promote, sell, or otherwise distribute software for use with a particular ASR.

16. What types of instrumentation can manufacturers promote for use with LDTs?

ASRs are intended to be sold as building blocks for use in design of a diagnostic test by the test developer. If an ASR is promoted as being intended for use with a particular instrument, FDA would not view the promoted product as an ASR. Use of the ASR with the particular instrument would be a design choice by the ASR manufacturer and not by the test developer. As a result, manufacturers should not promote specific laboratory instruments for use in conjunction with particular ASRs.

In contrast, open instruments that have user-defined capabilities, which allow the user to select an instrument and an ASR independently, and define, optimize, and validate the test performance characteristics and interpretation criteria, may be promoted for use generally in LDTs (e.g., spectrophotometers, HPLC). If instrumentation is used for an ASR-based test, the laboratory should be able to select the instrumentation and validate the performance of the LDT on that instrument.

17. Can the manufacturer of an ASR help with the validation and verification of performance specifications of a test that utilizes its ASR?

If a manufacturer or distributor wishes to market its product as an ASR, it should not assist with the development or validation of an LDT using its specific ASR. Under the CLIA regulations, the laboratory must conduct validation and verification of test performance specifications. 42 CFR 493.1213. This validation by the laboratory is the minimum requirement under CLIA for the laboratory to generate clinical results for tests of high complexity.

18. What type of information about a particular ASR can an ASR manufacturer provide to a laboratory?

An ASR manufacturer may provide laboratories with information, including peer-reviewed and published/presented literature, that establishes characteristics of the ASR itself, such as information describing the single ligand or target the ASR detects. Such information may not, however, describe the use of an ASR in a specific test, including information regarding an ASR's clinical utility and clinical performance as well as specific instructions-for-use and validation protocols. 21 CFR 809.30, 864.4020. When coupled with such information, the product would fall outside the definition of an ASR. See Section II, "What Meets the ASR Definition?".

IV. Research and Investigational Use of ASRs

19. Can ASRs be used for research?

Yes, ASRs can be used for research applications. The ASR requirements, including the need for the laboratory report disclaimer, apply only to clinical diagnostic use of these products and not to research applications. 21 CFR 864.4020(a)(2).

20. How is the ASR rule related to in vitro diagnostic products labeled for research or investigational use?

Products labeled for research use only (RUO) or investigational use only (IUO) are IVDs in different stages of development.

- FDA considers RUO products to be products that are in the laboratory research phase of development, that is, either basic research or the initial search for potential clinical utility, and not represented as an effective in vitro diagnostic product. During this phase, the focus of manufacturer-initiated studies is typically to evaluate limited-scale performance and potential clinical or informational usefulness of the test. These products must be labeled "For Research Use Only. Not for use in diagnostic procedures." as required under 21 CFR 809.10 (c)(2)(i).

- FDA considers IUO products to be products that are in the clinical investigation phase of development. They may be exempt from the investigational device (IDE) requirements of 21 CFR Part 812 (21 CFR 812.2(c)), or may be regulated under 21 CFR Part 812 as either a non-significant risk device or a significant risk device. Diagnostic devices exempt from IDE requirements cannot be used for human clinical diagnosis unless the diagnosis is being confirmed by another, medically-established diagnostic product or procedure (21 CFR 812.2(c)(3)(iv)). During this phase, the safety and effectiveness of the product are being studied; i.e., the clinical performance characteristics and expected values are being determined in the intended patient population(s). These products must be labeled, "For Investigational Use Only. The performance characteristics of this product have not been established." 21 CFR 809.10(c)(2)(ii).

21. What is the difference in GMP requirements for manufacturers of an ASR versus an RUO reagent?

Manufacturers establish and follow cGMPs, as established in the quality system regulation, to help ensure that their products are manufactured under controlled conditions that assure the devices meet consistent specifications across lots and over time. ASRs must be manufactured following cGMPs. 21 CFR 809.20. FDA does not expect RUO reagents to be manufactured in compliance with cGMPs because products labeled as RUO reagents cannot be used as clinical diagnostic products. 21 CFR 809.10(c)(2)(i).

Humanitarian Device Exemption (HDE) Regulation: Questions and Answers

Humanitarian Device Exemption (HDE) Regulation: Questions and Answers; Guidance for HDE Holders, Institutional Review Boards (IRBs), Clinical Investigators, and FDA Staff [†, 1]

U.S. Department of Health and Human Services
Food and Drug Administration
Center for Devices and Radiological Health
Center for Biologics Evaluation and Research

July 8, 2010

Contains Nonbinding Recommendations

This guidance document answers commonly asked questions about Humanitarian Use Devices (HUDs) and applications for Humanitarian Device Exemption (HDE) authorized by section 510(m)(2) of the Federal Food, Drug, and Cosmetic Act (the Act). It also reflects the additional requirements set forth in the Pediatric Medical Device Safety and Improvement Act of 2007.

Note: *This guidance represents the Food and Drug Administration's (FDA's) current thinking on this topic. It does not create or confer any rights for or on any person and does not operate to bind FDA or the public. You can use an alternative approach if the approach satisfies the requirements of the applicable statutes and regulations. If you want to discuss an alternative approach, contact the appropriate FDA staff. If you cannot identify the appropriate FDA staff, call the appropriate number listed on the title page of this guidance.*

Introduction

This guidance document answers commonly asked questions about Humanitarian Use Devices (HUDs) and applications for Humanitarian Device Exemption (HDE) authorized by section 510(m)(2) of the Federal Food, Drug, and Cosmetic Act (the Act). This guidance document reflects the additional requirements set forth in the Pediatric Medical Device Safety and Improvement Act of 2007.

For the purposes of this guidance, "you" refers to the HDE holder, the Institutional Review Board (IRB), or the clinical investigator depending upon how the question is asked and "we" refers to FDA.

FDA's guidance documents, including this guidance, do not establish legally enforceable responsibilities. Instead, guidances describe the Agency's current thinking on a topic and

---

† Available on the FDA website at: http://www.fda.gov/MedicalDevices/DeviceRegulationandGuidance/GuidanceDocuments/ucm110194.htm

1

should be viewed only as recommendations, unless specific regulatory or statutory requirements are cited. The use of the word should in Agency guidances means that something is suggested or recommended, but not required.

Definitions

1. What is a Humanitarian Use Device (HUD)?

As defined in 21 CFR 814.3(n), a HUD is a "medical device intended to benefit patients in the treatment or diagnosis of a disease or condition that affects or is manifested in fewer than 4,000 individuals in the United States per year."

2. What is a Humanitarian Device Exemption (HDE)?

A Humanitarian Device Exemption (HDE)is an application that is similar to a premarket approval (PMA) application, but is exempt from the effectiveness requirements of sections 514 and 515 of the Food, Drug, and Cosmetic Act (the Act). FDA approval of an HDE authorizes an applicant to market a Humanitarian Use Device (HUD), subject to certain profit and use restrictions set forth in section 520(m) of the Act. Specifically, as described below, HUDs cannot be sold for profit, except in narrow circumstances, and they can only be used in a facility after an IRB has approved their use in that facility, except in certain emergencies.

3. Who is an HDE holder?

An HDE holder is a person who obtains the approval of a Humanitarian Device Exemption (HDE) from FDA.

4. What does it mean to "use" a HUD?

The term "*use*" in this document, when unmodified, refer to the use of a HUD according to its approved labeling and indication(s) t o treat or diagnose patients. When a HUD is being used in a clinical investigation (i.e., collection of safety and effectiveness data), the terms "*investigational use*" or "*clinical investigation*" will be used. A HUD may be studied in a clinical investigation in accordance with its approved indication(s) for a different indication, subject to the requirements described below. For more information on "use" versus "investigational use"/"clinical investigation" of a HUD, see questions 40-42 and "Figure 1: Decision Tree for IRB Review of HUDs" at the end of this guidance

HUD Designations and HDE Applications

5. What is required in a request for HUD designation?

In accordance with 21 CFR 814.102(a), the applicant's request must include:

- a statement indicating that the applicant is requesting a HUD designation for a rare disease or condition, or a valid subset of the disease or condition

- the name and address of the applicant

- a description of the rare disease or condition for which the device is to be used, the proposed indication or indications for use of the device, and the reasons why such therapy is needed

- a description of the device and a discussion of the scientific rationale for the use of the device for the rare disease or condition and

- documentation, with appended authoritative references, to demonstrate that the device meets the definition of 21 CFR 814.3(n).

See 21 CFR 814.102(a) for additional information on each of the above items.

6. When does FDA determine whether a device is eligible for designation as a HUD?

After all supportive materials have been received along with the applicant's request for HUD designation, we determine whether the device is for a rare disease or condition that affects, or is manifested in fewer than 4,000 individuals in the United States (US) per year. In the case of a device used for diagnostic purposes, we also determine at that time whether the documentation demonstrates that fewer than 4,000 individuals per year would be subjected to diagnosis by the device in the United States (21 CFR 814.102(a)(5)).

The applicant should submit the request for a HUD designation before submitting an application for an HDE.

7. Can a device qualify for HUD designation if the affected patient population is fewer than 4,000 per year but there may be multiple contacts with the device for a single patient?

Yes. FDA recognizes that, in some cases, the number of contacts with the device may exceed one per patient. A device that involves multiple patient contacts may still qualify for HUD designation as long as the total number of patients affected, or in which the disease or condition is manifested, is less than 4,000 per year in the US. In the case of a device used for diagnostic purposes, it may also still qualify for HUD designation despite there being multiple contacts with the device by a single patient; the documentation must demonstrate that fewer than 4,000 individuals per year would be subjected to diagnosis by the device in the United States (21 CFR 814.102(a)(5)). That is, devices used in 4,000 or more patients a year to diagnose a subpopulation of less than 4,000 patients with a disease or condition would not be eligible for HUD designation (21 CFR 814.102(b)(3)(ii)).

8. What is required in an HDE application?

The applicant must include a copy of or reference to FDA's HUD designation letter with the HDE application (21 CFR 814.104(b)(1)). Other contents required in an HDE application are described in detail in 21 CFR 814.104. This information enables FDA to determine whether the device meets the statutory criteria for a HUD set forth in section 520(m)(2) of the Act.

The Pediatric Medical Device Safety and Improvement Act of 2007 (Public Law 110-85) requires additional information in all original HDE applications, if such information is readily available. Specifically, it requires: a description of any pediatric subpopulations that suffer from the disease or condition that the device is intended to treat, diagnose, or cure; and the number of affected pediatric patients. See section 515A(a)(2) of the Act.[2]

9. Can you submit an HDE application if another comparable device is available to treat or diagnose the disease or condition?

We will consider an HDE application for any of the following:

- no comparable device is available to treat or diagnose the disease or condition; or
- a comparable device is available under another approved HDE application; or
- a comparable device is being studied under an approved Investigational Device Exemption (IDE) (21 CFR 814.104(b)(2)).

However, we cannot approve an HDE for a HUD device once a comparable device with the same indications for use is marketed through either the premarket approval (PMA) process or the premarket notification (510(k)) process. See section 520(m)(2)(B) of the Act.

10. What does FDA consider a "comparable device"?

A "comparable device" need not be identical to the device submitted under the HDE application. In determining whether a comparable device exists, FDA will consider:

- the device's indications for use and technological characteristics
- the patient population to be treated or diagnosed with the device
- whether the device meets the needs of the identified patient population.

---

[2] Many of the statutory provisions cited throughout this guidance, including sections 515A(a)(2) and 520(m)(6) of the Act, were added by the Pediatric Medical Device Safety and Improvement Act of 2007.

Contact Information

11. Where do I submit a request for a HUD designation?

Submit 2 copies of your request for a HUD designation in accordance with 21 CFR 814.102 to:

> Office of Orphan Products Development (OOPD)
> Food and Drug Administration
> WO32-5271
> 10903 New Hampshire Avenue
> Silver Spring, Maryland 20993-0002

If you have questions about the HUD designation, FDA's Office of Orphan Products Development is available at (301) 796-8660.

12. Where do I submit an HDE application?

Submit 6 copies[3] of your HDE application in accordance with 21 814.104 to:

> For Products Regulated by CDRH
> U.S. Food and Drug Administration
> Center for Devices and Radiological Heath
> Document Mail Center – WO66-G609
> 10903 New Hampshire Avenue
> Silver Spring, MD 20993-0002.

For Products Regulated by CBER

> Document Control Center (HFM-99)
> Center for Biologics Evaluation and Research
> Food and Drug Administration
> 1401 Rockville Pike, Suite 200N
> Rockville, MD 20852-1448
> FDA's Review of HDE Applications

13. How long does FDA have to review an original HDE application?

FDA has 75 days from the date of receipt to approve or deny an HDE application under 21 CFR 814.114. This period includes a 30-day filing period during which we determine whether the HDE application is sufficiently complete to permit substantive review. If we notify the applicant that the application is incomplete and request additional information, the 75-day time

---

[3] We encourage submission of electronic copies. For more information on submission of electronic copies to CDRH, please see "Electronic Copies for Pre-Market Submissions." For electronic copies submitted to CBER, please see "Regulatory Submissions in Electronic Format for Biologic Products.".

frame will reset upon receipt of the additional information by FDA. See section 520(m)(2) of the Act; 21 CFR 814.114.

14. What are the review time frames for HDE amendments, supplements, and reports?

The review timeframe for HDE amendments, supplements, and reports is 75 days, the same as for HDE original applications, except for a supplement submitted as a 30-day notice (21 CFR 814.39(f)).

15. Are HDE amendments, supplements, and reports subject to the same regulations as those for PMAs?

Yes. HDE amendments, supplements, and reports are generally subject to the same regulations as those for PMAs. See 21 CFR 814.106, 814.108, 814.110, and 814.126 for specific HDE requirements.

16. Are HDEs subject to user fees?

No. User fees for HDEs are waived under the Medical Device User Fee and Modernization Act of 2002, as reauthorized and amended by the Medical Device User Fee Amendments of 2007.

17. Does the Quality Systems Regulation (QSR) (21 CFR Part 820) apply to HUDs?

Yes, however, we primarily focus on those manufacturing practices the agency deems most relevant to the safety of the device.

18. Can I request an exemption from the QSR?

Yes. If you believe that you cannot comply with or should not be held to the QSR requirements, you may request an exemption. As described in 21 CFR 820.1(e), the procedures for petitioning for an exemption are set forth in 21 CFR 10.30. In evaluating such a request, we will give overriding consideration to the risks posed by the device, the potential risks that a manufacturing defect might pose, and the public health need for the device.

HDEs and Pediatric Patients

19. If an HDE was approved for use in pediatric patients prior to the enactment of the Pediatric Medical Device Safety and Improvement Act of 2007, is the HDE holder prohibited from profiting from the sale of the device?

Yes, only original HDE applications for devices indicated for use in pediatric patients or in a pediatric subpopulation that are approved on or after September 27, 2007, are assigned an annual distribution number (ADN) and may be sold for profit (subject to restrictions described below). For example, an HDE supplement does not warrant eligibility for profit if the HDE was previously approved before September 27, 2007, for use in pediatric patients or in a pediatric subpopulation.

20. Are separate HDE applications required for a device indicated for pediatric and adult use?

No. Devices that are intended to treat both a pediatric population and an adult population may be included in a single HDE application, but the indications for use should specify use in pediatric patients, or pediatric subpopulation(s), as well as use in adults. In some cases, the safety and probable benefit profile for devices intended for use in a pediatric population, or in a pediatric subpopulation, may differ from its use in an adult population. Therefore, it is recommended that HDE applications for devices intended for use in pediatric populations and adult populations include data supporting the use in both pediatric and adult populations.

We note that the Act, as amended by the Pediatric Medical Device Safety and Improvement Act of 2007 (Public Law 110-85), requires us to establish the annual distribution number (ADN) by assessing projected use of the product in "individuals," a term that includes both pediatric and adult patients. See section 520(m)(6)(A)(ii) of the Act. This provision authorizes HDE holders to receive profit from the sale of HUDs that are indicated for pediatric use only, or for use in both pediatric and adult patients, subject to the upper limit of the ADN. In this way, when a device is potentially applicable to both pediatric and adult populations, the statute provides an incentive for an applicant to include in its HDE submission to FDA information establishing that the device will not expose pediatric patients to an unreasonable or significant risk of illness or injury and that the probable benefit to health from the use of the device outweighs the risk of injury or illness from its use. Such analysis should address the risks compared to the benefits, taking into account the probable risks and benefits of currently available devices or alternative forms of treatment. Only when a submission meets this standard for approval will FDA approve the product for use in pediatric patients, and only then will the HDE holder be eligible to receive profit from the sale of the device.

21. What is the annual distribution number (ADN) and how is it determined?

The Pediatric Medical Device Safety and Improvement Act of 2007 (Public Law 110-85) allows HUDs intended for use in pediatric patients or in a pediatric subpopulation and approved on or after September 27, 2007, to be sold for profit as long as the number of devices distributed in any calendar year does not exceed the annual distribution number (ADN). The ADN is determined by the agency when the agency approves the HDE. It is determined by estimating the number of individuals (pediatric and adult patients) affected by the disease or condition and likely to use the device each year multiplied by the number of devices reasonably necessary to treat each individual. If the number calculated is less than 4,000, then this number is the ADN. If the number calculated is equal to or more than 4,000, then the ADN is capped at 3,999 because the ADN must be less than 4,000 devices. See section 520(m)(6)(A)(ii) of the Act.

The applicant should provide supporting data for both the number of individuals likely to use the device each year, and the number of devices reasonably necessary to treat each such individual. The same principles that govern requests for a HUD designation, specifically documentation with appended authoritative references, should apply to requests for an ADN designation. See question 5 for more information on such documentation.

As stated in section 520(m)(8) of the Act, the agency's Pediatric Advisory Committee will annually review all HUDs intended for use in pediatric patients that are approved on or after September 27, 2007, to ensure that the HDE remains appropriate for the pediatric populations for which it is approved.

22. After an HDE is approved and an ADN has been assigned, can an HDE holder request to have the ADN modified?

Yes. An HDE holder may submit an HDE supplement (21 CFR 814.108) requesting modification of the ADN based on new information regarding the number of individuals affected by the disease or condition. Again, the ADN must be less than 4,000.

23. Do HDE holders with ADNs set by the agency have special reporting requirements?

HDE holders assigned an ADN must immediately notify the agency if the number of devices distributed in a year exceeds the ADN. See section 520(m)(6)(A)(iii) of the Act. FDA interprets this statutory requirement to mean that HDE holders must immediately notify the agency by submitting an HDE report whenever the number of devices shipped, or sold, in a year, however they are used, exceeds the ADN.[4] In this way, the new statutory notification requirement is generally consistent with the reporting requirement in 21 CFR 814.126(b)(1)(iii) discussed in the "After FDA Approves an HDE" section below (question 31): both concern the number of devices shipped or sold, however the devices are ultimately used (even if outside their approved indications). The only difference is that the new statutory provision requires immediate notification when the number shipped or sold in a year exceeds the ADN, whereas

---

[4] FDA recognizes that HDE holders may ship additional sizes to facilities to ensure that each device fits properly when used. These additional shipments may or may not count towards the annual ADN tally, depending on whether these additional sizes are used or are returned to the HDE holder.

the current regulations require periodic reports on a timeframe specified in the HDE approval order.

In those rare cases in which a device holds both an HDE approval for a certain indication, and a PMA approval for a different indication, sales or shipments of the device pursuant to the PMA are not subject to the ADN reporting requirement. The ADN relates only to those devices that are on the market through the HDE process for a disease or condition that occurs in pediatric patients or in a pediatric subpopulation. In that instance, the manufacturer is only required to notify FDA when sales or shipments tracked pursuant to the HDE exceed the ADN.

24. What happens when the number of devices shipped or sold in a year exceeds the ADN?

For HUDs labeled for use in pediatric patients or in a pediatric subpopulation and approved on or after September 27, 2007, FDA exempts a certain number of these devices each year -- known as the ADN -- from the prohibition on profit (see questions 29 and 30 for more on this prohibition). It is the HDE holder's responsibility to immediately notify the agency in the form of an HDE report (21 CFR 814.126) when the number of HUDs shipped or sold in a year, however they are used, exceeds the ADN. Once this notification occurs, or once FDA discovers through an inspection that the ADN has been exceeded, then the general prohibition on profit applies for the remainder of the year. See section 520(m)(6)(D) of the Act.

25. If a device is manufactured in various sizes depending on a patient's anatomy, the number of devices distributed may be more than the number of devices used in any year. Which number, the number used or the number distributed, is the ADN?

As described above, the ADN is the number of devices shipped or sold in a year that the agency exempts from the prohibition on profit. Once the HDE holder notifies the agency, or once the agency discovers through an inspection, that the ADN has been exceeded, sales of the device for the remainder of the year are subject to the general prohibition on profit. If the HDE holder ships multiple sizes, these shipments may or may not count toward the annual ADN tally, depending on whether these additional sizes are used or are returned to the HDE holder. (See footnote 3.)

26. What is the definition of pediatric patients?

As defined in section 520(m)(6)(E) of the Act, pediatric patients are patients who are 21 years of age or younger at the time of the diagnosis or treatment. A pediatric subpopulation means one of the following populations: neonates, infants, children, or adolescents. FDA reviews pediatric devices through all of its premarket pathways, including premarket notification (510(k)), premarket approval (PMA), biological license application (BLA), and humanitarian device exemption (HDE). Additional information about the definition of pediatric patients and pediatric use can be found in: "Guidance for Industry and FDA Staff: Premarket Assessment of Pediatric Medical Devices."[5]

After FDA Approves an HDE

27. Is the HDE holder required to submit to FDA the names and addresses of the IRBs that approved the use of a HUD?

No. The applicant is not required to submit the names and addresses of the reviewing IRBs to FDA. However, as required in 21 CFR 814.126(b)(2), the applicant must maintain records of:

- the names and addresses of the facilities to which the HUD was shipped
- correspondence with reviewing IRBs
- any other information required by a reviewing IRB or FDA.

28. Does the general prohibition on profit apply to HUDs even when used outside their approved indications?

HUDs, even when used outside their approved indications, are subject to the general prohibition on profit. See section 520(m)(3) of the Act; 21 CFR 814.104(b)(5).[6] As explained in the "HDEs and Pediatric Patients" section above, however, some HUDs are exempt from this prohibition if they are indicated for use in pediatric patients, or in a pediatric subpopulation, or for use in both pediatric and adult patients, subject to the upper limit of the ADN.

For devices that have both an HDE and a PMA approval for a different indication, there is no restriction on profit from sales pursuant to the PMA.

29. How should the HDE holder verify that the amount charged for the device does not exceed the costs of research and development, fabrication, and distribution?

If the HDE holder charges more than $250 for the device, FDA requires a report by an independent certified public accountant (CPA), or an attestation by a responsible individual of the HDE holder's organization, verifying that the amount does not exceed the costs of research, development, fabrication, and distribution (21 CFR 814.104(b)(5)). If the amount charged is $250 or less, this requirement is waived. HDEs for pediatric use approved on or after September 27, 2007, are exempt from the prohibition against profiting from the sale of the device up to ADN, as explained in the "HDEs and Pediatric Patients" section above.

---

[5] http://www.fda.gov/downloads/MedicalDevices/
DeviceRegulationandGuidance/GuidanceDocuments/UCM089742.pdf

[6] As discussed in a preamble to the HDE Regulation, "an applicant will not be considered in violation of [section 520(m)(3) of the Act] if [the applicant] receives incidental profits which exceed its good faith estimate of costs." 61 Fed. Reg. 33232, 33242 (June 26, 1996) (citing legislative history).

30. What adverse event reporting requirements apply to HUDs?

Device user facilities and manufacturers are required to submit medical device reports to FDA and to the "IRB of record" (i.e., the IRB approving the use of the HUD) (See sections 519(a) and (b) of the Act; 21 CFR 803.30, 803.50, and 814.126(a)). Among these requirements, manufacturers must submit reports to FDA and the IRB of record whenever a HUD may have caused or contributed to a death or serious injury, or has malfunctioned and would be likely to cause or contribute to a death or serious injury if the malfunction were to recur (21 CFR 803.50 and 814.126(a)). User facilities must submit reports to FDA, the IRB of record, and the manufacturer whenever a HUD may have caused or contributed to a death, and must submit reports to the manufacturer (or to FDA and the IRB of record if the manufacturer is unknown) whenever a HUD may have caused or contributed to a serious injury (21 CFR 803.30 and 814.126(a)). Serious injury means an injury or illness that (1) is life-threatening, (2) results in permanent impairment of a body function or permanent damage to a body structure, or (3) necessitates medical or surgical intervention to preclude permanent impairment of a body function or permanent damage to a body structure (21 CFR 803.3). Note: Pediatric adverse events will be reviewed periodically by the agency's Pediatric Advisory Committee. The specific requirements for this reporting are set forth in the Medical Device Reporting (MDR) Regulation, at 21 CFR Part 803

31. What does the HDE holder need to provide to FDA in its periodic report with respect to the HUD designation?

You must provide us with updated information on a periodic basis demonstrating that the HUD designation is still valid, based on the most current and authoritative information available (21 CFR 814.126(b)). As part of these reporting requirements, you must report the number of devices shipped or sold since initial HDE marketing approval (21 CFR 814.126(b)(1)(iii)). FDA interprets this regulation to require HDE holders to report the total number of devices shipped or sold, no matter how they are used (whether for the approved indication(s), emergency use, or otherwise). However, for devices that have both an HDE approval and a PMA approval for a different indication, you are only required to report on the number of devices that are shipped or sold pursuant to the HDE, unless specifically required by the PMA Approval Order. The required frequency for these periodic reports is specified in each HDE approval order, as explained in 63 Fed. Reg. 59217, 59218 (Nov. 3, 1998).

If, based on information contained in these reports, we believe that the HUD designation may no longer apply to your device, we may contact you for additional information. See 21 CFR 814.126(b)(1) for more information on these reports.

32. Can an HDE holder submit an HDE supplement for a new indication for use of an approved HUD?

No. If you are seeking a new indication for use of an approved HUD, you must first obtain a HUD designation for the new indication for use and then submit a new original HDE application. In the new application, any information or data submitted in the HDE for the original indication may be incorporated by reference. See 21 CFR 814.110.

33. What happens to an approved HDE if, subsequently, FDA makes the determination that the disease or condition affects or is manifested in 4,000 or more individuals in the US per year?

If we make the determination that 4,000 or more individuals in the US are affected or manifest a certain disease or condition per year, we may consider whether the HDE should be withdrawn. We intend to consider factors such as the number of patients with the disease or condition, the feasibility of conducting a pivotal clinical trial (to demonstrate reasonable assurance of safety and effectiveness), and the public health need for the device.

34. If a HUD is being investigated in an IDE study for a different indication, does it impact the number of allowable patients under the HDE?

No. Investigational use of a HUD in an IDE study for a different indication does not impact the HDE approval. The HUD is intended for use in the treatment or diagnosis of a disease or condition that affects or is manifested in fewer than 4,000 individuals in the United States per year. The device being investigated in the IDE study for possible subsequent PMA approval or 510(k) clearance will not be for the same indications for use as the HUD.

35. After FDA approves an HDE for a HUD, if FDA subsequently approves a PMA or clears a 510(k) for the device or another comparable device with the same indication, what is the status of the HDE approval?

If we subsequently approve a PMA or clear a 510(k) for the HUD or another comparable device with the same indication, we may withdraw the HDE. Once a comparable device becomes legally marketed through PMA approval or 510(k) clearance to treat or diagnose the disease or condition in question, there may no longer be a need for the HUD and so the HUD may no longer meet the requirements of section 520(m)(2)(B) of the Act.

The Role of Institutional Review Boards (IRBs)

36. What are the differences between an HDE and an IDE? They both use "device exemption" in their titles and can thus be confusing to IRBs.

Quite simply, the term "exemption" for the HDE means that certain statutes and regulations need not be followed in order to legally market a HUD. An HDE approval is based on safety and probable benefit; HDEs are exempt from the requirement to provide a reasonable assurance of effectiveness, as otherwise required in sections 514 and 515 of the Act.

The term "exemption" for the IDE means certain statutes and regulations need not be followed in order to study an unapproved or uncleared device (or an approved or cleared device for an unapproved or uncleared indication) in a research study involving humans (i.e., an IDE is an investigational exemption). With this exemption, the unapproved or uncleared device can be shipped and used in human research.

We remind IRBs that question 4 of this document makes a distinction between "use" of a HUD and "investigational use"/ "clinical investigation" of a HUD. The term "use" in this document, when unmodified, refers to the use of a HUD according to its approved labeling and indication(s). If a HUD is being used in a clinical investigation (i.e., collection of safety and effectiveness data), whether for its HDE-approved indication(s) or for a different indication, then this document refers to "investigational use" or "clinical investigation" of the HUD. Such investigational use is subject to the same requirements that apply to all FDA-regulated clinical studies, including 21 CFR Parts 50 (Protection of Human Subjects) and 56 (Institutional Review Boards). Additionally, if the HUD is being studied for a use other than its approved indication(s), the IDE regulations at 21 CFR Part 812 apply. See questions 40-42.

For a schematic view of the difference between "use" and "investigational use"/"clinical investigation" of a HUD, please refer to "Figure 1: Decision Tree for IRB Review of HUDs" at the end of this guidance.

37. Should an IRB be concerned if there is a HUD approved for one indication, while the same device is being studied or marketed for another indication that does not qualify for an HDE?

No. As stated above, a HUD may be used in accordance with its approved indication(s) for use while the same device is being studied under an IDE for a different indication. Additionally, the same device can be approved or cleared for another indication without impacting the HDE.

38. What are the differences between a PMA, 510(k) and an HDE?

Three regulatory paths to the market for devices are via Premarket Approval (PMA), Premarket Notification (510(k)), and HDE.

A device with an approved PMA is approved for marketing based on valid scientific evidence and reasonable assurance that the device is safe and effective for its intended use. Once approved, it can be marketed and sold within its approved labeling. There are no restrictions on the price, and it can be used by anyone qualified to use the device.

A 510(k) device is cleared for marketing when the agency finds that it is at least as safe and effective, that is, substantially equivalent, to a legally marketed device that is not required to have a PMA. Using valid scientific evidence, submitters compare their device to one or more similar legally marketed devices, comparing the indications for use and technological characteristics. Once cleared, it can be marketed and sold in accordance with its labeling. There are no restrictions on the price, and it can be used by anyone qualified to use the device.

A device with an approved HDE is approved for marketing, but the approval is based on evidence of safety and probable benefit. The Act and implementing regulations exempt HUDs from the requirement to establish a reasonable assurance of effectiveness. The HUD is intended for use in the treatment or diagnosis of a disease or condition that affects or is manifested in fewer than 4,000 individuals in the US per year. The manufacturer of a HUD can make a profit, subject to the limit of the ADN, only if it is indicated for use in a pediatric population or subpopulation or for use in both pediatric and adult patients, was approved on or after September 27, 2007, and with certain other restrictions. (See the "HDEs and Pediatric Patients" section above for further discussion of this profit allowance.) Another important difference is that HUDs require IRB approval before being used at a facility. See sections 520(m)(3), (4), (6) of the Act; 21 CFR 814.124.

39. How does an IRB distinguish between the use of a HUD and the study of a HUD in a clinical investigation (i.e., research)?

Prior to the approval of an HDE application for a device, any studies conducted using the device must be under the IDE regulations (21 CFR Part 812). Once the HDE is approved, the following information applies if a clinical investigator or the HDE holder wants to conduct a clinical investigation using the HUD.

An HDE holder may collect safety and effectiveness data in a clinical investigation for the HDE-approved indication(s) without an IDE. As long as the HUD is being studied in accordance with the approved indication(s) described in labeling, the HUD, as such, is legally marketed and can be lawfully shipped without an IDE. See 21 CFR 812.1. IRB approval (21 CFR Part 56) and protection of human subjects (21 CFR Part 50) are still required for these studies because they are FDA-regulated clinical studies.

Clinical investigation of a HUD for a different indication must be conducted in compliance with the IDE regulations at 21 CFR Part 812, in addition to requiring IRB approval (21 CFR Part 56) and protection of human subjects (21 CFR Part 50). If the device is a significant risk device, an FDA-approved IDE is required. See 21 CFR 812.1, 812.20. To date, all HUDs have been significant risk devices requiring FDA-approved IDEs. See question 42 for more discussion of significant risk devices.

In short, IRB approval, informed consent, and additional safeguards for children (if applicable) are required for the clinical investigation (investigational use) of a HUD, whether the HUD is being studied for its HDE-approved indication(s) or for a different indication. These requirements are separate and distinct from the requirements that apply to the use of a HUD at a facility: as described in questions 43 and 59, IRB approval is required before a HUD is used at a facility to treat or diagnose patients and the IRB may require informed consent as part of such approval. In other words, just because an IRB has approved use of a HUD at a facility to treat or diagnose patients does not mean that the IRB has approved investigational use of the HUD (i.e., in a clinical investigation), for the collection of safety and effectiveness data. For more information on the difference between "use" of a HUD and "investigational use"/"clinical investigation" of a HUD, see "Figure 1: Decision Tree for IRB Review of HUDs" at the end of this guidance.

40. What if the HDE holder decides to collect safety and effectiveness data in a study to support a PMA for the HDE-approved indications?

As stated above, you may collect safety and effectiveness data to support a PMA for the HDE-approved indication(s) without an IDE. While the work done to collect such safety and effectiveness data to support a PMA constitutes a clinical investigation, FDA considers the study exempt from the requirement for an IDE as long as the HUD is used in accordance with its approved indication(s). IRB approval (21 CFR Part 56) and protection of human subjects (21 CFR Part 50) are still needed, however, as required for all FDA-regulated clinical studies. As noted above, the IRB approval, informed consent, and additional safeguards for children (if applicable) required for the clinical investigation/investigational use of a HUD are separate and distinct from the IRB approval and any consent associated with the use of the HUD. That an IRB has approved use of a HUD at a facility to treat or diagnose patients does not mean the IRB has approved investigational use of the HUD (i.e., in a clinical investigation), for the collection of safety and effectiveness data.

If you want to collect safety and effectiveness data for a use other than the HDE-approved indication(s), you must comply with the IDE regulations at 21 CFR Part 812 in addition to complying with the requirements for IRB approval (21 CFR Part 56) and protection of human subjects (21 CFR Part 50).

41. Does an IRB have to make the determination of a significant risk (SR) or non-significant risk (NSR) device (21 CFR 812.66) when it reviews a HUD?

When an IRB is deciding whether to approve use of a HUD at a facility (see questions 43-52), its review does not include an SR/NSR determination. As noted above, use of a HUD at a facility to treat or diagnose patients is not a "clinical investigation"; the HUD as such is legally marketed for use within its HDE-approved indication(s).

If an IRB receives a request to review a clinical investigation of a HUD (i.e., collection of safety and effectiveness data), and that clinical investigation concerns the HDE-approved indication(s), then again the IRB does not have make an SR/NSR determination in its review. FDA considers such investigations exempt from the IDE requirements in 21 CFR Part 812, as noted above. Nonetheless, the IRB still has to approve the clinical investigation under 21 CFR Part 56 and informed consent and additional safeguards for children (if applicable) are required under 21 CFR Part 50, as for all FDA-regulated clinical studies.

In contrast, if the IRB receives a request to review an application for an investigational study of the HDE for a different indication, then the IRB should be alert that this type of clinical investigation is subject to the IDE regulations at 21 CFR Part 812. To date, all HUDs when studied for uses other than their approved indication(s) have been SR devices requiring an FDA-approved IDE. See 21 CFR 812.20(a). In practice, most sponsors have obtained an IDE from FDA before beginning such studies, and so IRBs have not needed to make the SR/NSR determination (i.e., the sponsors already knew their device was an SR device). However, in the event that a sponsor seeks IRB approval for research of a HUD for an indication other than its approved indication(s) without first obtaining an FDA-approved IDE, then the IRB should make the SR/NSR determination as described in 21 CFR 812.66.

42. Is IRB approval required before the use of a HUD at a facility?

Yes. As stated in section 520(m)(4) of the Act, IRB approval is required before a HUD is used at a facility, with the exception of emergency use (see question 65). The IRB must have among its members (or consultants) the appropriate experience and expertise to perform a complete and adequate review of the use of a HUD at that institution (21 CFR 56.107(a)). In addition, a local IRB may defer in writing to another similarly constituted IRB that has agreed to assume responsibility for review of the use of the HUD. This deferral letter must be sent to the HDE holder, because the HDE holder is responsible for ensuring that a HUD is administered only in facilities in which the reviewing IRB is constituted and acting in accordance with 21 CFR Part 56 (21 CFR 814.124(a)). See question 46 for further discussion of the scope of IRB approval.

43. Who is responsible for submitting materials to and obtaining approval from the IRB before the HUD is used at a facility?

As explained above, the HDE holder is responsible for ensuring that the HUD is administered only in facilities with properly constituted and functioning IRBs (see question 27). The health care provider at such facilities should be responsible for obtaining IRB approval before use of the HUD, except in certain emergencies where prior IRB approval is not required (see question 65). The IRB should have policies and procedures in place for receipt and evaluation of the materials necessary for initial approval and continuing review of the HUD.

44. How should an IRB evaluate requests for approval of the use of a HUD?

As stated in 21 CFR 814.124(a), an IRB that reviews and approves the use of a HUD must be constituted and act in accordance with the agency's regulation governing IRBs (21 CFR Part 56), which include initial and continuing review of the use of the device. FDA recommends that an IRB follow the review criteria at 21 CFR 56.111 and elsewhere in Part 56 as much as possible. For example, you should review the risks to patients that are found in the product labeling, ensure the risks are minimized, and evaluate whether the risks are reasonable in relation to the proposed use of the device.

Specifically, FDA recommends reviewing the following materials during initial review of the HUD: a copy of the HDE approval order; a description of the device; the product labeling; the patient information packet that may accompany the HUD; a sample consent form for the use of the HUD, if required by the IRB; and a summary of how the physician proposes to use the device, including a description of any screening procedures, the HUD procedure, and any patient follow-up visits, tests or procedures. A list of approved HDEs may be found at http://www.accessdata.fda.gov/scripts/cdrh/cfdocs/cfHDE/HDEInformation.cfm#2.

The approval order, labeling, and patient information may be found by selecting the number of the appropriate HDE. You should have policies and procedures in place for this review and approval, including whether your IRB requires a consent document for the use of the HUD.

45. To what extent should an IRB exercise oversight of clinician responsibilities in the use of a HUD?

In reviewing the use of the HUD, IRBs should be cognizant that the FDA has made a determination of safety and probable benefit for use of the HUD only within its approved indication(s). The IRB is not required to review and approve each individual use of a HUD. Rather, the IRB may use its discretion to determine how to approve use of a HUD. For example, if it so wishes, with the input of members with the appropriate expertise in the clinical area (21 CFR Part 56), an IRB may specify limitations on the use of the device based upon one or more measures of disease progression, prior use and failure of any alternative treatment modalities, reporting requirements to the IRB or IRB chairperson, appropriate follow-up precautions and evaluations, or any other criteria it determines to be appropriate.

46. What types of review functions are IRBs responsible for with respect to HUDs?

IRBs are responsible for initial as well as continuing review of the HUD. For initial review of a HUD, IRBs are required to perform their review at a convened meeting (21 CFR 56.108). For continuing review, IRBs may use the expedited review procedures (21 CFR 56.110). When applicable, review of the use of a HUD and review of the investigational use of a HUD in a clinical investigation may be done simultaneously.

47. Why does FDA suggest that an IRB perform the continuing review of a HUD using an expedited procedure?

FDA recommends the use of an expedited procedure because a HUD is a legally marketed device and no safety and effectiveness information is being collected systematically, as is required for a research protocol. An expedited review does not mean a less than substantive review. During the expedited review, the Chair or the Chair's designated member(s) should thoughtfully consider the risk and benefit information available and any Medical Device Reporting (MDR) reports (see question 50). IRBs may develop their own policies and procedures for continuing review of a HUD and may perform this review at a convened meeting.

48. Should other committees at an institution be involved in the review of a HUD?

There is no regulatory requirement for committees other than the IRB to approve the use of a HUD. However, the institution may require additional review. For example, the use of another committee to provide assessments of specific risk posed by the technology or software compatibility may supplement the IRB review.

49. What does an IRB have to know about Medical Device Reporting (MDR)?

The HDE regulation, 21 CFR 814.126(a), requires that MDR reports submitted to FDA, in accordance with 21 CFR Part 803 (see question 31) shall also be submitted to the "IRB of record" (i.e., the IRB approving the use of the HUD).

50. What should an IRB consider with respect to the health care provider(s) who will use the HUD?

The IRB may want to ensure that health care providers are qualified through training and expertise to use the device. For many HDEs, the HDE holder is required to provide training on the use of the device prior to the health care provider using the device. Such requirements would be specified in the HDE approval order, available at http://www.accessdata.fda.gov/scripts/cdrh/cfdocs/cfHDE/HDEInformation.cfm#2 (select the HDE number).

51. Must an IRB request a protocol to review before approving the use of the HUD?

When a HUD is used to treat or diagnose patients, i.e., not for research, we do not require submission of a protocol to the IRB for review. However, your IRB or institution may require one under its own policies and procedures.

52. Does FDA require an IRB to monitor the number of uses per year of a HUD?

No. It is the responsibility of the HDE holder to monitor how many devices are distributed each year, and if that number exceeds 4,000, to provide an explanation and estimate of how the device is being used by patients. See 21 CFR 814.126(b)(1)(iii).

53. Must an IRB review or audit the medical record of patients who received a HUD?

No, we do not require you to audit medical records of patients who receive a HUD.

54. Should an IRB ask for justification of the charges for the HUD?

No. There is no requirement for the IRB to request a justification of the charges for the HUD. FDA reviews the financial information in the HDE holder's initial application, and periodically thereafter.

55. Should an IRB be concerned if an HDE holder charges for a HUD?

HDE holders generally charge for the HUD that is used to treat or diagnose a patient. However, HUDs cannot be sold for a price that exceeds the costs of research and development, fabrication, and distribution of the device. The exception is if they are indicated for use in a pediatric population, or pediatric subpopulation, or for use in both pediatric and adult patients, were approved on or after September 27, 2007, and annual sales have not yet exceeded the ADN (as discussed in "HDEs and Pediatric Patients" section above). See sections 520(m)(4), (6) of the Act.

If a HUD is studied in a clinical investigation of a new indication , the sponsor of the clinical investigation may not charge subjects or investigators a price larger than necessary to recover the costs of manufacture, research, development, and handling (21 CFR 812.7(b)). Any costs for which a subject in a clinical investigation is responsible must when appropriate, be clearly explained in the informed consent document (21 CFR 50.25(b)(3)).

56. Does an IRB function as a Data Monitoring Committee for a HUD?

No. The IRB may, however, ask the HDE holder for copies of the safety information submitted to FDA in the periodic reports required by 21 CFR 814.126(b)(1). In this way, information that could have a bearing on human safety would be considered at the time of continuing review.

57. Do the requirements for review of a HUD change if an IRB has a Federal Wide Assurance (FWA) with the Department of Health and Human Services, Office for Human Research Protections?

No. The use of a HUD is not research; rather, it is use of a legally marketed device. We describe the IRBs responsibilities in section 520(m) of the Act and in the implementing regulations at 21 CFR 814.124. We also offer guidance to you in this document. If, however, a HUD is used in a clinical investigation (see question 41), IRBs should follow their FWA requirements and their written procedures for FDA-regulated research.

58. What information should be given to patients before they receive a HUD, and should patients consent to the HUD use?

Neither the Act nor the regulations require informed consent from patients for the use of a HUD. An IRB may, however, choose to require informed consent that is consistent with the approved labeling when the IRB approves use of the HUD in a facility.

Most HDE holders develop patient information packets that generally contain a discussion of the potential risks and benefits of the HUD and any procedures associated with its use. If patient information packets are available, the IRB should ensure that physicians distribute them to patients prior to their receiving the HUD. Even when an institution requires patients to sign a written consent document that describes the use of the HUD (and which may provide similar information found in the HDE holder's packet), the patient should always receive the HDE holder's patient information packet. For HUD patient information packets, go to http://www.accessdata.fda.gov/scripts/cdrh/cfdocs/cfHDE/HDEInformation.cfm#2 and select the HDE number. In addition to the above information, many institutions also require informed consent for the surgery or procedure related to the use of the HUD.

If a HUD is studied in a clinical investigation, the informed consent of the subject must be obtained in accordance with FDA regulations at 21 CFR Part 50 (see question 41).

59. If an IRB requires a written consent document for the use of a HUD, what information should be included?

It would be reasonable for the document to include much of the information found in the HDE holder's patient information packet. If no patient information packet is available, you may consider including the following: an explanation that the HUD is designed to diagnose or treat the disease or condition described in the HDE labeling and that no comparable device is available to treat the disease or condition; a description of any ancillary procedures associated with the use of the HUD; a description of the use of the HUD; all known risks or discomforts; and an explanation of the postulated mechanism of action of the HUD in relation to the disease or condition. You should also include information reflecting the HUD status of the device, such as a sentence indicating that the effectiveness of this device for this use has not been demonstrated. The IRB may decide to include other information.

If the HUD is studied in a clinical investigation, the elements included in the informed consent document must conform to the requirements found in 21 CFR 50.25.

60. Is it appropriate for the HUD labeling and materials to include the phrase "FDA approved"? What other information must the labeling contain?

HUD labeling and materials must be truthful and not misleading. See section 502(a) of the Act. The labeling may state that the device is approved as a HUD for its intended use, but the labeling must also include the following statement clarifying that effectiveness has not been demonstrated: "Humanitarian Device. Authorized by Federal law for use in the [treatment or diagnosis] of [specify disease or condition]. The effectiveness of this device for this use has not been demonstrated." See 21 CFR 814.104(b)(4)(ii) for more information on HUD labeling requirements.

61. What should IRBs tell physicians who want to study a HUD for a new indication?

Physicians who want to study a HUD for a new indication must submit an IDE application to FDA if the device is a significant risk device (see question 42). Physicians may be either the sponsor or investigator of the study or they may want to involve the HDE holder as the sponsor. The investigational use of a HUD under these circumstances is a clinical investigation and must be conducted in accordance with 21 CFR Parts 812, 50, 54, and 56.

62. Does the use of a HUD constitute treatment or research under the Health Insurance Portability and Accountability Act of 1996 (HIPAA)? Does the IRB need to waive a HIPAA authorization for the use or disclosure of protected health information related to the use of a HUD?

The Privacy Rule promulgated at 45 CFR Parts 160 and 164, Subparts A and E pursuant to HIPAA governs the use and disclosure of certain individually identifiable health information (protected health information). An entity that is covered by HIPAA (a covered entity) may use and disclose protected health information without the patient's authorization if the use or disclosure is for the purpose of treatment. If the use or disclosure of protected health information is for the purpose of research, then the covered entity generally must obtain the patient's authorization, unless an IRB or Privacy Board has determined that such an authorization is not necessary because the research satisfies certain waiver criteria.

The use of a HUD according to its approved labeling and indication is generally for treatment or diagnosis, even though such use requires IRB approval. If a HUD is being used according to its approved labeling and indication, and not in a clinical investigation, then protected health information about a patient may be used or disclosed for treatment or diagnostic purposes without the patient's authorization under HIPAA.

If a HUD is being used in a clinical investigation, whether or not the use of the HUD is the subject of the investigation, then protected health information about a patient that is used or disclosed for purposes of the clinical investigation requires the patient's authorization under the HIPAA Privacy Rule. The IRB may waive this authorization if certain waiver criteria are met.

63. Does reporting of safety and effectiveness data to the sponsor require a HIPAA authorization or does this activity fall under an FDA-related activity under 45 CFR 164.512(b) (public health reporting)?

Reporting HUD safety information to the sponsor does not require a HIPAA authorization since it falls under the permissive disclosure for FDA-related activities at 45 CFR 164.512(b)(iii).

Using HUDs in Emergency Use Situations

64. When can a HUD be used without prior IRB approval?

If a physician in an emergency situation determines that IRB approval for the use of the HUD at the facility cannot be obtained in time to prevent serious harm or death to a patient, a HUD may be used without prior IRB approval. The physician must report the emergency use within five days; provide written notification of the use to the IRB chair person including identification of the patient involved, the date of the use, and the reason for the use. See section 520(m)(4) of the Act; 21 CFR 814.124.

65. After an IRB approves the use of the HUD at the facility, can a physician use a HUD outside its approved indication(s) in an emergency or if the physician determines there is no alternative device for the patient's condition?

Physicians should be cognizant that FDA has made a determination of safety and probable benefit for use of the HUD only within its approved indication(s). If a physician wants to use a HUD outside its approved indication(s), FDA recommends that the physician obtain informed consent from the patient and ensure that reasonable patient protection measures are followed, such as devising schedules to monitor the patient, taking into consideration the patient's specific needs and the limited information available about the risks and benefits of the device. FDA further recommends that the physician submit a follow-up report on the patient's condition to the HDE holder and first check with the IRB before such use to review any institutional policy. The extent of IRB oversight in these circumstances is up to the IRB (see questions 45 and 46). Note: as discussed in question 30, MDR reports must be submitted to FDA and to the "IRB of record" (i.e., the IRB approving the use of the HUD) if the device may have caused or contributed to death or serious injury and for certain malfunctions.

Figure 1: Decision Tree for IRB Review of HUDs

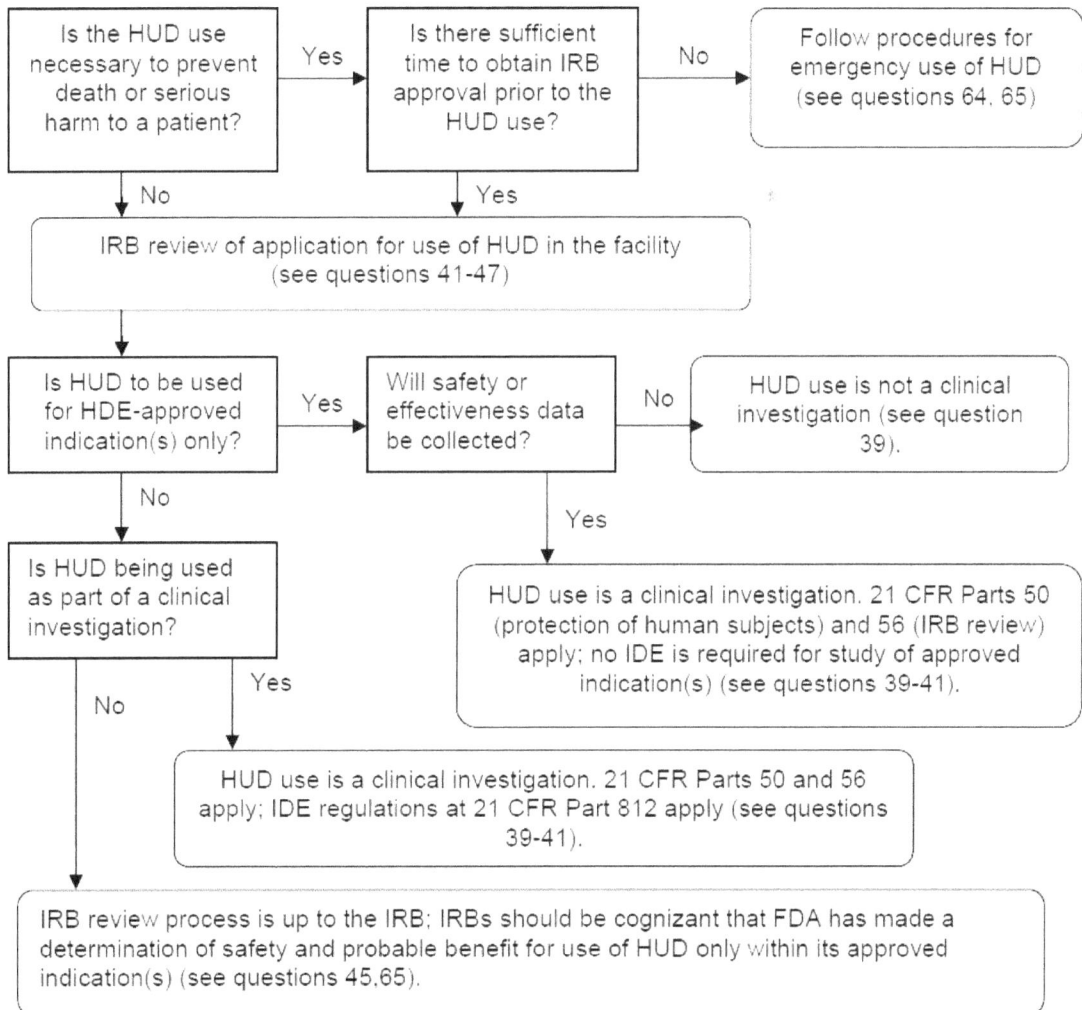

Note: *Medical device reporting is required under 21 CFR Part 803 whenever the use of a HUD may have caused or contributed to a death or serious injury, or has malfunctioned and would be likely to cause or contribute to a death or serious injury if the malfunction were to recur (see questions 30, 49, 65). For investigational use of a HUD under an IDE, reports of unanticipated adverse device effects must be reported under 21 CFR 812.150(a)(1) and 812.150(b)(1).*

Section 508 text for Figure 1.

66. Flowchart.

Is the HUD use necessary to prevent death or serious harm to a patient? If no, proceed to node 1; if yes, is there sufficient time to obtain IRB approval prior to the HUD use? If yes, proceed to node 1; if no, Follow procedures for emergency use of HUD (see questions 64, 65). Node 1, IRB review of application for use of HUD in the facility (see questions 41-47). Is HUD to be used for HDE-approved indication(s) only? If no, proceed to node 2; if yes, will safety or effectiveness data be collected? If yes, HUD use is a clinical investigation. 21 CFR Parts 50 (protection of human subjects) and 56 (IRB review) apply; no IDE is required for study of approved indication(s) (see questions 39-41). If no, HUD use is not a clinical investigation (see question 39). Node 2, is HUD being used as part of a clinical investigation? If yes, HUD use is a clinical investigation. 21 CFR Parts 50 and 56 apply; IDE regulations at 21 CFR Part 812 apply (see questions 39-41). If no, IRB review process is up to the IRB; IRBs should be cognizant that FDA has made a determination of safety and probable benefit for use of HUD only within its approved indication(s) (see questions 45, 65).

67. Paperwork Reduction Act of 1995

This guidance contains information collection provisions that are subject to review by the Office of Management and Budget (OMB) under the Paperwork Reduction Act of 1995 (44 U.S.C. 3501-3520). The time required to complete this information collection is estimated to average 100 hours per response, including the time to review instructions, search existing data sources, gather the data needed, and complete and review the information collection. Send comments regarding this burden estimate or suggestions for reducing this burden to:

> U.S. Food and Drug Administration
> Center for Devices and Radiological Health
> HDE Program WO66-1645
> 10903 New Hampshire Avenue
> Silver Spring, MD 20993-0002

This guidance also refers to previously approved collections of information found in FDA regulations. The collections of information in 21 CFR part 803 have been approved under OMB control number 0910-0437; the collections of information in 21 CFR part 812 have been approved under OMB control number 0910-0078; the collections of information in 21 CFR part 807, subpart E have been approved under OMB control number 0910-0120; the collections of information in 21 CFR part 814, subparts A, B, and C have been approved under OMB control number 0910-0231; the collections of information in 21 CFR parts 50 and 56 have been approved under OMB control number 0910-0130; the collections of information in 21 CFR part 820 have been approved under OMB control number 0910-0073; the collections of information in 21 CFR part 814, subpart H have been approved under OMB control number 0910-0332; and the collections of information in 21 CFR 10.30 have been approved under OMB control number 0910-0183.

An agency may not conduct or sponsor, and a person is not required to respond to, a collection of information unless it displays a currently valid OMB control number. The OMB control number for this information collection is 0910-0661, which expires on 05/31/2013.

Informed Consent for *In Vitro* Diagnostic Device Studies Using Leftover Human Specimens that are Not Individually Identifiable

Informed Consent for *In Vitro* Diagnostic Device Studies Using Leftover Human Specimens that are Not Individually Identifiable, Guidance for Sponsors, Institutional Review Boards, Clinical Investigators and FDA Staff [ŧ, 1]

U.S. Department of Health and Human Services
Food and Drug Administration
Center for Devices and Radiological Health
Office of In Vitro Diagnostic Device Evaluation and Safety
Center for Biologic Evaluation and Research
Office of Blood Research and Review

April 25, 2006

Contains Nonbinding Recommendations

This guidance informs sponsors, institutional review boards (IRBs), clinical investigators, and agency staff that the FDA intends to exercise enforcement discretion, under certain circumstances, with respect to its current regulations governing the requirement for informed consent when human specimens are used for FDA regulated in vitro diagnostic device investigations.

Note: *This guidance represents the Food and Drug Administration's (FDA's) current thinking on this topic. It does not create or confer any rights for or on any person and does not operate to bind FDA or the public. You can use an alternative approach if the approach satisfies the requirements of the applicable statutes and regulations. If you want to discuss an alternative approach, contact the appropriate FDA staff. If you cannot identify the appropriate FDA staff, call the appropriate number listed on the title page of this guidance.*

1. Introduction

FDA is issuing this guidance to inform sponsors, institutional review boards (IRBs), clinical investigators, and agency staff that the FDA intends to exercise enforcement discretion, under certain circumstances, with respect to its current regulations governing the requirement for

---

[ŧ] Available on the FDA website at: http://www.fda.gov/RegulatoryInformation/
Guidances/ucm127022.htm

[1]

informed consent when human specimens are used for FDA-regulated *in vitro* diagnostic (IVD)[2] device investigations. As described below, FDA does not intend to object to the use, without informed consent, of leftover human specimens -- remnants of specimens collected for routine clinical care or analysis that would otherwise have been discarded -- in investigations that meet the criteria for exemption from the Investigational Device Exemptions (IDE) regulation at 21 CFR 812.2(c)(3), as long as subject privacy is protected by using only specimens that are not individually identifiable. FDA also intends to include in this policy specimens obtained from specimen repositories[3] and specimens that are leftover from specimens previously collected for other unrelated research, as long as these specimens are not individually identifiable.

Under FDA's current regulations governing the conduct of IVD device studies, the definition of human subject includes individuals on whose specimens an investigational device is used [see 21 CFR 812.3(p)]. Because these regulations require informed consent for FDA-regulated human subject research, except in limited circumstances specified in the regulations,[4] informed consent is required before specimens can be used in FDA-regulated research [see 21 CFR part 50].

This aspect of FDA's human subject protection regulations has created confusion and difficulty for persons developing IVDs. Many clinicians, research hospitals, and companies have viewed the requirement for informed consent for IVD studies using leftover specimens as unnecessary for the protection of human subjects and as overly burdensome and costly.

FDA's recent "Critical Path[5]" initiative has also focused the agency's concern on unnecessary obstacles to medical product development. The agency has received comments from trade associations and research institutions that identify the challenge of obtaining informed consent for the use of leftover specimens as an unnecessary obstacle and expense for investigational efforts. When leftover specimens are available, it is often difficult, if not impossible, to locate the donor and obtain consent. This difficulty may deter a manufacturer's research efforts that would bring safe and effective IVDs to market more quickly. At the same time, many researchers maintain that for this particular type of study, the human subject protection values that informed consent is intended to ensure either are not implicated, or can be adequately safeguarded through less burdensome measures.

The confusion regarding the application of informed consent requirements to IVD studies and concerns about unnecessary obstacles to product development have prompted FDA to issue

---

[2] In vitro diagnostic products are those reagents, instruments, and systems intended for use in the diagnosis of disease or other conditions, including a determination of the state of health, in order to cure, mitigate, treat, or prevent disease or its sequelae. Such products are intended for use in the collection, preparation, and examination of specimens taken from the human body. 21 CFR 809.3(a).

[3] A specimen repository is a common site for storage of collections of human biological specimens available for study.

[4] See 21 CFR 50.23(a) and 50.24.

[5] " Innovation or Stagnation? -- Challenge and Opportunity on the Critical Path to New Medical Products" FDA Report issued on March 16, 2004. This document may be found at:
http://www.fda.gov/oc/initiatives/criticalpath/whitepaper.html

this guidance document. The agency believes this guidance will facilitate product development in a manner consistent with the values of human subject protection. FDA intends that the exercise of enforcement discretion expressed in this guidance begin immediately. In accordance with the agency's Good Guidance Practice regulations, 21 CFR 10.115, you may submit comments on this guidance at any time. The agency will consider your comments and determine whether to revise the guidance at a later date.

2. Scope

This document applies only to IVD device investigations regulated by FDA in accordance with section 520(g) of Federal Food, Drug, and Cosmetic Act (the Act), 21 USC 360j(g), that are exempt from most requirements of the IDE regulation (21 CFR 812) under 21 CFR 812.2(c)(3), and that use leftover specimens that are not individually identifiable. A leftover specimen is the remnant of a human specimen collected for routine clinical care or analysis that would otherwise have been discarded. A specimen is not individually identifiable when the identity of the subject is not known to or may not readily be ascertained by the investigator or any other individuals associated with the investigation, including the sponsor. (See Section 4, below.) This guidance also applies to specimens that were previously collected for other unrelated research and that are not individually identifiable.

This guidance will be implemented on the date it is issued. It applies to investigations using leftover specimens that are already existing on the date the guidance is issued and investigations using specimens that will be collected after the issuance of this guidance, so long as both the specimens and the investigations meet the circumstances outlined below (see section 4).

3. Background

FDA's investigational device regulations are intended to encourage the development of new, useful devices in a manner that is consistent with public health and safety and with ethical standards. (See 21 U.S.C. 360j(g)). Investigators should have freedom to pursue the least burdensome means of accomplishing this goal. However, to ensure that the balance is maintained between product development and the protection of public health and safety and ethical standards, FDA has established human subject protection regulations addressing requirements for informed consent and IRB review that apply to all FDA-regulated clinical investigations involving human subjects. In particular, informed consent requirements further both safety and ethical considerations by allowing potential subjects to consider both the physical and privacy risks they face if they agree to participate in a trial.

Under FDA regulations, clinical investigations using human specimens conducted in support of premarket submissions to FDA are considered human subject investigations [see 21 CFR 812.3(p)]. Many IVD studies are exempt from most provisions of 21 CFR part 812, Investigational Device Exemptions, under 21 CFR 812.2(c)(3), but FDA's regulations for the protection of human subjects (21 CFR parts 50 and 56) apply to all clinical investigations that are regulated by FDA [see 21 CFR 50.1; 21 CFR 56.1; 21 U.S.C. 360j(g)(3)(A) & (D)].

FDA does have narrow exceptions from the general requirements of informed consent for certain emergency and military research,[6] but FDA regulations do not contain exceptions from the requirements of informed consent on the grounds that the specimens are not identifiable and or that they are remnants of human specimens collected for routine clinical care or analysis that would otherwise have been discarded. Nor do FDA regulations allow IRBs to decide whether or not to waive informed consent for research involving leftover or unidentifiable specimens.

Leftover specimens are frequently used in feasibility studies and studies to characterize the performance of new in vitro diagnostic devices for several reasons. The evaluation of new devices often requires the use of specimens with specific laboratory characteristics, e.g., positive or negative for a particular disease marker, in order to meet the study inclusion criteria. Routine clinical care testing can provide information about the laboratory characteristics of the specimen that permit investigators to quickly ascertain whether the specimen will meet the study inclusion criteria. The remnants of these specimens thus become valuable to research at a point when they are of no value to the patient and are ready to be discarded. The lower cost of these specimens, compared to the cost of specimens collected prospectively for research, makes studies using leftover specimens more affordable, permitting manufacturers to conduct studies that otherwise may not be done. In addition, banked leftover specimens are a source for unique and possibly rare specimens in sufficient quantity to permit the rapid completion of investigations that would be difficult if not impossible to conduct in a reasonable timeframe without these specimens.

FDA believes that it is possible in certain circumstances for IVD device investigations to be conducted using leftover specimens obtained without informed consent while protecting the human subjects who are the source of such specimens. When IVD study sponsors use leftover specimens for which the subject cannot be identified and where results of the investigational test are not communicated to or otherwise associated with the identified subject, concerns associated with privacy are minimized. In addition, these studies do not pose new medical risks to subjects from whom the specimens were originally collected: Any risks from specimen collection were incurred prior to the involvement of the patient as a subject in an investigation, when the specimen was obtained for the patient's own clinical needs, and no risks from erroneous test results are presented because the results of the testing are not used for clinical management of the subject. Like leftover specimens that have been collected for routine clinical care, the investigational use of leftover specimens previously collected for other research purposes involves no additional medical risk, and privacy risks are mitigated by limiting the applicability of this guidance to specimens that are not identifiable.

---

[6] 21 CFR 50.23 and 50.24.

4. In what circumstances does FDA intend to exercise enforcement discretion as to the requirements for informed consent for use of specimens in FDA-regulated IVD studies ?

FDA intends to exercise enforcement discretion as to the informed consent requirements for clinical investigators, sponsors, and IRBs if an *in vitro* diagnostic device investigation is performed and all of the following are true:

a.   The investigation meets the IDE exemption criteria at 21 CFR 812.2(c) (3).

b.   The study uses leftover specimens, that is, remnants of specimens collected for routine clinical care or analysis that would have been discarded. The study may also use specimens obtained from specimen repositories or leftover specimens that were previously collected for other research purposes.

c.   The specimens are not individually identifiable, i.e., the identity of the subject is not known to and may not readily be ascertained by the investigator or any other individuals associated with the investigation, including the sponsor. If the specimen is coded,[7] it will be considered to be not individually identifiable if neither the investigator(s) nor any other individuals associated with the investigation or the sponsor can link the specimen to the subject from whom the specimen was collected, either directly or indirectly through coding systems.

d.   The specimens may be accompanied by clinical information as long as this information does not make the specimen source identifiable to the investigator or any other individual associated with the investigation, including the sponsor.

e.   The individuals caring for the patients are different from and do not share information about the patient with those conducting the investigation.

f.   The specimens are provided to the investigator(s) without identifiers and the supplier of the specimens has established policies and procedures to prevent the release of personal information.

g.   The study has been reviewed by an IRB in accordance with 21 CFR Part 56, except as described in section 7 of this guidance document.

Studies that do not fall within the intended enforcement discretion expressed in this guidance include (but are not limited to) studies where any of the following is true:

---

[7] For the purposes of this document, coded means that: 1) a number, letter, symbol, or combination thereof (i.e., the code) has replaced identifying information (such as name or social security number) that would enable the investigator or any other individuals associated with the investigation, including the sponsor to readily ascertain the identity of the individual to whom the specimen pertains; and 2) a key to decipher the code exists, enabling linkage of the identifying information to the specimen.

- The study does not meet the IDE exemption criteria at 21 CFR 812.2(c)(3);

- the specimens are individually identifiable, i.e., the identity of the subject is known to or may be readily ascertained by the investigator or any other individuals associated with the investigation, including the sponsor.

- the specimens were collected specifically for the proposed investigation. That is, the specimens are not leftover from routine clinical care or analysis or leftover from other research.

- the amount of specimen needed for the study is more than would be leftover from what is usually collected for routine clinical analysis

or,

- the test results will be reported to the subject's health care provider. For example, in the course of comparative studies involving B. anthracis detection devices, it would be inappropriate not to report positive results if they occur in the course of an investigation.

5. What type of records should be kept for these types of studies?

We recommend that sponsors maintain written documentation regarding the factors described in section 4 (a)-(g) of this guidance, including the policies and procedures followed by the specimen provider to ensure that the subject cannot be identified. FDA may request to inspect this documentation. FDA recommends that IRBs review this documentation before approving an investigation paying particular attention to privacy and confidentiality, and the potential for use of information from the investigation for clinical patient management.

6. Should sponsors consider anything else in deciding whether or not to conduct a study that may fall within the exercise of enforcement discretion contemplated by this guidance?

Sponsors should consider whether a study exhibiting the factors relevant to the exercise of enforcement discretion described in section 4 will generate sufficient data to support the product application they are considering. Although FDA does not intend to reject data from a study exhibiting the factors described in section 4 solely because it was conducted without complying with the informed consent requirements found in 21 CFR part 50, FDA also does not guarantee that the data generated from a study with those characteristics will be sufficient to support a premarket clearance or approval. FDA may determine that additional clinical information is important in order to evaluate test results. For some studies, masking of clinical information may be problematic and may bias data collection. Sponsors should understand that by choosing to conduct an investigation without informed consent, even in a manner consistent with this guidance, they accept the risk that they may not be able to provide sufficient information to satisfy FDA's premarket review needs.

7. What should IRBs do when reviewing the types of IVD studies that are the focus of this guidance?

To facilitate IVD device development, FDA intends to exercise enforcement discretion toward IRBs who approve IVD investigations that are consistent with the factors in section 4 of this guidance, with respect to the IRB's duties under 21 CFR part 56 regarding informed consent for those studies. (Noncompliance with requirements of 21 CFR part 56 not related to informed consent is not subject to enforcement discretion under this guidance.) We recommend that the IRB review the sponsor's documentation regarding the factors described in Section 4, (a) through (f), including the policies and procedures followed by the specimen provider to ensure that the subject cannot be identified. IRBs should apply existing FDA regulations, including all informed consent requirements, to any other investigational IVD study. We encourage IRBs to contact FDA if they have questions about the guidance or a specific study under review (see FDA contact information on the title page of this guidance.)

8. Paperwork Reduction Act of 1995

This guidance contains information collection provisions that are subject to review by the Office of Management and Budget (OMB) under the Paperwork Reduction Act of 1995 (44 U.S.C. 3501-3520).

The time required to complete this information collection is estimated to average 4 hours per response, including the time to review instructions, search existing data sources, gather the data needed, and complete and review the information collection. Send comments regarding this burden estimate or suggestions for reducing this burden to:

Denver Presley, Office of Information Resources Management (HFA-250), Food and Drug Administration, 5600 Fishers Lane, Room 16B-26; Rockville, MD 20857, 301-827-1472.

An agency may not conduct or sponsor, and a person is not required to respond to, a collection of information unless it displays a current valid OMB control number. the OMB control number for this information collection is 0910-0582, expires 09/30/2006

In Vitro Diagnostic (IVD) Device Studies - Frequently Asked Questions

In Vitro Diagnostic (IVD) Device Studies -Frequently Asked Questions, Guidance for Industry and FDA Staff †

U.S. Department of Health and Human Services
Food and Drug Administration
Center for Devices and Radiological Health
Center for Biologics Evaluation and Research

June 25, 2010

Contains Nonbinding Recommendations

This guidance document outlines FDA regulations applicable to studies for investigational IVD devices, including those regulations related to human subject protection. The guidance also explains data considerations that ultimately will affect the quality of the premarket submission. It includes a glossary, a reference list with related web addresses, and a quick-reference table.

Note: *This guidance represents the Food and Drug Administration's (FDA's) current thinking on this topic. It does not create or confer any rights for or on any person and does not operate to bind FDA or the public. You can use an alternative approach if the approach satisfies the requirements of the applicable statutes and regulations. If you want to discuss an alternative approach, contact the appropriate FDA staff. If you cannot identify the appropriate FDA staff, call the appropriate number listed on the title page of this guidance.*

I. Background

The Investigational Device Exemptions (IDE) regulation, Title 21, Code of Federal Regulations (21 CFR) Part 812, sets forth regulatory requirements for studies of investigational devices. Certain investigational IVD device studies (see the Glossary), however, are exempt from most of the provisions of 21 CFR Part 812 (21 CFR 812.2(c)(3)).[1] This guidance document, written in question and answer format, is intended to assist you[2] (the manufacturer, sponsor, applicant, investigator and the IVD device industry in general) in the development of

---

† Available on the FDA website at: http://www.fda.gov/downloads/RegulatoryInformation/ Guidances/UCM126834.pdf

[1] As explained below, even if a particular IVD study is exempt from most requirements of 21 CFR Part 812, studies that will support applications to FDA are subject to 21 CFR 812.119 (Disqualification of a Clinical Investigator), 21 CFR Part 50 (Informed Consent), and 21 CFR Part 56 (Institutional Review Boards).

[2] For the purpose of this document, "you" refers to the manufacturer, sponsor, applicant, investigator and the IVD device industry in general. If the text refers only to one or some of these entities, the appropriate entity is referenced by its name.

IVD studies, particularly those exempt from most of the requirements of the IDE regulation and to provide you with a broad view of the regulatory framework pertaining to the development phase of IVD devices. The information in this guidance document is also pertinent to investigators who participate in IVD studies and to institutional review boards (IRB) that review and approve such studies. The document is intended to facilitate the movement of new IVD technology from the investigational stage to the marketing stage.

This guidance document outlines FDA regulations applicable to studies for investigational IVD devices, including those regulations related to human subject protection. The guidance also explains data considerations that ultimately will affect the quality of the premarket submission. This document includes a glossary, a reference list with related web addresses, and a quick-reference table.

The Center for Devices and Radiological Health (CDRH) and the Center for Biologics Evaluation and Research (CBER) each have regulatory responsibilities for IVD devices; information included in this document applies to Class I, II, and III IVD devices regulated by either Center.

Note: *Some devices used to test blood donor suitability, and blood donor and recipient compatibility are licensed as biological products under Section 351 of the Public Health Service Act and are subject to the applicable regulations in 21 CFR Parts 600-680 . Examples of licensed biologics devices include blood donor screening tests for human immunodeficiency virus (HIV) and hepatitis B and C tests intended for blood screening and reagents used in blood grouping, antibody detection and identification, and crossmatching for pre-transfusion compatibility testing. This guidance is written to address only IVD devices that are approved or cleared under the Federal Food, Drug, and Cosmetic Act (the Act) and Part 800 of the device regulations, regardless of which Center reviews the submission. If you have questions about a device licensed by CBER, you may go to the CBER website for published guidance (http://www.fda.gov/BiologicsBloodVaccines/GuidanceComplianceRegulatoryInformation/default. htm) or contact CBER for further information on applicable guidance and regulations. (See Introduction, Section II, question # 4, of this guidance for a listing of CBER contact numbers).*

FDA's guidance documents, including this guidance, do not establish legally enforceable responsibilities. Instead, guidances describe the Agency's current thinking on a topic and should be viewed only as recommendations, unless specific regulatory or statutory requirements are cited. The use of the word should in Agency guidances means that something is suggested or recommended, but not required.

II. Introduction

1. What is the purpose of this guidance document and how does it differ from other guidance documents related to IVD products?

FDA prepared this comprehensive document as a resource for you and for its own staff to address issues concerning IVD studies. This guidance document contains information relevant to studies conducted during the development of a new IVD product, as well as other general considerations about applicable requirements and marketing of the new device. It addresses particularly those investigational studies that are exempt from the majority of requirements under 21 CFR Part 812. IVD study investigators and members of IRBs who review and approve such studies may also find it helpful. There are also device-specific guidance documents available for specific IVD products that can be found at http://www.accessdata.fda.gov/scripts/cdrh/cfdocs/cfggp/search.cfm. The use of investigational IVD devices in clinical studies designed to evaluate new drug products falls outside the scope of this guidance.

2. Why are in vitro diagnostics considered devices?

In vitro diagnostics (IVDs) meet the definition of a device under the Act. Section 201(h) of the Act defines a device as:

"an instrument, apparatus, implement, machine, contrivance, implant, in vitro reagent, or other similar or related article, including any component, part, or accessory, which is—

(1) recognized in the official National Formulary, or the United States Pharmacopeia, or any supplement to them,

(2) intended for use in the diagnosis of disease or other conditions, or in the cure, mitigation, treatment, or prevention of disease, in man or other animals, or

(3) intended to affect the structure or any function of the body of man or other animals, and

which does not achieve its primary intended purposes through chemical action within or on the body of man or other animals and which is not dependent upon being metabolized for the achievement of its primary intended purposes." 21 U.S.C. 321(h) (emphasis added).

3. How do IVD devices differ from other devices?

Most other devices function on or in a patient. In contrast, IVDs include products used to collect specimens, or to prepare or examine specimens (e.g., blood, serum, urine, spinal fluid, tissue samples) after they are removed from the human body.

4. Which Divisions at FDA are responsible for review of IVD products?

Center for Devices and Radiological Health (CDRH)

- Office of In Vitro Diagnostic Device Evaluation and Safety (OIVD)

- Division of Chemistry and Toxicology Devices – Phone: (301) 796-5470

- Division of Immunology and Hematology Devices – Phone: (301) 796-5481

- Division of Microbiology Devices – Phone: (301) 796-5461

Center for Biologics Evaluation and Research (CBER)

- Office of Cell, Tissues, and Gene Therapy (OCTGT) – Phone: (301) 827-5102

- Office of Blood Research and Review (OBRR)

- Division of Blood Applications (DBA) – Phone: (301) 827-3524

DBA schedules all review-related meetings for OBRR

- Division of Emerging and Transfusion Transmitted Diseases (DETTD) – Phone: (301) 827-3008

- Division of Hematology (DH) – Phone: (301) 496-4396

5. Whom should I consult when I have questions about the manufacturing regulations or the conduct of a study (e.g., human subject protection issues)?

Center for Devices and Radiological Health (CDRH)

For questions regarding manufacturing regulations and IVD-specific conduct of studies, contact:

Office of In Vitro Diagnostic Device Evaluation and Safety (OIVD)
Regulatory Staff, Patient Safety and Product Quality – Phone: (301) 796-5450

For questions regarding the conduct of studies, contact:

Office of Compliance (OC)
Division of Bioresearch Monitoring (DBM)
Phone: (301) 796-5490

or

Investigational Device Exemptions (IDE) Staff
Office of Device Evaluation
Phone: (301) 796-5640

Center for Biologics Evaluation and Research (CBER)

For questions regarding manufacturing regulations and IVD-specific conduct of studies, contact the appropriate reviewing division identified in the previous answer.

For questions regarding the conduct of studies, contact:

Office of Compliance and Biologics Quality (OCBQ)
Division of Inspections and Surveillance (DIS)
Bioresearch Monitoring Branch – Phone: (301) 827-6221

III. General Regulatory Issues

1. Which regulations contain provisions relevant to the IVD industry?

Listed below are some of the regulations that implement the Act and that are relevant to IVDs covered by this guidance. See Table 1 (Appendix 1) for additional information. This is not an all-inclusive list.

- Title 21, Code of Federal Regulations (21 CFR)
- Part 11, Electronic Records; Electronic Signatures
- Part 50, Protection of Human Subjects
- Part 54, Financial Disclosure by Clinical Investigators
- Part 56, Institutional Review Boards
- Part 801, Labeling
- Part 803, Medical Device Reporting
- Part 807, Establishment Registration and Device Listing for Manufacturers and Initial Importers of Devices
- Part 809, In Vitro Diagnostic Products for Human Use
- Part 810, Medical Device Recall Authority
- Part 812, Investigational Device Exemptions
- Part 814, Premarket Approval of Medical Devices
- Part 820, Quality System Regulation
- Part 860, Medical Device Classification Procedures
- Part 862, Clinical Chemistry and Clinical Toxicology Devices
- Part 864, Hematology and Pathology Devices
- Part 866, Immunology and Microbiology Devices

In addition, certain sections of Part 610 apply to devices that employ human blood components. For example:

- 610.40, Test Requirements (Testing Requirements for Communicable Disease Agents)
- 610.42, Restrictions on Use for Further Manufacture of Medical Devices

2. How do I determine the applicability of the IDE regulation to my IVD study?

We recommend that you begin with the exemptions in 21 CFR 812.2(c). Your proposed IVD study is exempt from most provisions of the IDE regulation if it fits any one of the following three categories:

a.   The IVD is a pre-amendments device (i.e., a device that was in commercial distribution prior to the enactment of the 1976 Medical Device Amendments to the Act), other than a transitional device (see the Glossary for definition), and is used or investigated according to the indications in the labeling at that time.

b.   The IVD is a device, other than a transitional device, that has been found to be substantially equivalent to a pre-amendments device and is used or investigated according to the indications in the labeling reviewed by FDA in determining substantial equivalence.

c.   The IVD

- is properly labeled in accordance with 21 CFR 809.10(c);

- is noninvasive (see question #5 below);

- does not require an invasive sampling procedure that presents significant risk (see question #4 below);

- does not by design or intention introduce energy into a subject; and

- is not used as a diagnostic procedure without confirmation of the diagnosis by another, medically established diagnostic product or procedure (see question # 6 below).

For your study to be exempt from most of the requirements of the IDE regulation under this third category, it must meet all of the conditions listed in "c" above. (See also the decision tree in Appendix 1.) You should refer to 21 CFR Parts 50 and 56 for applicable requirements relating to IRBs and informed consent, including for device studies that meet the criteria described in 21 CFR 812.2(c). Additionally, investigators for those studies are still subject to 21 CFR 812.119 (the provision entitled "Disqualification of a clinical investigator.")

If your proposed study does not fit into one of the three categories listed above, you, the sponsor, must have an approved IDE (21 CFR 812.2) before you may begin your investigation, including any shipment of your investigational IVD. (Note: A device that is approved under a premarket approval application (PMA) or cleared under a 510(k) and then used in a study in accordance with the approved or cleared labeling is not investigational and, therefore, is not subject to the IDE regulation.)

The requirements for an IDE depend on the level of risk that the study presents to subjects.

For a significant risk device (see the Glossary for definition), the sponsor must apply to FDA for an IDE approval (see 21 CFR 812.1, 812.20). For a non-significant risk device (see the Glossary for definition), the sponsor must meet the abbreviated requirements of 21 CFR 812.2(b), including review and approval of the investigation by an institutional review board

(IRB) and compliance with informed consent requirements. A non-significant risk study is considered to have an approved IDE when the abbreviated requirements are met.

Note: The requirements of the "Protection of Human Subjects" and "Institutional Review Boards" regulations (21 CFR Parts 50 and 56) apply to all clinical investigations regulated by FDA under section 520(g) of the Act, as well as other clinical investigations that support applications for research or marketing permits. (21 CFR 50.1, 56.101; see also Section V, Human Subject Protection, of this guidance.) Therefore, all studies of investigational IVDs that will support applications to FDA are subject to 21 CFR Parts 50 and 56, even if they are not subject to most requirements of 21 CFR Part 812.

3. How do I determine if the study is a significant or non-significant risk study under 21 CFR 812.2(b)?

A significant risk IVD device is generally one that is for a use of substantial importance in diagnosing, curing, mitigating, or treating disease, or otherwise preventing impairment of human health and presents a potential for serious risk to the health, safety, or welfare of a subject or otherwise presents a potential for serious risk to health, safety, or welfare of a subject. 21 CFR 812.3(m).

For IVDs, we interpret "potential for serious risk" in relation to the nature of the harm that may result to the subject. Misdiagnosis and/or error in treatment caused by inaccurate test results would be considered a significant risk if the potential harm to the subject could be life-threatening, or could result in permanent impairment of a body function or permanent damage to the body structure.

False positive results can lead to unnecessary confirmatory testing, unnecessary treatment that can be invasive or have harmful side effects, and/or unnecessary psychological trauma when serious or life-threatening diseases or conditions are involved. False negative results can lead to a delay in establishing the correct diagnosis, failure to start or continue needed treatment, false security that may prevent timely follow-up and retesting, and contribute to the potential spread of infectious agents to others. If the potential risk does not rise to the level described above, the study is not considered to pose a significant risk. FDA recommends the sponsor consider all these factors when determining the risk associated with your investigational IVD. (See 21 CFR 812.3(m) and also "Information Sheet Guidance for IRBs, Clinical Investigators, and Sponsors," available at http://www.fda.gov/ScienceResearch/SpecialTopics/RunningClinicalTrials/GuidancesInformationSheetsandNotices/ucm113709.htm, particularly the one on "Significant Risk and Nonsignificant Risk Medical Devices" at http://www.fda.gov/MedicalDevices/DeviceRegulationandGuidance/GuidanceDocuments/ucm126622.htm.)

4. How do I determine if an invasive sampling technique presents a significant risk under 21 CFR 812.2(c)(3)?

To determine whether an invasive sampling technique presents a serious risk, we recommend that you base your risk determination on the nature of the harm that may result from sampling.

For example, FDA considers sampling techniques that require biopsy of a major organ, use of general anesthesia, or placement of a blood access line into an artery or large vein (subclavian, femoral, or iliac) to present a significant risk.

5. What does noninvasive mean?

A noninvasive device is one that does not, by design or intention:

    a.   penetrate or pierce the skin or mucous membranes of the body, the ocular cavity, or the urethra; or

    b.   enter the ear beyond the external auditory canal, the nose beyond the nares, the mouth beyond the pharynx, the anal canal beyond the rectum, or the vagina beyond the cervical os.

    (21 CFR 812.3(k)).

Blood sampling that involves simple venipuncture is considered noninvasive, and the use of surplus samples of body fluids or tissues that are left over from samples taken for noninvestigational purposes is also considered noninvasive (21 CFR 812.3(k)).

6. What does it mean to have "confirmation of the diagnosis by another, medically established diagnostic product or procedure?"

For an investigational study to be exempt under 21 CFR 812.2(c)(3), clinical investigators must use a medically established means of diagnosis (e.g., another cleared or approved IVD or culture) of the disease or condition as the basis for decisions regarding treatment of all subjects participating in the study. 21 CFR 812.2(c)(3)(iv). Additionally, test results from the exempt IVD investigation should not influence patient treatment or clinical management decisions before the diagnosis is established by a medically established product or procedure.

If an investigational test uses a new technology or represents a significant technological advance, established diagnostic products or procedures may not be adequate to confirm the diagnosis provided by the investigational IVD. For example, if an investigational test is designed to identify an infection at the earliest stages of viral infection (before formation of antibodies), established diagnostic products or procedures that rely on the detection of antibodies to the virus would be inadequate to confirm diagnoses. Under these conditions the study would not meet the criteria for exemption under 812.2(c)(3) since the testing could not be confirmed with a medically established diagnostic product or procedure. You may consider whether the device is a non-significant risk device subject to abbreviated IDE requirements (21 CFR 812.2(b)).

7. What if no medically established means for diagnosing the disease or condition exists?

If there is no medically established diagnostic product or procedure and clinical investigators use the results from the investigational study to decide on treatment, FDA would not consider the study exempt from IDE requirements under 21 CFR 812.2. The sponsor would need to obtain FDA approval of an IDE if the results are used in diagnosis without confirmation (e.g., to assist in determining treatment) (21 CFR 812.1, 812.2) and if a significant risk device is involved.

8. Can an investigational IVD device be used outside of the study protocol, in anemergency situation?

Yes. (See also Chapter III, "Expanded Access to Unapproved Devices," of the guidance document "IDE Policies and Procedures.")[3] A physician may use an investigational IVD device in an emergency situation if:

    a.   the patient has a serious disease or condition;

    b.   no generally accepted alternative diagnostic device or treatment for the condition is available; and

    c.   there is no time to use existing procedures to get FDA approval for the emergency use.

FDA recommends that the physician make the determination that the patient's circumstances meet the above criteria, to assess the potential for benefit from the use of the unapproved device, and to have substantial reason to believe that benefits will exist. In the event that a device is used in circumstances meeting the criteria listed above, the physician should follow as many of the patient protection procedures listed below as possible. These include obtaining:

- Informed consent from the patient or a legally authorized representative;
- Clearance from the institution as specified by their policies;
- Concurrence of the IRB chairperson;
- An assessment from a physician who is not participating in the study; and
- Authorization from the IDE sponsor, if an approved IDE exists for the device;
- Authorization from the device company, if no IDE exists.

Although prior FDA approval for emergency use of the investigational device is not required, 21 U.S.C. § 360bbb(a), if an IDE exists, the use shall be reported to FDA in a supplemental IDE by the IDE sponsor within 5 working days from the time the sponsor learns of the use (21 CFR 812.35(a)(2)). The IDE supplement should contain a summary of the conditions

---

[3] This reference can be found at: http://www.fda.gov/downloads/MedicalDevices/ DeviceRegulationandGuidance/GuidanceDocuments/ucm080203.pdf.

constituting the emergency, patient outcome information, and the patient protection measures that were followed. If no IDE exists, the physician should follow the above procedures and report the emergency use to the sponsor and to CDRH or CBER, as appropriate.

For additional information on the procedures physicians and IRBs should follow in an emergency use situation, please see Chapter III, "Expanded Access to Unapproved Devices" of the guidance entitled, IDE Policies and Procedures.

9. Can an unapproved or uncleared investigational IVD device ever be used for nonemergency treatment of patients who do not meet the inclusion criteria of an investigational study?

Yes, in exceptional situations. FDA recognizes that there are circumstances when an unapproved or uncleared IVD is the only available option for a patient or small group of patients who do not meet the inclusion criteria and "compassionate use"/single patient use of the device may be appropriate. Section 561 of the Act. CBER refers to such situations as "single patient exemptions." Appropriate patient protection measures are needed for these studies.

Use of an investigational IVD device for one or a small group of patients who do not meet the study inclusion criteria would require a change to the investigational plan. 21 CFR 812.35(a). If the study is being conducted under an approved IDE, the sponsor should submit a supplement to the IDE requesting a change to the investigational plan for "compassionate use." 21 CFR 812.35(a). The review of this supplement can be facilitated by a phone call to the reviewing division and by the submission by facsimile of an advanced copy of the supplement. If the investigational IVD device would require an FDA-approved IDE, but one has not yet been submitted or approved, FDA intends to exercise enforcement discretion where the sponsor submits a compassionate use request to CDRH or CBER, as appropriate, and follows the patient protection measures listed above for emergency use.[4]

If a study is being conducted according to an exemption under 21 CFR 812.2(c)(3), or as a non-significant risk IDE under 21 CFR 812.2(b), the sponsor should obtain prior approval for the specific compassionate use/single patient use for the individual(s) in question from the FDA and the reviewing IRB at the site where the physician proposes to use the device. For CDRH regulated products, the required information can be submitted to the Director of the IDE Program:

> Attn: Director, IDE Program
> U.S. Food and Drug Administration
> Center for Devices and Radiological Health
> WO66 Room G609

---

[4] See Chapter III, "Expanded Access to Unapproved Devices," of the guidance document "IDE Policies and Procedures," http://www.fda.gov/downloads/MedicalDevices/DeviceRegulationandGuidance/GuidanceDocuments/ucm080203.pdf.

10903 New Hampshire Avenue
Silver Spring, MD 20993-0002

Compassionate use information is available on FDA's web site at
http://www.fda.gov/MedicalDevices/DeviceRegulationandGuidance/HowtoMarketYourDev
ice/InvestigationalDeviceExemptionIDE/ucm051345.htm#compassionateuse

For CBER regulated products, the required information should be submitted to the
appropriate reviewing division (see section II above):

Document Control Center (HFM-99)
Center for Biologics Evaluation and Research
Food and Drug Administration
1401 Rockville Pike, Suite 200N
Rockville, MD 20852-1448

10. Are treatment IDEs and continued access available for investigational IVDs under an IDE?

Yes, both are available. See 21 CFR 812.36 and the Glossary for definitions of treatment IDE
and continued access.

11. Can my IVD device be considered a humanitarian use device (HUD) and can I apply for marketing approval through a humanitarian device exemption (HDE)?

Yes, it is possible for an IVD device to be approved for marketing under the HDE. See the
Glossary for definitions, 21 CFR Part 814, Subpart H, and Appendix 1 for more information.

12. Can an IVD device qualify for HUD designation if the affected patient population is fewer than 4,000 per year but each patient may need to be tested multiple times?

IVD devices qualify for an HUD designation when the number of persons tested with the
device is fewer than 4,000 per year. FDA recognizes that the number of tests with the device
may exceed one per patient. A device that involves multiple patient uses may still qualify for
HUD designation as long as the IVD device is designed for diagnosis or treatment of a total of
fewer than 4,000 patients per year in the US.

If a device is being developed to diagnose or to help diagnose a disease or condition with an
incidence of fewer than 4,000 patients per year, but there are more than 4,000 patients a year
"at risk" who would be subject to testing using the device, then the device may not qualify as a
HUD. 21 CFR 814.102(a)(5).

13. Is there a regulation that specifically addresses labeling of IVD products?

Yes. The regulation, "Labeling for in vitro diagnostic products," (21 CFR 809.10), specifies the
information required on labeling and in package inserts of marketed products as well as

products in development that are distributed for use in studies. We recommend that you design IVD studies so that the results will support the proposed indications for use in the package insert and labeling.

14. Are there different goals for IVD studies compared to other device studies?

No. The goals for IVD studies are the same as the goals for other device studies, even if the IVD study is exempt from most IDE requirements under 21 CFR 812.2(c)(3). We recommend that the sponsor and the investigators conduct an IVD device study with the goals of

- producing valid scientific evidence (for a definition, see 21 CFR 860.7(c)(2) and answer #1 of section IV) demonstrating reasonable assurance of the safety and effectiveness of the product, as described below, and

- protecting the rights and welfare of study subjects. (See Human Subject Protection, Section V of this guidance).

15. What regulations describe the content requirements for IVD premarket submissions?

Regulations that describe the basic content requirements by submission type include:

- Investigational Device Exemption (IDE) – 21 CFR 812.20

- Premarket Notification (510(k)) – 21 CFR 807.87

- Premarket Approval (PMA) – 21 CFR 814.20

- Humanitarian Device Exemption (HDE) – 21 CFR 814.104

Currently, there is no regulation describing the contents for a Product Development Protocol (PDP). However, section 515(f)(1) of the Act and the CDRH website at http://www.fda.gov/MedicalDevices/DeviceRegulationandGuidance/HowtoMarketYourDev ice/PremarketSubmissions/PremarketApprovalPMA/ucm048168.htm#pdp describe PDP requirements.

In addition, the FDA 510(k) substantial equivalence determination summaries and FDA PMA summaries of safety and effectiveness are currently available on the CDRH OIVD web page at http://www.fda.gov/MedicalDevices/ProductsandMedicalProcedures/InVitroDiagnostics/La bTest/ucm126189.htm or for CBER products at http://www.fda.gov/BiologicsBloodVaccines/BloodBloodProducts/ApprovedProducts/Pre marketApprovalsPMAs/ucm089793.htm and http://www.fda.gov/BiologicsBloodVaccines/DevelopmentApprovalProcess/510kProcess/u cm133429.htm. We recommend that the sponsor structure submissions according to the relevant regulations and provide sufficient detail to give the reader an understanding of the scientific data and information supplied. OIVD has issued many device specific guidances that describe FDA's recommendations for premarket submissions for particular types of IVDs.

16. Can published literature be used to support an IVD premarket submission?

FDA has developed a guidance document entitled "Supplements to Approved Applications for Class III Medical Devices: Use of Published Literature, Use of Previously Submitted Materials, and Priority Review," which can be found on the CDRH website at http://www.fda.gov/MedicalDevices/DeviceRegulationandGuidance/GuidanceDocuments/ucm080183.htm. CDRH and CBER believe that the principles outlined in this guidance are applicable to other submissions, specifically those for a 510(k), PMA, and HDE. (See the Glossary for definitions of these submission types.)

17. Can data from studies performed outside of the United States (U.S.) be used to support an IVD premarket submission?

Yes. FDA recognizes that clinical investigations may be conducted outside of the U.S., for example, in order to find adequate numbers of subjects for certain disease states, conditions, or pathogens. The PMA regulation contains information regarding research conducted outside of the U.S. (21 CFR 814.15). FDA can also accept data from foreign studies in support of 510(k)s.

18. Can foreign/international data be used as the sole support of a marketing application?

Yes, but only if warranted. The PMA regulation, 21 CFR Part 814, allows foreign data to be used as the sole support of a marketing application but only if (1) the data are applicable to the U.S. population and to U.S. medical practices, including laboratory practices, (2) the studies have been performed by clinical investigators of recognized competence, and (3) the data may be considered valid without the need for an on-site FDA inspection or, if necessary, FDA can validate the data through an on-site inspection or other appropriate means (21 CFR 814.15(d)).

For IVD devices, FDA would consider differences in population demographics, disease prevalence, disease presentation, laboratory practices, and medical standards of care. If the sponsor plans to submit an application based solely on foreign data, FDA recommends that the sponsor consult with the reviewing division prior to submission of the application.

See Introduction, Section II, question # 4 of this guidance for a list of reviewing divisions in both CDRH and CBER.

19. What is a master file and how is one submitted?

A device master file (MAF) is a reference source that a person submits to FDA. In general, it is a file of trade secret or confidential commercial/financial information submitted by a third party (i.e., someone other than the applicant) for use as a reference source in support of at least one application. FDA will accept MAFs from organizations or persons who have not submitted or will not directly submit the information in a PMA, IDE, 510(k), or other device-related submission to FDA. MAFs may include information on the following:

• facilities and manufacturing procedures and controls;

- synthesis, formulation, purification and specifications for chemicals, materials (an alloy, plastic, etc.) or subassemblies for a device;

- packaging materials;

- contract packaging and other manufacturing (such as sterilization);

- nonclinical study data; and

- clinical study data.

We recommend that a MAF include a cover letter, preferably on company letterhead, signed by a responsible company official that identifies the submission as a MAF and provides the name of a contact person at the company or a designated agent. For more information concerning MAFs see the CDRH website at http://www.fda.gov/MedicalDevices/DeviceRegulationandGuidance/HowtoMarketYourDevice/PremarketSubmissions/PremarketApprovalPMA/ucm142714.htm

20. How do I arrange to reference a MAF?

You, the sponsor, should contact the company that owns the information you would like to incorporate by reference in your premarket submission to FDA, and find out if this information is currently in a master file. If it is, you should obtain a written authorization from the master file holder (or an authorized designated agent/representative) on company letterhead. You should include the original authorization letter in the original copy of the premarket submission to FDA, and a copy of the authorization letter in each subsequent copy of the premarket submission. The master file holder should not send the authorization letter directly to FDA for inclusion in the master file or for inclusion in your premarket submission.

If the information you, the sponsor, would like to incorporate by reference in your premarket submission to FDA is not already in a master file, you should request that the company that owns this information submit a master file to FDA.

For more information on referencing MAFs see the CDRH website at http://www.fda.gov/MedicalDevices/DeviceRegulationandGuidance/HowtoMarketYourDevice/PremarketSubmissions/PremarketApprovalPMA/ucm142714.htm.

IV. Investigational Studies

1. What does FDA consider to be valid scientific evidence?

Valid scientific evidence is defined in the "Medical Device Classification Procedures" regulation, 21 CFR Part 860, as:

Evidence from well-controlled investigations, partially controlled studies, studies and objective trials without matched controls, well-documented case histories conducted by qualified experts, and reports of significant human experience with a marketed device, from which it can fairly and responsibly be concluded by qualified experts that there is reasonable assurance of the safety and effectiveness of a device under its conditions of use (21 CFR 860.7(c)(2)).

The intended use of the IVD, the level and quality of information in the literature relevant to the device use, and FDA knowledge of the technology obtained from reviewing other premarket applications determine the type of study and the level of evidence you may need to demonstrate reasonable assurance of its safety and effectiveness. For example, if you are studying an IVD device that uses a well-characterized technology and has an intended use that falls within a type of device that has been classified into Class I or Class II, the study may consist of a comparison of analytic performance to that of a legally marketed (i.e., predicate) device. On the other hand, if your IVD uses novel or unproven technology or has a new intended use, you may need to conduct a well-planned clinical study of the device in the target population defined by your intended use. You may contact the Division in the appropriate Center if you have questions regarding the type of study you need to conduct for your device.

We recommend that sponsors and investigators of all studies, including exempt studies under 21 CFR 812.2(c)(3), conduct the studies with the following goals in mind: producing valid scientific evidence of the product's safety and effectiveness and protecting the rights and welfare of study subjects. Sponsors and investigators of significant and non-significant risk studies must comply with the regulation requirements in 21 CFR Part 812. FDA recommends that sponsors and investigators of studies exempt from the majority of requirements under 21 CFR Part 812 use the relevant sections of 21 CFR Part 812 regarding the general conduct of device studies as guidance. (General Regulatory Issues, Section III of this guidance, discusses how 21 CFR Part 812 may apply to a particular IVD study.)

2. Why should I review the information regarding the conduct of device studies found in the IDE regulation even if, after considering the exemption criteria in the regulation, I determine that my proposed studies are exempt from most IDE requirements?

Some studies are exempt from most of 21 CFR Part 812 because of the low risk they pose to study subjects. However, studies that will support a PMA or other premarket submission should have the same goals as all other device studies: 1) to produce valid scientific evidence to support reasonable assurance of a product's safety and effectiveness, and 2) to protect study subjects. Therefore, the information in 21 CFR Part 812 will also be useful to sponsors and investigators of device studies exempt under 21 CFR 812.2(c). In addition, all studies that will support applications to FDA are subject to 21 CFR 812.119(c) as well as to 21 CFR Parts 50 and 56.

3. Should I review the "International Conference on Harmonization; Good Clinical

Practice: Consolidated Guideline" ("ICHGCP") published in the Federal Register Vol. 62, No. 90, May 9, 1997, pp. 25691-25709 or the draft ISO 14155, "Clinical Investigation of Medical Devices for Human Subjects," when developing studies for devices that fall within the exemption at 21 CFR 812.2(c)?

Although the ICH document was written for studies of pharmaceuticals, sections of the guidance address study issues common to all investigational products. Thus, these sections of the ICH GCP provide a useful reference regarding the proper conduct of studies.

The draft ISO document specifically states that it does not apply to IVD devices. The draft ISO document is an international document intended to reflect basic practices appropriate to clinical trials worldwide. It does not include all of FDA's specific requirements for clinical studies and is not presently a standard that FDA has officially recognized; therefore, we do not recommend that you rely on it.

4. Is FDA willing to review and discuss a study protocol even if the study is exempt from most of the 21 CFR Part 812 requirements?

Yes. Both CDRH and CBER have developed processes that allow sponsors to obtain early FDA input and review of proposed studies by submission of the protocol and other study materials in the form of a "pre-IDE" document and/or a discussion in the form of a "pre-IDE" meeting. While we refer to this early input as a "pre-IDE" process, it is also available for studies that are exempt from most IDE requirements under 21 CFR 812.2(c)(3) or that will be conducted under the abbreviated IDE regulations for NSR studies (21 CFR 812.2). FDA encourages use of the pre-IDE submission and/or meeting whenever the sponsor desires early feedback for clinical studies, particularly those for novel or high risk (Class III) devices. If you (the sponsor) are interested in submitting a pre-IDE, we recommend that you contact the Division that will review your device before you initiate your studies (See Introduction, Section II, question # 4 of this guidance). Use of the pre-IDE process does not obligate you in any way to future submission of an IDE. FDA also encourages continued communication throughout the course of the study. This communication can be in the form of an informational meeting/telephone call or status reports to the pre-IDE file.

5. Can I obtain a more formal evaluation of my study design or investigational plan through a determination and/or agreement meeting?

Yes, for Class III IVDs. (See the Glossary for definition of terms.) A guidance document regarding these meetings, "Early Collaboration Meetings Under the FDA Modernization Act (FDAMA); Final Guidance for Industry and for CDRH Staff," is available at

http://www.fda.gov/MedicalDevices/DeviceRegulationandGuidance/GuidanceDocuments/ucm073604.htm. CBER also follows this meeting guidance document when determination and/or agreement meetings are requested.

6. Under 21 CFR Part 812, what are the sponsor's and investigator's responsibilities for studies of a non-significant risk device conducted under the abbreviated requirements in 21 CFR Part 812?

The majority of the sponsor's and investigator's responsibilities in a study of a nonsignificant risk device are found in 21 CFR 812.2(b)(1) of the IDE regulation and are summarized below:

a. Label the device in accordance with 21 CFR 812.5;

b. Obtain IRB approval of the investigation after presenting the reviewing IRB with a brief explanation of why the device is not a significant risk device, and maintain such approval.

c. Ensure that each investigator participating in an investigation of the device obtains from each subject under the investigator's care, informed consent under part 50 and documents it, unless documentation is waived by an IRB under 21 CFR 56.109(c).

d. Comply with the requirements of 21 CFR 812.46 with respect to monitoring investigations;

e. Maintain the records required under 21 CFR 812.140(b)(4) and (5) and make the reports required under 21 CFR 812.150 (b)(1) through (3) and (5) through (10);

f. Ensure that participating investigators maintain the records required by 21 CFR 812.140(a)(3)(i) and make the reports required under 21 CFR 812.150(a)(1), (2), (5), and (7);

g. Comply with the prohibitions in 21 CFR 812.7 against promotion and other practices.

All studies should have a written protocol as described in 21 CFR 812.25(b) and a risk analysis as described in 21 CFR 812.25(c), regardless of the status of the study under 21 CFR Part 812.

All sites participating in the study should use identical copies of the protocol and receive protocol amendments simultaneously so that data is collected in a consistent manner. Data collected from different sites otherwise may not be able to be pooled in the final analysis due to inconsistencies in how it was collected. We recommend that protocols describe the study objectives, design, methodology, subject populations, types of specimens, data to be collected and planned data analysis. (See also Data Considerations, Section VI, of this guidance). For information on how to monitor the study, you may refer to the FDA guidance document entitled "Guideline for the Monitoring of Clinical Investigations," which is available at http://www.fda.gov/downloads/ICECI/EnforcementActions/BioresearchMonitoring/UCM133752.pdf

7. What are my responsibilities as the sponsor or the investigator of a study of a significant risk device subject to 21 CFR Part 812?

The sponsor's responsibilities for significant risk device investigations are described in Appendix 3 of this guidance. This information is also included as an enclosure in all IDE approval letters.

The investigator's responsibilities for significant risk device investigations are described in Appendix 4 of this guidance. This information is also included as an enclosure in all IDE approval letters.

8. Is it appropriate to use a quality systems approach in the conduct of IVD studies?

Yes, we recommend that sponsors and investigators follow quality systems methodologies, including accountability and traceability of the investigational device, auditing of data collected and monitoring to make sure the protocol was followed, documentation of training of staff in the use of the device [21 CFR 812.43(a)], and notifying FDA of unanticipated adverse device effects [21 CFR 812.150(b)(1)] in the conduct of IVD studies. Also, 21 CFR 812.20(b)(3) requires "[a] description of the methods, facilities, and controls used for the manufacture, processing, packing, storage, and, where appropriate, installation of the device, in sufficient detail so that a person generally familiar with good manufacturing practices can make a knowledgeable judgment about the quality control used in the manufacture of the device." This suggestion is consistent with both the need to provide scientifically valid information in support of premarket submissions and the design control requirements. Manufacturers of Class II and Class III IVD devices, and some Class I devices, are required to follow design controls, as described in 21 CFR 820.30 of the "Quality System Regulation," during the development of investigational devices. 21 CFR 812.1(a). See the Glossary for the definition of device classes.

9. If a sponsor's 'in-house' laboratory participates in the study of an IVD is the laboratory considered to be a study site?

Yes. All locations involved in an IVD study are considered study sites whether they are located at a sponsor-owned facility or at an independently-owned laboratory. The sponsor should list the laboratory as a study site, and the study should be conducted under the same investigational plan. As a study site, this laboratory can be inspected as part of the FDA's bioresearch monitoring (BIMO) inspection program.

V. Human Subject Protection

1. When does an IVD study involve human subjects?

Under FDA's regulations governing the conduct of IVD device studies, the definition of "subject" includes individuals on whose specimens an investigational device is used [see 21 CFR 812.3(p)]. As a result, an IVD study using human specimens involves human subjects.

2. Am I required to follow the "Good Laboratory Practice for Nonclinical Studies" regulation (21 CFR Part 58) in my IVD study?

Part 58 applies only to nonclinical laboratory studies, which are defined as in vivo or in vitro experiments in which test articles are studied prospectively in test systems under laboratory conditions to determine their safety. The term does not include studies utilizing human subjects or clinical studies or field trials in animals. The term does not include basic exploratory studies carried out to determine whether a test article has any potential utility or to determine physical or chemical characteristics of a test article (21 CFR 58.3(d)).

Moreover, because the safety of IVDs is related to the accuracy of the result, most IVD studies that are intended to establish safety would necessarily use human specimens. As noted above, an IVD study using human specimens involves human subjects and thus is excluded from the definition of nonclinical laboratory studies. Such studies to establish safety are subject to 21 CFR Parts 50 and 56 and 21 CFR Part 812 as applicable, dealing with human subject research, rather than to 21 CFR Part 58.

3. What regulations apply regarding human subject protection in investigational IVD studies?

FDA regulations on "Protection of Human Subjects" and "Institutional Review Boards" (21 CFR Parts 50 and 56) apply to all clinical investigations (investigations involving human subjects) regulated by FDA under section 520(g) of the Federal Food, Drug and Cosmetic Act, as well as other clinical investigations that support applications for research or marketing permits for products regulated by FDA. (21 CFR 50.1, 56.101). As described above, any study using human specimens involves human subjects. FDA has expressed an intent to exercise enforcement discretion regarding informed consent for certain IVD studies, in the guidance entitled, "Guidance on Informed Consent for In Vitro Diagnostic Device Studies using Leftover Human Specimens that are not Individually Identifiable." http://www.fda.gov/MedicalDevices/DeviceRegulationandGuidance/GuidanceDocuments/ucm078384.htm

Some research studies involving human subjects that are conducted or supported by a federal department or agency will be required to follow the Federal Policy for the Protection of Human Subjects (the "Common Rule"). The Department of Health and Human Services (HHS) has codified the Common Rule at 45 CFR Part 46, subpart A. Research involving human subjects that is conducted or supported by HHS and that is not otherwise exempt must comply with 45 CFR Part 46, which contains additional protections for specific populations. For further information about the applicability of 45 CFR Part 46 to your study, refer to http://www.hhs.gov/ohrp/.

FDA regulation and the Common Rule contain some differing requirements. Although FDA and HHS seek to harmonize requirements related to informed consent where possible, certain requirements of the FDA regulation and the Common Rule are different. The same study may be subject to both sets of requirements, either one, or neither. It is the responsibility of sponsors and investigators to comply with all applicable requirements.

4. How do FDA regulations for protection of human subjects, 21 CFR Parts 50 and 56, differ from HHS regulations in 45 CFR Part 46 subpart A?

A general comparison appears in the FDA's Good Clinical Practice program website, Comparison of FDA and HHS Human Subject Protection Regulations found at http://www.fda.gov/ScienceResearch/SpecialTopics/RunningClinicalTrials/EducationalMaterials/ucm112910.htm. For additional information on 45 CFR Part 46 subpart A refer to the

Office for Human Research Protections at
http://www.hhs.gov/ohrp/humansubjects/guidance/45cfr46.htm.

5. Can investigational IVD studies receive expedited review (see Glossary for definition) by an IRB?

Yes, in many cases an investigational IVD study is eligible for IRB expedited review (see 21 CFR 56.110), for both initial approval and continuing review. The categories of research that may be reviewed by the IRB through an expedited review procedure are described in the Federal Register notice on expedited review, found at http://www.fda.gov/ScienceResearch/SpecialTopics/RunningClinicalTrials/GuidancesInfor mationSheetsandNotices/ucm118099.htm. As stated in a Federal Register notice, however, sponsors and investigators may not use the expedited review procedure where identification of the subjects and/or their responses would reasonably place the subjects at risk of criminal or civil liability or be damaging to the subjects' financial standing, employability, insurability, reputation, or be stigmatizing, unless reasonable and appropriate protections will be implemented so that risks related to invasion of privacy and breach of confidentiality are no greater than minimal. (63 FR 60353, November 9, 1998).

6. Can leftover specimens be used in IVD studies without informed consent?

The document entitled, "Guidance for Industry, Institutional Review Boards, Clinical Investigators, and Food and Drug Administration: Guidance on Informed Consent for In Vitro Diagnostic Device Studies using Leftover Human Specimens that are not Individually Identifiable," http://www.fda.gov/MedicalDevices/DeviceRegulationandGuidance/GuidanceDocuments/u cm078384.htm, describes the limited circumstances in which FDA intends to exercise enforcement discretion regarding requirements for informed consent. (See Glossary for definition of "leftover specimens".)

7. Can those who routinely conduct studies with IVDs (e.g., research hospitals) use a general informed consent to address future studies using samples collected in their own facility?

To fulfill FDA informed consent requirements for studies of IVDs, a site may develop an informed consent process to address the use of samples collected at the facility (see the Glossary for definition) in a specific study or for a broader category of future studies. This general informed consent process may be used for subjects seen at and/or admitted to a specific facility. The informed consent document must contain all of the required elements found in 21 CFR 50.25.

8. Can a human specimen that was initially collected in a study with the informed consent of the subject be used in a later study without a new consent process?

If the original informed consent document contains a statement that excess specimen(s) will be stored for future use in specified types of studies and the new study meets the criteria stated in that consent document, it is possible that no further consent is necessary. This assumes that the original informed consent document contains all of the other essential elements, including notice to the subject that FDA may review their files and an explanation of the purposes and benefits of the research. (See 21 CFR 50.25.) We recommend sponsors and investigators consult with the IRB regarding the need for a new informed consent process in such a case. The IRB decision should include consideration of any state and/or local requirements regarding informed consent and patient rights. If new testing could expose the subject to previously unanticipated risks (e.g., privacy concerns for the subject and/or his family related to testing for a genetic marker), a new consent may be needed. In addition, if the original informed consent did not address future research use at all, or did not cover the type of study now under consideration, it is likely a new consent will be needed.

Under certain circumstances, for human specimens leftover from specimens originally collected for a previous study, FDA intends to exercise enforcement discretion regarding informed consent requirements. See "Guidance for Industry, Institutional Review Boards, Clinical Investigators, and Food and Drug Administration: Guidance on Informed Consent for In Vitro Diagnostic Device Studies using Leftover Human Specimens that are not Individually Identifiable," http://www.fda.gov/MedicalDevices/DeviceRegulationandGuidance/GuidanceDocuments/ucm078384.htm,

VI. Data Considerations

1. What information should the protocol include to ensure that the investigational IVD study will be scientifically sound?

We recommend that the protocol include a clear description of study design; objectives, estimation of performance goals (e.g., desired confidence interval widths) that are directly related to the intended claims for the IVD device, or hypotheses; and a statistical plan to be applied to the data. (See the Glossary for definitions of protocol, statistical hypothesis, and confidence interval.)

2. Is it acceptable to develop new or to revise existing study hypotheses as the study progresses?

We generally believe it would be inappropriate to draw conclusions from after-the-fact hypotheses. We recommend that changes in study protocols be carefully documented and explained. FDA encourages sponsors to contact the appropriate review division to discuss studies before they are initiated and to consult FDA before changes in protocols are made mid-study. (For the FDA divisions responsible for review of IVD products, see Introduction, Section II, of this guidance.)

3. How should I determine appropriate sample size for a study?

The sponsor should formulate sample size based on standard statistical techniques and the sample size should account for any unique issues related to intended use(s), device technologies, and/or the biology of the condition being studied.

4. What guidance is available for sponsors to determine how to estimate IVD performance in terms of sensitivity and specificity, how to handle discrepant results, and what to do when a study is performed without a truth standard ("gold standard") (see the Glossary for definitions)?

FDA has recognized a number of Clinical and Laboratory Standards Institute [(CLSI), formerly National Committee on Clinical Laboratory Standards (NCCLS)] standards related to these issues. A list of these standards, but not the standards themselves, can be found through the database on the CDRH web page
http://www.accessdata.fda.gov/scrIpts/cdrh/cfdocs/cfStandards/search.cfm

National Committee on Clinical Laboratory Standards

The agency's guidance entitled "Statistical Guidance on Reporting Results from Studies Evaluating Diagnostic Tests" can be viewed at
http://www.fda.gov/MedicalDevices/DeviceRegulationandGuidance/GuidanceDocuments/u cm071148.htm

5. How much leeway is there in deciding on the populations from which human specimens are collected and under what conditions are data on simulated specimens (see the Glossary for definition) acceptable?

Studies should be performed in a representative sample of the intended use population (i.e., representation of both diseased and non-diseased cases, and controlling for subject demographics and morbidity factors that may affect the level of device performance). When a disease is rare or samples are needed specifically to challenge cut-off points, sponsors may use enriched samples, panels of credentialed samples (e.g., Center for Disease Control and Prevention (CDC) panels), and/or spiked or contrived samples. The acceptance of simulated specimens depends on how well they represent specimens from the intended-use population and whether their performance accurately reflects what the IVD device user can expect.

6. Is it acceptable to eliminate data that appear to be out of line with the main body of the dataset (i.e., "outliers")?

An outlier, an observation that lies an unexpected distance from other values, does not in itself prove that an error or violation has occurred. Therefore, the primary data analysis should include all such observations, including "outliers". A sponsor may perform some specific supplemental analyses on subsets of the data, if clinically and scientifically justified, that may exclude outlying data points. Excluding large amounts of data, regardless of the reasons for exclusion, will seriously bias the results.

7. Can I add additional testing on the same subject to the dataset, particularly when it is hard to find study subjects?

In most studies the sponsor should avoid multiple testings of the same subject because it may skew performance statistics and under-estimate standard deviations. When multiple testing is done, the sponsor should explain why it is necessary and choose methods of statistical analysis that allow adjustment for within-subject correlation.

It is appropriate to conduct repeated testing of the same sample to evaluate test reproducibility – i.e., the ability of the test to yield the same or similar readings when expected.

8. How much precision (see the Glossary for definition) is needed for measurement data, e.g., in terms of decimal places?

Study data should contain no more decimal places than the precision of the instrument allows, i.e., if the instrument is only precise to the second decimal place the sponsor should not analyze the data using three decimal places.

9. What records should help to ensure scientific soundness of an IVD investigational study?

Unless a study falls within the exemption at 21 CFR 812.2(c), specific record requirements are listed in 21 CFR 812.140. In general, the records that are needed are those that provide the data for testing the study hypotheses. Records should contain sufficient detail to allow the study to be reproduced when the same protocol is followed. We recommend that investigators maintain detailed records because a review of the study may indicate the need for other analyses of the collected data.

We also recommend that investigators:

    a.   Maintain records of all data elements captured in the study, including raw measurements and subject co-variables in the form of demographic and morbidity factors;

    b.   Link every observation recorded to the subject and that person's co-variable data;

    c.   Preserve information obtained for all subjects enrolled and for all specimens collected.

Additionally, electronic spreadsheets of study data are useful. Given the possible need to review or analyze study data at the most detailed level, electronic spreadsheets may help to minimize review time. For information on electronic records, see the guidance document, "Part 11, Electronic Records; Electronic Signatures -- Scope and Application," at http://www.fda.gov/downloads/Drugs/GuidanceComplianceRegulatoryInformation/Guidan ces/ucm072322.pdf. There is also a more general guidance document available on electronic records for clinical studies that is entitled "Computerized Systems Used in Clinical Trials," which can be found at http://www.fda.gov/RegulatoryInformation/Guidances/ucm126402.htm.

10. What does FDA recommend be included in the final report of the investigation from the sponsor to all reviewing IRBs (and to FDA for significant risk studies) (21 CFR 812.150(b)(7))?

A final report should be a basic scientific report of the studies conducted, including the results of testing the study hypotheses. This report can be a useful means of providing a simple account of the data collection and study outcome. Such a report can facilitate preparation of the eventual submission for regulatory action, particularly when accompanied by the information included in the investigational plan (see the Glossary for definition).

The suggested format for the IDE final report, which FDA includes as an enclosure in all IDE approval letters, is found in Appendix 5 of this guidance.

It should be noted that FDA will consider submission of a marketing application (510(k), PMA, or HDE) to serve as the final report for the IDE. When a study sponsor submits a marketing application in lieu of the final report, the sponsor should still submit a supplement to the IDE stating that the marketing application should be considered the final report for the study.

The final report for significant risk device investigations must be submitted to the IRBs and/or FDA within six months after termination or completion of the study. 21 CFR 812.150(b)(7).

VII. Glossary

Note: this glossary is written in plain language and is for use exclusively with this guidance document.

Definitions that have been taken from the Act, other pertinent laws, or in Federal regulations include the relevant citation.

510(k) – See Premarket Notification.

Agreement meeting – a meeting, under section 520(g)(7) of the Act (21 U.S.C. § 360j(g)(7)), that is available to anyone planning to investigate the safety or effectiveness of a class III device (see definition below) or any implant. The purpose of the meeting is to reach agreement on the key parameters of the investigational plan, including the study protocol. The meeting is to be held within 30 days of the receipt of a written request. FDA will document in writing any agreement reached and make it a part of the administrative record. The agreement is binding on FDA and can only be changed with the written agreement of the applicant or when there is a substantial scientific issue essential to determining the safety or effectiveness of the device. See 21 U.S.C. § 360j(g)(7). A guidance document regarding these meetings, "Early Collaboration Meetings Under the FDA Modernization Act (FDAMA); Final Guidance for Industry and for CDRH Staff," is available at http://www.fda.gov/MedicalDevices/DeviceRegulationandGuidance/GuidanceDocuments/ucm073604.htm.

Analyte specific reagent (ASR) - ASRs are defined as "antibodies, both polyclonal and monoclonal, specific receptor proteins, ligands, nucleic acid sequences, and similar reagents which, through specific binding or chemical reactions with substances in a specimen, are intended for use in a diagnostic application for identification and quantification of an individual chemical substance or ligand in biological specimens." 21 CFR 864.4020(a). ASRs are medical devices that are regulated by FDA. They are subject to general controls, including current Good Manufacturing Practices (cGMPs), 21 CFR Part 820, as well as the specific provisions of the ASR regulations (21 CFR 809.10(e), 809.30, 864.4020).

Class I devices – devices for which the general controls of the Act are sufficient to provide reasonable assurance of their safety and effectiveness. They typically present minimal potential for harm to the user and the person being tested. They are subject to general controls, which include registration and listing, labeling, and adverse event reporting requirements (section 513(a)(1)(A) of the Act). Most Class I devices are exempt from premarket notification (see definition below), subject to certain limitations found in section 510(l) of the Act and in 21 CFR 862.9, 864.9, and 866.9. Some are also exempt from the "Quality Systems Regulation" found in 21 CFR Part 820. IVD examples of Class I devices include complement reagent, phosphorus (inorganic) test systems (21 CFR 862.1580), and E. coli serological reagents (21 CFR 866.3255).

Class II devices – devices for which general controls alone are insufficient to provide reasonable assurance of their safety and effectiveness and for which establishment of special controls can provide such assurances. Special controls may include special labeling, mandatory performance standards, risk mitigation measures identified in guidance, and postmarket surveillance (section 513(a)(1)(B) of the Act). Some Class II devices are exempt from premarket notification (see definition below), subject to limitation in 21 CFR 862.9, 864.9, and 866.9. IVD examples of Class II devices include glucose test systems (21 CFR 862.1345), antinuclear antibody immunological test systems (21 CFR 866.5100), and coagulation instruments (21 CFR 864.5400).

Class III devices – devices for which insufficient information exists to provide reasonable assurance of safety and effectiveness through general or special controls. Class III devices are usually those that support or sustain human life, are of substantial importance in preventing impairment of human health, or which present a potential, unreasonable risk of illness or injury (section 513(a)(1)(C) of the Act). Most Class III devices require premarket approval (PMA, see definition below). IVD examples of these include automated PAP smear readers, nucleic acid amplification devices for tuberculosis, and total prostate specific antigen (PSA) for the detection of cancer. A limited number of Class III devices that are equivalent to devices legally marketed before enactment of the Medical Device Amendments of 1976 may be marketed through the premarket notification (510(k)) process (see definition below), until FDA has published a requirement for manufacturers of that generic type of device to submit PMA data.

Compassionate use – The compassionate use provision allows access for patients with a serious disease or condition who do not meet the requirements for inclusion in the clinical investigation but for whom the treating physician believes the device may provide a benefit in treating and/or diagnosing their disease or condition. There must be no feasible alternative therapies/diagnostics available. Compassionate use is typically available only for individual patients but also may be used to treat a small group. Prior FDA approval is needed before compassionate use occurs.

Further information can be found at http://www.fda.gov/MedicalDevices/DeviceRegulationandGuidance/HowtoMarketYourDev ice/InvestigationalDeviceExemptionIDE/ucm051345.htm.

Confidence interval – the range of plausible values for a statistical parameter, consistent with the observed data, which is computed with a sample estimate parameter (e.g., mean) and its standard deviation. For example, a "95% Confidence Interval" is computed such that, if the parameter was determined for repeated experiments, the resulting values would include the true parameter value 95% of the time.

Continued access to investigational devices – FDA may allow sponsors to continue to enroll subjects under an IDE, after the trial has been completed, while a marketing application is prepared by the sponsor and/or reviewed by FDA. To continue enrolling subjects, a sponsor should show that there is a public health need for the device, that preliminary evidence indicates that the device is likely to be effective for the indications proposed, and that no significant safety concerns have been identified for the proposed indication. A guidance document, entitled "Continued Access to Investigational Devices During PMA Preparation and Review," can be found at http://www.fda.gov/MedicalDevices/DeviceRegulationandGuidance/ GuidanceDocuments/ucm080260.htm .

Co-variables – data elements relating to a subject that might affect how well a diagnostic test works, such as demographic status (age, gender, etc.), morbidity, or concurrent therapy.

Determination Meeting – A Determination meeting under section 513(a)(3)(D) of the Act is available to anyone anticipating submitting a PMA or PDP and is intended to provide the applicant with the FDA's determination of the type of valid scientific evidence that will be necessary to demonstrate that the device is effective for its intended use. As a result of this meeting, FDA will determine whether clinical studies are needed to establish effectiveness and, in consultation with the applicant, determine the least burdensome way of evaluating device effectiveness that has a reasonable likelihood of success. The applicant can expect that FDA will determine if concurrent randomized controls, concurrent non-randomized controls, historical controls, or other types of evidence will be acceptable. FDA's determination is to be written, shared with the applicant within 30 days following the meeting, and is binding upon the FDA, unless it could be contrary to public health. 21 U.S.C. § 360c(a)(3)(D). A guidance document regarding these meetings, "Early Collaboration Meetings Under the FDA Modernization Act (FDAMA); Final Guidance for Industry and for CDRH Staff," is available at http://www.fda.gov/MedicalDevices/DeviceRegulationandGuidance/ GuidanceDocuments/ucm073604.htm.

Device – as defined in the Act, section 201(h): an instrument, apparatus, implement, machine, contrivance, implant, in vitro reagent, or other similar or related article, including any component, part, or accessory, which is a) recognized in the official National Formulary, or the

United States Pharmacopeia, or any supplement to them; b) intended for use in the diagnosis of disease or other conditions, or in the cure, mitigation, treatment, or prevention of disease, in man or other animals; or c) intended to affect the structure or any function of the body of man or other animals, and which does not achieve its primary intended purposes through chemical action within or on the body of man or other animals and which is not dependent upon being metabolized for the achievement of its primary intended purposes.

Excess samples – remnants of human specimens collected for routine clinical care or analysis that would otherwise have been discarded, as well as specimens leftover from specimens previously collected for other unrelated research or investigations. Excess samples are also referred to as "surplus samples," "residual", "reserved samples," "library samples," and "leftover specimens."

Expedited review – review by an institutional review board (IRB) that does not require full board review or a convened meeting. Such a review may be carried out by the IRB chairperson or one or more experienced reviewers assigned by the IRB chairperson from among the members of the IRB. Reviewers may exercise all of the authorities of the IRB except they may not disapprove the study. Disapproval may only result through the IRB's non-expedited review process. Expedited review is reserved for minimal risk studies. (See 21 CFR 56.110.)

General purpose reagents – chemical reagents that have general laboratory application and that are not labeled or otherwise intended for a specific diagnostic application. They are used to collect, prepare, and/or examine specimens from the human body for diagnostic purposes. (Example: reagents used for general staining in microscopic procedures.) General purpose reagents do not include laboratory machinery, automated or powered systems (21 CFR 864.4010(a)).

Gold standard – see truth standard.

Humanitarian use devices (HUDs) –HUDs are devices intended to diagnose a disease or condition in fewer than 4,000 patients in the U.S. per year. Such devices are regulated under 21 CFR Part 814, Subpart H. If a device receives a designation as an HUD, a Humanitarian Device Exemption request (HDE) can be submitted. HUDs that are approved for marketing under an HDE have specific labeling requirements. IRB approval is required for use of a HUD (21 CFR 814.124).

Investigation – a clinical investigation or research involving one or more subjects to determine the safety or effectiveness of a device (21 CFR 812.3(h)). It is often referred to as a clinical trial and is sometimes referred to as a field trial.

Investigational Device Exemption (IDE) – application which, when approved, allows the device to be used lawfully for the purpose of conducting studies regarding the safety and effectiveness of the device, without complying with certain requirements of the Act. (See 21 CFR 812.1 for specific exemptions.) For significant risk (SR) device studies (see definition below), a sponsor must apply to FDA to obtain approval for an IDE. (See 21 CFR 812.20.) For non-significant risk (NSR) device studies (see definition below), an IDE is considered approved when a sponsor meets the abbreviated requirements found in 21 CFR 812.2(b), which include approval from the reviewing Institutional Review Board(s) (IRB(s)).

Investigational plan – sponsor's overall plan regarding the conduct of an investigational study. It includes, but is not limited to, the purpose of the study, a written protocol, a risk analysis, device description, labeling, written monitoring procedures, informed consent materials, and Institutional Review Board (IRB) information. (21 CFR 812.25.) For IVD studies, protocols should describe the study objectives, design, methodology, subject populations, types of specimens, data to be collected and planned data analysis.

Investigational Use in vitro diagnostic (IVD) product – an IVD product being used for product testing prior to full commercial marketing (e.g., for use on specimens derived from humans to compare the usefulness of the product with other products or procedures in current use or recognized as useful). These products must be labeled according to 21 CFR 812.5 for non-significant risk or significant risk devices and according to 21 CFR 809.10(c)(2)(ii) for devices that are exempt under 21 CFR 812.2(c).

Investigator – an individual who actually conducts a clinical investigation, i.e., a person under whose immediate direction the investigational product is administered, dispensed, or used, provided that the investigation involves a subject. In the event of an investigation conducted by a team of individuals, the investigator is the responsible leader of that team (21 CFR 812.3(i)).

In vitro diagnostic (IVD) products – those reagents, instruments, and systems intended for use in the diagnosis of disease or other conditions, including a determination of the state of health, in order to cure, mitigate, treat, or prevent disease or its sequelae. Such products are intended for use in the collection, preparation, and examination of specimens taken from the human body. IVD products are devices as defined in section 201(h) of the Act and may also be biological products subject to section 351 of the Public Health Service Act. The regulatory definition of in vitro diagnostic products is found in 21 CFR 809.3(a).

Leftover specimens -- remnants of specimens collected for routine clinical care or analysis that would otherwise have been discarded, or remnants of specimens previously collected for other unrelated research. These specimens may be obtained from a specimen repository -- a common site for storage of collections of human biological specimens available for study. See also Excess samples.

Non-significant risk (NSR) device – a device that does not meet the definition of significant risk (SR) device (see definition below). An IDE is considered approved for a NSR investigational device study once sponsors meet the abbreviated requirements found in the "Investigational Device Exemptions" regulation at 21 CFR 812.2(b). The risk determination for an investigational device study should be based on the proposed use of the device in the investigation in addition to the characteristics of the device.

Outlier – a data observation whose value appears to be out of line with the main body of data that has been collected.

Pre-amendment in vitro diagnostic (IVD) tests – IVD tests that were in commercial distribution before May 28, 1976.

Precision – the closeness of agreement between independent diagnostic test results obtained under stipulated conditions. For additional information refer to the Harmonized Technology Database, Clinical and Laboratory Standards Institute, available at http://www.clsi.org.

Premarket Approval Application (PMA) – the application for approval required prior to the marketing of most Class III medical devices (section 515 of the Act, 21 U.S.C. 360e). (See definitions of Class I, II, and III devices above.) PMA approval is based on a determination by FDA that the applicant's submission provides sufficient valid scientific evidence to provide reasonable assurance that the device is safe and effective for its intended use(s). The PMA regulation is 21 CFR Part 814. PMA information is available at http://www.fda.gov/MedicalDevices/DeviceRegulationandGuidance/HowtoMarketYourDev ice/PremarketSubmissions/PremarketApprovalPMA/default.htm.

Premarket Notification – also referred to as a 510(k), is a submission to FDA that contains information to demonstrate that a device is substantially equivalent (SE) to a legally marketed (predicate) device. Governing regulations regarding premarket notification procedures are found in Subpart E of the "Establishment Registration and Device Listing for Manufacturers and Initial Importers of Devices" regulation (21 CFR Part 807). The 510(k) device advice page is available at http://www.fda.gov/MedicalDevices/DeviceRegulationandGuidance/HowtoMarketYourDev ice/PremarketSubmissions/PremarketNotification510k/default.htm.

Product Development Protocol (PDP) – FDA process of approval for marketing of medical devices, usually reserved for Class III devices (see definitions of device classes above), by which the sponsor and FDA agree on the product design and testing early in the concept and planning stages of a product (section 515(f) of the Act).

Protocol – a document that contains a description of the objectives and design of an investigational study, methodology(s) to be used, and data to be collected. It may also contain information regarding the planned data analysis and study monitoring. For most studies in the development of an IVD product, it also contains information regarding types of specimens and subject populations.

Reserved samples – see excess samples.

Sensitivity – the probability that a diagnostic test will yield a positive result when the disease or the target analyte is present.

Significant risk (SR) device – an investigational device that presents a potential for serious risk to the health, safety, or welfare of a subject and:

1. is intended as an implant;

2. is purported or represented to be for use in supporting or sustaining human life;

3. is for a use of substantial importance in diagnosing, curing, mitigating, or treating disease, or otherwise preventing impairment of human health; or

4. otherwise presents a potential for serious risk to the health, safety, or welfare of a subject.

The risk determination for an investigational device study should be based on the proposed use of the device in the investigation in addition to the device characteristics. Sponsors of significant risk device studies must apply to FDA for an Investigational Device Exemption (IDE) (see definition above). (21 CFR 812.3(a), 812.3(m); 812.20.)

Simulated specimens – specimens made in the laboratory by adding the analyte of interest in known concentrations to a medium that simulates the natural matrix.

Specificity – the probability that a diagnostic test will yield a negative result when the disease or target analyte is absent.

Sponsor – a person who initiates, but who does not actually conduct, the investigation, i.e., the investigational device is administered, dispensed, or used under the immediate direction of another individual. A person other than an individual that uses one or more of its own employees to conduct an investigation that it has initiated is a sponsor, not a sponsor-investigator (see next definition), and the employees are investigators (see definition above) (21 CFR 812.3(n)).

Sponsor-investigator – an individual who both initiates and actually conducts, alone or with others, an investigation, i.e., under whose immediate direction the investigational device is administered, dispensed, or used. The term does not include any person other than an individual. The obligations of a sponsor-investigator include both those of an investigator and those of a sponsor (21 CFR 812.3(o)).

Statistical hypothesis – a statement about some state of nature that a proposed study or set of studies will either accept or reject on the basis of the experimental data. The hypothesis is usually broken down into a null hypothesis (a statement of what the testing results will hopefully reject) and an alternative hypothesis (a statement of what the testing results will hopefully accept).

Study – as used in this document, covers the systematic evaluations conducted in the development of an IVD product, including the feasibility, analytical assessments, method comparison, and evaluations to determine clinical utility of a product.

Subject – a human who participates in an investigation, either as an individual on whom or on whose specimen an investigational device is used or as a control. A subject may be in normal health or may have a medical condition or disease (21 CFR 812.3(p)).

Surplus samples – see excess samples.

Transitional device – a product defined as a device as of May 28, 1976, but previously considered by FDA to be a new drug or an antibiotic drug (21 CFR 812.3(r)).

Treatment IDE – use of an unapproved investigational device for the treatment or diagnosis of patients during the clinical trial or prior to final FDA action on the marketing application, if during the course of the clinical trial the data suggest that the device is effective. A treatment IDE may cover a large number of patients that exceeds the number of clinical sites and patients stipulated in the original IDE. The device must be for treatment or diagnosis of a serious or immediately life-threatening disease or condition; there must be no comparable or satisfactory alternative device or therapy available; the device must be under investigation in a controlled clinical study for the same use under an approved IDE, or such clinical studies have been completed; and the sponsor must be actively pursuing marketing approval or clearance of the device. Requirements for an application for a treatment IDE are found in the Investigational Device Exemptions regulation at 21 CFR 812.36.

Truth Standard ("Gold" Standard) – any medical procedure or laboratory method or combination of procedures and methods that the clinical community relies upon for diagnosis, that is accepted by FDA, and that is regarded as having negligible risk of either a false positive or a false negative result. The truth standard result should be definitive (positive/negative, present/absent, or diseased/non-diseased), and should not give an indeterminate result. As science and technology improve, newer, more reliable standards may replace previous standards, particularly in the case of new disease markers.

VIII. **References**

Note: *this listing is presented in the order that the documents are first referred to in this guidance document.*

1. 21 CFR Part 812, Investigational Device Exemptions, found at http://www.accessdata.fda.gov/scripts/cdrh/cfdocs/cfCFR/ShowCFR.cfm?CFRPart=812.

2. 21 CFR Part 312, Investigational New Drug Application, found at http://www.accessdata.fda.gov/scripts/cdrh/cfdocs/cfCFR/ShowCFR.cfm?CFRPart=312.

3. 21 CFR Part 809, In Vitro Diagnostic Products for Human Use, found at http://www.accessdata.fda.gov/scripts/cdrh/cfdocs/cfCFR/CFRSearch.cfm?CFRPart=809

4. 21 CFR 820.30, Subpart C of the Quality System Regulation, found at http://www.accessdata.fda.gov/scripts/cdrh/cfdocs/cfcfr/CFRSearch.cfm?fr=820.30

5. 21 CFR 860.7, Determination of safety and effectiveness, in Medical Device Classification Procedures, found at http://www.accessdata.fda.gov/scripts/cdrh/cfdocs/cfCFR/ShowCFR.cfm?FR=860.7.

6. 21 CFR Part 54, Financial Disclosure by Clinical Investigators, found at http://www.accessdata.fda.gov/scripts/cdrh/cfdocs/cfCFR/ShowCFR.cfm?CFRPart=54

7. "Expanded Access to Unapproved Devices," Chapter III, of the guidance document IDE Policies and Procedures. Guidance document found at http://www.fda.gov/MedicalDevices/DeviceRegulationandGuidance/GuidanceDocuments/ucm080202.htm.

8. "Test Requirements," (21 CFR 610.40), found at
   http://www.accessdata.fda.gov/scripts/cdrh/cfdocs/cfcfr/CFRSearch.cfm?FR=610.40 and the "Restrictions on Use for Further Manufacture of Medical Devices, " (21 CFR 610.42), found at
   http://www.accessdata.fda.gov/scripts/cdrh/cfdocs/cfcfr/CFRSearch.cfm?FR=610.42

9. Information concerning Master Files for Devices (MAFs) is found on the CDRH website at
   http://www.fda.gov/MedicalDevices/DeviceRegulationandGuidance/HowtoMarket YourDevice/PremarketSubmissions/PremarketApprovalPMA/ucm142714.htm

10. "Supplements to Approved Applications for Class III Medical Devices: Use of Published Literature, Use of Previously Submitted Materials, and Priority Review," which can be found on the CDRH website at
    http://www.fda.gov/MedicalDevices/DeviceRegulationandGuidance/GuidanceDoc uments/ucm080183.htm.

11. Guidance regarding Product Development Protocol (PDP) applications, found at
    http://www.fda.gov/MedicalDevices/DeviceRegulationandGuidance/HowtoMarket YourDevice/PremarketSubmissions/PremarketApprovalPMA/ucm048168.htm

12. FDA premarket final review summaries and FDA PMA summaries of safety and effectiveness, found at
    http://www.fda.gov/MedicalDevices/ProductsandMedicalProcedures/InVitroDiagn ostics/LabTest/ucm126189.htm and
    http://www.fda.gov/BiologicsBloodVaccines/BloodBloodProducts/ApprovedProdu cts/PremarketApprovalsPMAs/ucm089793.htm.

13. Guideline for the Monitoring of Clinical Investigations, found at
    http://www.fda.gov/ICECI/EnforcementActions/BioresearchMonitoring/ucm1350 75.htm

14. "Preparing Notices of Availability of Investigational Medical Devices and for Recruiting Study Subjects" found at
    http://www.fda.gov/MedicalDevices/DeviceRegulationandGuidance/GuidanceDoc uments/ucm073568.htm.

15. International Conference on Harmonization: "Good Clinical Practice" Guideline published in the Federal Register Vol.62, No.90, May 9, 1997, pp. 25691-25709, found at
    http://www.fda.gov/downloads/RegulatoryInformation/Guidances/UCM129515.pd f.

16. "Early Collaboration Meetings Under the FDA Modernization Act (FDMA); Final Guidance for Industry and for CDRH Staff," is available at
    http://www.fda.gov/MedicalDevices/DeviceRegulationandGuidance/GuidanceDoc uments/ucm073604.htm

17. 21 CFR 814.15, Research conducted outside of the United States, in Premarket Approval of Medical Devices, found at http://www.accessdata.fda.gov/scripts/cdrh/cfdocs/cfCFR/ShowCFR.cfm?FR=814.15

18. FDA's Good Clinical Practice program website, Comparison of FDA and HHS Human Subject Protection Regulations found at http://www.fda.gov/ScienceResearch/SpecialTopics/RunningClinicalTrials/EducationalMaterials/ucm112910.htm.

19. "Expedited Review of Premarket Submissions for Devices," found at http://www.fda.gov/MedicalDevices/DeviceRegulationandGuidance/GuidanceDocuments/ucm089643.htm

20. 21 CFR Part 50, Protection of Human Subjects, found at http://www.accessdata.fda.gov/scripts/cdrh/cfdocs/cfCFR/CFRSearch.cfm?CFRPart=50

21. 21 CFR Part 56, Institutional Review Boards, found at http://www.accessdata.fda.gov/scripts/cdrh/cfdocs/cfCFR/CFRSearch.cfm?CFRPart=56

22. 21 CFR 50.25, Elements of informed consent, in Protection of Human Subjects, found at http://www.accessdata.fda.gov/scripts/cdrh/cfdocs/cfCFR/ShowCFR.cfm?FR=50.25

23. "Statistical Guidance on Reporting Results from Studies Evaluating Diagnostic Test" found at http://www.fda.gov/MedicalDevices/DeviceRegulationandGuidance/GuidanceDocuments/ucm071148.htm

24. 21 CFR Part 11, Electronic Records; Electronic Signatures, found at http://www.accessdata.fda.gov/scripts/cdrh/cfdocs/cfCFR/ShowCFR.cfm?CFRPart=11.

25. Guidance for Industry: Part 11, Electronic Records; Electronic Signatures – Scope and Application, guidance document found at http://www.fda.gov/downloads/Drugs/GuidanceComplianceRegulatoryInformation/Guidances/ucm072322.pdf.

26. Draft Guidance for Institutional Review Boards, Clinical Investigators, and Sponsors: Exception from Informed Consent Requirements for Emergency Research at http://edocket.access.gpo.gov/2006/E6-14262.htm.

27. The PMA information, found at http://www.fda.gov/MedicalDevices/ProductsandMedicalProcedures/DeviceApprovalsandClearances/PMAApprovals/default.htm

28. Food and Drug Administration Modernization Act of 1997; List of Documents Issued by the Food and Drug Administration That Apply to Medical Devices Regulated by the

Center for Biologics Evaluation and Research (4/26/99; 64 FR20312) found at http://www.fda.gov/ohrms/dockets/98fr/042699d.pdf.

Appendix 1: Regulatory Decision Tree (21 CFR PART 812) for IVD Investigational Studies

Is it a Pre-amendments device (other than transitional) used according to the labeling in effect at the time, or is it a device, determined by FDA as substantially equivalent (SE) to a preamendments device, used according to the labeling reviewed as part of the SE determination?

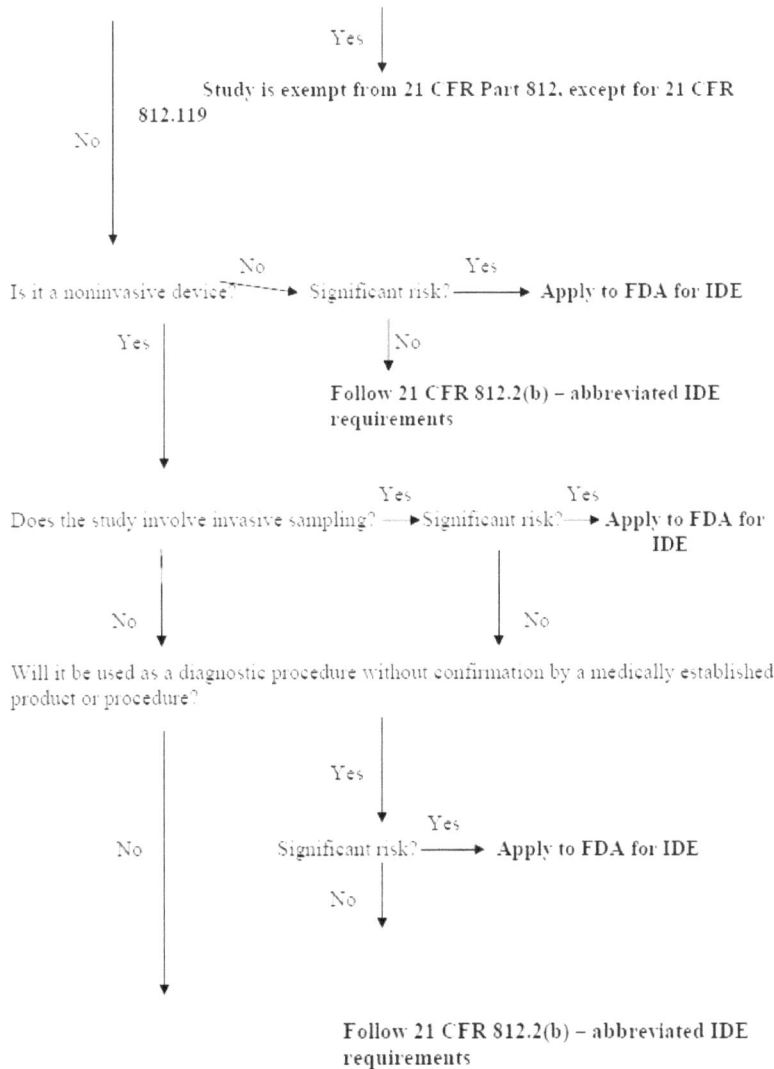

Yes

Study is exempt from 21 CFR Part 812, except for 21 CFR 812.119

No

Is it a noninvasive device? ------No------> Significant risk? ------Yes------> Apply to FDA for IDE

Yes

No

Follow 21 CFR 812.2(b) – abbreviated IDE requirements

Does the study involve invasive sampling? ---Yes---> Significant risk? ---Yes---> Apply to FDA for IDE

No

No

Will it be used as a diagnostic procedure without confirmation by a medically established product or procedure?

Yes

No

Significant risk? ------Yes------> Apply to FDA for IDE

No

Follow 21 CFR 812.2(b) – abbreviated IDE requirements

If the sponsor complies with the applicable requirements of 21 CFR 809.10(c), the study is exempt from 21 CFR Part 812, with the exception of 21 CFR 812.119.

Appendix 2. Table 1. Regulatory Framework for IVD Products Regulated as Devices

	Registration & Listing	Application	Applicable Labeling 801-Subpart A & 801.119 to all	Limitations	QSRs	IRB & Informed Consent	Adverse Event Reporting
General Purpose Reagents (Class I)	No 807.65(c)	No 864.4010(b)	809.10(d)	For Laboratory Use 809.10(d)(1)(iv)	Only 820.180 and 820.198 (if not sterile) 864.4010(b)	N.A.	Yes Part 803
ASR's Class I	Yes 807.20(a)	No 864.4020(b)(1)	809.10(e) 809.30(c)	Restrictions on sale, use, distribution, & reporting 809.30	Yes 809.20(b)	N.A.	Yes Part 803
Class II	Yes 807.20(a)	510(k) 807.81 864.4020(b)(2)					
Class III	Yes 807.20(a)	PMA 814.1 864.4020(b)(3)					
Investigational Use:							
• Significant Risk (SR)	No 812.1	IDE 812.20	812.5 [exempt from 809.10(a) & (b)]	Prohibition on promotion and commercialization 812.7	No (except for 820.30) 812.1 But see 812.20(b)(3)	Yes Parts 50 & 56	For SR and NSR: Investigators report to IRB 812.150(a)(1) Sponsors may need to report to FDA 812.150(b)(1)
• Non-Significant Risk (NSR)	No 812.1	No 812.2(b)	809.10(c)(1)(iii)	For Investigational Use Only. The performance characteristics of this product have not been established 809.10(c)(2)(ii)	No*	Yes Parts 50 & 56	
• Exempt from most requirements of Part 812 (investigational devices are not in commercial distribution) Note: 21 CFR 812.119 applies to all investigational devices	No 807.20(a)	No 812.2(c)(3)			No*		
Research Use	No 807.65(f)	No	809.10(c)(2)(ii)	Research Use Only 809.10(c)(2)(i)	No*	See Parts 50.1 & 56.101 for applicable requirements for clinical research	No 803.19(a)(2)
For non investigational commercially marketed *in vitro* diagnostic devices:							
• Pre-Amendment device	Yes 807.20(a)	No [unless 807.81(a)(3) and unless rule requires PMA. 21 U.S.C. 360e(b)]	809.10 "For In Vitro Diagnostic Use" 809.10(a)(4)	See "Labeling"	Yes 809.20(b)	N.A.	Yes 803.1
• Premarket Notification* (Class I, II, or III)	Yes 807.20(a)	510(k) 807.81 (Exemptions subject to limitations in 862.9, 864.9 & 866.9)			Yes. 809.20(b)		
• PMA PDP (Class III)	Yes 807.20(a)	PMA 814.20			Yes 809.20(b)		
*See 21CFR 862-866 (generic device class regulations) for specific-device exemptions • HDE	Yes 807.20(a)	HUD 814.102 HDE 814.104	809.10 814.104(b)(4)(iii)	Labeling & Cost 814.104(b)(4)(iii) 814.104(b)(5)	Yes 809.20(b)	IRB approval 814.124 [no informed consent]	Yes 814.126(a) 803.1

+ All references in table are to Title 21 of the Code of Federal Regulations.

* While investigational IVDs exempt from most of the provisions of 21 CFR Part 812 are not exempt from the QSR requirements, we generally do not intend to enforce such requirements for investigational IVDs that are exempt from most 21 CFR Part 812 requirements; except for design controls.

Appendix 3: Sponsor's Responsibilities for Significant Risk Device Investigations

Sponsors are required to comply with all applicable duties under the regulations. We summarize them below.

1. General Duties (21 CFR 812.40)

a. Submitting the IDE application to FDA

b. Obtaining both FDA and IRB approval for the investigation

c. Selecting qualified investigators and providing them with the information they need to conduct the investigation properly

d. Ensuring proper monitoring of the investigation

e. Ensuring that any reviewing IRB and FDA are promptly informed of significant new information about an investigation

2. Selection of Investigators (21 CFR 812.43)

a. Assuring selection of investigators qualified by training and experience

b. Permitting only participating investigators to use the investigational device

c. Obtaining a signed investigator's agreement from each investigator containing:

(1) investigator's curriculum vitae

(2) statement of investigator's relevant experience, including dates, location, extent, and type of experience

(3) if an investigator was involved in an investigation or other research that was terminated, an explanation of the circumstances that led to the termination

(4) statement of the investigator's commitment to:

- conduct the investigation in accordance with the agreement, the investigational plan, 21 CFR Part 812 and other applicable regulations, and any conditions of approval imposed by the IRB or FDA

- supervise all testing of the device involving human subjects

- ensure that the requirements for obtaining informed consent are met (21 CFR Part 50)

    d.   Selecting monitor(s) qualified by training and experience to monitor the progress of the investigation in accordance with FDA regulations.

    e.   Providing investigators with the investigational plan and report of prior investigations of the device. (21 CFR 812.45)

3. Monitoring (21 CFR 812.46)

a. S ecuring compliance of all investigators in accordance with the signed investigator's agreement, the investigational plan, the requirements of 21 CFR Part 812 or other applicable FDA regulations, or any condition of approval imposed by the reviewing IRB or FDA. If compliance cannot be secured, use of the device by the investigator and the investigator's participation in the investigation must be discontinued.

    b.   Evaluating all unanticipated adverse device effects and terminating the investigation, or portions of it, as soon as possible if that effect presents an unreasonable risk to subjects (Reporting requirements are listed below.)

    c.   Resuming terminated investigations only after IRB and/or FDA approvals are obtained, as required by this regulation.

4. Controlling Distribution and Disposition of Devices that are Shipped

Although investigators are responsible for ensuring that investigational devices are made available only to persons who are legally authorized to receive them (see 21 CFR 812.110(c)), sponsors also bear responsibility for taking proper measures to ensure that devices are not diverted outside of legally authorized channels. Sponsors may ship investigational devices only to qualified investigators participating in the clinical investigation (21 CFR 812.43(b)). Sponsors must also maintain complete, current, and accurate records pertaining to the shipment and disposition of the investigational device (21 CFR 812.140(b)(2)). Records of shipment shall include the name and address of the consignee, type and quantity of device, date of shipment, and batch number or code mark. Records of disposition shall describe the batch number or code marks of any devices returned to the sponsor, repaired, or disposed of in other ways by the investigator or another person, and the reasons for and method of disposal.

To further ensure compliance with these requirements, sponsors should take appropriate measures to instruct investigators regarding their responsibilities with respect to recordkeeping and device disposition. The specific recordkeeping requirements for investigators are set forth at 21 CFR 812.140(a). Upon completion or termination of a clinical investigation (or the investigator's part of an investigation), or at the sponsor's request, an investigator is required to return to the sponsor any remaining supply of the device or otherwise to dispose of the device as the sponsor directs (21 CFR 812.110(e)).

5. Prohibition of Promotion and Other Practices (21 CFR 812.7)

The IDE regulation prohibits the promotion and commercialization of a device that has not been first cleared or approved for marketing by FDA. This prohibition is applicable to sponsors and investigators (or any person acting on behalf of a sponsor or investigator), and encompasses the following activities:

a. Promotion or test marketing of the investigational device

b. Charging subjects or investigators for the device a price larger than is necessary to recover the costs of manufacture, research, development, and handling

c. Unduly prolonging an investigation beyond the point needed to collect data required to determine whether the device is safe and effective, and

d. Representing that the device is safe or effective for the purposes for which it is being investigated.

6. Supplemental Applications (21 CFR 812.35(a) and (b))

Supplemental applications are required to be submitted to, and approved by, FDA in the following situations:

a. *Changes in the investigational plan:* FDA approval is required for any change that may affect the scientific soundness of the investigation or the rights, safety or welfare of the subjects. IRB approval is also required for changes that may affect the rights, safety or welfare of the subjects. The change in the investigational plan may not be implemented until FDA approval (and IRB approval, if required) is obtained.

b. *Addition of new institutions:* IRB approval is also required for new institutions.

The investigation at the new institution(s) may not begin until both FDA and IRB approval(s) are obtained, and certification of IRB approval is submitted to FDA.

7. Maintaining Records (21 CFR 812.140(b))

A sponsor shall maintain the following accurate, complete, and current records relating to an investigation (also See Table I, next page):

a. Correspondence (including reports) with another sponsor, monitor, investigators, an IRB or FDA

b. Records of any shipment, including:

   (1) name and address of consignee

   (2) type and quantity of device

   (3) date of shipment

   (4) batch numbers or code marks

c. Records of disposition, describing:

   (1) Batch number or code mark of devices returned, repaired, or disposed of by the investigator or other persons

   (2) Reasons for and method of disposal

d. Signed investigator agreements

e. Adverse device effects (whether anticipated or unanticipated) and complaints

f.   Any other records that FDA requires by regulation or by specific requirement for a category of investigation or a particular investigation

Table I: Responsibilities for Maintaining Records for a Significant Risk Device Study

Records	Maintained by Investigator	Maintained by Sponsor
All Correspondence Pertaining to the Investigation	X	X
Shipment. Receipt. Disposition	X	X
Device Administration and Use	X	-
Subject Case Histories	X	-
Informed Consent	X	-
Protocols and Reasons for Deviations from Protocol	X	-
Adverse Device Effects and Complaints	X	X
Signed Investigator Agreements	-	X
Membership Employment/Conflicts of Interest	-	X
Minutes of Meetings	-	-

8. Submitting Reports (21 CFR 812.150(b))

A sponsor shall prepare and submit the following complete, accurate, and timely reports (also see Table II, next page):

a.   Unanticipated adverse device effects (with evaluation) to FDA, all IRBs, and investigators within 10 working days after notification by the investigator.

Subsequent reports on the effect may be required by FDA

b.   Withdrawal of IRB approval

c.   Withdrawal of FDA approval

d.   Current 6-month investigator list

e.   Progress reports (at least annual) - see attached suggested format for IDE progress report

f. Recall and device disposition (within 30 working days after the request was made)

g. Final report - see attached suggested format for progress reports

h. Use of device without obtaining patient informed consent within 5 working days of receipt of notice of such use

i. Significant risk determinations by the IRB when sponsor had proposed nonsignificant risk within 5 working days of receipt of such IRB determination

j. Other reports requested by the IRB or FDA about any aspect of the investigation

Table II: Responsibilities for Preparing and Submitting Reports for Significant Risk Device Studies

Type of Report	Prepared by Investigators for	Prepared by Sponsors for
Unanticipated Adverse Effect Evaluation	Sponsors and IRBs	FDA, IRBs and Investigators
Withdrawal of IDE Approval	Sponsors	FDA, IRBs, and Investigators
Progress Report	Sponsors, Monitors and IRBs	FDA and IRBs
Final Report	Sponsors and IRBs	FDA, IRBs, and Investigators
Emergencies (Protocol Deviations)	Sponsors and IRBs	FDA
Inability to Obtain Informed Consent	Sponsors and IRBs	FDA
Withdrawal of FDA Approval	N A	IRBs and Investigators
Current Investigator List	N A	FDA

Recall and Device Disposition	N A	FDA and IRBs
Records Maintenance Transfer	Sponsors	FDA
Significant Risk Determinations	N A	FDA

9. Inspections (21 CFR 812.145)

Sponsors are required to permit FDA to enter and inspect (at reasonable times and in a reasonable manner) any establishment where devices are held (including any establishment where devices are manufactured, processed, packed, installed, used, or implanted or where records or results from use of devices are kept). FDA may also inspect and copy all records relating to an investigation including, in certain situations, records which identify subjects.

Appendix 4: Investigator's Responsibilities for Significant Risk Device Investigations

Investigators are required to comply with all applicable duties under the regulations. We summarize them below.

1. General Responsibilities of Investigators (21 CFR 812.100)

a. Ensuring that the investigation is conducted according to the signed agreement, the investigational plan and applicable FDA regulations

b. Protecting the rights, safety, and welfare of subjects under the investigator's care

c. Controlling devices under investigation

d. Ensuring that informed consent is obtained from each subject in accordance with 21 CFR Part 50

2. Specific Responsibilities of Investigators (21 CFR 812.110)

a. Awaiting IRB approval and any necessary FDA approval before requesting written informed consent or permitting subject participation

b. Conducting the investigation in accordance with:

    (1) the signed agreement with the sponsor

    (2) the investigational plan

    (3) the regulations set forth in 21 CFR Part 812 and all other applicable FDA regulations, and

    (4) any conditions of approval imposed by an IRB or FDA

c. Supervising the use of the investigational device. An investigator shall permit an investigational device to be used only with subjects under the investigator's supervision. An investigator shall not supply an investigational device to any person not authorized under 21 CFR Part 812 to receive it.

d. Financial disclosure. A clinical investigator shall disclose to the sponsor sufficient accurate financial information to allow the applicant to submit complete and accurate certification or disclosure statements under Part 54.

e. Disposing of the device properly. Upon completion or termination of a clinical investigation or the investigator's part of an investigation, or at the sponsor's request, an investigator shall return to the sponsor any remaining supply of the device or otherwise dispose of the device as the sponsor directs.

3. Maintaining Records (21 CFR 812.140)

An investigator shall maintain the following accurate, complete, and current records relating to the investigator's participation in an investigation:

    a.   Correspondence with another investigator, an IRB, the sponsor, a monitor, or FDA

    b.   Records of receipt, use or disposition of a device that relate to:

        (1)   the type and quantity of the device, dates of receipt, and batch numbers or code marks

        (2)   names of all persons who received, used, or disposed of each device

        (3)   the number of units of the device returned to the sponsor, repaired, or otherwise disposed of, and the reason(s) therefore

    c.   Records of each subject's case history and exposure to the device, including:

        (1)   documents evidencing informed consent and, for any use of a device by the investigator without informed consent, any written concurrence of a licensed physician and a brief description of the circumstances

        (2)   justifying the failure to obtain informed consent

        (3)   document all relevant observations, including records concerning adverse device effects (whether anticipated or not), information and data on the condition of each subject upon entering, and during the course of, the investigation, including information about relevant previous medical history and the results of all diagnostic tests

        (4)   a record of the exposure of each subject to the investigational device, including the date and time of each use, and any other therapy

    d.   The protocol, with documents showing the dates of and reasons for each deviation from the protocol

    e.   Any other records that FDA requires to be maintained by regulation or by specific requirement for a category of investigations or a particular investigation.

4. Inspections (21 CFR 812.145)

Investigators are required to permit FDA to inspect and copy any records pertaining to the investigation including, in certain situations, those which identify subjects.

5. Submitting Reports (21 CFR 812.150)

An investigator shall prepare and submit the following complete, accurate, and timely reports:

    a.   To the sponsor and the IRB:

        (1)   Any unanticipated adverse device effect occurring during an investigation. (Due no later than 10 working days after the investigator first learns of the effect.)

        (2)   Progress reports on the investigation. (These reports must be provided at regular intervals, but in no event less often than yearly. If there is a study monitor, a copy of the report should also be sent to the monitor.)

(3) Any deviation from the investigational plan made to protect the life or physical well-being of a subject in an emergency. (Report is due as soon as possible but no later than 5 working days after the emergency occurs. Except in emergency situations, a protocol deviation requires prior sponsor approval; and if the deviation may affect the scientific soundness of the plan or the rights, safety, or welfare of subjects, prior FDA and IRB approval are required.)

(4) Any use of the device without obtaining informed consent. (Due within 5 working days after such use.)

(5) A final report. (Due within 3 months following termination or completion of the investigation or the investigator's part of the investigation. For additional guidance, see the discussion under the section entitled "Annual Progress Reports and Final Reports.")

(6) Any further information requested by FDA or the IRB about any aspect of the investigation.

b. To the Sponsor:

(1) Withdrawal of IRB approval of the investigator's part of an investigation. (Due within 5 working days of such action).

6. Investigational Device Distribution and Tracking

The IDE regulations prohibit an investigator from providing an investigational device to any person not authorized to receive it (21 CFR 812.110(c)). The best strategy for reducing the risk that an investigational device could be improperly dispensed (whether purposely or inadvertently) is for the sponsor and the investigators to closely monitor the shipping, use, and final disposal of the device(s). Upon completion or termination of a clinical investigation (or the investigator's part of an investigation), or at the sponsor's request, an investigator is required to return to the sponsor any remaining supply of the device or otherwise to dispose of the device as the sponsor directs (21 CFR 812.110(e)). Investigators must also maintain complete, current and accurate records of the receipt, use, or disposition of investigational devices (21 CFR 812.140(a)(2)). Specific investigator recordkeeping requirements are set forth at 21 CFR 812.140(a).

7. Prohibition of Promotion and Other Practices (21 CFR 812.7)

The IDE regulations prohibit the promotion and commercialization of a device that has not been first cleared or approved for marketing by FDA. This prohibition is applicable to sponsors and investigators (or any person acting on behalf of a sponsor or investigator), and encompasses the following activities:

a. Promotion or test marketing of the investigational device

b. Charging subjects or investigators for the device a price larger than is necessary to recover the costs of manufacture, research, development, and handling

    c.   Unduly prolonging an investigation beyond the point needed to collect data required to determine whether the device is safe and effective, and

    d.   Representing that the device is safe or effective for the purposes for which it is being investigated.

8. Annual Progress Reports and Final Reports

The IDE regulations do not specify the content of the annual progress or final reports. With respect to reports to the IRB, the IRB itself may specify what information it wishes to be included in these reports. Because FDA does require the information listed below, it is suggested that, at a minimum, the annual progress and final reports to the sponsor and the IRB also include the following items:

    a.   IDE number

    b.   Device name

    c.   Indications for use

    d.   Brief summary of study progress in relation to investigational plan

    e.   Number of investigators and investigational sites

    f.   Number of subjects enrolled

    g.   Number of devices received, used, and the final disposition of unused devices

    h.   Brief summary of results and conclusions

    i.   Summary of anticipated and unanticipated adverse device effects

    j.   Description of any deviations from investigational plan

    k.   Reprints of any articles published by the investigator in relation to the study

Appendix 5: Suggested Format for IDE Final Report

1. The Basics

    a.   IDE Number

    b.   Device name and indication for use

    c.   Sponsor's name, address and phone number

    d.   Contact person

2. Study Progress (Data from beginning of the study should be reported, unless otherwise indicated.)

    a.   Brief summary of study progress in relation to investigational plan

    b.   Number of investigators/investigational sites (attach list of investigators)

    c.   Number of subjects enrolled (by indication or model)

    d.   Number of devices shipped

    e.   Disposition of all devices shipped

    f.   Brief summary of results

    g.   Summary of anticipated and unanticipated adverse effects

    h.   Description of any deviations from the investigational plan by investigators (since last progress report)

3. Risk Analysis

    a.   Summary of any new adverse information (since last progress report) that may affect the risk analysis; this includes preclinical data, animal studies, foreign data, clinical studies, etc.

    b.   Reprints of any articles published from data collected from this study

4. Other Changes

    a.   Summary of any changes in manufacturing practices and quality control (including changes not reported in a supplemental application)

    b.   Summary of all changes in investigational plan not required to be submitted in a supplemental application

5. Marketing Application or Future Plans

    a.   Progress toward product approval, with date (or projected date) of PMA or 510(k) submission; or indication that marketing of device is not planned.

    b.   Any plans to submit another IDE application for this device or a modification of this device.

General Principles of Software Validation

General Principles of Software Validation, Guidance for Industry and FDA Staff [T, 1]

U.S. Department Of Health and Human Services
Food and Drug Administration
Center for Devices and Radiological Health
Center for Biologics Evaluation and Research

January 11, 2002

Contains Nonbinding Recommendations

The primary purpose of this guidance is to outline general validation principles that the FDA considers applicable to the validation of medical device software or the validation of software used to design, develop, or manufacture medical devices. In addition, it contains useful validation principles applicable to software used in the conduct of clinical trials.

Note: *This guidance represents the Food and Drug Administration's (FDA's) current thinking on this topic. It does not create or confer any rights for or on any person and does not operate to bind FDA or the public. You can use an alternative approach if the approach satisfies the requirements of the applicable statutes and regulations. If you want to discuss an alternative approach, contact the appropriate FDA staff. If you cannot identify the appropriate FDA staff, call the appropriate number listed on the title page of this guidance.*

Section 1. Purpose

This guidance outlines general validation principles that the Food and Drug Administration (FDA) considers to be applicable to the validation of medical device software or the validation of software used to design, develop, or manufacture medical devices. This final guidance document, Version 2.0, supersedes the draft document, *General Principles of Software Validation, Version 1.1*, dated June 9, 1997.

Section 2. Scope

This guidance describes how certain provisions of the medical device Quality System regulation apply to software and the agency's current approach to evaluating a software validation system. For example, this document lists elements that are acceptable to the FDA

[T] Available on the FDA website at: http://www.fda.gov/RegulatoryInformation/Guidances/ucm126954.htm

[1]

for the validation of software; however, it does not list all of the activities and tasks that must, in all instances, be used to comply with the law.

The scope of this guidance is somewhat broader than the scope of validation in the strictest definition of that term. Planning, verification, testing, traceability, configuration management, and many other aspects of good software engineering discussed in this guidance are important activities that together help to support a final conclusion that software is validated.

This guidance recommends an integration of software life cycle management and risk management activities. Based on the intended use and the safety risk associated with the software to be developed, the software developer should determine the specific approach, the combination of techniques to be used, and the level of effort to be applied. While this guidance does not recommend any specific life cycle model or any specific technique or method, it does recommend that software validation and verification activities be conducted throughout the entire software life cycle.

Where the software is developed by someone other than the device manufacturer (e.g., off-the-shelf software) the software developer may not be directly responsible for compliance with FDA regulations. In that case, the party with regulatory responsibility (i.e., the device manufacturer) needs to assess the adequacy of the off-the-shelf software developer's activities and determine what additional efforts are needed to establish that the software is validated for the device manufacturer's intended use.

2.1. Applicability

This guidance applies to:

- Software used as a component, part, or accessory of a medical device;
- Software that is itself a medical device (e.g., blood establishment software);
- Software used in the production of a device (e.g., programmable logic controllers in manufacturing equipment); and
- Software used in implementation of the device manufacturer's quality system (e.g., software that records and maintains the device history record).

This document is based on generally recognized software validation principles and, therefore, can be applied to any software. For FDA purposes, this guidance applies to any software related to a regulated medical device, as defined by Section 201(h) of the Federal Food, Drug, and Cosmetic Act (the Act) and by current FDA software and regulatory policy. This document does not specifically identify which software is or is not regulated.

2.2. Audience

This guidance provides useful information and recommendations to the following individuals:

- Persons subject to the medical device Quality System regulation
- Persons responsible for the design, development, or production of medical device software

- Persons responsible for the design, development, production, or procurement of automated tools used for the design, development, or manufacture of medical devices or software tools used to implement the quality system itself

- FDA Investigators

- FDA Compliance Officers

- FDA Scientific Reviewers

2.3. The Least Burdensome Approach

We believe we should consider the least burdensome approach in all areas of medical device regulation. This guidance reflects our careful review of the relevant scientific and legal requirements and what we believe is the least burdensome way for you to comply with those requirements. However, if you believe that an alternative approach would be less burdensome, please contact us so we can consider your point of view. You may send your written comments to the contact person listed in the preface to this guidance or to the CDRH Ombudsman. Comprehensive information on CDRH's Ombudsman, including ways to contact him, can be found on the Internet at:

http://www.fda.gov/cdrh/resolvingdisputes/ombudsman.html.

2.4. Regulatory Requirements for Software Validation

The FDA's analysis of 3140 medical device recalls conducted between 1992 and 1998 reveals that 242 of them (7.7%) are attributable to software failures. Of those software related recalls, 192 (or 79%) were caused by software defects that were introduced when changes were made to the software after its initial production and distribution. Software validation and other related good software engineering practices discussed in this guidance are a principal means of avoiding such defects and resultant recalls.

Software validation is a requirement of the Quality System regulation, which was published in the Federal Register on October 7, 1996 and took effect on June 1, 1997. (See Title 21 Code of Federal Regulations (CFR) Part 820, and 61 Federal Register (FR) 52602, respectively.) Validation requirements apply to software used as components in medical devices, to software that is itself a medical device, and to software used in production of the device or in implementation of the device manufacturer's quality system.

Unless specifically exempted in a classification regulation, any medical device software product developed after June 1, 1997, regardless of its device class, is subject to applicable design control provisions. (See of 21 CFR §820.30.) This requirement includes the completion of current development projects, all new development projects, and all changes made to existing medical device software. Specific requirements for validation of device software are found in 21 CFR §820.30(g). Other design controls, such as planning, input, verification, and reviews, are required for medical device software. (See 21 CFR §820.30.) The corresponding documented results from these activities can provide additional support for a conclusion that medical device software is validated.

Any software used to automate any part of the device production process or any part of the quality system must be validated for its intended use, as required by 21 CFR §820.70(i). This requirement applies to any software used to automate device design, testing, component acceptance, manufacturing, labeling, packaging, distribution, complaint handling, or to automate any other aspect of the quality system.

In addition, computer systems used to create, modify, and maintain electronic records and to manage electronic signatures are also subject to the validation requirements. (See 21 CFR §11.10(a).) Such computer systems must be validated to ensure accuracy, reliability, consistent intended performance, and the ability to discern invalid or altered records.

Software for the above applications may be developed in-house or under contract. However, software is frequently purchased off-the-shelf for a particular intended use. All production and/or quality system software, even if purchased off-the-shelf, should have documented requirements that fully define its intended use, and information against which testing results and other evidence can be compared, to show that the software is validated for its intended use.

The use of off-the-shelf software in automated medical devices and in automated manufacturing and quality system operations is increasing. Off-the-shelf software may have many capabilities, only a few of which are needed by the device manufacturer. Device manufacturers are responsible for the adequacy of the software used in their devices, and used to produce devices. When device manufacturers purchase "off-the-shelf" software, they must ensure that it will perform as intended in their chosen application. For off-the-shelf software used in manufacturing or in the quality system, additional guidance is included in Section 6.3 of this document. For device software, additional useful information may be found in FDA's *Guidance for Industry, FDA Reviewers, and Compliance on Off-The-Shelf Software Use in Medical Devices.*

2.4. Quality System Regulation vs Pre-Market Submissions

This document addresses Quality System regulation issues that involve the implementation of software validation. It provides guidance for the management and control of the software validation process. The management and control of the software validation process should not be confused with any other validation requirements, such as process validation for an automated manufacturing process.

Device manufacturers may use the same procedures and records for compliance with quality system and design control requirements, as well as for pre-market submissions to FDA. This document does not cover any specific safety or efficacy issues related to software validation. Design issues and documentation requirements for pre-market submissions of regulated software are not addressed by this document. Specific issues related to safety and efficacy, and the documentation required in pre-market submissions, should be addressed to the Office of Device Evaluation (ODE), Center for Devices and Radiological Health (CDRH) or to the Office of Blood Research and Review, Center for Biologics Evaluation and Research (CBER). See the references in Appendix A for applicable FDA guidance documents for pre-market submissions.

Section 3. Context for Software Validation

Many people have asked for specific guidance on what FDA expects them to do to ensure compliance with the Quality System regulation with regard to software validation. Information on software validation presented in this document is not new. Validation of software, using the principles and tasks listed in Sections 4 and 5, has been conducted in many segments of the software industry for well over 20 years.

Due to the great variety of medical devices, processes, and manufacturing facilities, it is not possible to state in one document all of the specific validation elements that are applicable. However, a general application of several broad concepts can be used successfully as guidance for software validation. These broad concepts provide an acceptable framework for building a comprehensive approach to software validation. Additional specific information is available from many of the references listed in Appendix A.

3.1. Definitions and Terminology

Unless defined in the Quality System regulation, or otherwise specified below, all other terms used in this guidance are as defined in the current edition of the FDA *Glossary of Computerized System and Software Development Terminology*.

The medical device Quality System regulation (21 CFR 820.3(k)) defines "**establish**" to mean "define, document, and implement." Where it appears in this guidance, the words "establish" and "established" should be interpreted to have this same meaning.

Some definitions found in the medical device Quality System regulation can be confusing when compared to commonly used terminology in the software industry. Examples are requirements, specification, verification, and validation.

3.1.1 Requirements and Specifications

While the Quality System regulation states that design input requirements must be documented, and that specified requirements must be verified, the regulation does not further clarify the distinction between the terms "requirement" and "specification." A **requirement** can be any need or expectation for a system or for its software. Requirements reflect the stated or implied needs of the customer, and may be market-based, contractual, or statutory, as well as an organization's internal requirements. There can be many different kinds of requirements (e.g., design, functional, implementation, interface, performance, or physical requirements). Software requirements are typically derived from the system requirements for those aspects of system functionality that have been allocated to software. Software requirements are typically stated in functional terms and are defined, refined, and updated as a development project progresses. Success in accurately and completely documenting software requirements is a crucial factor in successful validation of the resulting software.

A **specification** is defined as "a document that states requirements." (See 21 CFR §820.3(y).) It may refer to or include drawings, patterns, or other relevant documents and usually indicates the means and the criteria whereby conformity with the requirement can be checked. There are many different kinds of written specifications, e.g., system requirements specification, software requirements specification, software design specification, software test specification, software integration specification, etc. All of these documents establish "specified requirements" and are design outputs for which various forms of verification are necessary.

3.1.2 Verification and Validation

The Quality System regulation is harmonized with ISO 8402:1994, which treats "verification" and "validation" as separate and distinct terms. On the other hand, many software engineering journal articles and textbooks use the terms "verification" and "validation" interchangeably, or in some cases refer to software "verification, validation, and testing (VV&T)" as if it is a single concept, with no distinction among the three terms.

Software verification provides objective evidence that the design outputs of a particular phase of the software development life cycle meet all of the specified requirements for that phase. Software verification looks for consistency, completeness, and correctness of the software and its supporting documentation, as it is being developed, and provides support for a subsequent conclusion that software is validated. Software testing is one of many verification activities intended to confirm that software development output meets its input requirements. Other verification activities include various static and dynamic analyses, code and document inspections, walkthroughs, and other techniques.

Software validation is a part of the design validation for a finished device, but is not separately defined in the Quality System regulation. For purposes of this guidance, FDA considers software validation to be "**confirmation by examination and provision of objective evidence that software specifications conform to user needs and intended uses, and that the particular requirements implemented through software can be consistently fulfilled.**" In practice, software validation activities may occur both during, as well as at the end of the software development life cycle to ensure that all requirements have been fulfilled. Since software is usually part of a larger hardware system, the validation of software typically includes evidence that all software requirements have been implemented correctly and completely and are traceable to system requirements. A conclusion that software is validated is highly dependent upon comprehensive software testing, inspections, analyses, and other verification tasks performed at each stage of the software development life cycle. Testing of device software functionality in a simulated use environment, and user site testing are typically included as components of an overall design validation program for a software automated device.

Software verification and validation are difficult because a developer cannot test forever, and it is hard to know how much evidence is enough. In large measure, software validation is a matter of developing a "level of confidence" that the device meets all requirements and user expectations for the software automated functions and features of the device. Measures such as defects found in specifications documents, estimates of defects remaining, testing coverage, and other techniques are all used to develop an acceptable level of confidence before shipping the product. The level of confidence, and therefore the level of software validation, verification, and testing effort needed, will vary depending upon the safety risk (hazard) posed by the automated functions of the device. Additional guidance regarding safety risk management for software may be found in Section 4 of FDA's *Guidance for the Content of Pre-*

market Submissions for Software Contained in Medical Devices, and in the international standards ISO/IEC 14971-1 and IEC 60601-1-4 referenced in Appendix A.

3.1.3 IQ/OQ/PQ

For many years, both FDA and regulated industry have attempted to understand and define software validation within the context of process validation terminology. For example, industry documents and other FDA validation guidance sometimes describe user site software validation in terms of installation qualification (IQ), operational qualification (OQ) and performance qualification (PQ). Definitions of these terms and additional information regarding IQ/OQ/PQ may be found in FDA's *Guideline on General Principles of Process Validation*, dated May 11, 1987, and in FDA's *Glossary of Computerized System and Software Development Terminology*, dated August 1995.

While IQ/OQ/PQ terminology has served its purpose well and is one of many legitimate ways to organize software validation tasks at the user site, this terminology may not be well understood among many software professionals, and it is not used elsewhere in this document. However, both FDA personnel and device manufacturers need to be aware of these differences in terminology as they ask for and provide information regarding software validation.

3.2. Software Development As Part Of System Design

The decision to implement system functionality using software is one that is typically made during system design. Software requirements are typically derived from the overall system requirements and design for those aspects in the system that are to be implemented using software. There are user needs and intended uses for a finished device, but users typically do not specify whether those requirements are to be met by hardware, software, or some combination of both. Therefore, software validation must be considered within the context of the overall design validation for the system.

A documented requirements specification represents the user's needs and intended uses from which the product is developed. A primary goal of software validation is to then demonstrate that all completed software products comply with all documented software and system requirements. The correctness and completeness of both the system requirements and the software requirements should be addressed as part of the design validation process for the device. Software validation includes confirmation of conformance to all software specifications and confirmation that all software requirements are traceable to the system specifications. Confirmation is an important part of the overall design validation to ensure that all aspects of the medical device conform to user needs and intended uses.

3.3. Software is Different From Hardware

While software shares many of the same engineering tasks as hardware, it has some very important differences. For example:

- The vast majority of software problems are traceable to errors made during the design and development process. While the quality of a hardware product is highly dependent on design, development and manufacture, the quality of a software product is dependent primarily on design and development with a minimum concern for software manufacture. Software manufacturing consists of reproduction that can be easily verified. It is not difficult to manufacture thousands of program copies that function exactly the same as the original; the difficulty comes in getting the original program to meet all specifications.

- One of the most significant features of software is branching, i.e., the ability to execute alternative series of commands, based on differing inputs. This feature is a major contributing factor for another characteristic of software - its complexity. Even short programs can be very complex and difficult to fully understand.

- Typically, testing alone cannot fully verify that software is complete and correct. In addition to testing, other verification techniques and a structured and documented development process should be combined to ensure a comprehensive validation approach.

- Unlike hardware, software is not a physical entity and does not wear out. In fact, software may improve with age, as latent defects are discovered and removed. However, as software is constantly updated and changed, such improvements are sometimes countered by new defects introduced into the software during the change.

- Unlike some hardware failures, software failures occur without advanced warning. The software's branching that allows it to follow differing paths during execution, may hide some latent defects until long after a software product has been introduced into the marketplace.

- Another related characteristic of software is the speed and ease with which it can be changed. This factor can cause both software and non-software professionals to believe that software problems can be corrected easily. Combined with a lack of understanding of software, it can lead managers to believe that tightly controlled engineering is not needed as much for software as it is for hardware. In fact, the opposite is true. **Because of its complexity, the development process for software should be even more tightly controlled than for hardware, in order to prevent problems that cannot be easily detected later in the development process.**

- Seemingly insignificant changes in software code can create unexpected and very significant problems elsewhere in the software program. The software development process should be sufficiently well planned, controlled, and documented to detect and correct unexpected results from software changes.

- Given the high demand for software professionals and the highly mobile workforce, the software personnel who make maintenance changes to software may not have been involved in the original software development. Therefore, accurate and thorough documentation is essential.

- Historically, software components have not been as frequently standardized and interchangeable as hardware components. However, medical device software developers are beginning to use component-based development tools and techniques. Object-oriented methodologies and the use of off-the-shelf software components hold promise for faster and less expensive software development. However, component-based approaches require very careful attention during integration. Prior to integration, time is needed to fully define and develop reusable software code and to fully understand the behavior of off-the-shelf components.

For these and other reasons, software engineering needs an even greater level of managerial scrutiny and control than does hardware engineering.

3.4. Benefits of Software Validation

Software validation is a critical tool used to assure the quality of device software and software automated operations. Software validation can increase the usability and reliability of the device, resulting in decreased failure rates, fewer recalls and corrective actions, less risk to patients and users, and reduced liability to device manufacturers. Software validation can also reduce long term costs by making it easier and less costly to reliably modify software and revalidate software changes. Software maintenance can represent a very large percentage of the total cost of software over its entire life cycle. An established comprehensive software validation process helps to reduce the long-term cost of software by reducing the cost of validation for each subsequent release of the software.

3.5 Design Review

Design reviews are documented, comprehensive, and systematic examinations of a design to evaluate the adequacy of the design requirements, to evaluate the capability of the design to meet these requirements, and to identify problems. While there may be many informal technical reviews that occur within the development team during a software project, a formal design review is more structured and includes participation from others outside the development team. Formal design reviews may reference or include results from other formal and informal reviews. Design reviews may be conducted separately for the software, after the software is integrated with the hardware into the system, or both. Design reviews should include examination of development plans, requirements specifications, design specifications, testing plans and procedures, all other documents and activities associated with the project, verification results from each stage of the defined life cycle, and validation results for the overall device.

Design review is a primary tool for managing and evaluating development projects. For example, formal design reviews allow management to confirm that all goals defined in the software validation plan have been achieved. The Quality System regulation requires that at least one formal design review be conducted during the device design process. However, it is recommended that multiple design reviews be conducted (e.g., at the end of each software life cycle activity, in preparation for proceeding to the next activity). Formal design review is especially important at or near the end of the requirements activity, before major resources have been committed to specific design solutions. Problems found at this point can be resolved more easily, save time and money, and reduce the likelihood of missing a critical issue.

Answers to some key questions should be documented during formal design reviews. These include:

- Have the appropriate tasks and expected results, outputs, or products been established for each software life cycle activity?

- Do the tasks and expected results, outputs, or products of each software life cycle activity:

  - Comply with the requirements of other software life cycle activities in terms of correctness, completeness, consistency, and accuracy?

  - Satisfy the standards, practices, and conventions of that activity?

  - Establish a proper basis for initiating tasks for the next software life cycle activity?

Section 4. Principles Of Software Validation

This section lists the general principles that should be considered for the validation of software.

4.1. Requirements

A documented software requirements specification provides a baseline for both validation and verification. The software validation process cannot be completed without an established software requirements specification (Ref: 21 CFR 820.3(z) and (aa) and 820.30(f) and (g)).

4.2. Defect Prevention

Software quality assurance needs to focus on preventing the introduction of defects into the software development process and not on trying to "test quality into" the software code after it is written. Software testing is very limited in its ability to surface all latent defects in software code. For example, the complexity of most software prevents it from being exhaustively tested. **Software testing is a necessary activity. However, in most cases software testing by itself is not sufficient to establish confidence that the software is fit for its intended use.** In order to establish that confidence, software developers should use a mixture of methods and techniques to prevent software errors and to detect software errors that do occur. The "best mix" of methods depends on many factors including the development environment, application, size of project, language, and risk.

4.3. Time and Effort

To build a case that the software is validated requires time and effort. Preparation for software validation should begin early, i.e., during design and development planning and design input. The final conclusion that the software is validated should be based on evidence collected from planned efforts conducted throughout the software lifecycle.

4.4. Software Life Cycle

Software validation takes place within the environment of an established software life cycle. The software life cycle contains software engineering tasks and documentation necessary to support the software validation effort. In addition, the software life cycle contains specific verification and validation tasks that are appropriate for the intended use of the software. This guidance does not recommend any particular life cycle models - only that they should be selected and used for a software development project.

4.5. Plans

The software validation process is defined and controlled through the use of a plan. The software validation plan defines "what" is to be accomplished through the software validation effort. Software validation plans are a significant quality system tool. Software validation plans specify areas such as scope, approach, resources, schedules and the types and extent of activities, tasks, and work items.

4.6. Procedures

The software validation process is executed through the use of procedures. These procedures establish "how" to conduct the software validation effort. The procedures should identify the specific actions or sequence of actions that must be taken to complete individual validation activities, tasks, and work items.

4.7. Software Validation After a Change

Due to the complexity of software, a seemingly small local change may have a significant global system impact. When any change (even a small change) is made to the software, the validation status of the software needs to be re-established. **Whenever software is changed, a validation analysis should be conducted not just for validation of the individual change, but also to determine the extent and impact of that change on the entire software system.** Based on this analysis, the software developer should then conduct an appropriate level of software regression testing to show that unchanged but vulnerable portions of the system have not been adversely affected. Design controls and appropriate regression testing provide the confidence that the software is validated after a software change.

4.8. Validation Coverage

Validation coverage should be based on the software's complexity and safety risk - not on firm size or resource constraints. The selection of validation activities, tasks, and work items should be commensurate with the complexity of the software design and the risk associated with the use of the software for the specified intended use. For lower risk devices, only baseline validation activities may be conducted. As the risk increases additional validation activities should be added to cover the additional risk. Validation documentation should be sufficient to demonstrate that all software validation plans and procedures have been completed successfully.

4.9. Independence of Review

Validation activities should be conducted using the basic quality assurance precept of "independence of review." Self-validation is extremely difficult. When possible, an independent evaluation is always better, especially for higher risk applications. Some firms contract out for a third-party independent verification and validation, but this solution may not always be feasible. Another approach is to assign internal staff members that are not involved in a particular design or its implementation, but who have sufficient knowledge to evaluate the project and conduct the verification and validation activities. Smaller firms may need to be creative in how tasks are organized and assigned in order to maintain internal independence of review.

4.10. Flexibility and Responsibility

Specific implementation of these software validation principles may be quite different from one application to another. The device manufacturer has flexibility in choosing how to apply these validation principles, but retains ultimate responsibility for demonstrating that the software has been validated.

Software is designed, developed, validated, and regulated in a wide spectrum of environments, and for a wide variety of devices with varying levels of risk. FDA regulated medical device applications include software that:

- Is a component, part, or accessory of a medical device;

- Is itself a medical device; or

- Is used in manufacturing, design and development, or other parts of the quality system.

In each environment, software components from many sources may be used to create the application (e.g., in-house developed software, off-the-shelf software, contract software, shareware). In addition, software components come in many different forms (e.g., application software, operating systems, compilers, debuggers, configuration management tools, and many more). The validation of software in these environments can be a complex undertaking; therefore, it is appropriate that all of these software validation principles be considered when designing the software validation process. The resultant software validation process should be commensurate with the safety risk associated with the system, device, or process.

Software validation activities and tasks may be dispersed, occurring at different locations and being conducted by different organizations. However, regardless of the distribution of tasks, contractual relations, source of components, or the development environment, the device manufacturer or specification developer retains ultimate responsibility for ensuring that the software is validated.

Section 5. Activities and Tasks

Software validation is accomplished through a series of activities and tasks that are planned and executed at various stages of the software development life cycle. These tasks may be one time occurrences or may be iterated many times, depending on the life cycle model used and the scope of changes made as the software project progresses.

5.1. Software Life Cycle Activities

This guidance does not recommend the use of any specific software life cycle model. Software developers should establish a software life cycle model that is appropriate for their product and organization. The software life cycle model that is selected should cover the software from its birth to its retirement. Activities in a typical software life cycle model include the following:

- Quality Planning

- System Requirements Definition

- Detailed Software Requirements Specification

- Software Design Specification

- Construction or Coding

- Testing

- Installation

- Operation and Support

- Maintenance

- Retirement

Verification, testing, and other tasks that support software validation occur during each of these activities. A life cycle model organizes these software development activities in various ways and provides a framework for monitoring and controlling the software development project. Several software life cycle models (e.g., waterfall, spiral, rapid prototyping, incremental development, etc.) are defined in FDA's Glossary of Computerized System and Software Development Terminology, dated August 1995. These and many other life cycle models are described in various references listed in Appendix A.

5.2. Typical Tasks Supporting Validation

For each of the software life cycle activities, there are certain "typical" tasks that support a conclusion that the software is validated. However, the specific tasks to be performed, their order of performance, and the iteration and timing of their performance will be dictated by the specific software life cycle model that is selected and the safety risk associated with the software application. For very low risk applications, certain tasks may not be needed at all. However, the software developer should at least consider each of these tasks and should define and document which tasks are or are not appropriate for their specific application. The following discussion is generic and is not intended to prescribe any particular software life cycle model or any particular order in which tasks are to be performed.

5.2.1. Quality Planning

Design and development planning should culminate in a plan that identifies necessary tasks, procedures for anomaly reporting and resolution, necessary resources, and management review requirements, including formal design reviews. A software life cycle model and associated activities should be identified, as well as those tasks necessary for each software life cycle activity. The plan should include:

- The specific tasks for each life cycle activity;
- Enumeration of important quality factors (e.g., reliability, maintainability, and usability);
- Methods and procedures for each task;
- Task acceptance criteria;
- Criteria for defining and documenting outputs in terms that will allow evaluation of their conformance to input requirements;
- Inputs for each task;
- Outputs from each task;
- Roles, resources, and responsibilities for each task;
- Risks and assumptions; and
- Documentation of user needs.

Management must identify and provide the appropriate software development environment and resources. (See 21 CFR §820.20(b)(1) and (2).) Typically, each task requires personnel as well as physical resources. The plan should identify the personnel, the facility and equipment resources for each task, and the role that risk (hazard) management will play. A configuration management plan should be developed that will guide and control multiple parallel development activities and ensure proper communications and documentation. Controls are necessary to ensure positive and correct correspondence among all approved versions of the specifications documents, source code, object code, and test suites that comprise a software system. The controls also should ensure accurate identification of, and access to, the currently approved versions.

Procedures should be created for reporting and resolving software anomalies found through validation or other activities. Management should identify the reports and specify the contents, format, and responsible organizational elements for each report. Procedures also are necessary for the review and approval of software development results, including the responsible organizational elements for such reviews and approvals.

Typical Tasks - Quality Planning

- Risk (Hazard) Management Plan
- Configuration Management Plan
- Software Quality Assurance Plan

- Software Verification and Validation Plan

• Verification and Validation Tasks, and Acceptance Criteria

• Schedule and Resource Allocation (for software verification and validation activities)

• Reporting Requirements

- Formal Design Review Requirements

- Other Technical Review Requirements

• Problem Reporting and Resolution Procedures

• Other Support Activities

5.2.2. Requirements

Requirements development includes the identification, analysis, and documentation of information about the device and its intended use. Areas of special importance include allocation of system functions to hardware/software, operating conditions, user characteristics, potential hazards, and anticipated tasks. In addition, the requirements should state clearly the intended use of the software.

The software requirements specification document should contain a written definition of the software functions. It is not possible to validate software without predetermined and documented software requirements. Typical software requirements specify the following:

• All software system inputs;

• All software system outputs;

• All functions that the software system will perform;

• All performance requirements that the software will meet, (e.g., data throughput, reliability, and timing);

• The definition of all external and user interfaces, as well as any internal software-to-system interfaces;

• How users will interact with the system;

• What constitutes an error and how errors should be handled;

• Required response times;

• The intended operating environment for the software, if this is a design constraint (e.g., hardware platform, operating system);

• All ranges, limits, defaults, and specific values that the software will accept; and

• All safety related requirements, specifications, features, or functions that will be implemented in software.

Software safety requirements are derived from a technical risk management process that is closely integrated with the system requirements development process. Software requirement specifications should identify clearly the potential hazards that can result from a software failure in the system as well as any safety requirements to be implemented in software. The consequences of software failure should be evaluated, along with means of mitigating such failures (e.g., hardware mitigation, defensive programming, etc.). From this analysis, it should be possible to identify the most appropriate measures necessary to prevent harm.

The Quality System regulation requires a mechanism for addressing incomplete, ambiguous, or conflicting requirements. (See 21 CFR 820.30(c).) Each requirement (e.g., hardware, software, user, operator interface, and safety) identified in the software requirements specification should be evaluated for accuracy, completeness, consistency, testability, correctness, and clarity. For example, software requirements should be evaluated to verify that:

- There are no internal inconsistencies among requirements;
- All of the performance requirements for the system have been spelled out;
- Fault tolerance, safety, and security requirements are complete and correct;
- Allocation of software functions is accurate and complete;
- Software requirements are appropriate for the system hazards; and
- All requirements are expressed in terms that are measurable or objectively verifiable.

A software requirements traceability analysis should be conducted to trace software requirements to (and from) system requirements and to risk analysis results. In addition to any other analyses and documentation used to verify software requirements, a formal design review is recommended to confirm that requirements are fully specified and appropriate before extensive software design efforts begin. Requirements can be approved and released incrementally, but care should be taken that interactions and interfaces among software (and hardware) requirements are properly reviewed, analyzed, and controlled.

Typical Tasks - Requirements

- Preliminary Risk Analysis
- Traceability Analysis
  - Software Requirements to System Requirements (and vice versa)
  - Software Requirements to Risk Analysis
- Description of User Characteristics
- Listing of Characteristics and Limitations of Primary and Secondary Memory
- Software Requirements Evaluation
- Software User Interface Requirements Analysis
- System Test Plan Generation
- Acceptance Test Plan Generation
- Ambiguity Review or Analysis

5.2.3. Design

In the design process, the software requirements specification is translated into a logical and physical representation of the software to be implemented. The software design specification is a description of what the software should do and how it should do it. Due to complexity of the project or to enable persons with varying levels of technical responsibilities to clearly understand design information, the design specification may contain both a high level summary of the design and detailed design information. The completed software design specification constrains the programmer/coder to stay within the intent of the agreed upon requirements and design. A complete software design specification will relieve the programmer from the need to make ad hoc design decisions.

The software design needs to address human factors. Use error caused by designs that are either overly complex or contrary to users' intuitive expectations for operation is one of the most persistent and critical problems encountered by FDA. Frequently, the design of the software is a factor in such use errors. Human factors engineering should be woven into the entire design and development process, including the device design requirements, analyses, and tests. Device safety and usability issues should be considered when developing flowcharts, state diagrams, prototyping tools, and test plans. Also, task and function analyses, risk analyses, prototype tests and reviews, and full usability tests should be performed. Participants from the user population should be included when applying these methodologies.

The software design specification should include:

- Software requirements specification, including predetermined criteria for acceptance of the software;

- Software risk analysis;

- Development procedures and coding guidelines (or other programming procedures);

- Systems documentation (e.g., a narrative or a context diagram) that describes the systems context in which the program is intended to function, including the relationship of hardware, software, and the physical environment;

- Hardware to be used;

- Parameters to be measured or recorded;

- Logical structure (including control logic) and logical processing steps (e.g., algorithms);

- Data structures and data flow diagrams;

- Definitions of variables (control and data) and description of where they are used;

- Error, alarm, and warning messages;

- Supporting software (e.g., operating systems, drivers, other application software);

- Communication links (links among internal modules of the software, links with the supporting software, links with the hardware, and links with the user);

- Security measures (both physical and logical security); and

- Any additional constraints not identified in the above elements.

The first four of the elements noted above usually are separate pre-existing documents that are included by reference in the software design specification. Software requirements specification was discussed in the preceding section, as was software risk analysis. Written development procedures serve as a guide to the organization, and written programming procedures serve as a guide to individual programmers. As software cannot be validated without knowledge of the context in which it is intended to function, systems documentation is referenced. If some of the above elements are not included in the software, it may be helpful to future reviewers and maintainers of the software if that is clearly stated (e.g., There are no error messages in this program).

The activities that occur during software design have several purposes. Software design evaluations are conducted to determine if the design is complete, correct, consistent, unambiguous, feasible, and maintainable. Appropriate consideration of software architecture (e.g., modular structure) during design can reduce the magnitude of future validation efforts when software changes are needed. Software design evaluations may include analyses of control flow, data flow, complexity, timing, sizing, memory allocation, criticality analysis, and many other aspects of the design. A traceability analysis should be conducted to verify that the software design implements all of the software requirements. As a technique for identifying where requirements are not sufficient, the traceability analysis should also verify that all aspects of the design are traceable to software requirements. An analysis of communication links should be conducted to evaluate the proposed design with respect to hardware, user, and related software requirements. The software risk analysis should be re-examined to determine whether any additional hazards have been identified and whether any new hazards have been introduced by the design.

At the end of the software design activity, a Formal Design Review should be conducted to verify that the design is correct, consistent, complete, accurate, and testable, before moving to implement the design. Portions of the design can be approved and released incrementally for implementation; but care should be taken that interactions and communication links among various elements are properly reviewed, analyzed, and controlled.

Most software development models will be iterative. This is likely to result in several versions of both the software requirement specification and the software design specification. All approved versions should be archived and controlled in accordance with established configuration management procedures.

Typical Tasks - Design

- Updated Software Risk Analysis

- Traceability Analysis - Design Specification to Software Requirements (and vice versa)

- Software Design Evaluation

- Design Communication Link Analysis

- Module Test Plan Generation

- Integration Test Plan Generation
- Test Design Generation (module, integration, system, and acceptance)

5.2.4. Construction or Coding

Software may be constructed either by coding (i.e., programming) or by assembling together previously coded software components (e.g., from code libraries, off-the-shelf software, etc.) for use in a new application. Coding is the software activity where the detailed design specification is implemented as source code. Coding is the lowest level of abstraction for the software development process. It is the last stage in decomposition of the software requirements where module specifications are translated into a programming language.

Coding usually involves the use of a high-level programming language, but may also entail the use of assembly language (or microcode) for time-critical operations. The source code may be either compiled or interpreted for use on a target hardware platform. Decisions on the selection of programming languages and software build tools (assemblers, linkers, and compilers) should include consideration of the impact on subsequent quality evaluation tasks (e.g., availability of debugging and testing tools for the chosen language). Some compilers offer optional levels and commands for error checking to assist in debugging the code. Different levels of error checking may be used throughout the coding process, and warnings or other messages from the compiler may or may not be recorded. However, at the end of the coding and debugging process, the most rigorous level of error checking is normally used to document what compilation errors still remain in the software. If the most rigorous level of error checking is not used for final translation of the source code, then justification for use of the less rigorous translation error checking should be documented. Also, for the final compilation, there should be documentation of the compilation process and its outcome, including any warnings or other messages from the compiler and their resolution, or justification for the decision to leave issues unresolved.

Firms frequently adopt specific coding guidelines that establish quality policies and procedures related to the software coding process. Source code should be evaluated to verify its compliance with specified coding guidelines. Such guidelines should include coding conventions regarding clarity, style, complexity management, and commenting. Code comments should provide useful and descriptive information for a module, including expected inputs and outputs, variables referenced, expected data types, and operations to be performed. Source code should also be evaluated to verify its compliance with the corresponding detailed design specification. Modules ready for integration and test should have documentation of compliance with coding guidelines and any other applicable quality policies and procedures.

Source code evaluations are often implemented as code inspections and code walkthroughs. Such static analyses provide a very effective means to detect errors before execution of the code. They allow for examination of each error in isolation and can also help in focusing later dynamic testing of the software. Firms may use manual (desk) checking with appropriate controls to ensure consistency and independence. Source code evaluations should be extended to verification of internal linkages between modules and layers (horizontal and vertical interfaces), and compliance with their design specifications. Documentation of the procedures used and the results of source code evaluations should be maintained as part of design verification.

A source code traceability analysis is an important tool to verify that all code is linked to established specifications and established test procedures. A source code traceability analysis should be conducted and documented to verify that:

• Each element of the software design specification has been implemented in code;

• Modules and functions implemented in code can be traced back to an element in the software design specification and to the risk analysis;

• Tests for modules and functions can be traced back to an element in the software design specification and to the risk analysis; and

• Tests for modules and functions can be traced to source code for the same modules and functions.

Typical Tasks - Construction or Coding

• Traceability Analyses

 - Source Code to Design Specification (and vice versa)

 - Test Cases to Source Code and to Design Specification

• Source Code and Source Code Documentation Evaluation

• Source Code Interface Analysis

• Test Procedure and Test Case Generation (module, integration, system, and acceptance)

5.2.5. Testing by the Software Developer

Software testing entails running software products under known conditions with defined inputs and documented outcomes that can be compared to their predefined expectations. It is a time consuming, difficult, and imperfect activity. As such, it requires early planning in order to be effective and efficient.

Test plans and test cases should be created as early in the software development process as feasible. They should identify the schedules, environments, resources (personnel, tools, etc.), methodologies, cases (inputs, procedures, outputs, expected results), documentation, and reporting criteria. The magnitude of effort to be applied throughout the testing process can be linked to complexity, criticality, reliability, and/or safety issues (e.g., requiring functions or modules that produce critical outcomes to be challenged with intensive testing of their fault tolerance features). Descriptions of categories of software and software testing effort appear in the literature, for example:

• NIST Special Publication 500-235, Structured Testing: A Testing Methodology Using the Cyclomatic Complexity Metric;

• NUREG/CR-6293, Verification and Validation Guidelines for High Integrity Systems; and

- IEEE Computer Society Press, Handbook of Software Reliability Engineering.

Software test plans should identify the particular tasks to be conducted at each stage of development and include justification of the level of effort represented by their corresponding completion criteria.

Software testing has limitations that must be recognized and considered when planning the testing of a particular software product. Except for the simplest of programs, software cannot be exhaustively tested. Generally it is not feasible to test a software product with all possible inputs, nor is it possible to test all possible data processing paths that can occur during program execution. There is no one type of testing or testing methodology that can ensure a particular software product has been thoroughly tested. Testing of all program functionality does not mean all of the program has been tested. Testing of all of a program's code does not mean all necessary functionality is present in the program. Testing of all program functionality and all program code does not mean the program is 100% correct! Software testing that finds no errors should not be interpreted to mean that errors do not exist in the software product; it may mean the testing was superficial.

An essential element of a software test case is the expected result. It is the key detail that permits objective evaluation of the actual test result. This necessary testing information is obtained from the corresponding, predefined definition or specification. A software specification document must identify what, when, how, why, etc., is to be achieved with an engineering (i.e., measurable or objectively verifiable) level of detail in order for it to be confirmed through testing. The real effort of effective software testing lies in the definition of what is to be tested rather than in the performance of the test.

A software testing process should be based on principles that foster effective examinations of a software product. Applicable software testing tenets include:

- The expected test outcome is predefined;

- A good test case has a high probability of exposing an error;

- A successful test is one that finds an error;

- There is independence from coding;

- Both application (user) and software (programming) expertise are employed;

- Testers use different tools from coders;

- Examining only the usual case is insufficient;

- Test documentation permits its reuse and an independent confirmation of the pass/fail status of a test outcome during subsequent review.

Once the prerequisite tasks (e.g., code inspection) have been successfully completed, software testing begins. It starts with unit level testing and concludes with system level testing. There may be a distinct integration level of testing. A software product should be challenged with test cases based on its internal structure and with test cases based on its external specification. These tests should provide a thorough and rigorous examination of the software product's compliance with its functional, performance, and interface definitions and requirements.

Code-based testing is also known as structural testing or "white-box" testing. It identifies test cases based on knowledge obtained from the source code, detailed design specification, and other development documents. These test cases challenge the control decisions made by the program; and the program's data structures including configuration tables. Structural testing can identify "dead" code that is never executed when the program is run. Structural testing is accomplished primarily with unit (module) level testing, but can be extended to other levels of software testing.

The level of structural testing can be evaluated using metrics that are designed to show what percentage of the software structure has been evaluated during structural testing. These metrics are typically referred to as "coverage" and are a measure of completeness with respect to test selection criteria. The amount of structural coverage should be commensurate with the level of risk posed by the software. Use of the term "coverage" usually means 100% coverage. For example, if a testing program has achieved "statement coverage," it means that 100% of the statements in the software have been executed at least once. Common structural coverage metrics include:

- **Statement Coverage** - This criteria requires sufficient test cases for each program statement to be executed at least once; however, its achievement is insufficient to provide confidence in a software product's behavior.

- **Decision (Branch) Coverage** - This criteria requires sufficient test cases for each program decision or branch to be executed so that each possible outcome occurs at least once. It is considered to be a minimum level of coverage for most software products, but decision coverage alone is insufficient for high-integrity applications.

- **Condition Coverage** - This criteria requires sufficient test cases for each condition in a program decision to take on all possible outcomes at least once. It differs from branch coverage only when multiple conditions must be evaluated to reach a decision.

- **Multi-Condition Coverage** - This criteria requires sufficient test cases to exercise all possible combinations of conditions in a program decision.

- **Loop Coverage** - This criteria requires sufficient test cases for all program loops to be executed for zero, one, two, and many iterations covering initialization, typical running and termination (boundary) conditions.

- **Path Coverage** - This criteria requires sufficient test cases for each feasible path, basis path, etc., from start to exit of a defined program segment, to be executed at least once. Because of the very large number of possible paths through a software program, path coverage is generally not achievable. The amount of path coverage is normally established based on the risk or criticality of the software under test.

- **Data Flow Coverage** - This criteria requires sufficient test cases for each feasible data flow to be executed at least once. A number of data flow testing strategies are available.

Definition-based or specification-based testing is also known as functional testing or "black-box" testing. It identifies test cases based on the definition of what the software product (whether it be a unit (module) or a complete program) is intended to do. These test cases challenge the intended use or functionality of a program, and the program's internal and external interfaces. Functional testing can be applied at all levels of software testing, from unit to system level testing.

The following types of functional software testing involve generally increasing levels of effort:

- *Normal Case* - Testing with usual inputs is necessary. However, testing a software product only with expected, valid inputs does not thoroughly test that software product. By itself, normal case testing cannot provide sufficient confidence in the dependability of the software product.

- *Output Forcing* - Choosing test inputs to ensure that selected (or all) software outputs are generated by testing.

- *Robustness* - Software testing should demonstrate that a software product behaves correctly when given unexpected, invalid inputs. Methods for identifying a sufficient set of such test cases include Equivalence Class Partitioning, Boundary Value Analysis, and Special Case Identification (Error Guessing). While important and necessary, these techniques do not ensure that all of the most appropriate challenges to a software product have been identified for testing.

- *Combinations of Inputs* - The functional testing methods identified above all emphasize individual or single test inputs. Most software products operate with multiple inputs under their conditions of use. Thorough software product testing should consider the combinations of inputs a software unit or system may encounter during operation. Error guessing can be extended to identify combinations of inputs, but it is an ad hoc technique. Cause-effect graphing is one functional software testing technique that systematically identifies combinations of inputs to a software product for inclusion in test cases.

Functional and structural software test case identification techniques provide specific inputs for testing, rather than random test inputs. One weakness of these techniques is the difficulty in linking structural and functional test completion criteria to a software product's reliability. Advanced software testing methods, such as statistical testing, can be employed to provide further assurance that a software product is dependable. Statistical testing uses randomly generated test data from defined distributions based on an operational profile (e.g., expected use, hazardous use, or malicious use of the software product). Large amounts of test data are generated and can be targeted to cover particular areas or concerns, providing an increased possibility of identifying individual and multiple rare operating conditions that were not anticipated by either the software product's designers or its testers. Statistical testing also provides high structural coverage. It does require a stable software product. Thus, structural and functional testing are prerequisites for statistical testing of a software product.

Another aspect of software testing is the testing of software changes. Changes occur frequently during software development. These changes are the result of 1) debugging that finds an error and it is corrected, 2) new or changed requirements ("requirements creep"), and 3) modified designs as more effective or efficient implementations are found. Once a software product has been baselined (approved), any change to that product should have its own "mini life cycle," including testing. Testing of a changed software product requires additional effort. Not only should it demonstrate that the change was implemented correctly, testing should also demonstrate that the change did not adversely impact other parts of the software product.

Regression analysis and testing are employed to provide assurance that a change has not created problems elsewhere in the software product. Regression analysis is the determination of the impact of a change based on review of the relevant documentation (e.g., software requirements specification, software design specification, source code, test plans, test cases, test scripts, etc.) in order to identify the necessary regression tests to be run. Regression testing is the rerunning of test cases that a program has previously executed correctly and comparing the current result to the previous result in order to detect unintended effects of a software change. Regression analysis and regression testing should also be employed when using integration methods to build a software product to ensure that newly integrated modules do not adversely impact the operation of previously integrated modules.

In order to provide a thorough and rigorous examination of a software product, development testing is typically organized into levels. As an example, a software product's testing can be organized into unit, integration, and system levels of testing.

1) Unit (module or component) level testing focuses on the early examination of sub-program functionality and ensures that functionality not visible at the system level is examined by testing. Unit testing ensures that quality software units are furnished for integration into the finished software product.

2) Integration level testing focuses on the transfer of data and control across a program's internal and external interfaces. External interfaces are those with other software (including operating system software), system hardware, and the users and can be described as communications links.

3) System level testing demonstrates that all specified functionality exists and that the software product is trustworthy. This testing verifies the as-built program's functionality and performance with respect to the requirements for the software product as exhibited on the specified operating platform(s). System level software testing addresses functional concerns and the following elements of a device's software that are related to the intended use(s):

    • Performance issues (e.g., response times, reliability measurements);

    • Responses to stress conditions, e.g., behavior under maximum load, continuous use;

    • Operation of internal and external security features;

    • Effectiveness of recovery procedures, including disaster recovery;

    • Usability;

    • Compatibility with other software products;

    • Behavior in each of the defined hardware configurations; and

    • Accuracy of documentation.

    • Control measures (e.g., a traceability analysis) should be used to ensure that the intended coverage is achieved.

System level testing also exhibits the software product's behavior in the intended operating environment. The location of such testing is dependent upon the software developer's ability to produce the target operating environment(s). Depending upon the circumstances, simulation and/or testing at (potential) customer locations may be utilized. Test plans should identify the controls needed to ensure that the intended coverage is achieved and that proper documentation is prepared when planned system level testing is conducted at sites not directly controlled by the software developer. Also, for a software product that is a medical device or a component of a medical device that is to be used on humans prior to FDA clearance, testing involving human subjects may require an Investigational Device Exemption (IDE) or Institutional Review Board (IRB) approval.

Test procedures, test data, and test results should be documented in a manner permitting objective pass/fail decisions to be reached. They should also be suitable for review and objective decision making subsequent to running the test, and they should be suitable for use in any subsequent regression testing. Errors detected during testing should be logged, classified, reviewed, and resolved prior to release of the software. Software error data that is collected and analyzed during a development life cycle may be used to determine the suitability of the software product for release for commercial distribution. Test reports should comply with the requirements of the corresponding test plans.

Software products that perform useful functions in medical devices or their production are often complex. Software testing tools are frequently used to ensure consistency, thoroughness, and efficiency in the testing of such software products and to fulfill the requirements of the planned testing activities. These tools may include supporting software built in-house to facilitate unit (module) testing and subsequent integration testing (e.g., drivers and stubs) as well as commercial software testing tools. Such tools should have a degree of quality no less than the software product they are used to develop. Appropriate documentation providing evidence of the validation of these software tools for their intended use should be maintained (see section 6 of this guidance).

Typical Tasks - Testing by the Software Developer

- Test Planning
- Structural Test Case Identification
- Functional Test Case Identification
- Traceability Analysis - Testing
    - Unit (Module) Tests to Detailed Design
    - Integration Tests to High Level Design
    - System Tests to Software Requirements
- Unit (Module) Test Execution
- Integration Test Execution
- Functional Test Execution

- System Test Execution

- Acceptance Test Execution

- Test Results Evaluation

- Error Evaluation/Resolution

- Final Test Report

5.2.6. User Site Testing

Testing at the user site is an essential part of software validation. The Quality System regulation requires installation and inspection procedures (including testing where appropriate) as well as documentation of inspection and testing to demonstrate proper installation. (See 21 CFR §820.170.) Likewise, manufacturing equipment must meet specified requirements, and automated systems must be validated for their intended use. (See 21 CFR §820.70(g) and 21 CFR §820.70(i) respectively.)

Terminology regarding user site testing can be confusing. Terms such as beta test, site validation, user acceptance test, installation verification, and installation testing have all been used to describe user site testing. For purposes of this guidance, the term "user site testing" encompasses all of these and any other testing that takes place outside of the developer's controlled environment. This testing should take place at a user's site with the actual hardware and software that will be part of the installed system configuration. The testing is accomplished through either actual or simulated use of the software being tested within the context in which it is intended to function.

Guidance contained here is general in nature and is applicable to any user site testing. However, in some areas (e.g., blood establishment systems) there may be specific site validation issues that need to be considered in the planning of user site testing. Test planners should check with the FDA Center(s) with the corresponding product jurisdiction to determine whether there are any additional regulatory requirements for user site testing.

User site testing should follow a pre-defined written plan with a formal summary of testing and a record of formal acceptance. Documented evidence of all testing procedures, test input data, and test results should be retained.

There should be evidence that hardware and software are installed and configured as specified. Measures should ensure that all system components are exercised during the testing and that the versions of these components are those specified. The testing plan should specify testing throughout the full range of operating conditions and should specify continuation for a sufficient time to allow the system to encounter a wide spectrum of conditions and events in an effort to detect any latent faults that are not apparent during more normal activities.

Some of the evaluations that have been performed earlier by the software developer at the developer's site should be repeated at the site of actual use. These may include tests for a high volume of data, heavy loads or stresses, security, fault testing (avoidance, detection, tolerance, and recovery), error messages, and implementation of safety requirements. The developer may be able to furnish the user with some of the test data sets to be used for this purpose.

In addition to an evaluation of the system's ability to properly perform its intended functions, there should be an evaluation of the ability of the users of the system to understand and correctly interface with it. Operators should be able to perform the intended functions and respond in an appropriate and timely manner to all alarms, warnings, and error messages.

During user site testing, records should be maintained of both proper system performance and any system failures that are encountered. The revision of the system to compensate for faults detected during this user site testing should follow the same procedures and controls as for any other software change.

The developers of the software may or may not be involved in the user site testing. If the developers are involved, they may seamlessly carry over to the user's site the last portions of design-level systems testing. If the developers are not involved, it is all the more important that the user have persons who understand the importance of careful test planning, the definition of expected test results, and the recording of all test outputs.

Typical Tasks - User Site Testing

- Acceptance Test Execution
- Test Results Evaluation
- Error Evaluation/Resolution
- Final Test Report

5.2.7. Maintenance and Software Changes

As applied to software, the term maintenance does not mean the same as when applied to hardware. The operational maintenance of hardware and software are different because their failure/error mechanisms are different. Hardware maintenance typically includes preventive hardware maintenance actions, component replacement, and corrective changes. Software maintenance includes corrective, perfective, and adaptive maintenance but does not include preventive maintenance actions or software component replacement.

Changes made to correct errors and faults in the software are corrective maintenance. Changes made to the software to improve the performance, maintainability, or other attributes of the software system are perfective maintenance. Software changes to make the software system usable in a changed environment are adaptive maintenance.

When changes are made to a software system, either during initial development or during post release maintenance, sufficient regression analysis and testing should be conducted to demonstrate that portions of the software not involved in the change were not adversely impacted. This is in addition to testing that evaluates the correctness of the implemented change(s).

The specific validation effort necessary for each software change is determined by the type of change, the development products affected, and the impact of those products on the operation of the software. Careful and complete documentation of the design structure and interrelationships of various modules, interfaces, etc., can limit the validation effort needed when a change is made. The level of effort needed to fully validate a change is also dependent upon the degree to which validation of the original software was documented and archived. For example, test documentation, test cases, and results of previous verification and validation testing need to be archived if they are to be available for performing subsequent regression testing. Failure to archive this information for later use can significantly increase the level of effort and expense of revalidating the software after a change is made.

In addition to software verification and validation tasks that are part of the standard software development process, the following additional maintenance tasks should be addressed:

- *Software Validation Plan Revision* - For software that was previously validated, the existing software validation plan should be revised to support the validation of the revised software. If no previous software validation plan exists, such a plan should be established to support the validation of the revised software.

- *Anomaly Evaluation* - Software organizations frequently maintain documentation, such as software problem reports that describe software anomalies discovered and the specific corrective action taken to fix each anomaly. Too often, however, mistakes are repeated because software developers do not take the next step to determine the root causes of problems and make the process and procedural changes needed to avoid recurrence of the problem. Software anomalies should be evaluated in terms of their severity and their effects on system operation and safety, but they should also be treated as symptoms of process deficiencies in the quality system. A root cause analysis of anomalies can identify specific quality system deficiencies. Where trends are identified (e.g., recurrence of similar software anomalies), appropriate corrective and preventive actions must be implemented and documented to avoid further recurrence of similar quality problems. (See 21 CFR 820.100.)

- *Problem Identification and Resolution Tracking* - All problems discovered during maintenance of the software should be documented. The resolution of each problem should be tracked to ensure it is fixed, for historical reference, and for trending.

- *Proposed Change Assessment* - All proposed modifications, enhancements, or additions should be assessed to determine the effect each change would have on the system. This information should determine the extent to which verification and/or validation tasks need to be iterated.

- *Task Iteration* - For approved software changes, all necessary verification and validation tasks should be performed to ensure that planned changes are implemented correctly, all documentation is complete and up to date, and no unacceptable changes have occurred in software performance.

- *Documentation Updating* - Documentation should be carefully reviewed to determine which documents have been impacted by a change. All approved documents (e.g., specifications, test procedures, user manuals, etc.) that have been affected should be updated in accordance with configuration management procedures. Specifications should be updated before any maintenance and software changes are made.

Section 6. Validation of Automated Process Equipment and Quality System Software

The Quality System regulation requires that "when computers or automated data processing systems are used as part of production or the quality system, the [device] manufacturer shall validate computer software for its intended use according to an established protocol." (See 21 CFR §820.70(i)). This has been a regulatory requirement of FDA's medical device Good Manufacturing Practice (GMP) regulations since 1978.

In addition to the above validation requirement, computer systems that implement part of a device manufacturer's production processes or quality system (or that are used to create and maintain records required by any other FDA regulation) are subject to the Electronic Records; Electronic Signatures regulation. (See 21 CFR Part 11.) This regulation establishes additional security, data integrity, and validation requirements when records are created or maintained electronically. These additional Part 11 requirements should be carefully considered and included in system requirements and software requirements for any automated record `keeping systems. System validation and software validation should demonstrate that all Part 11 requirements have been met.

Computers and automated equipment are used extensively throughout all aspects of medical device design, laboratory testing and analysis, product inspection and acceptance, production and process control, environmental controls, packaging, labeling, traceability, document control, complaint management, and many other aspects of the quality system. Increasingly, automated plant floor operations can involve extensive use of embedded systems in:

- programmable logic controllers;
- digital function controllers;
- statistical process control;
- supervisory control and data acquisition;
- robotics;
- human-machine interfaces;
- input/output devices; and
- computer operating systems.

Software tools are frequently used to design, build, and test the software that goes into an automated medical device. Many other commercial software applications, such as word processors, spreadsheets, databases, and flowcharting software are used to implement the quality system. All of these applications are subject to the requirement for software validation, but the validation approach used for each application can vary widely.

Whether production or quality system software is developed in-house by the device manufacturer, developed by a contractor, or purchased off-the-shelf, it should be developed using the basic principles outlined elsewhere in this guidance. The device manufacturer has latitude and flexibility in defining how validation of that software will be accomplished, but validation should be a key consideration in deciding how and by whom the software will be developed or from whom it will be purchased. The software developer defines a life cycle model. Validation is typically supported by:

- verifications of the outputs from each stage of that software development life cycle; and

- checking for proper operation of the finished software in the device manufacturer's intended use environment.

6.1. How Much Validation Evidence is Needed?

The level of validation effort should be commensurate with the risk posed by the automated operation. In addition to risk other factors, such as the complexity of the process software and the degree to which the device manufacturer is dependent upon that automated process to produce a safe and effective device, determine the nature and extent of testing needed as part of the validation effort. Documented requirements and risk analysis of the automated process help to define the scope of the evidence needed to show that the software is validated for its intended use. For example, an automated milling machine may require very little testing if the device manufacturer can show that the output of the operation is subsequently fully verified against the specification before release. On the other hand, extensive testing may be needed for:

- a plant-wide electronic record and electronic signature system;

- an automated controller for a sterilization cycle; or

- automated test equipment used for inspection and acceptance of finished circuit boards in a life-sustaining / life-supporting device.

Numerous commercial software applications may be used as part of the quality system (e.g., a spreadsheet or statistical package used for quality system calculations, a graphics package used for trend analysis, or a commercial database used for recording device history records or for complaint management). The extent of validation evidence needed for such software depends on the device manufacturer's documented intended use of that software. For example, a device manufacturer who chooses not to use all the vendor-supplied capabilities of the software only needs to validate those functions that will be used and for which the device manufacturer is dependent upon the software results as part of production or the quality system. However, high risk applications should not be running in the same operating environment with non-validated software functions, even if those software functions are not used. Risk mitigation techniques such as memory partitioning or other approaches to resource protection may need to be considered when high risk applications and lower risk applications are to be used in the same operating environment. When software is upgraded or any changes are made to the software, the device manufacturer should consider how those changes may impact the "used portions" of the software and must reconfirm the validation of those portions of the software that are used. (See 21 CFR §820.70(i).)

6.2. Defined User Requirements

A very important key to software validation is a documented user requirements specification that defines:

- the "intended use" of the software or automated equipment; and
- the extent to which the device manufacturer is dependent upon that software or equipment for production of a quality medical device.

The device manufacturer (user) needs to define the expected operating environment including any required hardware and software configurations, software versions, utilities, etc. The user also needs to:

- document requirements for system performance, quality, error handling, startup, shutdown, security, etc.;
- identify any safety related functions or features, such as sensors, alarms, interlocks, logical processing steps, or command sequences; and
- define objective criteria for determining acceptable performance.

The validation must be conducted in accordance with a documented protocol, and the validation results must also be documented. (See 21 CFR §820.70(i).) Test cases should be documented that will exercise the system to challenge its performance against the pre-determined criteria, especially for its most critical parameters. Test cases should address error and alarm conditions, startup, shutdown, all applicable user functions and operator controls, potential operator errors, maximum and minimum ranges of allowed values, and stress conditions applicable to the intended use of the equipment. The test cases should be executed and the results should be recorded and evaluated to determine whether the results support a conclusion that the software is validated for its intended use.

A device manufacturer may conduct a validation using their own personnel or may depend on a third party such as the equipment/software vendor or a consultant. In any case, the device manufacturer retains the ultimate responsibility for ensuring that the production and quality system software:

- is validated according to a written procedure for the particular intended use; and
- will perform as intended in the chosen application.

The device manufacturer should have documentation including:

- defined user requirements;
- validation protocol used;
- acceptance criteria;
- test cases and results; and
- a validation summary

that objectively confirms that the software is validated for its intended use.

6.3. Validation of Off-The-Shelf Software and Automated Equipment

Most of the automated equipment and systems used by device manufacturers are supplied by third-party vendors and are purchased off-the-shelf (OTS). The device manufacturer is responsible for ensuring that the product development methodologies used by the OTS software developer are appropriate and sufficient for the device manufacturer's intended use of that OTS software. For OTS software and equipment, the device manufacturer may or may not have access to the vendor's software validation documentation. If the vendor can provide information about their system requirements, software requirements, validation process, and the results of their validation, the medical device manufacturer can use that information as a beginning point for their required validation documentation. The vendor's life cycle documentation, such as testing protocols and results, source code, design specification, and requirements specification, can be useful in establishing that the software has been validated. However, such documentation is frequently not available from commercial equipment vendors, or the vendor may refuse to share their proprietary information.

Where possible and depending upon the device risk involved, the device manufacturer should consider auditing the vendor's design and development methodologies used in the construction of the OTS software and should assess the development and validation documentation generated for the OTS software. Such audits can be conducted by the device manufacturer or by a qualified third party. The audit should demonstrate that the vendor's procedures for and results of the verification and validation activities performed the OTS software are appropriate and sufficient for the safety and effectiveness requirements of the medical device to be produced using that software.

Some vendors who are not accustomed to operating in a regulated environment may not have a documented life cycle process that can support the device manufacturer's validation requirement. Other vendors may not permit an audit. Where necessary validation information is not available from the vendor, the device manufacturer will need to perform sufficient system level "black box" testing to establish that the software meets their "user needs and intended uses." For many applications black box testing alone is not sufficient. Depending upon the risk of the device produced, the role of the OTS software in the process, the ability to audit the vendor, and the sufficiency of vendor-supplied information, the use of OTS software or equipment may or may not be appropriate, especially if there are suitable alternatives available. The device manufacturer should also consider the implications (if any) for continued maintenance and support of the OTS software should the vendor terminate their support.

For some off-the-shelf software development tools, such as software compilers, linkers, editors, and operating systems, exhaustive black-box testing by the device manufacturer may be impractical. Without such testing - a key element of the validation effort - it may not be possible to validate these software tools. However, their proper operation may be satisfactorily inferred by other means. For example, compilers are frequently certified by independent third-party testing, and commercial software products may have "bug lists", system requirements and other operational information available from the vendor that can be compared to the device manufacturer's intended use to help focus the "black-box" testing effort. Off-the-shelf operating systems need not be validated as a separate program. However, system-level validation testing of the application software should address all the operating system services used, including maximum loading conditions, file operations, handling of system error conditions, and memory constraints that may be applicable to the intended use of the application program.

For more detailed information, see the production and process software references in Appendix A.

Appendix A - References

Food and Drug Administration References

Design Control Guidance for Medical Device Manufacturers, Center for Devices and Radiological Health, Food and Drug Administration, March 1997.

Do It by Design, An Introduction to Human Factors in Medical Devices, Center for Devices and Radiological Health, Food and Drug Administration, March 1997.

Electronic Records; Electronic Signatures Final Rule, 62 Federal Register 13430 (March 20, 1997).

Glossary of Computerized System and Software Development Terminology, Division of Field Investigations, Office of Regional Operations, Office of Regulatory Affairs, Food and Drug Administration, August 1995.

Guidance for the Content of Pre-market Submissions for Software Contained in Medical Devices, Office of Device Evaluation, Center for Devices and Radiological Health, Food and Drug Administration, May 1998.

Guidance for Industry, FDA Reviewers and Compliance on Off-the-Shelf Software Use in Medical Devices, Office of Device Evaluation, Center for Devices and Radiological Health, Food and Drug Administration, September 1999.

Guideline on General Principles of Process Validation, Center for Drugs and Biologics, & Center For Devices and Radiological Health, Food and Drug Administration, May 1987.

Medical Devices; Current Good Manufacturing Practice (CGMP) Final Rule; Quality System Regulation, 61 Federal Register 52602 (October 7, 1996).

Reviewer Guidance for a Pre-Market Notification Submission for Blood Establishment Computer Software, Center for Biologics Evaluation and Research, Food and Drug Administration, January 1997

Student Manual 1, Course INV545, Computer System Validation, Division of Human Resource Development, Office of Regulatory Affairs, Food and Drug Administration, 1997.

Technical Report, Software Development Activities, Division of Field Investigations, Office of Regional Operations, Office of Regulatory Affairs, Food and Drug Administration, July 1987.

Other Government References

W. Richards Adrion, Martha A. Branstad, John C. Cherniavsky. NBS Special Publication 500-75, Validation, Verification, and Testing of Computer Software, Center for Programming Science and Technology, Institute for Computer Sciences and Technology, National Bureau of Standards, U.S. Department of Commerce, February 1981.

Martha A. Branstad, John C Cherniavsky, W. Richards Adrion, NBS Special Publication 500-56, Validation, Verification, and Testing for the Individual Programmer, Center for Programming Science and Technology, Institute for Computer Sciences and Technology, National Bureau of Standards, U.S. Department of Commerce, February 1980.

J.L. Bryant, N.P. Wilburn, Handbook of Software Quality Assurance Techniques Applicable to the Nuclear Industry, NUREG/CR-4640, U.S. Nuclear Regulatory Commission, 1987.

H. Hecht, et.al., Verification and Validation Guidelines for High Integrity Systems. NUREG/CR-6293. Prepared for U.S. Nuclear Regulatory Commission, 1995.

H. Hecht, et.al., Review Guidelines on Software Languages for Use in Nuclear Power Plant Safety Systems, Final Report. NUREG/CR-6463. Prepared for U.S. Nuclear Regulatory Commission, 1996.

J.D. Lawrence, W.L. Persons, Survey of Industry Methods for Producing Highly Reliable Software, NUREG/CR-6278, U.S. Nuclear Regulatory Commission, 1994.

J.D. Lawrence, G.G. Preckshot, Design Factors for Safety-Critical Software, NUREG/CR-6294, U.S. Nuclear Regulatory Commission, 1994.

Patricia B. Powell, Editor. NBS Special Publication 500-98, Planning for Software Validation, Verification, and Testing, Center for Programming Science and Technology, Institute for Computer Sciences and Technology, National Bureau of Standards, U.S. Department of Commerce, November 1982.

Patricia B. Powell, Editor. NBS Special Publication 500-93, Software Validation, Verification, and Testing Technique and Tool Reference Guide, Center for Programming Science and Technology, Institute for Computer Sciences and Technology, National Bureau of Standards, U.S. Department of Commerce, September 1982.

Delores R. Wallace, Roger U. Fujii, NIST Special Publication 500-165, Software Verification and Validation: Its Role in Computer Assurance and Its Relationship with Software Project Management Standards, National Computer Systems Laboratory, National Institute of Standards and Technology, U.S. Department of Commerce, September 1995.

Delores R. Wallace, Laura M. Ippolito, D. Richard Kuhn, NIST Special Publication 500-204, High Integrity Software, Standards and Guidelines, Computer Systems Laboratory, National Institute of Standards and Technology, U.S. Department of Commerce, September 1992.

Delores R. Wallace, et.al. NIST Special Publication 500-234, Reference Information for the Software Verification and Validation Process. Computer Systems Laboratory, National Institute of Standards and Technology, U.S. Department of Commerce, March 1996.

Delores R. Wallace, Editor. NIST Special Publication 500-235, Structured Testing: A Testing Methodology Using the Cyclomatic Complexity Metric. Computer Systems Laboratory, National Institute of Standards and Technology, U.S. Department of Commerce, August 1996.

International and National Consensus Standards

ANSI / ANS-10.4-1987, Guidelines for the Verification and Validation of Scientific and Engineering Computer Programs for the Nuclear Industry, American National Standards Institute, 1987.

ANSI / ASQC Standard D1160-1995, Formal Design Reviews, American Society for Quality Control, 1995.

ANSI / UL 1998:1998, Standard for Safety for Software in Programmable Components, Underwriters Laboratories, Inc., 1998.

AS 3563.1-1991, Software Quality Management System, Part 1: Requirements. Published by Standards Australia [Standards Association of Australia], 1 The Crescent, Homebush, NSW 2140.

AS 3563.2-1991, Software Quality Management System, Part 2: Implementation Guide. Published by Standards Australia [Standards Association of Australia], 1 The Crescent, Homebush, NSW 2140.

IEC 60601-1-4:1996, Medical electrical equipment, Part 1: General requirements for safety, 4. Collateral Standard: Programmable electrical medical systems. International Electrotechnical Commission, 1996.

IEC 61506:1997, Industrial process measurement and control - Documentation of application software. International Electrotechnical Commission, 1997.

IEC 61508:1998, Functional safety of electrical/electronic/programmable electronic safety-related systems. International Electrotechnical Commission, 1998.

IEEE Std 1012-1986, Software Verification and Validation Plans, Institute for Electrical and Electronics Engineers, 1986.

IEEE Standards Collection, Software Engineering, Institute of Electrical and Electronics Engineers, Inc., 1994. ISBN 1-55937-442-X.

ISO 8402:1994, Quality management and quality assurance - Vocabulary. International Organization for Standardization, 1994.

ISO 9000-3:1997, Quality management and quality assurance standards - Part 3: Guidelines for the application of ISO 9001:1994 to the development, supply, installation and maintenance of computer software. International Organization for Standardization, 1997.

ISO 9001:1994, Quality systems - Model for quality assurance in design, development, production, installation, and servicing. International Organization for Standardization, 1994.

ISO 13485:1996, Quality systems - Medical devices - Particular requirements for the application of ISO 9001. International Organization for Standardization, 1996.

ISO/IEC 12119:1994, Information technology - Software packages - Quality requirements and testing, Joint Technical Committee ISO/IEC JTC 1, International Organization for Standardization and International Electrotechnical Commission, 1994.

ISO/IEC 12207:1995, Information technology - Software life cycle processes, Joint Technical Committee ISO/IEC JTC 1, Subcommittee SC 7, International Organization for Standardization and International Electrotechnical Commission, 1995.

ISO/IEC 14598:1999, Information technology - Software product evaluation, Joint Technical Committee ISO/IEC JTC 1, Subcommittee SC 7, International Organization for Standardization and International Electrotechnical Commission, 1999.

ISO 14971-1:1998, Medical Devices - Risk Management - Part 1: Application of Risk Analysis. International Organization for Standardization, 1998.

Software Considerations in Airborne Systems and Equipment Certification. Special Committee 167 of RTCA. RTCA Inc., Washington, D.C. Tel: 202-833-9339. Document No. RTCA/DO-178B, December 1992.

Production Process Software References

The Application of the Principles of GLP to Computerized Systems, Environmental Monograph #116, Organization for Economic Cooperation and Development (OECD), 1995.

George J. Grigonis, Jr., Edward J. Subak, Jr., and Michael Wyrick, "Validation Key Practices for Computer Systems Used in Regulated Operations," Pharmaceutical Technology, June 1997.

Guide to Inspection of Computerized Systems in Drug Processing, Reference Materials and Training Aids for Investigators, Division of Drug Quality Compliance, Associate Director for Compliance, Office of Drugs, National Center for Drugs and Biologics, & Division of Field Investigations, Associate Director for Field Support, Executive Director of Regional Operations, Food and Drug Administration, February 1983.

Daniel P. Olivier, "Validating Process Software", FDA Investigator Course: Medical Device Process Validation, Food and Drug Administration.

GAMP Guide For Validation of Automated Systems in Pharmaceutical Manufacture,Version V3.0, Good Automated Manufacturing Practice (GAMP) Forum, March 1998:

Volume 1, Part 1: User Guide; Part 2: Supplier Guide; Volume 2: Best Practice for User and Suppliers.

Technical Report No. 18, Validation of Computer-Related Systems. PDA Committee on Validation of Computer-Related Systems. PDA Journal of Pharmaceutical Science and Technology, Volume 49, Number 1, January-February 1995 Supplement.

Validation Compliance Annual 1995, International Validation Forum, Inc.

General Software Quality References

Boris Beizer, Black Box Testing, Techniques for Functional Testing of Software and Systems, John Wiley & Sons, 1995. ISBN 0-471-12094-4.

Boris Beizer, Software System Testing and Quality Assurance, International Thomson Computer Press, 1996. ISBN 1-85032-821-8.

Boris Beizer, Software Testing Techniques, Second Edition, Van Nostrand Reinhold, 1990. ISBN 0-442-20672-0.

Richard Bender, Writing Testable Requirements, Version 1.0, Bender & Associates, Inc., Larkspur, CA 94777, 1996.

Frederick P. Brooks, Jr., The Mythical Man-Month, Essays on Software Engineering, Addison-Wesley Longman, Anniversary Edition, 1995. ISBN 0-201-83595-9.

Silvana Castano, et.al., Database Security, ACM Press, Addison-Wesley Publishing Company, 1995. ISBN 0-201-59375-0.

Computerized Data Systems for Nonclinical Safety Assessment, Current Concepts and Quality Assurance, Drug Information Association, Maple Glen, PA, September 1988.

M. S. Deutsch, Software Verification and Validation, Realistic Project Approaches, Prentice Hall, 1982.

Robert H. Dunn and Richard S. Ullman, TQM for Computer Software, Second Edition, McGraw-Hill, Inc., 1994. ISBN 0-07-018314-7.

Elfriede Dustin, Jeff Rashka, and John Paul, Automated Software Testing - Introduction, Management and Performance, Addison Wesley Longman, Inc., 1999. ISBN 0-201-43287-0.

Robert G. Ebenau and Susan H. Strauss, Software Inspection Process, McGraw-Hill, 1994. ISBN 0-07-062166-7.

Richard E. Fairley, Software Engineering Concepts, McGraw-Hill Publishing Company, 1985. ISBN 0-07-019902-7.

Michael A. Friedman and Jeffrey M. Voas, Software Assessment - Reliability, Safety, Testability, Wiley-Interscience, John Wiley & Sons Inc., 1995. ISBN 0-471-01009-X.

Tom Gilb, Dorothy Graham, Software Inspection, Addison-Wesley Publishing Company, 1993. ISBN 0-201-63181-4.

Robert B. Grady, Practical Software Metrics for Project Management and Process Improvement, PTR Prentice-Hall Inc., 1992. ISBN 0-13-720384-5.

Les Hatton, Safer C: Developing Software for High-integrity and Safety-critical Systems, McGraw-Hill Book Company, 1994. ISBN 0-07-707640-0.

Janis V. Halvorsen, A Software Requirements Specification Document Model for the Medical Device Industry, Proceedings IEEE SOUTHEASTCON '93, Banking on Technology, April 4th -7th, 1993, Charlotte, North Carolina.

Debra S. Herrmann, Software Safety and Reliability: Techniques, Approaches and Standards of Key Industrial Sectors, IEEE Computer Society, 1999. ISBN 0-7695-0299-7.

Bill Hetzel, The Complete Guide to Software Testing, Second Edition, A Wiley-QED Publication, John Wiley & Sons, Inc., 1988. ISBN 0-471-56567-9.

Watts S. Humphrey, A Discipline for Software Engineering. Addison-Wesley Longman, 1995. ISBN 0-201-54610-8.

Watts S. Humphrey, Managing the Software Process, Addison-Wesley Publishing Company, 1989. ISBN 0-201-18095-2.

Capers Jones, Software Quality, Analysis and Guidelines for Success, International Thomson Computer Press, 1997. ISBN 1-85032-867-6.

J.M. Juran, Frank M. Gryna, Quality Planning and Analysis, Third Edition, , McGraw-Hill, 1993. ISBN 0-07-033183-9.

Stephen H. Kan, Metrics and Models in Software Quality Engineering, Addison-Wesley Publishing Company, 1995. ISBN 0-201-63339-6.

Cem Kaner, Jack Falk, Hung Quoc Nguyen, Testing Computer Software, Second Edition, Vsn Nostrand Reinhold, 1993. ISBN 0-442-01361-2.

Craig Kaplan, Ralph Clark, Victor Tang, Secrets of Software Quality, 40 Innovations from IBM, McGraw-Hill, 1995. ISBN 0-07-911795-3.

Edward Kit, Software Testing in the Real World, Addison-Wesley Longman, 1995. ISBN 0-201-87756-2.

Alan Kusinitz, "Software Validation", Current Issues in Medical Device Quality Systems, Association for the Advancement of Medical Instrumentation, 1997. ISBN 1-57020-075-0.

Nancy G. Leveson, Safeware, System Safety and Computers, Addison-Wesley Publishing Company, 1995. ISBN 0-201-11972-2.

Michael R. Lyu, Editor, Handbook of Software Reliability Engineering, IEEE Computer Society Press, McGraw-Hill, 1996. ISBN 0-07-039400-8.

Steven R. Mallory, Software Development and Quality Assurance for the Healthcare Manufacturing Industries, Interpharm Press,Inc., 1994. ISBN 0-935184-58-9.

Brian Marick, The Craft of Software Testing, Prentice Hall PTR, 1995. ISBN 0-13-177411-5.

Steve McConnell, Rapid Development, Microsoft Press, 1996. ISBN 1-55615-900-5.

Glenford J. Myers, The Art of Software Testing, John Wiley & Sons, 1979. ISBN 0-471-04328-1.

Peter G. Neumann, Computer Related Risks, ACM Press/Addison-Wesley Publishing Co., 1995. ISBN 0-201-55805-X.

Daniel Olivier, Conducting Software Audits, Auditing Software for Conformance to FDA Requirements, Computer Application Specialists, San Diego, CA, 1994.

William Perry, Effective Methods for Software Testing, John Wiley & Sons, Inc. 1995. ISBN 0-471-06097-6.

William E. Perry, Randall W. Rice, Surviving the Top Ten Challenges of Software Testing, Dorset House Publishing, 1997. ISBN 0-932633-38-2.

Roger S. Pressman, Software Engineering, A Practitioner's Approach, Third Edition, McGraw-Hill Inc., 1992. ISBN 0-07-050814-3.

Roger S. Pressman, A Manager's Guide to Software Engineering, McGraw-Hill Inc., 1993 ISBN 0-07-050820-8.

A. P. Sage, J. D. Palmer, Software Systems Engineering, John Wiley & Sons, 1990.

Joc Sanders, Eugene Curran, Software Quality, Addison-Wesley Publishing Co., 1994. ISBN 0-201-63198-9.

Ken Shumate, Marilyn Keller, Software Specification and Design, A Disciplined Approach for Real-Time Systems, John Wiley & Sons, 1992. ISBN 0-471-53296-7.

Dennis D. Smith, Designing Maintainable Software, Springer-Verlag, 1999. ISBN 0-387-98783-5.

Ian Sommerville, Software Engineering, Third Edition, Addison Wesley Publishing Co., 1989. ISBN 0-201-17568-1.

Karl E. Wiegers, Creating a Software Engineering Culture, Dorset House Publishing, 1996. ISBN 0-932633-33-1.

Karl E. Wiegers, Software Inspection, Improving Quality with Software Inspections, Software Development, April 1995, pages 55-64.

Karl E. Wiegers, Software Requirements, Microsoft Press, 1999. ISBN 0-7356-0631-5.

Appendix B - Development Team

Center for Devices and Radiological Health

Office of Compliance: Stewart Crumpler

Office of Device Evaluation: James Cheng, Donna-Bea Tillman

Office of Health and Industry Programs: Bryan Benesch, Dick Sawyer

Office of Science and Technology: John Murray

Office of Surveillance and Biometrics: Howard Press

Center for Drug Evaluation and Research

Office of Medical Policy: Charles Snipes

Center for Biologics Evaluation and Research

Office of Compliance and Biologics Quality: Alice Godziemski

Office of Regulatory Affairs

Office of Regional Operations: David Bergeson, Joan Loreng

Part V

Selected FDA GCP/Clinical Trial Guidance Documents:
Manufacturing Requirements for Investigational Products

Current Good Manufacturing Practice for Phase 1 Investigational Drugs

Current Good Manufacturing Practice for Phase 1 Investigational Drugs, Guidance for Industry [†, 1]

U.S. Department of Health and Human Services
Food and Drug Administration
Center for Drug Evaluation and Research (CDER)
Center for Biologics Evaluation and Research (CBER)
Office of Regulatory Affairs (ORA)

July 2008

CGMP

Contains Nonbinding Recommendations

This guidance is intended to assist in applying current good manufacturing practice (CGMP) required under section 501(a)(2)(B) of the Federal Food, Drug, and Cosmetic Act (FD&C Act) in the manufacture of most investigational new drugs (IND) used in phase 1 clinical trials. These drugs, which include biological drugs, are exempt from complying with 21 CFR part 211 under 21 CFR 210.2(c) (referred to as phase 1 investigational drugs).

Note: *This guidance represents the Food and Drug Administration's (FDA's) current thinking on this topic. It does not create or confer any rights for or on any person and does not operate to bind FDA or the public. You can use an alternative approach if the approach satisfies the requirements of the applicable statutes and regulations. If you want to discuss an alternative approach, contact the appropriate FDA staff. If you cannot identify the appropriate FDA staff, call the appropriate number listed on the title page of this guidance.*

I. Introduction

This guidance is intended to assist in applying current good manufacturing practice (CGMP) required under section 501(a)(2)(B) of the Federal Food, Drug, and Cosmetic Act (FD&C Act) in the manufacture of most investigational new drugs (IND) used in phase 1 clinical trials.[2] These drugs, which include biological drugs, are exempt from complying with 21 CFR part 211 under 21 CFR 210.2(c) (referred to as phase 1 investigational drugs).

[†] Available on the FDA website at: http://www.fda.gov/downloads/Drugs/GuidanceComplianceRegulatoryInformation/Guidances/UCM070273.pdf

[1] This guidance has been prepared by an Agency working group with representatives from the Center for Drug Evaluation and Research (CDER), Center for Biologics Evaluation and Research (CBER), and the Office of Regulatory Affairs (ORA), at the Food and Drug Administration.

[2] See 21 CFR 312.21(a)

Because a phase 1 clinical trial initially introduces an investigational new drug into human subjects, appropriate CGMP help ensure subject safety. This guidance applies, as part of CGMP, quality control (QC) principles to the manufacture of phase 1 investigational drugs (i.e., interpreting and implementing CGMP consistent with good scientific methodology), which foster CGMP activities that are more appropriate for phase 1 clinical trials, improve the quality of phase 1 investigational drugs, and facilitate the initiation of investigational clinical trials in humans while continuing to protect trial subjects.

This guidance replaces the guidance issued in 1991 titled *Preparation of Investigational New Drug Products (Human and Animal)* (referred to as the 1991 guidance) (Ref. 1) for the manufacture of phase 1 investigational drugs described in this guidance (see section III). However, the 1991 guidance still applies to the manufacture of investigational new products (human and animal) used in phase 2 and phase 3 clinical trials.

The guidance finalizes the draft guidance entitled "INDs—Approaches to Complying with CGMP During Phase 1" dated January 2006; and is being issued concurrently with a final rule that specifies that 21 CFR part 211 no longer applies for most investigational products (see section III), including certain exploratory products (Ref. 2) that are manufactured for use in phase 1 clinical trials. The agency recommends using the approaches outlined in this guidance for complying with § 501(a)(2)(B) of the FD&C Act.

FDA's guidance documents, including this guidance, do not establish legally enforceable responsibilities. Instead, guidances describe the Agency's current thinking on a topic and should be viewed only as recommendations, unless specific regulatory or statutory requirements are cited. The use of the word should in Agency guidances means that something is suggested or recommended, but not required.

II. Background

A. Statutory and Regulatory Requirements

Section 501(a)(2)(B) of the FD&C Act (21 U.S.C. 351 (a)(2)(B)) requires drugs, which include IND products, to comply with current good manufacturing practice as follows:

> A drug...shall be deemed adulterated...if...the methods used in, or the facilities or controls used for, its manufacture, processing, packing, or holding do not conform to or are not operated or administered in conformity with current good manufacturing practice to assure that such drug meets the requirements of this Act as to safety and has the identity and strength, and meets the quality and purity characteristics, which it purports or is represented to possess.

Based on the statutory requirement for manufacturers to follow CGMP, FDA issued CGMP regulations for drug and biological products (see 21 CFR parts 210 and 211). Although FDA stated at the time of issuance that the regulations applied to all types of pharmaceutical production,[3] the preamble to the regulations indicated that FDA was considering proposing additional regulations governing drugs used in investigational clinical trials.

Because certain requirements in part 211, which implement § 501(a)(2)(B) of the FD&C Act, were directed at the commercial manufacture of products typically characterized by large, repetitive, commercial batch production (e.g., those regulations that address validation of manufacturing processes (§ 211.110(a)), and warehousing (§ 211.142)), they may not be appropriate to the manufacture of most investigational drugs used for phase 1 clinical trials.

Section 505(i) of the FD&C Act (21 U.S.C. 355(i)) directs the Secretary of Health and Human Services to promulgate regulations for exempting from the operation of section 505 "drugs intended solely for investigational use by experts qualified by scientific training and experience to investigate the safety and effectiveness of drugs." Based on this statutory mandate, among others, FDA has issued regulations governing IND products to protect human subjects enrolled in clinical trials. For example, in part 312 (21 CFR part 312), sponsors must submit chemistry, manufacturing and control (CMC) information on a drug or biological product as part of an IND application (§ 312.23(a)(7)) (Refs. 1 through 6). FDA reviews the submitted IND to determine whether the phase 1 investigational drug to be used in the clinical trial is sufficiently safe to permit the trial to proceed. This determination is based, in part on whether the investigational product has the identity, strength, quality, and purity, and purported effect described in the IND application. In certain circumstances, FDA also may choose to conduct an inspection (e.g., if there is insufficient information to assess the risks to subjects or if the subjects would be exposed to unreasonable and significant risk). Finally, FDA could decide to place a proposed or ongoing phase 1 clinical trial on clinical hold or terminate the IND. FDA can also take any of these actions if there is evidence of inadequate QC procedures that would compromise the safety of an investigational product.

B. Guidance

The 1991 guidance (reprinted in 1992) did not discuss all manufacturing situations, including, for example, small- or laboratory-scale manufacture of investigational products. In addition, the 1991 guidance did not address fully FDA's expectation for an appropriate approach to manufacturing controls during different phases of investigational product development, which for most products includes a change in manufacturing scale.

This guidance describes an approach manufacturers may use to implement manufacturing controls that are appropriate for the phase 1 clinical trial stage of development. The approach described in this guidance reflects the fact that some manufacturing controls and the extent of manufacturing controls needed to achieve appropriate product quality differ not only between

---

[3] Preamble to the CGMP 1978, comment #49. "The Commissioner finds that, as stated in 211.1, these CGMP regulations apply to the preparation of any drug product for administration to humans or animals, including those still in investigational stages. It is appropriate that the process by which a drug product is manufactured in the development phase be well documented and controlled in order to assure the reproducibility of the product for further testing and for ultimate commercial production. The Commissioner is considering proposing additional CGMP regulations specifically designed to cover drugs in research stages."

investigational and commercial manufacture, but also among the various phases of clinical trials. Consistent with FDA's CGMP for the 21 Century initiative,[4] where applicable, manufacturers are also expected to implement manufacturing controls that reflect product and manufacturing considerations, evolving process and product knowledge, and manufacturing experience.[5]

The 1991 guidance will continue to be relevant for the manufacture of IND investigational drugs for use during phase 2 and 3 clinical trials (Ref. 1) and for the manufacture of non-exempt phase 1 investigational drugs. Phase 2 and 3 manufacturing will continue to be subject to parts 210 and 211.

III. Scope

This guidance applies to investigational new drug and biological products (including finished dosage forms used as placebos) intended for human use during phase 1 development that are subject to CGMP requirements of section 501(a)(2)(B) of the FD&C Act, and are excluded from complying with the CGMP regulations in 21 CFR part 211, by operation of 21 CFR 210.2(c). These include but are not limited to:

- Investigational recombinant and non-recombinant therapeutic products

- Vaccine products

- Allergenic products

- In vivo diagnostics

- Plasma derivative products

- Blood and blood components[6]

- Gene therapy products

- Somatic cellular therapy products (including xenotransplantation products).

This guidance applies to phase 1 investigational drugs whether they are manufactured in small- or large-scale environments because phase 1 clinical trials (21 CFR 312.21(a)) are typically designed to assess tolerability, or feasibility, for further development of a specific drug or biological product. Furthermore, if an investigational drug has already been manufactured by an IND sponsor for use during phase 2 or phase 3 clinical trials or has been lawfully marketed, manufacture of such a drug must comply (21 CFR 211.1) with 21 CFR part 211 for the drug to be used in any subsequent phase 1 clinical trials, irrespective of the trial size or duration of dosing. See 21 CFR 210.2(c).

---

[4] See http://www.fda.gov/cder/gmp/21stcenturysummary.htm.
[5] We are considering issuing additional guidance and/or regulations to clarify FDA's expectations with regard to fulfilling the CGMP requirements when producing investigational drugs for phase 2 and phase 3 clinical trials.
[6] Manufacture of such a drug must comply with the appropriate sections of 21 CFR part 211 for the drug to be used in any subsequent phase 1 clinical trial, irrespective of the trial size or duration of dosing.

This guidance does not apply to the following phase 1 investigational products:

- Human cell or tissue products regulated solely under § 361 of the Public Health Service Act

- Clinical trials for products subject to the device approval or clearance provisions of the FD&C Act

- Investigational products manufactured for phase 2 and phase 3 clinical trials[7]

- Already approved products that are being used during phase 1 clinical trials (e.g., for a new indication)[6]

- Positron Emission Topography (PET) drugs that are subject to § 501(a)(2)(C) of the FD&C Act and/or the new PET CGMP in 21 CFR part 212 when finalized

If you need clarification on the applicability of this guidance to a specific clinical trial, contact the appropriate FDA Center with responsibility for review of the IND.

This guidance is applicable to all manufacturers of phase 1 investigational drugs, including contractors and other specialized service providers as well as IND sponsors who participate in any aspect of manufacturing.

We (FDA) recommend that you use this guidance as a companion to other FDA guidance documents describing the chemistry, manufacturing, and control (CMC) information submitted and reviewed in an IND application for phase 1 clinical trials (Refs. 1 through 6). In many cases, at this stage of development manufacture of the active pharmaceutical ingredient and the phase 1 investigational drug will be accomplished through a series of steps within a single facility. Manufacturers of new active pharmaceutical ingredients (also referred to as API or drug substance) must also conform with CGMP as required in § 501(a)(2)(B) of the FD&C Act. Limited guidance is available on CGMP for the manufacture of new API in some IND products (Ref. 3). Manufacturers of APIs should implement CGMP appropriate to the stage of clinical development and consider the recommendations described in this guidance for the manufacture of APIs used in phase 1 investigational drugs.

IV. General Guidance for Complying With The Statute

This guidance provides recommendations that manufacturers of phase 1 investigational drugs can use to comply with the statutory requirement for CGMP under § 501(a)(2)(B) of the FD&C Act. Manufacturers should also consult other resources, such as literature and technical bulletins for additional detailed information on CGMP that complement the approaches and recommendations in this guidance.

During product development, the quality and safety of phase 1 investigational drugs are maintained, in part, by having appropriate QC procedures in effect. Using established or standardized QC procedures and following appropriate CGMP will also facilitate the manufacture of equivalent or comparable IND product for future clinical trials as needed.

---

[7] Manufacture of such a drug must comply with the appropriate sections of 21 CFR part 211 for the drug to be used in any subsequent phase 1 clinical trial, irrespective of the trial size or duration of dosing.

Adherence to CGMP during manufacture of phase 1 investigational drugs occurs mostly through:

- Well-defined, written procedures

- Adequately controlled equipment and manufacturing environment

- Accurately and consistently recorded data from manufacturing (including testing)

Manufacturers may have acceptable alternatives to meet the objectives described in this guidance. It is the manufacturer's responsibility to provide and use such methods, facilities, and manufacturing controls to ensure that the phase 1 investigational drug meets appropriate standards of safety, identity, strength, quality, and purity. Manufacturers of phase 1 investigational drugs should consider carefully how to best ensure the implementation of standards, practices, and procedures that conform to CGMP for their specific product and manufacturing operation.

In applying appropriate CGMP, we recommend that manufacturers consider carefully the hazards and associated risks from the manufacturing environment that might adversely affect the quality of a phase 1 investigational drug, especially when the phase 1 investigational drug is manufactured in laboratory facilities that are not expressly or solely designed for their manufacture. For example, of particular importance is the susceptibility of a phase 1 investigational drug to contamination or cross contamination with other substances (e.g. chemicals, biologicals, adventitious agents) that may be present from previous or concurrent research or manufacturing activities.

We recommend the following steps to establish the appropriate manufacturing environment for phase 1 investigational drugs:

- A comprehensive and systematic evaluation of the manufacturing setting (i.e., product environment, equipment, process, personnel, materials) to identify potential hazards

- Appropriate actions prior to and during manufacturing to eliminate or mitigate potential hazards to safeguard the quality of the phase 1 investigational drug

Any manufacturing environment, including a laboratory, should have adequate work areas that are properly equipped and controlled for the specific operation(s) needed to manufacture a phase 1 investigational drug. Given the diversity of products and requisite manufacturing operations, not all environments may be acceptable for the manufacture of the specific phase 1 investigational drug under consideration. In these situations, you should use more suitable facilities.

A number of technologies and resources are available that can facilitate conformance with CGMP and streamline product development. Some examples include:

- Use of disposable equipment and process aids to reduce cleaning burden and chances of contamination

- Use of commercial, prepackaged materials (e.g., Water For Injection (WFI), pre-sterilized containers and closures) to eliminate the need for additional equipment or for demonstrating CGMP control of existing equipment

- Use of closed process equipment (i.e., the phase 1 investigational drug is not exposed to the environment during processing) to alleviate the need for stricter room classification for air quality

- Use of contract or shared CGMP manufacturing facilities and testing laboratories (including specialized services). For example, some academic institutions have developed shared manufacturing and testing facilities that can be used by institutional sponsors.

Under CGMP, if a sponsor or manufacturer initiates a contract with another party to perform part or all of the phase 1 investigational drug manufacturing, the sponsor or manufacturer, and contractor are both responsible for assuring that the phase 1 investigational drug is manufactured in compliance with CGMP. This assurance is achieved, in part, by having effective quality control functions (see section V.B). We recommend that the manufacturer or sponsor assess the contractor to ensure that effective quality control functions are in place.

V. Recommended CGMP for Phase 1 Investigational Drugs

Consistent with the FD&C Act (§ 501(a) (2) (B)), CGMP must be in effect for the manufacture of each batch of investigational drug used during phase 1 clinical trials. Manufacturers should establish manufacturing controls based on identified hazards for the manufacturing setting that follow good scientific and QC principles. The following manufacturing controls are applicable to the manufacture of phase 1 investigational drugs and in some specific manufacturing situations. These recommendations provide flexibility to the manufacturers in implementing CGMP controls appropriate to their specific situation and application.

A. Personnel

All personnel should have the education, experience, and training or any combination thereof to enable each individual to perform their assigned function. In particular, personnel should have the appropriate experience to prepare the phase 1 investigational drug and be familiar with QC principles and acceptable methods for complying with the statutory requirement of CGMP, such as the recommendations described in this guidance.

B. QC Function

Every manufacturer should establish a written plan that describes the role of and responsibilities for QC functions.[8] For example, a written plan should provide, at a minimum, for the following functions.

- Responsibility for examining the various materials used in the manufacture of a phase 1 investigational drug (e.g., containers, closures, in-process materials, raw materials, packaging materials, and labeling) to ensure that they are appropriate and meet defined, relevant quality standards

---

[8] For some manufacturers, the Quality Control Function as described in this guidance may be assigned between a quality control and quality assurance group and may be integrated into a more comprehensive quality system.

- Responsibility for review and approval of manufacturing procedures, testing procedures, and acceptance criteria

- Responsibility for releasing or rejecting each batch of phase 1 investigational drug based on a cumulative review of completed manufacturing records and other relevant information (e.g., procedures were followed, product tests performed appropriately, acceptance criteria met)

- Responsibility for investigating unexpected results or errors that occur during manufacturing or from complaints received and initiation of corrective action, if appropriate.

Although quality is the responsibility of all personnel involved in manufacturing, we recommend that you assign an individual(s) to perform QC functions independent of manufacturing responsibilities, especially for the cumulative review and release of phase 1 investigational drug batches.

However, in very limited circumstances and depending on the size and structure of an organization, all QC functions may be performed by the same individual(s) performing manufacturing. For example, in some small operations, it may be necessary to have the same individual perform both manufacturing and QC functions, including release or rejection of each batch. However, in such circumstances, we strongly recommend that another qualified individual not involved in the manufacturing operation conduct an additional periodic review of manufacturing records and other QC activities.

When activities such as testing, commonly performed by dedicated QC personnel in commercial manufacture, are performed by manufacturing personnel in phase 1 studies, adequate controls should be in place (e.g., segregation of testing from manufacturing) so as to not contaminate testing or negatively affect test results.

C. Facility and Equipment

Any facility used for manufacturing phase 1 investigational drugs should have adequate work areas and equipment for the intended task.

Each facility should provide the following described work area and equipment:

- Sufficient space, clean environment, appropriate construction

- Appropriate lighting, ventilation, and heating

- Appropriate cooling, plumbing, washing, and sanitation

- Appropriate equipment to maintain an air cleanliness classification suitable to the operation performed in the area. For example, appropriate air handling systems (e.g., laminar flow hoods) to aid in preventing contamination and cross-contamination of the phase 1 investigational drug.

- Appropriate equipment that will not contaminate the phase 1 investigational drug or otherwise react with, add to, or be absorbed by the phase 1 investigational drug; and that is properly maintained, calibrated, cleaned, and sanitized at appropriate intervals following written procedures.

We recommend that you identify all equipment used for a particular process and document such use in the manufacturing record. You should follow the provisions described under Sterile Products/Aseptically Processed Products (see section VI.C) for phase 1 investigational drugs prepared using aseptic processing.

Use of procedural controls in a facility promotes orderly manufacturing and aids in preventing contamination, cross contamination and mix-ups (see section VI.A).

D. Control of Components, and Containers and Closures

You should establish written procedures describing the handling, review, acceptance, and control of material (i.e., components, containers, closures) used in the manufacture of a phase 1 investigational drug. Materials should be controlled (e.g., segregated, labeled) until you have examined or tested the materials, as appropriate, and released them for use in manufacturing. It is important to handle and store such materials in a manner that prevents degradation or contamination.

The manufacturer should be able to identify and trace all materials used in the manufacture of a phase 1 investigational drug from receipt to use in the manufacture of each batch. We recommend that you keep a record (e.g., log book) containing relevant information on all materials. At a minimum, recorded relevant information would include receipt date, quantity of the shipment, supplier's name, material lot number, storage conditions, and corresponding expiration date.

The manufacturer should establish acceptance criteria for specified attributes on each material. For some materials, all relevant attributes or acceptance criteria may not be known at the phase 1 stage of product development. However, attributes and acceptance criteria selected for assessment should be based on scientific knowledge and experience for use in the specific phase 1 investigational drug. The material attributes and acceptance criteria will be reviewed in the IND application (Refs. 1 through 6).

We recommend that you examine the certificate of analysis (COA) and/or other documentation on each lot of material to ensure that it meets established acceptance criteria for specified attributes. For some (e.g., human and animal derived material), documentation should include information on sourcing and/or test results for adventitious agents, as appropriate. If documentation for a material is incomplete for a specified attribute, we recommend that you test for the incomplete specified attribute of the material. For each batch of the API (or drug substance), you should perform confirmatory identity testing.

E. Manufacturing and Records

The manufacture of phase 1 investigational drugs should follow written manufacturing and process control procedures that provide for the following records.

• A record of manufacturing data that details the materials, equipment, procedures used, and any problems encountered during manufacturing. We recommend that manufacturers retain records sufficient to replicate the manufacturing process. Similarly, if the manufacture of a phase 1 investigational drug batch is initiated but not completed, we recommend that the record include an explanation of why manufacturing was terminated.

- A record of changes in procedures and processes used for subsequent batches along with the rationale for any changes

- A record of the microbiological controls that have been implemented (including written procedures) for the production of sterile-processed phase 1 investigational drugs that are covered by this guidance. You should follow the recommendations for use of aseptic techniques and the control of in-process materials, components, and container closures designed to prevent microbial and endotoxin contamination (see section VI.C).

F. Laboratory Controls

1. Testing

Laboratory tests used in manufacturing (e.g., testing of materials, in-process material, packaging, drug product) should be scientifically sound (e.g., specific, sensitive, and accurate), suitable and reliable for the specified purpose. You should perform tests under controlled conditions and follow written procedures describing the testing methodology. You should maintain records of all test results, procedures, and changes in procedures.

You should perform laboratory testing of the phase 1 investigational drug to evaluate quality attributes including those that define the identity, strength, potency, purity, as appropriate. Specified attributes should be monitored, and acceptance criteria applied appropriately. For known safety-related concerns, specifications should be established and met. For some phase 1 investigational drug attributes, all relevant acceptance criteria may not be known at this stage of development. This information will be reviewed in the IND submission (Refs. 1 through 6).

To ensure reliability of test results, we recommend that you calibrate laboratory equipment at appropriate intervals and maintain the equipment according to established written procedures. We recommend that personnel verify that the equipment is in good working condition when samples are analyzed (e.g., system suitability).

You should retain a representative sample from each batch of phase 1 investigational drug. We recommend retention of both the API and phase 1 investigational drug in containers used in the clinical trials. When feasible, we recommend that the sample consist of a quantity adequate to perform additional testing or investigation if required at a later date (e.g., twice the quantity necessary to conduct release testing, excluding testing for pyrogenicity and sterility). We recommend that you appropriately store and retain the samples for at least two years following clinical trial termination, or withdrawal of the IND application.

2. Stability

We recommend initiation of a stability study using representative samples of the phase 1 investigational drug to monitor the stability and quality of the phase 1 investigational drug during the clinical trial (i.e., date of manufacture through date of last administration).[9]

G. Packaging, Labeling and Distributing

The phase 1 investigational drug should be suitably packaged to protect it from alteration, contamination, and damage during storage, handling, and shipping. You should establish written procedures for controlling packaging, labeling, and distribution operations. We recommend the use of appropriate measures (e.g., product segregation, label reconciliation, verify operations by a second person, confirmatory laboratory testing, QC review) to achieve effective control especially in situations where the potential for mix-ups is more likely (e.g., use of placebo, blinded trials, multiple strengths).

As it relates to phase 1 clinical trials, distribution includes the transport of a phase 1 investigational drug covered by this guidance to clinical investigators. You should handle phase 1 investigational drugs in accordance with labeled conditions (e.g., temperature) to ensure retention of the quality of the product. A distribution record of each batch of phase 1 investigational drug must be sufficiently detailed to allow traceability and facilitate recall of the phase 1 investigational drug if necessary (§ 312.57(a)).[10]

H. Recordkeeping

As indicated in previous sections, manufacturers should keep complete records relating to the quality and operation of the manufacturing processes, including but not limited to:

- Equipment maintenance and calibration
- Manufacturing records and related analytical test records
- Distribution records
- QC functions (as defined in section V.B)
- Component records
- Deviations and investigations
- Complaints

Under § 312.57(c), sponsors must retain records for at least two years after a marketing application is approved for the drug, or if an application is not approved for the drug, until two years after shipment and delivery of the drug for investigational use is discontinued and FDA is notified.

---

[9] IND regulations require information sufficient to assure the drug product's stability during the planned studies (see 21CFR 312.23(a)(7)(iv) (b)).

[10] IND regulation 21 CFR 312.57 governs the retention of all records required by Part 312 (see 21 CFR 312.57).

VI. Special Manufacturing Situations

A. Multi-Product Facilities

We recommend that you manufacture only one phase 1 investigational drug at any given time, in an area or room separate from unrelated activities. However, you could use the same area or room for multiple purposes, including manufacture of other investigational products or laboratory research, provided that appropriate cleaning and procedural controls are in place to ensure that there is no carry-over of materials or products, or mix-ups. In such cases, the design or layout of an area should promote the orderly handling of materials and equipment, the prevention of mix-ups, and the prevention of contamination of equipment or product by substances, previously manufactured products, personnel, or environmental conditions.

Examples of procedural controls could include procedures for clearing the room of previous product materials, product segregation, component segregation, and use of unique product identifiers. We recommend that you periodically evaluate the implemented procedural controls for their effectiveness. You should take appropriate corrective action when indicated by the evaluation or when other events warrant.

B. Biological and Biotechnological Products

1. General Considerations

The manufacturing process is critical to ensure the correct composition, quality, and safety of biological and biotechnology products. For these products, it can be difficult to distinguish changes in quality attributes, or predict the impact of observed changes in quality attributes on safety. This is especially true for phase 1 clinical trials where knowledge and understanding of a phase 1 investigational drug is limited and where comprehensive product characterization is often unavailable, especially for products that are difficult to characterize. Therefore, it is critical to carefully control and record the manufacturing process in conjunction with appropriate testing to reproduce a comparable phase 1 investigational drug as may be necessary. Properly stored retained samples (e.g., API or drug substance, in-process material, phase 1 investigational drug) that can be subsequently analyzed for comparison, can provide important links in reproducing comparable biological and biotechnological products.

You should have in place appropriate equipment and controls in manufacturing to ensure that unit operations with safety-related functions (e.g., viral clearance, virus/toxin attenuation, pasteurization) perform their function with a high degree of assurance. Specific testing may also serve to complement these functions. In manufacturing, you should use testing for safety-related purposes such as viral loads, bioburden, detoxification of bacterial toxins, virus clearance (i.e., removal or inactivation), and removal of residual substances (e.g., antibiotics, chemicals) as appropriate (see section VI.B.2).

2. Adventitious Agent Control

When evaluating the manufacturing environment for biological and biotechnology phase 1 investigational drugs, it is of particular importance to evaluate for susceptibility to contaminate the environment with biological substances, including microbial adventitious agents (e.g., bacterial, viral, mycoplasm), that may remain from previous research or manufacturing activities.

Some biological and biotechnology phase 1 investigational drugs, including those made from pathogenic microorganisms, spore-forming microorganisms, transgenic animals and plants, live viral vaccines, and gene therapy vectors, warrant additional containment considerations. We encourage you to discuss such containment issues with the applicable Center within FDA (i.e., product and facility group with responsibility for the product) prior to engaging in manufacturing.

In addition to the recommendation in section VI.A, multi-product facilities should have in place cleaning and testing procedures that ensure prevention and/or detection of contamination by adventitious agents. To the extent possible, we recommend the use of dedicated equipment and/or disposable parts (e.g., tubing) for this reason. For multi-product areas, you should establish procedures to prevent cross-contamination, and to demonstrate removal of the previously manufactured product from shared equipment and work surfaces, especially if live viral and vector processing occurs in a manufacturing area.

3. Gene Therapy and Cellular Therapy Products

Due to the wide variety and unique manufacturing aspects of investigational gene and cellular therapy products, manufacturers should consider the appropriateness of additional or specialized controls. Although you should manufacture phase 1 investigational cell and gene therapy products following the recommendations in this guidance, we recognize that it may not be possible to follow each recommendation. For example, with some cellular products, it may be impossible to retain samples of the final cellular product due to the limited amounts of material available. Therefore, we recommend that you include your justification for adopting additional controls or alternative approaches to the recommendations in this guidance in the records on the phase 1 investigational drug.

In some cases, investigational gene and cellular therapy products may be manufactured as one batch per subject in phase 1 clinical trials (e.g., gene vector modified autologous cell products, autologous cell products). Manufacture of multiple batches will allow manufacturing and testing information to accumulate in an accelerated manner as compared to more conventional products. As manufacturing methods and assays used for testing can be novel for these products, it is important to monitor manufacturing performance to ensure product safety and quality.

When manufacturing multiple batches of the same phase 1 investigational drug, we recommend that manufacturers periodically conduct and document internal performance reviews. We recommend that this review assess whether the manufacturing process is optimal to ensure overall product quality. Based on the review, appropriate modifications and corrective actions can be taken to control procedures and manufacturing operations.

C. Sterile Products/Aseptically Processed Products

Because product sterility is a critical element of human subject safety, you should take special precautions for phase 1 investigational drugs that are intended to be sterile. You should give thorough consideration to implementing appropriate controls for aseptic processing to ensure a sterile phase 1 investigational drug. The guidance issued by FDA on aseptic processing is a good reference when using aseptic processing (Ref. 7). Particular manufacturing controls include:

- Conducting aseptic manipulation in an aseptic workstation (e.g., laminar air flow workbench, biosafety cabinets, or barrier isolator system) under laminar airflow conditions that meet Class A, ISO 5. You should perform all manipulations of sterile products and materials under aseptic conditions.

- Conducting a process simulation using bacterial growth media to demonstrate that the aseptic processing/controls and production environment are capable of producing a sterile drug

- Performing environmental monitoring of the aseptic workstation during processing to ensure appropriate microbiological control. Such monitoring may include microbial monitoring by settling plates or by active air monitoring.

- Disinfecting the entire aseptic workstation as appropriate to the nature of the operations (e.g., before aseptic manipulation, or between different operations during the same day)

- Ensuring proper workstation installation and placement (e.g., workstations sufficiently separated to allow appropriate airflow) and ensuring that items within a laminar airflow aseptic workstation do not interrupt the unidirectional airflow

- Disinfecting gloves or changing them frequently when working in the laminar flow hood

- Disinfecting the surface of nonsterile items (e.g., test tube rack, and the overwrap for sterile syringes and filters) with sterile disinfectant solution before placing them in the laminar flow hood

- Documenting and following all procedures intended to maintain the sterility of the components, in-process materials, and final phase 1 investigational drug

- Demonstrating that the test article does not interfere with the sterility tests (e.g., USP <71>)

- Employing aseptic technique and control of microbiological impurities in components designed to prevent microbial and endotoxin contamination

- Training personnel in using aseptic techniques

- Verifying that the equipment is suitable for its intended use, i.e., sterilization (e.g., autoclave, hot air oven); performing appropriate calibration of temperature probes used to monitor the sterilization cycle; using suitable and qualified biological indicators; and retaining maintenance and cycle run log records

- Demonstrating that the sterilization method for sterile components and disposable equipment (e.g., filters, bags, containers/ stoppers) is suitable and creating documentation that supports the appropriate use and shelf life of sterile components and equipment

- Ensuring that release of the final phase 1 investigational drug by the QC unit, or designated individual, includes an acceptable review of manufacturing records that demonstrate aseptic procedures and precautions were followed

- Ensuring that final phase 1 investigational drugs are not released until acceptable results of sterility testing are known. We understand that you may have to release a phase 1 investigational drug with a short shelf-life (e.g., radiopharmaceutical, cellular product) based on results from relevant tests (e.g., assessment of sterile filtration by bubble point filter integrity test, cell product — an in-process test result and a negative gram stain, or other rapid microbial detection test and negative endotoxin test on the final product) while results of the sterility test are pending. When you obtain positive results or other relevant results from sterility testing, we recommend that you perform an investigation to determine the cause of contamination followed by corrective action, if warranted. You should notify the person responsible for the associated clinical trials so appropriate action can be taken.

Glossary

Acceptance Criteria means numerical limits, ranges, or other suitable measures of test results necessary to determine acceptance of the drug substance, drug products, or materials at stages of their manufacture.

Active Pharmaceutical Ingredient (API) (or drug substance) means any substance or mixture of substances intended to be used in the manufacture of a drug (medicinal) product and that, when used in the production of a drug, becomes an active ingredient of the drug product. Such substances are intended to furnish pharmacological activity or other direct effect in the diagnosis, cure, mitigation, treatment, or prevention of disease or to affect the structure and function of the body.

Batch means a specific quantity of a drug or other material intended to have uniform character and quality, within specified limits, and is manufactured according to a single manufacturing order during the same cycle of manufacture.

Component means any ingredient intended for use in the manufacture of a drug product, including those that may not appear in such drug product.

Contamination means the undesired introduction of impurities of a chemical or microbiological nature, or of foreign matter, into or onto a raw material, in-process material, or phase 1 investigational drug during manufacturing, sampling, packaging or repackaging, storage, or transport.

Cross-Contamination means contamination of a material or phase 1 investigational drug with another material or product.

Drug product means a finished dosage form (e.g., tablet, capsule, solution) that contains an active drug ingredient generally, but not necessarily, in association with inactive ingredients. The term also includes a finished dosage form that does not contain an active ingredient, but is intended to be used as a placebo.

In-process material means any material fabricated, compounded, blended, or derived by chemical reaction (e.g., intermediate) that is manufactured for, and used in, the preparation of the phase 1 investigational drug.

Phase 1 investigational drug – a new drug or biological drug that is used in phase 1 of a clinical investigation. The term also includes a biological product that is used in vitro for diagnostic purposes.

Multiproduct means more than one approved product, licensed product, IND drug; or separate process.

Manufacture (production) means all operations involved in the preparation of a phase 1 investigational drug from receipt of materials through distribution including processing, storage, packaging, labeling, laboratory testing, and QC.

Manufacturer means a person who takes responsibility for and is involved in any aspect of the manufacture of a phase 1 investigational drug.

Procedural control means manufacturing methodologies executed in such a manner as to prevent or minimize contamination.

Specification means a list of tests, references to analytical procedures, and appropriate acceptance criteria or other criteria for the tests. It establishes the set of criteria to which a drug substance or drug product should conform to be considered acceptable for its intended use. Conformance to specification means that the material, when tested according to the listed analytical procedures, will meet the listed acceptance criteria.

Sponsor means a person who takes responsibility for and initiates a clinical investigation.

References

1. FDA guidance on the "Preparation of Investigational New Drug Products (Human and Animal)" 1991 (reprinted November 1992).

2. FDA "Guidance for Industry, Investigators, and Reviewers: Exploratory IND Studies."

3. FDA "Guidance for Industry: Q7A Good Manufacturing Practice Guidance for Active Pharmaceutical Ingredients," Section 19.

4. FDA "Guidance for Industry: Content and Format of Investigational New Drug Applications (INDs) for Phase 1 Studies of Drugs, Including Well-Characterized, Therapeutic, Biotechnology-derived Products."

5. FDA "Draft Guidance for Industry: Instructions and Template for Chemistry, Manufacturing, and Control (CMC) Reviewers of Human Somatic Cell Therapy Investigational New Drug Applications (INDs)," August 2003.

6. FDA "Draft Guidance for FDA Review Staff and Sponsors: Content and Review of Chemistry, Manufacturing, and Control (CMC) Information for Human Gene Therapy Investigational New Drug Applications (INDs)," November 2004.

7. FDA "Guidance for Industry: Sterile Drug Products Produced by Aseptic Processing – Current Good Manufacturing Practices."

Design Control Guidance for Medical Device Manufacturers

Design Control Guidance for Medical Device Manufacturers [T, 1]

This Guidance relates to FDA 21 CFR 820.30 and Sub-clause 4.4 of ISO 9001

March 11, 1997

Contains Nonbinding Recommendations

This guidance is intended to assist manufacturers in understanding quality system requirements concerning design controls. Assistance is provided by interpreting the language of the quality systems requirements and explaining the underlying concepts in practical terms. As discussed under Device Advice, devices approved under an investigational device exemption (IDE) are exempt from the Quality System (QS) regulation, except for the design control requirements under §820.30. However, the sponsor may state an intention to comply with other parts of the QS regulation. The extent to which the Quality System regulation will be followed in manufacturing the device must be documented in the sponsor's IDE records [§812.140(b)(4)(v)].

Note: *This guidance represents the Food and Drug Administration's (FDA's) current thinking on this topic. It does not create or confer any rights for or on any person and does not operate to bind FDA or the public. You can use an alternative approach if the approach satisfies the requirements of the applicable statutes and regulations. If you want to discuss an alternative approach, contact the appropriate FDA staff. If you cannot identify the appropriate FDA staff, call the appropriate number listed on the title page of this guidance.*

Foreword

To ensure that good quality assurance practices are used for the design of medical devices and that they are consistent with quality system requirements worldwide, the Food and Drug Administration revised the Current Good Manufacturing Practice (CGMP) requirements by incorporating them into the Quality System Regulation, 21 CFR Part 820. An important component of the revision is the addition of design controls.

Because design controls must apply to a wide variety of devices, the regulation does not prescribe the practices that must be used. Instead, it establishes a framework that manufacturers must use when developing and implementing design controls. The framework provides manufacturers with the flexibility needed to develop design controls that both comply with the regulation and are most appropriate for their own design and development processes.

[T] Available on the FDA website at: http://www.fda.gov/downloads/MedicalDevices/ DeviceRegulationandGuidance/GuidanceDocuments/UCM070642.pdf

1

This guidance is intended to assist manufacturers in understanding the intent of the regulation. Design controls are based upon quality assurance and engineering principles. This guidance complements the regulation by describing its intent from a technical perspective using practical terms and examples.

Draft guidance was made publicly available in March, 1996. We appreciate the many comments, suggestions for improvement, and encouragement we received from industry, interested parties, and the Global Harmonization Task Force (GHTF) Study Group 3. The comments were systematically reviewed, and revisions made in response to those comments and suggestions are incorporated in this version. As experience is gained with the guidance, FDA will consider the need for additional revisions within the next six to eighteen months.

The Center publishes the results of its work in scientific journals and in its own technical reports. Through these reports, CDRH also provides assistance to industry and to the medical and healthcare professional communities in complying with the laws and regulations mandated by Congress. These reports are sold by the Government Printing Office (GPO) and by the National Technical Information Service (NTIS). Many reports, including this guidance document, are also available via Internet on the World Wide Web at www.fda.gov.

We welcome your comments and suggestions for future revisions.

D. Bruce Burlington, M.D.
Director
Center for Devices and Radiological Health

Preface

Effective implementation of design controls requires that the regulation and its intent be well understood. The Office of Compliance within CDRH is using several methods to assist manufacturers in developing this understanding. Methods include the use of presentations, teleconferences, practice audits, and written guidance.

Those persons in medical device companies charged with responsibility for developing, implementing, or applying design controls come from a wide variety of technical and non-technical backgrounds—engineering, business administration, life sciences, computer science, and the arts. Therefore, it is important that a tool be provided that conveys the intent of the regulation using practical terminology and examples. That is the purpose of his guidance.

The response of medical device manufacturers and other interested parties to the March, 1996 draft version of this guidance has significantly influenced this latest version. Most comments centered on the complaint that the guidance was too prescriptive. Therefore, it has been rewritten to be more pragmatic, focusing on principles rather than specific practices.

It is noteworthy that many comments offered suggestions for improving the guidance, and that the authors of the comments often acknowledged the value of design controls and the potential benefit of good guidance to the medical device industry, the public, and the FDA. Some comments even included examples of past experiences with the implementation of controls.

Finally, there are several people within CDRH that deserve recognition for their contributions to the development of this guidance. Al Taylor and Bill Midgette of the Office of Science and Technology led the development effort and served as co-chairs of the CDRH Design Control Guidance Team that reviewed the comments received last spring. Team members included Ashley Boulware, Bob Cangelosi, Andrew Lowrey, Deborah Lumbardo, Jack McCracken, Greg O'Connell, and Walter Scott. As the lead person within CDRH with responsibility for implementing the Quality System Regulation, Kim Trautman reviewed the guidance and coordinated its development with the many other concurrent and related activities. Their contributions are gratefully acknowledged.

FDA would also like to acknowledge the significant contributions made by the Global Harmonization Task Force (GHTF) Study Group 3. The Study Group reviewed and revised this guidance at multiple stages during its development. It is hoped that this cooperative effort will lead to this guidance being accepted as an internationally recognized guidance document through the GHTF later this year.

Lillian J. Gill
Director
Office of Compliance

Acknowledgement

FDA wishes to acknowledge the contributions of the Global Harmonization Task Force (GHTF) Study Group 3 to the development of this guidance. As has been stated in the past, FDA is firmly committed to the international harmonization of standards and regulations governing medical devices. The GHTF was formed in 1992 to further this effort. The GHTF includes representatives of the Canadian Ministry of Health and Welfare; the Japanese Ministry of Health and Welfare; FDA; industry members from the European Union, Australia, Canada, Japan, and the United States; and a few delegates from observing countries.

Among other efforts, the GHTF Study Group 3 started developing guidance on the application of design controls to medical devices in the spring of 1995. Study Group 3 has recognized FDA's need to publish timely guidance on this topic in conjunction with promulgation of its new Quality System Regulation. The Study Group has therefore devoted considerable time and effort to combine its draft document with the FDA's efforts as well as to review and comment on FDA's subsequent revisions. FDA, for its part, delayed final release of its guidance pending final review by the Study Group. As a result, it is hoped that this document, with some minor editorial revisions to make the guidance global to several regulatory schemes, will be recognized through the GHTF as an international guidance document.

Introduction

I. Purpose

This guidance is intended to assist manufacturers in understanding quality system requirements concerning design controls. Assistance is provided by interpreting the language of the quality systems requirements and explaining the underlying concepts in practical terms.

Design controls are an interrelated set of practices and procedures that are incorporated into the design and development process, i.e., a system of checks and balances. Design controls make systematic assessment of the design an integral part of development. As a result, deficiencies in design input requirements, and discrepancies between the proposed designs and requirements, are made evident and corrected earlier in the development process. Design controls increase the likelihood that the design transferred to production will translate into a device that is appropriate for its intended use.

In practice, design controls provide managers and designers with improved visibility of the design process. With improved visibility, managers are empowered to more effectively direct the design process—that is, to recognize problems earlier, make corrections, and adjust resource allocations. Designers benefit both by enhanced understanding of the degree of conformance of a design to user and patient needs, and by improved communications and coordination among all participants in the process.

The medical device industry encompasses a wide range of technologies and applications, ranging from simple hand tools to complex computer-controlled surgical machines, from implantable screws to artificial organs, from blood-glucose test strips to diagnostic imaging systems and laboratory test equipment. These devices are manufactured by companies varying in size and structure, methods of design and development, and methods of management. These factors significantly influence how design controls are actually applied. Given this diversity, this guidance does not suggest particular methods of implementation, and therefore, must not be used to assess compliance with the quality system requirements. Rather, the intent is to expand upon the distilled language of the quality system requirements with practical explanations and examples of design control principles. Armed with this basic knowledge, manufacturers can and should seek out technology-specific guidance on applying design controls to their particular situation.

When using this guidance, there could be a tendency to focus only on the time and effort required in developing and incorporating the controls into the design process. However, readers should keep in mind the intrinsic value of design controls as well. It is a well-established fact that the cost to correct design errors is lower when errors are detected early in the design and development process. Large and small companies that have achieved quality systems certification under ISO 9001 cite improvements in productivity, product quality, customer satisfaction, and company competitiveness. Additional benefits are described in comments received from a quality assurance manager of a medical device firm regarding the value of a properly documented design control system:

> "...there are benefits to an organization and the quality improvement of an organization by having a written design control system. By defining this

system on paper, a corporation allows all its employees to understand the requirements, the process, and expectations of design and how the quality of design is assured and perceived by the system. It also provides a baseline to review the system periodically for further improvements based on history, problems, and failures of the system (not the product)."

II. Scope

The guidance applies to the design of medical devices as well as the design of the associated manufacturing processes. The guidance is applicable to new designs as well as modifications or improvements to existing device designs. The guidance discusses subjects in the order in which they appear in FDA's Quality System regulation and is crossreferenced to International Organization for Standards (ISO) 9001:1994, *Quality Systems—Model for Quality Assurance in Design, Development, Production, Installation, and Servicing,* and the ISO draft international standard ISO/DIS 13485, *Quality Systems—Medical Devices—Particular Requirements for the Application of ISO 9001,* dated April 1996.

Design controls are a component of a comprehensive quality system that covers the life of a device. The assurance process is a total systems approach that extends from the development of device requirements through design, production, distribution, use, maintenance, and eventually, obsolescence. Design control begins with development and approval of design inputs, and includes the design of a device and the associated manufacturing processes.

Design control does not end with the transfer of a design to production. Design control applies to all changes to the device or manufacturing process design, including those occurring long after a device has been introduced to the market. This includes evolutionary changes such as performance enhancements as well as revolutionary changes such as corrective actions resulting from the analysis of failed product. The changes are part of a continuous, ongoing effort to design and develop a device that meets the needs of the user and/or patient. Thus, the design control process is revisited many times during the life of a product.

Some tools and techniques are described in the guidance. Although aspects of their utility are sometimes described, they are included in the guidance for illustrative purposes only. Including them does not mean that they are preferred. There may be alternative ways that are better suited to a particular manufacturer and design activity. The literature contains an abundance of information on tools and techniques. Such topics as project management, design review, process capability, and many others referred to in this guidance are available in textbooks, periodicals, and journals. As a manufacturer applies design controls to a particular task, the appropriate tools and techniques used by competent personnel should be applied to meet the needs of the unique product or process for that manufacturer.

III. Application of Design Controls

Design controls may be applied to any product development process. The simple example shown in Figure 1 illustrates the influence of design controls on a design process.

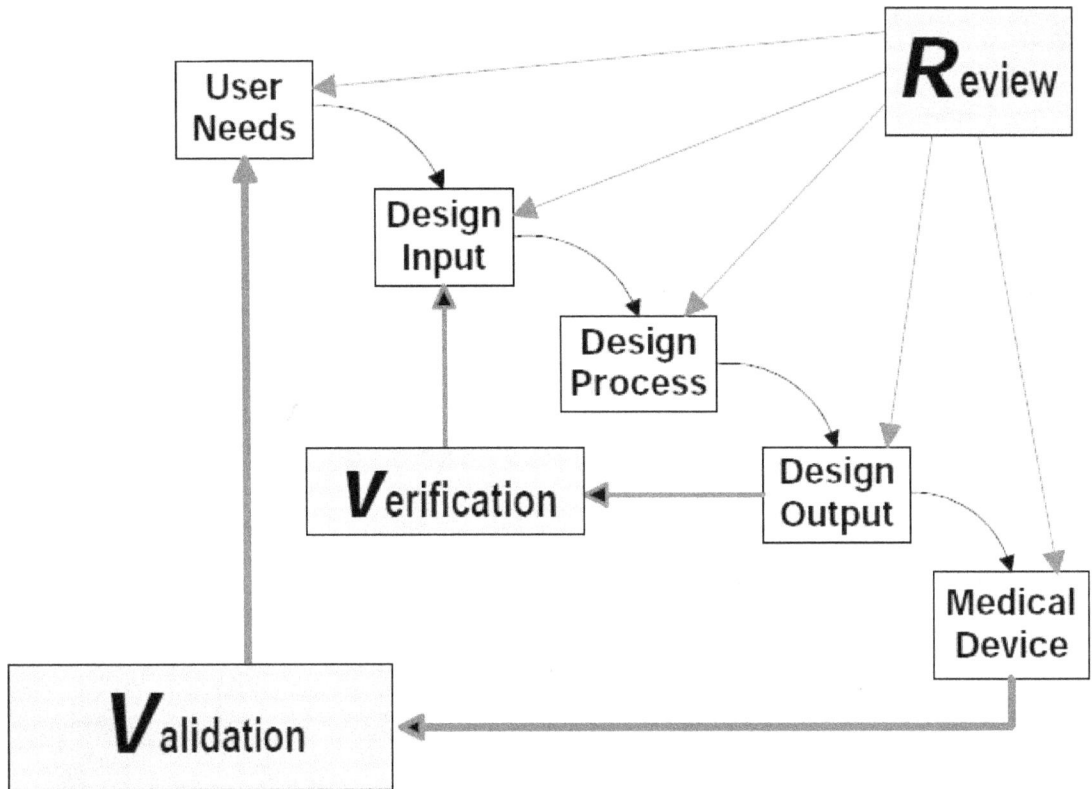

Figure 1 – Application of Design Controls to Waterfall Design Process (figure used with permission of Medical Devices Bureau, Health Canada)

The development process depicted in the example is a traditional waterfall model. The design proceeds in a logical sequence of phases or stages. Basically, requirements are developed, and a device is designed to meet those requirements. The design is then evaluated, transferred to production, and the device is manufactured. In practice, feedback paths would be required between each phase of the process and previous phases, representing the iterative nature of product development. However, this detail has been omitted from the figure to make the influence of the design controls on the design process more distinct.

The importance of the design input and verification of design outputs is illustrated by this example. When the design input has been reviewed and the design input requirements are determined to be acceptable, an iterative process of translating those requirements into a device design begins. The first step is conversion of the requirements into system or highlevel specifications. Thus, these specifications are a design output. Upon verification that the high-level specifications conform to the design input requirements, they become the design input for the next step in the design process, and so on.

This basic technique is used repeatedly throughout the design process. Each design input is converted into a new design output; each output is verified as conforming to its input; and it then becomes the design input for another step in the design process. In this manner, the design input requirements are translated into a device design conforming to those requirements.

The importance of design reviews is also illustrated by the example. The design reviews are conducted at strategic points in the design process. For example, a review is conducted to assure that the design input requirements are adequate before they are converted into the design specifications. Another is used to assure that the device design is adequate before prototypes are produced for simulated use testing and clinical evaluation. Another, a validation review, is conducted prior to transfer of the design to production. Generally, they are used to provide assurance that an activity or phase has been completed in an acceptable manner, and that the next activity or phase can begin.

As the figure illustrates, design validation encompasses verification and extends the assessment to address whether devices produced in accordance with the design actually satisfy user needs and intended uses.

An analogy to automobile design and development may help to clarify these concepts. Fuel efficiency is a common design requirement. This requirement might be expressed as the number of miles-per-gallon of a particular grade of gasoline for a specified set of driving conditions. As the design of the car proceeds, the requirements, including the one for fuel efficiency, are converted into the many layers of system and subsystem specifications needed for design. As these various systems and subsystems are designed, design verification methods are used to establish conformance of each design to its own specifications. Because several specifications directly affect fuel efficiency, many of the verification activities help to provide confirmation that the overall design will meet the fuel efficiency requirement. This might include simulated road testing of prototypes or actual road testing. This is establishing by objective evidence that the design output conforms to the fuel efficiency requirement. However, these verification activities alone are not sufficient to validate the design. The design may be validated when a representative sample of users have driven production vehicles under a specified range of driving conditions and judged the fuel efficiency to be adequate. This is providing objective evidence that the particular requirement for a specific intended use can be consistently fulfilled.

Concurrent Engineering.

Although the waterfall model is a useful tool for introducing design controls, its usefulness in practice is limited. The model does apply to the development of some simpler devices. However, for more complex devices, a concurrent engineering model is more representative of the design processes in use in the industry.

In a traditional waterfall development scenario, the engineering department completes the product design and formally transfers the design to production. Subsequently, other departments or organizations develop processes to manufacture and service the product. Historically, there has frequently been a divergence between the intent of the designer and the reality of the factory floor, resulting in such undesirable outcomes as low manufacturing yields, rework or redesign of the product, or unexpectedly high cost to service the product.

One benefit of concurrent engineering is the involvement of production and service personnel throughout the design process, assuring the mutual optimization of the characteristics of a device and its related processes. While the primary motivations of concurrent engineering are shorter development time and reduced production cost, the practical result is often improved product quality.

Concurrent engineering encompasses a range of practices and techniques. From a design control standpoint, it is sufficient to note that concurrent engineering may blur the line between development and production. On the one hand, the concurrent engineering model properly emphasizes that the development of production processes is a design rather than a manufacturing activity. On the other hand, various components of a design may enter production before the design as a whole has been approved. Thus, concurrent engineering and other more complex models of development usually require a comprehensive matrix of reviews and approvals to ensure that each component and process design is validated prior to entering production, and the product as a whole is validated prior to design release.

Risk Management and Design Controls.

Risk management is the systematic application of management policies, procedures, and practices to the tasks of identifying, analyzing, controlling, and monitoring risk. It is intended to be a framework within which experience, insight, and judgment are applied to successfully manage risk. It is included in this guidance because of its effect on the design process.

Risk management begins with the development of the design input requirements. As the design evolves, new risks may become evident. To systematically identify and, when necessary, reduce these risks, the risk management process is integrated into the design process. In this way, unacceptable risks can be identified and managed earlier in the design process when changes are easier to make and less costly.

An example of this is an exposure control system for a general purpose x-ray system. The control function was allocated to software. Late in the development process, risk analysis of the system uncovered several failure modes that could result in overexposure to the patient. Because the problem was not identified until the design was near completion, an expensive, independent, back-up timer had to be added to monitor exposure times.

The Quality System and Design Controls.

In addition to procedures and work instructions necessary for the implementation of design controls, policies and procedures may also be needed for other determinants of device quality that should be considered during the design process. The need for policies and procedures for these factors is dependent upon the types of devices manufactured by a company and the risks associated with their use. Management with executive responsibility has the responsibility for determining what is needed.

Example of topics for which policies and procedures may be appropriate are:

- risk management

- device reliability

- device durability

- device maintainability

- device serviceability

- human factors engineering

- software engineering

- use of standards

- configuration management

- compliance with regulatory requirements

- device evaluation (which may include third party product certification or approval)

- clinical evaluations

- document controls

- use of consultants

- use of subcontractors

- use of company historical data

Section A. General

I. Requirements

§ 820.30(a) General.

(1) Each manufacturer of any class III or class II device, and the class I devices listed in paragraph (a) (2) of this section, shall establish and maintain procedures to control the design of the device in order to ensure that specified design requirements are met.

(2) The following class I devices are subject to design controls:

   (i) Devices automated with computer software; and

   (ii) The devices listed in the chart below.

Section	Device
868.6810	Catheter, Tracheobronchial Suction
878.4460	Glove, Surgeon's
880.6760	Restraint, Protective
892.5650	System, Applicator, Radionuclide, Manual
892.5740	Source, Radionuclide Teletherapy

II. Definitions

§ 820.3 (n) *Management with executive responsibility* means those senior employees of a manufacturer who have the authority to establish or make changes to the manufacturer's quality policy and quality system.

§ 820.3 (s) *Quality* means the totality of features and characteristics that bear on the ability of a device to satisfy fitness-for-use, including safety and performance.

§ 820.3 (v) *Quality system* means the organizational structure, responsibilities, procedures, processes, and resources for implementing quality management.

Cross reference to ISO 9001:1994 and ISO/DIS 13485 Section 4.4.1 General.

III. Discussion and Points to Consider

The essential quality aspects and the regulatory requirements, such as safety, performance, and dependability of a product (whether hardware, software, services, or processed materials) are established during the design and development phase. Deficient design can be a major cause of quality problems.

The context within which product design is to be carried out should be set by the manufacturer's senior management. It is their responsibility to establish a design and development plan which sets the targets to be met. This plan defines the constraints within which the design is to be implemented.

The quality system requirements do not dictate the types of design process that a manufacturer must use. Manufacturers should use processes best suited to their needs. However, whatever the processes may be, it is important that the design controls are applied in an appropriate manner. This guidance document contains examples of how this might be achieved in a variety of situations.

It is important to note that the design function may apply to various facets of the operation having differing styles and time scales. Such facets are related to products, including services and software, as well as to their manufacturing processes.

Senior management needs to decide how the design function is to be managed and by whom. Senior management should also ensure that internal policies are established for design issues such as:

- assessing new product ideas

- training and retraining of design managers and design staff

- use of consultants

- evaluation of the design process

- product evaluation, including third party product certification and approvals

- patenting or other means of design protection

It is for senior management to ensure that adequate resources are available to carry out the design in the required time. This may involve reinforcing the skills and equipment available internally and/or obtaining external resources.

Section B. Design and Development Planning

I. Requirements

§ 820.30(b) Design and development planning.

- Each manufacturer shall establish and maintain plans that describe or reference the design and development activities and define responsibility for implementation.

- The plans shall identify and describe the interfaces with different groups or activities that provide, or result in, input to the design and development process.

- The plans shall be reviewed, updated, and approved as design and development evolves.

Cross-reference to ISO 9001:1994 and ISO/DIS 13485 sections 4.4.2 Design and development planning and 4.4.3 Organizational and technical interfaces.

II. Discussion and Points to Consider

Design and development planning is needed to ensure that the design process is appropriately controlled and that device quality objectives are met. The plans must be consistent with the remainder of the design control requirements. The following elements would typically be addressed in the design and development plan or plans:

- Description of the goals and objectives of the design and development program; i.e., what is to be developed;

- Delineation of organizational responsibilities with respect to assuring quality during the design and development phase, to include interface with any contractors;

- Identification of the major tasks to be undertaken, deliverables for each task, and individual or organizational responsibilities (staff and resources) for completing each task;

- Scheduling of major tasks to meet overall program time constraints;

- Identification of major reviews and decision points;

- Selection of reviewers, the composition of review teams, and procedures to be followed by reviewers;

- Controls for design documentation;

- Notification activities.

Planning enables management to exercise greater control over the design and development process by clearly communicating policies, procedures, and goals to members of the design and development team, and providing a basis for measuring conformance to quality system objectives.

Design activities should be specified at the level of detail necessary for carrying out the design process. The extent of design and development planning is dependent on the size of the developing organization and the size and complexity of the product to be developed. Some manufacturers may have documented policies and procedures which apply to all design and development activities. For each specific development program, such manufacturers may also prepare a plan which spells out the project-dependent elements in detail, and incorporates the general policies and procedures by reference. Other manufacturers may develop a comprehensive design and development plan which is specifically tailored to each individual project.

In summary, the form and organization of the planning documents are less important than their content. The following paragraphs discuss the key elements of design and development planning.

Organizational Responsibilities.

The management responsibility section of the quality system requirements[2] requires management to establish a quality policy and implement an organizational structure to ensure quality. These are typically documented in a quality manual or similarly named document. In some cases, however, the design and development plan, rather than the quality manual, is the best vehicle for describing organizational responsibilities relative to design and development activities. The importance of defining responsibilities with clarity and without ambiguity should be recognized. When input to the design is from a variety of sources, their interrelationships and interfaces (as well as the pertinent responsibilities and authorities) should be defined, documented, coordinated, and controlled. This might be the case, for example, if a multidisciplinary product development team is assembled for a specific project, or if the team includes suppliers, contract manufacturers, users, outside consultants, or independent auditors.

Task Breakdown.

The plan establishes, to the extent possible:

---

[2] § 820.20 of the FDA Quality System Regulation; section 4.1 of ISO 9001 and ISO/DIS 13485.

- The major tasks required to develop the product

- The time involved for each major task

- The resources and personnel required

- The allocation of responsibilities for completing each major task

- The prerequisite information necessary to start each major task and the interrelationship between tasks

- The form of each task output or deliverable

- Constraints, such as applicable codes, standards, and regulations

Tasks for all significant design activities, including verification and validation tasks, should be included in the design and development plan. For example, if clinical trials are anticipated, there may be tasks associated with appropriate regulatory requirements.

For complex projects, rough estimates may be provided initially, with the details left for the responsible organizations to develop. As development proceeds, the plan should evolve to incorporate more and better information.

The relationships between tasks should be presented in such a way that they are easily understood. It should be clear which tasks depend on others, and which tasks need to be performed concurrently. Planning should reflect the degree of perceived development risk; for example, tasks involving new technology or processes should be spelled out in greater detail, and perhaps be subjected to more reviews and checks, than tasks which are perceived as routine or straightforward.

The design and development plan may include a schedule showing starting and completion dates for each major task, project milestone, or key decision points. The method chosen and the detail will vary depending on the complexity of the project and the level of risk associated with the device. For small projects, the plan may consist of only a simple flow diagram or computer spreadsheet. For larger projects, there are a number of project management tools that are used to develop plans. Three of the most commonly used are the Program Evaluation and Review Technique (PERT), the Critical Path Method (CPM), and the Gantt chart. Software is available in many forms for these methods. When selecting these tools, be careful to choose one that best fits the needs of the project. Some of the software programs are far more complex than may be necessary.

Unless a manufacturer has experience with the same type of device, the plan will initially be limited in scope and detail. As work proceeds, the plan is refined. Lack of experience in planning often leads to optimistic schedules, but slippage may also occur for reasons beyond the control of planners, for example, personnel turnover, materiel shortage, or unexpected problems with a design element or process. Sometimes the schedule can be compressed by using additional resources, such as diverting staff or equipment from another project, hiring a contractor, or leasing equipment.

It is important that the schedule be updated to reflect current knowledge. At all times, the plan should be specified at a level of detail enabling management to make informed decisions, and provide confidence in meeting overall schedule and performance objectives. This is important because scheduling pressures have historically been a contributing factor in many design defects which caused injury. To the extent that good planning can prevent schedule pressures, the potential for design errors is reduced.

However, no amount of planning can eliminate all development risk. There is inherent conflict between the desire to maximize performance and the need to meet business objectives, including development deadlines. In some corporate cultures, impending deadlines create enormous pressure to cut corners. Planning helps to combat this dilemma by ensuring management awareness of pressure points. With awareness, decisions are more likely to be made with appropriate oversight and consideration of all relevant factors. Thus, when concessions to the clock must be made, they can be justified and supported.

Section C. Design Input

I. Requirements

§ 820.30(c) Design input.

- Each manufacturer shall establish and maintain procedures to ensure that the design requirements relating to a device are appropriate and address the intended use of the device, including the needs of the user and patient.

- The procedures shall include a mechanism for addressing incomplete, ambiguous, or conflicting requirements.

- The design input requirements shall be documented and shall be reviewed and approved by designated individual(s).

- The approval, including the date and signature of the individual(s) approving the requirements, shall be documented.

Cross reference to ISO 9001:1994 and ISO/DIS 13485 section 4.4.4 Design input.

II. Definitions

§ 820.3(f) *Design input* means the physical and performance requirements of a device that are used as a basis for device design.

III. Discussion and Points to Consider

Design input is the starting point for product design. The requirements which form the design input establish a basis for performing subsequent design tasks and validating the design. Therefore, development of a solid foundation of requirements is the single most important design control activity.

Many medical device manufacturers have experience with the adverse effects that incomplete requirements can have on the design process. A frequent complaint of developers is that "there's never time to do it right, but there's always time to do it over." If essential requirements are not identified until validation, expensive redesign and rework may be necessary before a design can be released to production.

By comparison, the experience of companies that have designed devices using clear-cut, comprehensive sets of requirements is that rework and redesign are significantly reduced and product quality is improved. They know that the development of requirements for a medical device of even moderate complexity is a formidable, time-consuming task. They accept the investment in time and resources required to develop the requirements because they know the advantages to be gained in the long run.

Unfortunately, there are a number of common misconceptions regarding the meaning and practical application of the quality system requirements for design input. Many seem to arise from interpreting the requirements as a literal prescription, rather than a set of principles to be followed. In this guidance document, the focus is on explaining the principles and providing examples of how they may be applied in typical situations.

Concept Documents Versus Design Input

In some cases, the marketing staff, who maintain close contact with customers and users, determine a need for a new product, or enhancements to an existing product. Alternatively, the idea for a new product may evolve out of a research or clinical activity. In any case, the result is a concept document specifying some of the desired characteristics of the new product.

Some members of the medical device community view these marketing memoranda, or the equivalent, as the design input. However, that is not the intent of the quality system requirements. Such concept documents are rarely comprehensive, and should not be expected to be so. Rather, the intent of the quality system requirements is that the product conceptual description be elaborated, expanded, and transformed into a complete set of design input requirements which are written to an engineering level of detail.

This is an important concept. The use of qualitative terms in a concept document is both appropriate and practical. This is often not the case for a document to be used as a basis for design. Even the simplest of terms can have enormous design implications. For example, the term "must be portable" in a concept document raises questions in the minds of product developers about issues such as size and weight limitations, resistance to shock and vibration, the need for protection from moisture and corrosion, the capability of operating over a wide temperature range, and many others. Thus, a concept document may be the starting point for development, but it is not the design input requirement. This is a key principle—the design input requirements are the result of the first stage of the design control process.

Research and Development.

Some manufacturers have difficulty in determining where research ends and development begins. Research activities may be undertaken in an effort to determine new business opportunities or basic characteristics for a new product. It may be reasonable to develop a rapid prototype to explore the feasibility of an idea or design approach, for example, prior to developing design input requirements. But manufacturers should avoid falling into the trap of equating the prototype design with a finished product design. Prototypes at this stage lack safety features and ancillary functions necessary for a finished product, and are developed under conditions which preclude adequate consideration of product variability due to manufacturing.

Responsibility for Design Input Development.

Regardless of who developed the initial product concept, product developers play a key role in developing the design input requirements. When presented with a set of important characteristics, it is the product developers who understand the auxiliary issues that must be addressed, as well as the level of detail necessary to design a product. Therefore, a second key principle is that the product developer(s) ultimately bear responsibility for translating user and/or patient needs into a set of requirements which can be validated prior to implementation. While this is primarily an engineering function, the support or full participation of production and service personnel, key suppliers, etc., may be required to assure that the design input requirements are complete.

Care must be exercised in applying this principle. Effective development of design input requirements encompasses input from both the product developer as well as those representing the needs of the user, such as marketing. Terminology can be a problem. In some cases, the product conceptual description may be expressed in medical terms. Medical terminology is appropriate in requirements when the developers and reviewers are familiar with the language, but it is often preferable to translate the concepts into engineering terms at the requirements stage to minimize miscommunication with the development staff.

Another problem is incorrect assumptions. Product developers make incorrect assumptions about user needs, and marketing personnel make incorrect assumptions about the needs of the product designers. Incorrect assumptions can have serious consequences that may not be detected until late in the development process. Therefore, both product developers and those representing the user must take responsibility for critically examining proposed requirements, exploring stated and implied assumptions, and uncovering problems.

Some examples should clarify this point. A basic principle is that design input requirements should specify what the design is intended to do while carefully avoiding specific design solutions at this stage. For example, a concept document might dictate that the product be housed in a machined aluminum case. It would be prudent for product developers to explore why this type of housing was specified. Perhaps there is a valid reason—superior electrical shielding, mechanical strength, or reduced time to market as compared to a cast housing. Or perhaps machined aluminum was specified because a competitor's product is made that way, or simply because the user didn't think plastic would be strong enough.

Not all incorrect assumptions are made by users. Incorrect assumptions made by product developers may be equally damaging. Failure to understand the abuse to which a portable instrument would be subjected might result in the selection of housing materials inadequate for the intended use of the product.

There are occasions when it may be appropriate to specify part of the design solution in the design input requirements. For example, a manufacturer may want to share components or manufacturing processes across a family of products in order to realize economies of scale, or simply to help establish a corporate identity. In the case of a product upgrade, there may be clear consensus regarding the features to be retained. However, it is important to realize that every such design constraint reduces implementation flexibility and should therefore be documented and identified as a possible conflicting requirement for subsequent resolution.

Scope And Level Of Detail.

Design input requirements must be comprehensive. This may be quite difficult for manufacturers who are implementing a system of design controls for the first time. Fortunately, the process gets easier with practice. It may be helpful to realize that design input requirements fall into three categories. Virtually every product will have requirements of all three types.

- Functional requirements specify what the device does, focusing on the operational capabilities of the device and processing of inputs and the resultant outputs.

- Performance requirements specify how much or how well the device must perform, addressing issues such as speed, strength, response times, accuracy, limits of operation, etc. This includes a quantitative characterization of the use environment, including, for example, temperature, humidity, shock, vibration, and electromagnetic compatibility. Requirements concerning device reliability and safety also fit into this category.

- Interface requirements specify characteristics of the device which are critical to compatibility with external systems; specifically, those characteristics which are mandated by external systems and outside the control of the developers. One interface which is important in every case is the user and/or patient interface.

What is the scope of the design input requirements development process and how much detail must be provided? The scope is dependent upon the complexity of a device and the risk associated with its use. For most medical devices, numerous requirements encompassing functions, performance, safety, and regulatory concerns are implied by the application. These implied requirements should be explicitly stated, in engineering terms, in the design input requirements.

Determining the appropriate level of detail requires experience. However, some general guidance is possible. The marketing literature contains product specifications, but these are superficial. The operator and service manuals may contain more detailed specifications and performance limits, but these also fall short of being comprehensive. Some insight as to what is necessary is provided by examining the requirements for a very common external interface. For the power requirements for AC-powered equipment, it is not sufficient to simply say that a unit shall be AC-powered. It is better to say that the unit shall be operable from AC power in North America, Europe, and Japan, but that is still insufficient detail to implement or validate the design. If one considers the situation just in North America, where the line voltage is typically 120 volts, many systems are specified to operate over the range of 108 to 132 volts. However, to account for the possibility of brownout, critical devices may be specified to operate from 95 to 132 volts or even wider ranges. Based on the intended use of the device, the manufacturer must choose appropriate performance limits.

There are many cases when it is impractical to establish every functional and performance characteristic at the design input stage. But in most cases, the form of the requirement can be determined, and the requirement can be stated with a to-be-determined (TBD) numerical value or a range of possible values. This makes it possible for reviewers to assess whether the requirements completely characterize the intended use of the device, judge the impact of omissions, and track incomplete requirements to ensure resolution.

For complex designs, it is not uncommon for the design input stage to consume as much as thirty percent of the total project time. Unfortunately, some managers and developers have been trained to measure design progress in terms of hardware built, or lines of software code written. They fail to realize that building a solid foundation saves time during the implementation. Part of the solution is to structure the requirements documents and reviews such that tangible measures of progress are provided.

At the other extreme, many medical devices have very simple requirements. For example, many new devices are simply replacement parts for a product, or are kits of commodity items. Typically, only the packaging and labeling distinguishes these products from existing products. In such cases, there is no need to recreate the detailed design input requirements of the item. It is acceptable to simply cite the predecessor product documentation, add any new product information, and establish the unique packaging and labeling requirements.

Assessing Design Input Requirements for Adequacy.

Eventually, the design input must be reviewed for adequacy. After review and approval, the design input becomes a controlled document. All future changes will be subject to the change control procedures, as discussed in Section I (Design Changes).

Any assessment of design input requirements boils down to a matter of judgment. As discussed in Section E (Design Review), it is important for the review team to be multidisciplinary and to have the appropriate authority. A number of criteria may be employed by the review team.

- Design input requirements should be unambiguous. That is, each requirement should be able to be verified by an objective method of analysis, inspection, or testing. For example, it is insufficient to state that a catheter must be able to withstand repeated flexing. A better requirement would state that the catheter should be formed into a 50 mm diameter coil and straightened out for a total of fifty times with no evidence of cracking or deformity. A qualified reviewer could then make a judgment whether this specified test method is representative of the conditions of use.

- Quantitative limits should be expressed with a measurement tolerance. For example, a diameter of 3.5 mm is an incomplete specification. If the diameter is specified as 3.500 ± 0.005 mm, designers have a basis for determining how accurate the manufacturing processes have to be to produce compliant parts, and reviewers have a basis for determining whether the parts will be suitable for the intended use.

- The set of design input requirements for a product should be self-consistent. It is not unusual for requirements to conflict with one another or with a referenced industry standard due to a simple oversight. Such conflicts should be resolved early in the development process.

• The environment in which the product is intended to be used should be properly characterized. For example, manufacturers frequently make the mistake of specifying "laboratory" conditions for devices which are intended for use in the home. Yet, even within a single country, relative humidity in a home may range from 20 percent to 100 percent (condensing) due to climactic and seasonal variations. Household temperatures in many climates routinely exceed 40 °C during the hot season. Altitudes may exceed 3,000 m, and the resultant low atmospheric pressure may adversely affect some kinds of medical equipment. If environmental conditions are fully specified, a qualified reviewer can make a determination of whether the specified conditions are representative of the intended use.

• When industry standards are cited, the citations should be reviewed for completeness and relevance. For example, one medical device manufacturer claimed compliance with an industry standard covering mechanical shock and vibration. However, when the referenced standard was examined by a reviewer, it was found to prescribe only the method of testing, omitting any mention of pass/fail criteria. It was incumbent on the manufacturer in this case to specify appropriate performance limits for the device being tested, as well as the test method.

Evolution of the Design Input Requirements.

Large development projects often are implemented in stages. When this occurs, the design input requirements at each stage should be developed and reviewed following the principles set forth in this section. Fortunately, the initial set of requirements, covering the overall product, is by far the most difficult to develop. As the design proceeds, the output from the early stages forms the basis for the subsequent stages, and the information available to designers is inherently more extensive and detailed.

It is almost inevitable that verification activities will uncover discrepancies which result in changes to the design input requirements. There are two points to be made about this. One is that the change control process for design input requirements must be carefully managed. Often, a design change to correct one problem may create a new problem which must be addressed. Throughout the development process, it is important that any changes are documented and communicated to developers so that the total impact of the change can be determined. The change control process is crucial to device quality.

The second point is that extensive rework of the design input requirements suggests that the design input requirements may not be elaborated to a suitable level of detail, or insufficient resources are being devoted to defining and reviewing the requirements. Managers can use this insight to improve the design control process. From a design control perspective, the number of requirements changes made is less important than the thoroughness of the change control process.

Section D. Design Output

I. Requirements

§ 820.30(d) Design output.

- Each manufacturer shall establish and maintain procedures for defining and documenting design output in terms that allow an adequate evaluation of conformance to design input requirements.

- Design output procedures shall contain or make reference to acceptance criteria and shall ensure that those design outputs that are essential for the proper functioning of the device are identified.

- Design output shall be documented, reviewed, and approved before release.

- The approval, including the date and signature of the individual(s) approving the output, shall be documented.

Cross-reference to ISO 9001:1994 and ISO/DIS 13485 section 4.4.5 Design output.

II. Definitions

§ **820.3(g)** *Design output* means the results of a design effort at each design phase and at the end of the total design effort. The finished design output is the basis for the device master record. The total finished design output consists of the device, its packaging and labeling, and the device master record.

§ **820.3(y)** *Specification* means any requirement with which a product, process, service, or other activity must conform.

III. Discussion and Points to Consider

The quality system requirements for design output can be separated into two elements: Design output should be expressed in terms that allow adequate assessment of conformance to design input requirements and should identify the characteristics of the design that are crucial to the safety and proper functioning of the device. This raises two fundamental issues for developers:

- What constitutes design output?

- Are the form and content of the design output suitable?

The first issue is important because the typical development project produces voluminous records, some of which may not be categorized as design output. On the other hand, design output must be reasonably comprehensive to be effective. As a general rule, an item is design output if it is a work product, or deliverable item, of a design task listed in the design and development plan, and the item defines, describes, or elaborates an element of the design implementation. Examples include block diagrams, flow charts, software high-level code, and system or subsystem design specifications. The design output in one stage is often part of the design input in subsequent stages.

Design output includes production specifications as well as descriptive materials which define and characterize the design.

Production Specifications.

Production specifications include drawings and documents used to procure components, fabricate, test, inspect, install, maintain, and service the device, such as the following:

- assembly drawings
- component and material specifications
- production and process specifications
- software machine code (e.g., diskette or master EPROM)
- work instructions
- quality assurance specifications and procedures
- installation and servicing procedures
- packaging and labeling specifications, including methods and processes used

In addition, as discussed in Section H (Design Transfer), production specifications may take on other forms. For example, some manufacturers produce assembly instructions on videotapes rather than written instructions. Similarly, a program diskette, used by a computer-aided milling machine to fabricate a part, would be considered a production specification. The videotape and the software on the program diskette are part of the device master record.

Other Descriptive Materials.

Other design output items might be produced which are necessary to establish conformance to design input requirements, but are not used in its production. For example, for each part which is fabricated by computer-aided machine, there should be an assembly drawing which specifies the dimensions and characteristics of the part. It is a part of the design output because it establishes the basis for the machine tool program used to fabricate the part. Other examples of design output include the following:

- the results of risk analysis
- software source code
- results of verification activities
- biocompatibility test results
- bioburden test results

Form and Content.

Manufacturers must take steps to assure that the design output characterizes all important aspects of the design and is expressed in terms which allow adequate verification and validation. Two basic mechanisms are available to manufacturers to accomplish these objectives.

- First, the manufacturer proactively can specify the form and content of design output at the planning stage. For some types of design output, form and content may be codified in a consensus standard which can be referenced. In other cases, a manufacturer could specify the desired characteristics, or even simply specify that the form and content of an existing document be followed.

- Second, form and content can be reviewed retroactively as a part of the design verification process. For example, the verification of design output could include assessing whether specified documentation standards have been adhered to.

As these examples illustrate, conformance with the quality system requirements concerning design output generally requires no "extra" effort on the part of the manufacturer, but simply the application of some common sense procedures during the planning, execution, and review of design tasks.

Section E. Design Review

I. Requirements

§ 820.30(e) Design review.

- Each manufacturer shall establish and maintain procedures to ensure that formal documented reviews of the design results are planned and conducted at appropriate stages of the device's design development.

- The procedures shall ensure that participants at each design review include representatives of all functions concerned with the design stage being reviewed and an individual(s) who does not have direct responsibility for the design stage being reviewed, as well as any specialists needed.

- The results of a design review, including identification of the design, the date, and the individual(s) performing the review, shall be documented in the design history file (the DHF).

Cross-reference to ISO 9001:1994 and ISO/DIS 13485 section 4.4.6 Design review.

II. Definitions

§ 820.3(h) *Design review* means a documented, comprehensive, systematic examination of a design to evaluate the adequacy of the design requirements, to evaluate the capability of the design to meet these requirements, and to identify problems.

III. Discussion and Points to Consider

In general, formal design reviews are intended to:

- provide a systematic assessment of design results, including the device design and the associated designs for production and support processes;

- provide feedback to designers on existing or emerging problems;

- assess project progress; and/or

- provide confirmation that the project is ready to move on to the next stage of development.

Many types of reviews occur during the course of developing a product. Reviews may have both an internal and external focus. The internal focus is on the feasibility of the design and the produceability of the design with respect to manufacturing and support capabilities. The external focus is on the user requirements; that is, the device design is viewed from the perspective of the user.

The nature of reviews changes as the design progresses. During the initial stages, issues related to design input requirements will predominate. Next, the main function of the reviews may be to evaluate or confirm the choice of solutions being offered by the design team. Then, issues such as the choice of materials and the methods of manufacture become more important. During the final stages, issues related to the verification, validation, and production may predominate.

The term "review" is commonly used by manufacturers to describe a variety of design assessment activities. Most, but not all, of these activities meet the definition of formal design reviews. The following exceptions may help to clarify the distinguishing characteristics of design reviews.

- Each design document which constitutes the formal output, or deliverable, of a design task is normally subject to evaluation activities, sometimes referred to as informal peer review, supervisory review, or technical assessment. These activities, while they may be called reviews, are often better described as verification activities, because they are not intended to be comprehensive, definitive, and multidisciplinary in their scope. Rather, their purpose is to confirm that design output meets design input. Verification activities affect and add to the design output, and are themselves subject to subsequent design review.

- Developers may conduct routine or ad hoc meetings to discuss an issue, coordinate activities, or assess development progress. Decisions from such meetings may not require formal documentation; however, if a significant issue is resolved, this should be documented. If the outcome results in change to an approved design document, then applicable change control procedures should be followed, as discussed in Section I (Design Changes).

Control of the design review process is achieved by developing and implementing a formal design review program consistent with quality system requirements. The following issues should be addressed and documented in the design and development plan(s).

Number and Type of Reviews.

It is a well-accepted fact that the cost to correct design errors increases as the design nears completion, and the flexibility to implement an optimal solution decreases. When an error is discovered at the end of the development cycle, difficult decisions have to be made regarding an acceptable corrective action. When that corrective action is implemented in haste, the result is often an unintended consequence leading to a new problem. Thus, formal design reviews should be planned to detect problems early. A corollary is that planners should presume that problems will be detected, and allocate a reasonable amount of time to implement corrective actions. Typically, formal reviews are conducted at the end of each phase and at important milestones in the design process.

As discussed in Section C (Design Input), it is beneficial in almost every case to conduct a formal review of the design input requirements early in the development process. The number of reviews depends upon the complexity of the device.

- For a simple design, or a minor upgrade to an existing product, it might be appropriate to conduct a single review at the conclusion of the design process.

- For a product involving multiple subsystems, an early design task is to allocate the design input requirements among the various subsystems. For example, in a microprocessor-based system, designers must decide which functions will be performed by hardware and which by software. In another case, tolerance buildup from several components may combine to create a clearance problem. System designers must establish tolerance specifications for each component to meet the overall dimensional specification. In cases like these, a formal design review is a prudent step to ensure that all such system-level requirements have been allocated satisfactorily prior to engaging in detailed design of each subsystem.

- For complex systems, additional reviews are often built into the development plan. For example, engineering sketches may be developed for prototyping purposes prior to development of production drawings. Evaluation of the prototype would typically culminate in a formal design review. Similarly, software development commonly includes a high-level design phase, during which requirements are elaborated to a greater level of detail and algorithms are developed to implement key functions. A formal design review would typically be conducted to review this work prior to beginning detailed coding.

There are a number of approaches to conducting formal design reviews at the end of the design process. In some organizations, engineering essentially completes the design, tests an engineering prototype, and conducts a formal design review prior to turning the design over to manufacturing. In such cases, an additional review will be needed after the design has been validated using production devices.

In some instances, components having long lead times may enter production prior to completion of the overall device design. The primary motivation for early production is to reduce time to market. The manufacturer runs the business risk that the design review at the end of the design process will uncover a defect that must be corrected in production devices before any devices are distributed.

All of these approaches to scheduling formal design reviews are valid. What is important is that the manufacturer establish a reasonable rationale for the number and type of reviews, based on sound judgment.

Selection of Reviewers.

In determining who should participate in a formal design review, planners should consider the qualifications of reviewers, the types of expertise required to make an adequate assessment, and the independence of the reviewers. Each of these concerns is discussed briefly in the following paragraphs.

Qualifications.

Formal design reviews should be conducted by person(s) having technical competence and experience at least comparable to the developers. For a small manufacturer, this may require that an outside consultant be retained to participate in the evaluation of the design.

A manufacturer will often employ one or more specialists to conduct certain types of specialized assessments which are beyond the capabilities of the designers. For example, a mechanical engineer may be retained to perform a structural analysis of a design, and perhaps conduct vibration testing to verify its performance under stress. Such specialists may be assigned to participate in the formal design review. Alternatively, they may be assigned to make an independent assessment and submit observations and recommendations to the reviewers. Either approach is valid.

Types of expertise required.

Many medical device designs involve a number of technologies, such as electronics, mechanics, software, materials science, or pneumatics. In addition, a variety of clinical and manufacturing issues may influence the design. Manufacturers should carefully consider which interests should be represented at formal design reviews. Subtle distinctions in reviewer perspective may have dramatic impact on device quality. For example, the marketing department of a small manufacturer shared a new design with several surgeons on their advisory board. The surgeons all thought the design was terrific. Subsequently, the manufacturer invited two experienced operating room nurses to participate in the final design review. During the course of the review, it became apparent that while surgeons may be the customers, nurses are the primary users of the device, and no one up to that point had consulted with any nurses. The nurses at the design review didn't like some of the features of the design. After some further market survey, the manufacturer decided to make changes to the design to accommodate these concerns. It was unfortunate (and expensive) in this case that the user requirements were not considered until late in the development cycle, but the design review was ultimately very successful.

Independence.

The formal design review should include at least one individual who does not have direct responsibility for the design stage under review. In a small company, complete independence is very difficult to obtain. Within the context of formal design reviews, the practical solution is simply to ensure a fresh perspective, based on the principle that those who are too close to the design may overlook design errors. Thus, reviewers will often be from the same organization as the developers, but they should not have been significantly involved in the activities under review. As discussed in the following section, the formal design review procedures play a large role in assuring independent and objective reviews.

Design Review Procedures.

The manufacturer should have documented formal design review procedures addressing the following:

- Evaluation of the design (including identification of concerns, i.e., issues and potential problems with the design)
- Resolution of concerns
- Implementation of corrective actions

Evaluation of the design.

Many formal design reviews take the form of a meeting. At this meeting, the designer(s) may make presentations to explain the design implementation, and persons responsible for verification activities may present their findings to the reviewers. Reviewers may ask for clarification or additional information on any topic, and add their concerns to any raised by the presenters. This portion of the review is focused on finding problems, not resolving them.

There are many approaches to conducting design review meetings. In simple cases, the technical assessor and reviewer may be the same person, often a project manager or engineering supervisor, and the review meeting is a simple affair in the manager's office. For more elaborate reviews, detailed written procedures are desirable to ensure that all pertinent topics are discussed, conclusions accurately recorded, and action items documented and tracked.

There is a dangerous tendency for design review meetings to become adversarial affairs. The reputation of the designers tends to be linked to the number of discrepancies found, causing the designers to become defensive, while the reviewers score points by finding weaknesses in the design. The resulting contest can be counterproductive. An added complication is the presence of invited guests, often clinicians, who are expected to provide the user perspective. These reviewers are often very reluctant to ask probing questions, especially if they sense that they may become involved in a conflict where all the rules and relationships are not evident.

These difficulties can be avoided by stating the goals and ground rules for conducting the formal design review clearly at the outset. While the designers are in the best position to explain the best features of the design, they are also most likely to be aware of the design's weaknesses. If the designers and reviewers are encouraged to work together to systematically explore problems and find solutions, the resultant design will be improved and all parties will benefit from the process. Participants must be encouraged to ask questions, avoid making assumptions, and think critically. The focus must be on the design, not the participants.

Not all formal design reviews involve meetings. For extremely simple designs or design changes, it may be appropriate to specify a procedure in which review materials are distributed or circulated among the reviewers for independent assessment and approval. However, such a procedure negates the benefits of synergy and teamwork, and should be considered only in cases where the design issues are limited in scope and well defined.

Resolution of concerns.

The reviewers consider concerns raised during the evaluation portion of the formal design review and decide on an appropriate disposition for each one. There is wide variation in the way companies implement decision-making processes. In some cases, the reviewers play an advisory role to the engineering manager or other company official, who directs the formal design review and ultimately selects a course of action. In other cases, the reviewers are given limited or broad authority to make decisions and commit resources to resolve problems. The approach used should be documented.

In the real world, reviews often leave unresolved issues. Therefore, review procedures should include a process for resolving differences, and provide reviewers with enough leeway to make practical decisions while protecting the integrity of the process.

Implementation of corrective actions.

Not all identified concerns result in corrective actions. The reviewers may decide that the issue is erroneous or immaterial. In most cases, however, resolution involves a design change, a requirements change, or a combination of the two. If the solution is evident, the reviewers may specify the appropriate corrective action; otherwise, an action item will be assigned to study the problem further. In any case, action items and corrective actions are normally tracked under the manufacturer's change control procedures.

Relationship of Design Review to Verification and Validation.

In practice, design review, verification, and validation overlap one another, and the relationship among them may be confusing. As a general rule, the sequence is: verification, review, validation, review.

In most cases, verification activities are completed prior to the design review, and the verification results are submitted to the reviewers along with the other design output to be reviewed. Alternatively, some verification activities may be treated as components of the design review, particularly if the verification activity is complex and requires multidisciplinary review.

Similarly, validation typically involves a variety of activities, including a determination that the appropriate verifications and reviews have been completed. Thus, at the conclusion of the validation effort, a review is usually warranted to assure that the validation is complete and adequate.

Section F. Design Verification

I. Requirements

§ 820.30(f) Design verification.

- Each manufacturer shall establish and maintain procedures for verifying the device design.

- Design verification shall confirm that the design output meets the design input requirements.

- The results of the design verification, including identification of the design, method(s), the date, and the individual(s) performing the verification, shall be documented in the Design History File.

Cross-reference to ISO 9001:1994 and ISO/DIS 13485 section 4.4.7 Design verification.

I. Definitions

§820.3(y) *Specification* means any requirement with which a product, process, service, or other activity must conform.

§ 820.3(z) *Validation* means confirmation by examination and provision of objective evidence that the particular requirements for a specific intended use can be consistently fulfilled.

(1) *Process Validation* means establishing by objective evidence that a process consistently produces a result or product meeting its predetermined specifications.

(2) *Design Validation* means establishing by objective evidence that device specifications conform with user needs and intended use(s).

§820.3(aa) *Verification* means confirmation by examination and provision of objective evidence that specified requirements have been fulfilled.

III. Discussion and Points to Consider

Verification and validation are associated concepts with very important differences. Various organizations have different definitions for these terms. Medical device manufacturers are encouraged to use the terminology of the quality system requirements in their internal procedures.

To illustrate the concepts, consider a building design analogy. In a typical scenario, the senior architect establishes the design input requirements and sketches the general appearance and construction of the building, but associates or contractors typically elaborate the details of the various mechanical systems. Verification is the process of checking at each stage whether the output conforms to requirements for that stage. For example: does the air conditioning system deliver the specified cooling capacity to each room? Is the roof rated to withstand so many newtons per square meter of wind loading? Is a fire alarm located within 50 meters of each location in the building?

At the same time, the architect has to keep in mind the broader question of whether the results are consistent with the ultimate user requirements. Does the air conditioning system keep the occupants comfortable throughout the building? Will the roof withstand weather extremes expected at the building site? Can the fire alarm be heard throughout the building? Those broader concerns are the essence of validation.

In the initial stages of design, verification is a key quality assurance technique. As the design effort progresses, verification activities become progressively more comprehensive. For example, heat or cooling delivery can be calculated and verified by the air conditioning designer, but the resultant air temperature can only be estimated. Occupant comfort is a function not only of delivered air temperature, but also humidity, heat radiation to or from nearby thermal masses, heat gain or loss through adjacent windows, etc. During the latter design phases, the interaction of these complex factors may be considered during verification of the design.

Validation follows successful verification, and ensures that each requirement for a particular use is fulfilled. Validation of user needs is possible only after the building is built. The air conditioning and fire alarm performance may be validated by testing and inspection, while the strength of the roof will probably be validated by some sort of analysis linked to building codes which are accepted as meeting the needs of the user—subject to possible confirmation during a subsequent severe storm.

Validation is the topic of Section G of this guidance document. The remainder of this section focuses on verification principles.

Types of Verification Activities.

Verification activities are conducted at all stages and levels of device design. The basis of verification is a three-pronged approach involving tests, inspections, and analyses. Any approach which establishes conformance with a design input requirement is an acceptable means of verifying the design with respect to that requirement. In many cases, a variety of approaches are possible.

Complex designs require more and different types of verification activities. The nature of verification activities varies according to the type of design output. The intent of this guidance document is not to suggest or recommend verification techniques which should be performed by device manufacturers. Rather, the manufacturer should select and apply appropriate verification techniques based on the generally accepted practices for the technologies employed in their products. Many of these practices are an integral part of the development process, and are routinely performed by developers. The objective of design controls is to ensure adequate oversight by making verification activities explicit and measuring the thoroughness of their execution. Following are a few examples of verification methods and activities.

- Worst case analysis of an assembly to verify that components are derated properly and not subject to overstress during handling and use.

- Thermal analysis of an assembly to assure that internal or surface temperatures do not exceed specified limits.

- Fault tree analysis of a process or design.

- Failure modes and effects analysis.

- Package integrity tests.

- Biocompatibility testing of materials.

- Bioburden testing of products to be sterilized.

- Comparison of a design to a previous product having an established history of successful use.

For some technologies, verification methods may be highly standardized. In other cases, the manufacturer may choose from a variety of applicable methods. In a few cases, the manufacturer must be creative in devising ways to verify a particular aspect of a design.

Some manufacturers erroneously equate production testing with verification. Whereas verification testing establishes conformance of design output with design input, the aim of production testing is to determine whether the unit under test has been correctly manufactured. In other words, production testing is designed to efficiently screen out manufacturing process errors and perhaps also to detect infant mortality failures. Typically, a small subset of functional and performance tests accomplish this objective with a high degree of accuracy. Therefore, production testing is rarely, if ever, comprehensive enough to verify the design. For example, a leakage test may be used during production to ensure that a hermetically-sealed enclosure was properly assembled. However, the leakage test may not be sensitive enough to detect long-term diffusion of gas through the packaging material. Permeability of the packaging material is an intrinsic property of the material rather than an assembly issue, and would likely be verified using a more specialized test than is used during production.

Documentation of Verification Activities.

Some verification methods result in a document by their nature. For example, a failure modes and effects analysis produces a table listing each system component, its postulated failure modes, and the effect of such failures on system operation.

Another self-documenting verification method is the traceability matrix. This method is particularly useful when the design input and output are both documents; it also has great utility in software development. In the most common form of the traceability matrix, the input requirements are enumerated in a table, and references are provided to each section in the output documents (or software modules) which address or satisfy each input requirement. The matrix can also be constructed "backwards," listing each feature in the design output and tracing which input requirement bears on that feature. This reverse approach is especially useful for detecting hidden assumptions. Hidden assumptions are dangerous because they often lead to overdesign, adding unnecessary cost and complexity to the design. In other cases, hidden assumptions turn out to be undocumented design input requirements which, once exposed, can be properly tracked and verified.

However, many verification activities are simply some sort of structured assessment of the design output relative to the design input. When this is the case, manufacturers may document completion of verification activities by linking these activities with the signoff procedures for documents. This may be accomplished by establishing a procedure whereby each design output document must be verified and signed by designated persons. The presence of the reviewers' signatures on the document signifies that the design output has been verified in accordance with the signoff procedure.

Section G. Design Validation

I. Requirements

§ 820.30(g) Design validation.

- Each manufacturer shall establish and maintain procedures for validating the device design.

- Design validation shall be performed under defined operating conditions on initial production units, lots, or batches, or their equivalents.

- Design validation shall ensure that devices conform to defined user needs and intended uses and shall include testing of production units under actual or simulated use conditions.

- Design validation shall include software validation and risk analysis, where appropriate.

- The results of the design validation, including identification of the design, method(s), the date, and the individual(s) performing the validation, shall be documented in the Design History File.

Cross-reference to ISO 9001:1994 and ISO/DIS 13485 section 4.4.8 Design validation.

II. Definitions

§820.3(y) *Specification* means any requirement with which a product, process, service, or other activity must conform.

§ 820.3(z) *Validation* means confirmation by examination and provision of objective evidence that the particular requirements for a specific intended use can be consistently fulfilled.

(1) Process Validation means establishing by objective evidence that a process consistently produces a result or product meeting its predetermined specifications.

(2) Design Validation means establishing by objective evidence that device specifications conform with user needs and intended use(s).

§820.3(aa) *Verification* means confirmation by examination and provision of objective evidence that specified requirements have been fulfilled.

III. Discussion and Points to Consider

Whereas verification is a detailed examination of aspects of a design at various stages in the development, design validation is a cumulative summation of all efforts to assure that the design will conform with user needs and intended use(s), given expected variations in components, materials, manufacturing processes, and the use environment.

Validation Planning.

Planning for validation should begin early in the design process. The performance characteristics that are to be assessed should be identified, and validation methods and acceptance criteria should be established. For complex designs, a schedule of validation activities and organizational or individual responsibilities will facilitate maintaining control over the process. The validation plan should be reviewed for appropriateness, completeness, and to ensure that user needs and intended uses are addressed.

Validation Review.

Validation may expose deficiencies in the original assumptions concerning user needs and intended uses. A formal review process should be used to resolve any such deficiencies. As with verification, the perception of a deficiency might be judged insignificant or erroneous, or a corrective action may be required.

Validation Methods.

Many medical devices do not require clinical trials. However, all devices require clinical evaluation and should be tested in the actual or simulated use environment as a part of validation. This testing should involve devices which are manufactured using the same methods and procedures expected to be used for ongoing production. While testing is always a part of validation, additional validation methods are often used in conjunction with testing, including analysis and inspection methods, compilation of relevant scientific literature, provision of historical evidence that similar designs and/or materials are clinically safe, and full clinical investigations or clinical trials.

Some manufacturers have historically used their best assembly workers or skilled lab technicians to fabricate test articles, but this practice can obscure problems in the manufacturing process. It may be beneficial to ask the best workers to evaluate and critique the manufacturing process by trying it out, but pilot production should simulate as closely as possible the actual manufacturing conditions.

Validation should also address product packaging and labeling. These components of the design may have significant human factors implications, and may affect product performance in unexpected ways. For example, packaging materials have been known to cause electrostatic discharge (ESD) failures in electronic devices. If the unit under test is delivered to the test site in the test engineer's briefcase, the packaging problem may not become evident until after release to market.

Validation should include simulation of the expected environmental conditions, such as temperature, humidity, shock and vibration, corrosive atmospheres, etc. For some classes of device, the environmental stresses encountered during shipment and installation far exceed those encountered during actual use, and should be addressed during validation.

Particular care should be taken to distinguish among customers, users, and patients to ensure that validation addresses the needs of all relevant parties. For a consumer device, the customer, user, and patient may all be the same person. At the other extreme, the person who buys the device may be different from the person who routinely uses it on patients in a clinical setting. Hospital administrators, biomedical engineers, health insurance underwriters, physicians, nurses, medical technicians, and patients have distinct and sometimes competing needs with respect to a device design.

Validation Documentation.

Validation is a compilation of the results of all validation activities. For a complex design, the detailed results may be contained in a variety of separate documents and summarized in a validation report. Supporting information should be explicitly referenced in the validation report and either included as an appendix or available in the design history file.

Section H. Design Transfer

I. Requirements

§ 820.30(h) Design transfer.

- Each manufacturer shall establish and maintain procedures to ensure that the device design is correctly translated into production specifications.

Cross reference to ISO 9001:1994 and ISO/DIS 13485 section 4.2.3(c) Quality planning.

II. Discussion and Points to Consider

Production specifications must ensure that manufactured devices are repeatedly and reliably produced within product and process capabilities. If a manufactured device deviates outside those capabilities, performance may be compromised. Thus, the process of encapsulating knowledge about the device into production specifications is critical to device quality.

The level of detail necessary to accomplish this objective varies widely, based on the type of device, the relationship between the design and manufacturing organizations, and the knowledge, experience, and skills of production workers. In some cases, devices are produced by contract manufacturers who have no involvement in the development and little or no contact with the designers. At the other extreme, some devices are handcrafted by skilled artisans with extensive knowledge about the use of the product.

One normally associates the term "production specifications" with written documents, such as assembly drawings, component procurement specifications, workmanship standards, manufacturing instructions, and inspection and test specifications. While these types of documents are widely employed in medical device production, other equally acceptable means of conveying design information exist, and manufacturers have the flexibility to employ these alternate means of communication as appropriate. For example, each of the following could constitute "production specifications" within the meaning of the quality system requirements:

- documentation (in electronic format as well as paper)

- training materials, e.g., manufacturing processes, test and inspection methods

- digital data files, e.g., programmable device files, master EPROM, computer-aided manufacturing (CAM) programming files

- manufacturing jigs and aids, e.g., molds, sample wiring harness to be duplicated

Historically, shortcomings in the production specifications tend to be manifested late in the product life cycle. When the design is new, there is often intensive interaction between the design and production teams, providing ample opportunity for undocumented information flow. Later, as production experience is gained, some decoupling often occurs between design and production teams. In addition, key personnel may leave, and their replacements may lack comparable training, experience, or institutional knowledge.

Particular care should be taken when the product involves new and unproved manufacturing processes, or established processes which are new to the manufacturer. It may not be possible to determine the adequacy of full-scale manufacturing on the basis of successfully building prototypes or models in a laboratory and testing these prototypes or models. The engineering feasibility and production feasibility may be different because the equipment, tools, personnel, operating procedures, supervision and motivation could be different when a manufacturer scales up for routine production.

No design team can anticipate all factors bearing on the success of the design, but procedures for design transfer should address at least the following basic elements.

- First, the design and development procedures should include a qualitative assessment of the completeness and adequacy of the production specifications.

- Second, the procedures should ensure that all documents and articles which constitute the production specifications are reviewed and approved.

- Third, the procedures should ensure that only approved specifications are used to manufacture production devices.

The first item in the preceding list may be addressed during design transfer. The second and third elements are among the basic principles of document control and configuration management. As long as the production specifications are traditional paper documents, there is ample information available to guide manufacturers in implementing suitable procedures. When the production specifications include non-traditional means, flexibility and creativity may be needed to achieve comparable rigor.

Section I. Design Changes

I. Requirements

§ 820.30(i) Design changes.

- Each manufacturer shall establish and maintain procedures for the identification, documentation, validation or where appropriate verification, review, and approval of design changes before their implementation.

Cross-reference to ISO 9001:1994 and ISO/DIS 13485 section 4.4.9 Design changes.

II. Discussion And Points To Consider

There are two principal administrative elements involved in controlling design changes:

- Document control—enumeration of design documents, and tracking their status and revision history. Throughout this section, the term "document" is used in an inclusive sense to mean all design documents, drawings, and other items of design input or output which characterize the design or some aspect of it.

- Change control—enumeration of deficiencies and corrective actions arising from verification and review of the design, and tracking their resolution prior to design transfer.

For a small development project, an adequate process for managing change involves little more than documenting the design change, performing appropriate verification and validation, and keeping records of reviews. The main objectives are ensuring that:

- corrective actions are tracked to completion;

- changes are implemented in such a manner that the original problem is resolved and no new problems are created; or if new problems are created, they are also tracked to resolution; and

- design documentation is updated to accurately reflect the revised design.

For projects involving more than two persons, coordination and communication of design changes become vitally important. In other words, manufacturers should take steps to avoid the common situation where, for example, Jon and Marie agree to a make a change but neglect to inform Pat of their decision.

Medical device manufacturers are usually quite comfortable with the processes of document control and change control with respect to managing manufacturing documents. The principles of these processes are reviewed in the following paragraphs. Subsequently, we will explore how these may be applied to design activities.

Document Control.

The features of a manufacturing document control system typically include the following:

- Documents should be identified (i.e., named and numbered) in accordance with some logical scheme which links the documents to the product or component they describe or depict and illuminates the drawing hierarchy.

- A master list or index of documents should be maintained which presents a comprehensive overview of the documentation which collectively defines the product and/or process.

- Approval procedures should be prescribed which govern entry of documents into the document control system.

- A history of document revisions should be maintained.

- Procedures for distributing copies of controlled documents and tracking their location should be prescribed.

- Files of controlled documents should be periodically inventoried to ensure that the contents are up to date.

- A person or persons should be assigned specific responsibility to oversee and carry out these procedures. It is desirable that the document control system be administered by a person who is not directly involved with developing or using the documents. For a small manufacturer, document control might be a part-time job for a technician or clerical staff person. More typically, one or more librarians or full-time clerical or paraprofessional employees are required to administer the system.

- There should be a procedure for removal and deletion of obsolete documents.

Change Control.

Manufacturing change control is usually implemented using a set of standardized procedures similar to the following:

- A change request might be originated by a developer, manager, reviewer, marketing representative, user, customer, quality assurance representative, or production personnel, and identifies a design problem which the requester believes should be corrected. Change requests are typically reviewed following the manufacturer's prescribed review process, and the request might be rejected, deferred, or accepted.

- If a change request is accepted and corrective action is straightforward, a change order might be issued on the spot to implement the change. The change order pertains to an explicitly identified document or group of documents, and specifies the detailed revision of the document content which will fix the identified problem.

- Often, the change request results in an assignment to developers to further study the problem and develop a suitable corrective action. If the change is extensive, wholesale revision of affected documents may be warranted in lieu of issuing change orders.

- Change requests and change orders should be communicated to all persons whose work might be impacted by the change.

- It may not be practical to immediately revise documents affected by a change order. Instead, the common practice is to distribute and attach a copy of the change order to each controlled copy of the original document.

- Change control procedures should incorporate review and assessment of the impact of the design change on the design input requirements and intended uses.

- A mechanism should be established to track all change requests and change orders to ensure proper disposition.

- Change control procedures are usually administered by the document control staff.

Application of Document and Change Controls to Design.

The design control system has to be concerned with the creation and revision of documents, as well as the management of finished documents. Additional mechanisms are required to provide needed flexibility while preserving the integrity of design documentation. These additional mechanisms are embodied in the procedures for review and approval of various documents.

It is important that the design change procedures always include re-verifying and revalidating the design. Fortunately, most design changes occur early in the design process, prior to extensive design validation. Thus, for most design changes, a simple inspection is all that is required. The later in the development cycle that the change occurs, the more important the validation review becomes. There are numerous cases when seemingly innocuous design changes made late in the design phase or following release of the design to market have had disastrous consequences.

For example, a manufacturer encountered problems in the field with a valve sticking in a ventilator due to moisture in the breathing circuit. The problem was resolved by slightly increasing the weight of the disc. Since the change was minor, minimal testing was performed to verify the change. Subsequently, when the revised valves entered production, significant numbers of valves began failing. Investigation revealed that the heavier disc was causing the valve cage to separate due to higher inertia. This failure mode was more serious than the original sticking problem, and resulted in a safety recall.

Section J. Design History File (DHF)

I. Requirements

§ 820.30(j) Design history file.

- Each manufacturer shall establish and maintain a DHF for each type of device.

- The DHF shall contain or reference the records necessary to demonstrate that the design was developed in accordance with the approved design plan and the requirements of this part.

Cross-reference to ISO 9001:1994 and ISO/DIS 13485 section 4.16 Control of quality records.

II. Definitions

§ 820.3(e) *Design history file (DHF)* means a compilation of records which describes the design history of a finished device.

III. Discussion and Points to Consider

There is no specific requirement in ISO 9001 or ISO 13485 for a design history file. However, in order to market a medical device in the United States, a manufacturer must comply with the U. S. Food and Drug Administration (FDA) quality system regulation, which requires a design history file. For this reason, some guidance is provided on the U. S. FDA design history file.

Other national regulations require some form of documentation and records. Product documentation required by Canada, Europe, and Japan contain certain elements of the U. S. FDA design history file requirements without requiring all the elements to be compiled in a file.

Virtually every section of the design control requirements specifies information which should be recorded. The compilation of these records is sometimes referred to as the design history file. Throughout this guidance document, suggestions are made when warranted as to the form and content of documents contained in the design history file.

The primary beneficiary of the device history file is the device manufacturer. For example, in one case, a microprocessor-controlled enteral feeding pump was reported to be behaving erratically in the field. Some of the symptoms pointed to software problems. But the manufacturer admitted that they did not possess a copy of the software source code for the product. The software had been developed by a contractor who had delivered only a master EPROM (memory chip) which was duplicated by the manufacturer to install the software in each machine. The contractor had subsequently withdrawn following a contractual dispute, leaving the manufacturer with no rights to the source code developed by the contractor, and no practical way to maintain the software. For this and other reasons, the product was the subject of a mandatory recall and all known units were collected and destroyed.

This is admittedly an extreme case, but many similar cases have been documented in which the manufacturer lacked design information necessary to validate a design and maintain it throughout the product life cycle. This occurs for the most innocent of reasons—contracts expire, companies reorganize, employees move on to new projects or new jobs. Even when the designer is available, he or she may forget why a particular decision was made years, months, or even weeks before. Since design decisions often directly affect the well-being of device users and patients, it is to the manufacturer's benefit to maintain the knowledge base which forms a basis for the product design.

Except for small projects, it is unusual for all design history documents to be filed in a single location. For example, many design engineers maintain laboratory notebooks which are typically retained in the engineers' personal files. In addition, the design history may include memoranda and electronic mail correspondence which are stored at various physical locations. Quality system plans applicable to a development project may reside in the quality assurance department, while the chief engineer may be responsible for maintaining design and development plans. These diverse records need not be consolidated at a single location. The intent is simply that manufacturers have access to the information when it is needed. If a manufacturer has established procedures for multiple filing systems which together satisfy that intent, there is no need to create additional procedures or records.

As an example of the level of detail which may be entailed, some manufacturers have policies covering laboratory notebooks. Manufacturers typically find that without such written procedures, a breakdown in communications eventually occurs, resulting in a loss of control. These procedures might address the following points.

- Laboratory notebooks are the property of the manufacturer, not the individual.

- A separate notebook is to be maintained for each project, and surrendered to the engineering librarian at the conclusion of the engineer's active participation in the project.

- Laboratory notebooks are to be surrendered if the employee leaves the company.

- Product development supervisors shall review employees' laboratory notebooks at specified intervals to ensure that records are complete, accurate, and legible.

There are no requirements on the location or organization of the design history file. In some cases, especially for simple designs, the designer will assemble and maintain the entire design history file. For larger projects, a document control system will likely be established for design documents, and these files will likely be maintained in some central location, usually within the product development department.

Based on the structure (or lack thereof) of the product development organization, more or less extensive controls will be required. For example, company policy should state unequivocally that all design history documentation is the property of the manufacturer, not the employee or contractor. Design and development contracts should explicitly specify the manufacturer's right to design information and establish standards for the form and content of design documentation. Finally, certain basic design information may be maintained in a single project file in a specified location. This may include the following:

- Detailed design and development plan specifying design tasks and deliverables.

- Copies of approved design input documents and design output documents.

- Documentation of design reviews.

- Validation documentation.

- When applicable, copies of controlled design documents and change control records.

INDs for Phase 2 and Phase 3 Studies: Chemistry, Manufacturing, and Controls Information

INDs for Phase 2 and Phase 3 Studies: Chemistry, Manufacturing, and Controls Information, Guidance for Industry [T, 1]

U.S. Department of Health and Human Services
Food and Drug Administration
Center for Drug Evaluation and Research (CDER)

May 2003

CMC

Contains Nonbinding Recommendations

This guidance provides recommendations to sponsors of investigational new drug applications (INDs) on the chemistry, manufacturing, and controls (CMC) information that would be submitted for phase 2 and phase 3 studies conducted under INDs. This document applies to human drugs (as defined in the Federal Food, Drug, and Cosmetic Act). It does not apply to botanical drug products, protein drug products derived from natural sources or produced by the use of biotechnology, or other biologics. The goals of this document are to (1) ensure that sufficient data will be submitted to the Agency to assess the safety, as well as the quality of the proposed clinical studies from the CMC perspective, (2) expedite the entry of new drug products into the marketplace by clarifying the type, extent, and reporting of CMC information for phase 2 and phase 3 studies, and (3) facilitate drug discovery and development.

Note: *This guidance represents the Food and Drug Administration's (FDA's) current thinking on this topic. It does not create or confer any rights for or on any person and does not operate to bind FDA or the public. You can use an alternative approach if the approach satisfies the requirements of the applicable statutes and regulations. If you want to discuss an alternative approach, contact the appropriate FDA staff. If you cannot identify the appropriate FDA staff, call the appropriate number listed on the title page of this guidance.*

[T] Available on the FDA website at: http://www.fda.gov/downloads/RegulatoryInformation/ Guidances/UCM126834.pdf

[1] This guidance has been prepared by IND Reform Committee of the Chemistry, Manufacturing, and Controls Coordinating Committee (CMCCC) in the Center for Drug Evaluation and Research (CDER) at the FDA.

I. Introduction

This guidance provides recommendations to sponsors of investigational new drug applications (INDs) on the chemistry, manufacturing, and controls (CMC) information that would be submitted for phase 2 and phase 3 studies conducted under INDs.[2] This document applies to human drugs (as defined in the Federal Food, Drug, and Cosmetic Act). The guidance does not apply to botanical drug products,[3] protein drug products derived from natural sources or produced by the use of biotechnology, or other biologics. The goals of the guidance are to (1) ensure that sufficient data will be submitted to the Agency to assess the safety, as well as the quality of the proposed clinical studies from the CMC perspective, (2) expedite the entry of new drug products into the marketplace by clarifying the type, extent, and reporting of CMC information for phase 2 and phase 3 studies, and (3) facilitate drug discovery and development.

The amount and depth of CMC information that would be submitted to the Agency depends, in large part, on the phase of the investigation, the testing proposed in humans, and whether the information is safety related. This guidance identifies CMC information that would be presented in information amendments (i.e., CMC safety information) and annual reports (i.e., corroborating information).

The recommendations in this guidance are intended to provide regulatory relief for IND sponsors by providing greater flexibility in the collecting and reporting of data and by avoiding redundant submissions. Four areas of regulatory relief are as follows:

- Certain information that traditionally has been submitted in information amendments would be identified as corroborating information (see section II.B.2) and can be submitted in an annual report.

- The limited phase 2 corroborating information recommended in section III need not be submitted before initiation of phase 2 studies and can be generated during phase 2 drug development.

- The phase 3 corroborating information recommended in section IV need not be submitted before the initiation of phase 3 studies and can be generated during phase 3 drug development.

- The corroborating information and a summary of CMC safety information submitted during a subject-reporting period would be included in the annual report. Therefore, there should be no need for *general CMC updates* at the end of phase 1 or phase 2.

---

[2] Recommendations are provided in other guidances for CMC issues relating to pre-IND, end-of-phase 2 (EOP2), and pre-new drug application (NDA) meetings (e.g., guidance for industry on *IND Meetings for Human Drugs and Biologics; Chemistry, Manufacturing, and Controls Information*), and pre-new drug application (NDA) rolling submissions (guidance for industry on *Fast Track Drug Development Programs – Designation, Development, and Application Review*).

[3] Information on INDs for botanical drug products will be provided in FDA's forthcoming guidance for industry on Botanical Drug Products (draft published August 2000; 65 FR 49247).

Although applicable to INDs that are sponsored by both commercial establishments and individual investigators, the guidance's greater value and relevance will be for commercial INDs.

For phase 1 submissions, sponsors can refer to the guidance for industry on *Content and Format of Investigational New Drug Applications (INDs) for Phase 1 Studies of Drugs, Including Well-Characterized, Therapeutic, Biotechnology-derived Products* (phase 1 guidance).

FDA's guidance documents, including this guidance, do not establish legally enforceable responsibilities. Instead, guidances describe the Agency's current thinking on a topic and should be viewed only as recommendations, unless specific regulatory or statutory requirements are cited. The use of the word *should* in Agency guidances means that something is suggested or recommended, but not required.

II. Background

A. Current Requirements

Under current regulations in the United States, use of a human drug product not previously authorized for marketing in the United States requires the submission of an IND to the gency. FDA's regulations at 21 CFR 312.22 and 312.23, respectively, contain the general principles underlying the IND submission and the general requirements for content and format. Section 312.23(a)(7)(i) requires that an IND for each phase of investigation include sufficient CMC information to ensure the proper identity, strength or potency, quality, and purity of the drug substance and drug product. The type of information submitted will depend on the phase of the investigation, the extent of the human study, the duration of the investigation, the nature and source of the drug substance, and the drug product dosage form.

As clinical development of the drug product proceeds, sponsors can discuss with the Agency the type of CMC information that would be submitted to support the use of the drug in all investigational phases. The Agency encourages sponsors to meet with the CMC review team, if appropriate, before the initiation of or during phase 3 clinical trials to discuss issues and protocols that might affect the approvability of the NDA. The Agency will grant CMC-specific meetings when justified (see CDER's guidance on *IND Meetings for Human Drugs and Biologics; Chemistry, Manufacturing, and Controls Information*).

B. General Principles

This guidance provides recommendations on CMC safety information and the limited corroborating information that should be submitted to support phase 2 and phase 3 studies. The scope of the guidance covers many different types of drug substances and drug products. Therefore, every recommendation may not be applicable to a particular drug substance or drug product. CMC safety information and corroborating information should be submitted in information amendments (21 CFR 312.31) and annual reports (21 CFR 312.33), respectively.

Under § 312.33 annual reports must be submitted during the ongoing development of the drug. With respect to CMC information, each annual report to the IND should include a summary of CMC safety information submitted in information amendments during the past year (i.e., subject-reporting period) and, when applicable, corroborating information. The annual report should also include updates of corroborating information or corrections to information previously provided to the IND that cannot be considered significant enough to warrant an information amendment.

FDA recommends that the sponsor carefully document its drug development program. This more-detailed information is often used to establish correlations between data generated during IND studies and the to-be-marketed product and to support other aspects of the NDA (e.g., process controls, justification of specifications) even when the submission of this information was not warranted during the IND studies.

1. CMC Safety Information

CMC safety information should be submitted to support the safe use of the drug. FDA reviews the safety information to determine whether a clinical hold on the IND is warranted. A summary of CMC safety information submitted during a subject-reporting period should be included in the annual report.

CMC safety information, as recommended in this guidance, and any other CMC information available to the sponsor that relates to the safe use of drug should be submitted in information amendments as follows:

- The CMC safety information identified in section III (Phase 2 Studies) should be submitted before initiation of the phase 2 studies. This information can be submitted during phase 1 or before the initiation of phase 2 studies.

- The CMC safety information identified in section IV (Phase 3 Studies) should be submitted before initiation of the phase 3 studies. This information can be submitted during phase 1 or phase 2 or before the initiation of phase 3 studies.

- When new information becomes available that relates to the safe use of the drug or when there are changes in previously submitted CMC safety information, the information should be submitted during IND clinical trials in an information amendment as the information becomes available.

FDA recommends that the sponsor carefully assess any changes in the drug substance and drug product manufacturing process or drug product formulation at any phase of clinical development to determine if the changes can directly or indirectly affect the safety of the product. For changes with a significant potential to affect the safety of the product (see examples below), an information amendment should be submitted that describes the changes and contains relevant information at a level of detail sufficient for an adequate review and assessment. When appropriate, this information should include data from tests on the drug substance and/or drug product produced from the previous manufacturing process and the changed manufacturing process to evaluate product equivalency, quality, and safety. In addition, when analytical data from tests on the drug substance and/or drug product demonstrate that the materials manufactured before and after are not comparable, sponsors should perform additional qualification and/or bridging studies to support the safety and bioavailability of the material to be used in the proposed trials and, when applicable, to support the quality of the trials.

The CMC safety concerns identified in the phase 1 guidance are equally applicable to phase 2 and phase 3.

CMC modifications throughout the IND process that can affect safety include, but are not limited to, a change in:

- the synthetic pathway used to manufacture the drug substance

    -- material change in one of the bond forming steps

    -- change in a solvent used for the last reaction and/or crystallization step

    -- change resulting in a different impurity profile

- the manufacturing process that can affect the quality of a drug substance produced by fermentation or derived from a natural source (plant, animal, or human)

- the manufacturing process that can directly or indirectly affect viral or impurity clearance for a drug substance produced by fermentation or derived from a natural source

- the manufacturing method from one manufacturing method (chemical synthesis, fermentation, or derivation from a natural source) to another

- source material (e.g., plant to animal, species, part used) or country of origin for a drug substance derived from a natural source

- species and/or strain of microorganism for a drug substance produced by fermentation

- certain aspects of specifications (see sections III.A.4, III.B.4, IV.A.4, and IV.B.4)

- the method of sterilization of the drug substance or drug product

- the route of administration

- the composition and/or dosage form of the drug product

- the drug product manufacturing process that can affect product quality

- the drug product container closure system that can affect product quality (e.g., metering capability, dose delivery)

2. Corroborating Information

Corroborating information is used to assess the scientific quality of the drug substance and drug product used in the clinical investigations to ensure that the clinical investigations will yield reliable and interpretable data and to corroborate the quality and safety of clinical materials used in earlier investigational phases.[4] Corroborating information is less likely to affect the safe use of the drug but should be submitted to ensure the proper identity, strength or potency, quality, and purity of the investigational drug (21 CFR 312.23(a)(7)(i)).

---

[4] Although FDA's review of phase 1 submissions will focus on assessing the safety of phase 1 investigations, FDA's review of phases 2 and 3 submissions will include an assessment of the scientific quality of the clinical investigations and the likelihood that the investigations will yield data capable of meeting statutory standards for market approvals (21 CFR 312.22(a)).

Corroborating information should be submitted in annual reports. In general, the corroborating information should focus on summaries and analyses of data rather than extensive compilations of data. However, there are some exceptions when compilations of data should be submitted (e.g., stability data). Occasionally, CDER may request more detailed corroborating information when warranted to assess the scientific quality of the investigations.

Submission of corroborating information in annual reports should occur as follows:

- Corroborating information specified in section III (Phase 2 Studies) that is generated during phase 1 need not be submitted until the first annual report after initiation of phase 2 studies. However, a sponsor can choose to submit the information in annual reports during phase 1 studies.

- Corroborating information specified in section III (Phase 2 Studies) that is generated during phase 2 studies should be submitted in the next annual report after the information becomes available. Corroborating information specified in section III need not be submitted before initiating phase 2 clinical trials.

- Corroborating information specified in section IV (Phase 3 Studies) that is generated earlier during phase 1 and phase 2 need not be submitted until the first annual report after initiation of phase 3 studies. However, a sponsor can choose to submit the information in annual reports during phase 1 and phase 2 studies.

- Corroborating information specified in section IV (Phase 3 Studies) that is generated during phase 3 studies should be submitted in the next annual report after the information becomes available. Corroborating information specified in section IV need not be submitted before initiating phase 3 clinical trials.

III. Phase 2 Studies

The CMC information provided to support the phase 2 studies should focus on additional CMC information to maintain the continued safety of the patients enrolled in these studies. Corroborating information should be provided to ensure that the clinical investigations will yield reliable and interpretable data. During or before phase 2, CMC safety information that has previously been submitted to the IND may have changed and consequently must be updated as required under 21 CFR 312.31.[5] For information amendments submitted to the IND during ongoing development, the emphasis should be on reporting significant changes that can have a safety-related impact. These include, but are not limited to, those changes specified in section II.B.1. In cases where studies begin with phase 2 clinical studies, CMC safety information should be submitted before initiation of the phase 2 studies as specified in the phase 1 guidance and in this section.

---

[5] Throughout section III, the guidance indicates that updates of or changes in CMC safety information from that provided for phase 1 should be submitted. This statement refers to updates that are warranted because changes are being made for materials intended for phase 2 studies. If changes relate to materials used in phase 1 studies, the changes should have been reported during the phase 1 studies.

The information recommended in sections III.A and III.B is considered CMC safety information that should be submitted in an information amendment before initiation of phase 2 studies, except where particular information is identified for submission in an annual report (i.e., corroborating information). Information that is considered corroborating information that can be submitted in an annual report is indicated in sections III.A.5, 6, and 7 and III.B.5 and 6.

Sponsors can reference an official compendium or other FDA recognized standard reference (e.g., *AOAC International Book of Methods*) to provide certain recommended CMC information (e.g., general methods, monograph standard) for an investigational drug substance or drug product, when applicable.[6] Reference to drug master files (DMFs) or other existing INDs or NDAs, with an authorization letter from the holder, sponsor, or applicant, can also be used to provide CMC information in support of the IND submission (21 CFR 312.23(b)).

A. Drug Substance

1. General Information

The Agency recommends that sponsors provide updates on the brief description of the drug substance, which was provided to support the phase 1 studies, and a more detailed description of the configuration and chemical structure for complex organic compounds (e.g., paclitaxel, polyketides). This information will be helpful in predicting the structure of possible metabolites. For peptides, characterization should include data on the amino acid sequence, and when relevant, peptide map. For DNA products, characterization should include nucleic acid sequence, DNA melting point, and side chain modifications when applicable.

2. Manufacture

a. Manufacturers

The addition, deletion, or change of any manufacturer of the drug substance from that specified during phase 1 should be reported.

b. Description of Manufacturing Process and Process Controls

An updated flow diagram for the synthesis or manufacturing process should be provided, if applicable. This information can be a general description. However, the flow diagram should contain the chemical structures and configurations, including stereochemical information of the starting materials, intermediates (either in situ or isolated), and, when feasible, significant side products.

Reagents, solvents and auxiliary materials, equipment (e.g., fermenters, columns), and provisions for monitoring and controlling critical conditions should be identified. Furthermore, a manufacturing step should be described in more detail if it is unique or critical to the synthetic or manufacturing process. For example, for fermentation and natural source drug substances, identification of model number or manufacturer of the fermenter is not warranted,

---

[6] Official compendium is defined in section 201(j) of the Federal Food, Drug, and Cosmetic Act (21 U.S.C. 321(j)).

but process controls that ensure performance of safety-related manufacturing steps (e.g., viral or impurity clearance) should be clearly described.

The general description of the synthetic and manufacturing process (e.g., fermentation, purification) provided to support the phase 1 studies should be updated from a safety perspective if changes or modifications have been introduced. Reprocessing procedures and controls need not be described except for natural source drug substances when the steps affect safety (e.g., virus or impurity clearance).

For sterile drug substances, updates on the manufacturing process from that provided for phase 1 studies should be submitted. The phase 2 information should include changes in the drug substance sterilization process (e.g., terminal sterilization to aseptic processing). Information related to the validation of the sterilization process need not be submitted at this time.

c. Control of Materials

The structures of the proposed starting materials should be provided if they have not been previously submitted. The source, analytical procedures, and test results for the starting materials should be submitted upon request. A list of any new reagents, solvents, auxiliary materials, or biological raw materials should be provided. For critical, complex materials (e.g., monoclonal antibodies configured in affinity matrices), a full description of the manufacturing process and acceptance criteria for the material should be provided.

For fermentation products or natural substances extracted from plant, human, or animal sources, the following information would have been provided in phase 1: (1) origin (e.g., country), source (e.g., pancreas), and taxonomy (e.g., family, genus, species, variety) of the starting materials or strain of the microorganism; (2) details of appropriate screening procedures for adventitious agents, if relevant; and (3) information to support the safe use of any materials of microbial, plant, human, or animal origin (e.g., certification, screening, testing). Any updates to the information submitted in phase 1 and any new information to support the safety of materials of human or animal origin should be provided.

d. Controls of Critical Steps and Intermediates

To the extent possible in phase 2, sponsors should provide information on controls of critical steps and intermediates and tentative acceptance criteria to ensure that the manufacturing process is controlled at predetermined points. Although controls of critical steps and intermediates can still be in development, information on controls for monitoring adventitious agents should be provided for fermentation and natural source (human or animal) drug substances, as appropriate.

3. Characterization

Evidence to reasonably support the proposed chemical structure of the drug substance should be provided. Data on particle size distribution and other physical properties (e.g., polymorphic or solid state form) should be generated so that relevant correlations can be established between data generated during early and late drug development. Data on the particle size distribution and/or physical properties should be submitted, when appropriate (e.g., inhalation, suspension, modified release solid dosage forms).

4. Control of Drug Substance

A *specification sheet* is a list of tests, analytical procedures, and acceptance criteria (i.e., numerical limits, ranges, or other criteria for the tests described). Critical quality attributes include, but are not limited to, identity, purity, quality, potency or strength, and impurities. During the clinical investigation process, the sponsor would establish tentative acceptance criteria that are continually refined based on data obtained from analysis of batches of drug substance and new information that becomes available. In the course of product development, the analytical technology or methodology often evolves parallel to the clinical investigations. In setting subsequent acceptance criteria, relevant correlations should be established between data generated during early and late drug development.

Any change in the tentative specification, including the tentative acceptance criteria, should be reported. This includes changes in the sponsor's drug substance specification and, if different, drug product manufacturer's acceptance testing for the drug substance. Test results and analytical data (e.g., infrared spectra, chromatograms) from batch release of representative clinical trial materials should be provided initially and when any change is made in the specification.

The analytical procedure (e.g., high-pressure liquid chromatography) used to perform a test and to support the tentative acceptance criteria should be briefly described and changes reported when the changes are such that an update of the brief description is warranted. A complete description of analytical procedures and appropriate validation data should be available for the analytical procedures that are not from an FDArecognized standard reference (e.g., official compendium, *AOAC International Book of Methods*), and this information should be submitted upon request.

New impurities (e.g., from a change in synthetic pathway) should be qualified, quantified, and reported, as appropriate. Procedures to evaluate impurities to support an NDA (e.g., recommended identification levels) may not be practical at this point in drug development. Suitable limits should be established based on manufacturing experience, stability data, and safety considerations.

5. Reference Standards or Materials

Where a recognized national or international standard (such as a standard from the World Health Organization (WHO)) is available, the manufacturer's reference material and/or working standard should be qualified against this standard. A national or international reference standard may not be available because many INDs will be for new molecular entities. In this case, the sponsor can select a batch of drug substance to be used as a reference material, against which initial clinical batches would be tested before their release. Preferably, the sponsor would establish a working standard even at the initial stage of drug development. For the purpose of this guidance, a *working standard* is a reference material that has been further characterized beyond the standard batch release tests. The protocol for establishing the working standard should be submitted in an information amendment. However, the results from the testing to establish the working standard can be reported in an annual report.

When a reference material is fully characterized, it would become the manufacturer's primary reference material. The manufacturer can continue to establish new working standards that are qualified against that primary reference material.

6. Container Closure System

The *container closure system* is defined as the sum of packaging components that together contain and protect the drug substance. A brief description of the container closure system (also referred to as the packaging system) and any subsequent changes should be provided in an annual report.

7. Stability

A description of the stability program to support the drug substance under clinical investigation in phase 2 should be submitted that includes a list of tests, analytical procedures, acceptance criteria, test time points for each of the tests, storage conditions, and the duration of the study.

Any available stability data for the clinical material used in the phase 1 study that were not reported during phase 1 should be provided in an information amendment. Stability data from representative clinical trial materials used in phase 2 should be provided in annual reports as the data become available.

If degradation of the drug substance occurs during manufacture or storage, this change should be considered when establishing acceptance criteria and monitoring quality.

Because of the inherent complexity of many drug substances, there may be no single stability-indicating assay or parameter that profiles all the stability characteristics of the drug substance. Consequently, the manufacturer should consider the development of stability-indicating analytical procedures that will detect significant changes in the quality of the drug substance. The nature of the particular drug substance will determine which tests should be included in the stability program. Performance of stability stress studies with the drug substance early in drug development is encouraged, as these studies provide information crucial to the selection of stability indicating analytical procedures for real time studies.

B. Drug Product

1. Description and Composition of the Drug Product

Any changes from the information specified for phase 1 (i.e., table listing of all components) should be provided. All components used in the manufacture of the drug product, regardless of whether or not they appear in the finished drug product, should be identified by their established names and a reference to a quality standard (e.g., *United States Pharmacopeia (USP)*, *National Formulary (NF)*) included. The quantitative composition on a per unit basis (e.g., milligram (mg)/milliliter (mL), mg/tablet) should be provided. However, quantitative values need not be reported for components that are removed during manufacturing and do not appear in the final drug product.

2. Manufacture

a. Manufacturers

The addition, deletion, or change of any manufacturer of the drug product from that specified during phase 1 should be reported.

b. Batch Formula

A representative batch formula should be provided if not already submitted. Quantitative information should be reported for all components in the batch formula whether or not the component appears in the final drug product.

c. Description of Manufacturing Process and Process Controls

An updated flow diagram and a brief step-by-step description of the manufacturing process should be provided. The description can focus on the unit operation (e.g., blending) rather than the individual manufacturing steps of the unit operation. Information, such as the following, need not be provided in either the flow diagram or description: (1) equipment used (e.g., V-blender); (2) the packaging and labeling process; (3) controls, except for sterile products (e.g., injectables, implants, ophthalmics) or atypical dosage forms (e.g., metered dose inhalation (MDI), liposomal encapsulation, implants, injectable microspheres); and (4) information on reprocessing procedures and controls, unless it is safety related. Where the qualitative formulation does not change, a single description of the manufacture of different strength unit doses can be provided.

For sterile products, updates on the manufacturing process from that provided for phase 1 studies should be submitted. The phase 2 information should include changes in the drug product sterilization process (e.g., terminal sterilization to aseptic processing). Information related to the validation of the sterilization process need not be submitted at this time.

3. Control of Excipients

For compendial excipients, references to quality standards (e.g., *USP, NF*) should be provided if changed from phase 1.

For noncompendial excipients, a specification sheet should be provided that identifies the tests and acceptance criteria and indicates the types of analytical procedure (e.g., HPLC) used. A complete description of the analytical procedures should be submitted upon request. A brief description of the manufacture and control of these components or an appropriate reference should be provided (e.g., DMF, NDA). Information for excipients not used in previously approved drug products in the United States (e.g., novel excipients) should be equivalent to that submitted for drug substances.

4. Control of Drug Product

Physicochemical tests (e.g., identity, assay, content uniformity, degradants, impurities, dissolution, viscosity, particle size), biological (e.g., potency), and microbiological tests (e.g., sterility and pyrogens or bacterial endotoxins for sterile products, antimicrobial preservative for multiple-dose sterile and nonsterile dosage forms, and microbial limits for nonsterile dosage forms) that have been added or deleted from the specification should be reported. Data on the particle size distribution and/or polymorphic form of the drug substance used in clinical trial materials should be included, when appropriate (e.g., inhalation, suspension, modified release solid dosage forms) so that relevant correlations can be established between data generated during early and late drug development and in vivo product performance. Relaxation of acceptance criteria or any change that affects safety should be reported. Test results and analytical data (e.g., chromatograms) from batch release of representative clinical trial materials should be provided initially and when any changes are made in the specification.

The analytical procedure (e.g., HPLC) used to perform a test should be briefly described and changes reported when the change is such that an update of the brief description is warranted. A complete description of analytical procedures and appropriate validation data should be available for analytical procedures that are not from an FDA-recognized standard reference (e.g., official compendium, *AOAC International Book of Methods*), and this information should be submitted upon request.

Data updates on the degradation profile should be provided so safety assessments can be made.

5. Container Closure System

The *container closure system* is defined as the sum of packaging components that together contain and protect the drug product. A brief description of the container closure system (also referred to as packaging system) should be provided in an information amendment. When changes are made in the container closure system, information should be submitted in an information amendment if there can be an effect on product quality. Otherwise, the changes can be reported in an annual report. Additional information may be requested for atypical delivery systems (e.g., MDIs, disposable injection devices).

6. Stability

A description of the stability program to support phase 2 clinical studies should be submitted that includes a list of the tests, analytical procedures, acceptance criteria, test time points for each of the tests, storage conditions, and the duration of the study, which should be long enough to cover the expected duration of the clinical studies.

Any stability data for the clinical material used in the phase 1 study that were not reported during phase 1 should be provided in an information amendment. The stability of reconstituted products should be studied and data submitted if not already provided or when there are formulation or diluent changes. Stability data from representative clinical trial materials used in phase 2 should be provided in annual reports as the data become available.

If degradation of the drug product occurs during manufacture or storage, this change should be considered when establishing acceptance criteria and monitoring quality.

Because of the inherent complexity of many dosage forms, there may be no single stability-indicating assay or parameter that profiles all the stability characteristics of the drug product. Consequently, the manufacturer should consider the development of stability-indicating analytical procedures that will detect significant changes in the quality of the drug product. The nature of the particular drug product will determine which tests should be included in the stability program.

IV. Phase 3 Studies

CMC development continues in parallel with the clinical development during phase 3 studies. The CMC safety information provided to support phase 3 studies should focus on the information that is warranted in maintaining the continued safety of the patients enrolled in these studies. For information amendments submitted to the IND during ongoing development, the emphasis should be on reporting significant changes that can have a safety-related impact. During or before phase 3, CMC safety information that has previously been submitted to the IND may have changed and, consequently, should be updated as required under § 312.31.[7] These changes include, but are not limited to, those specified in section II.B.1. The corroborating information should be provided in the annual report to ensure that the clinical investigations will yield reliable and interpretable data.

Sponsors can reference an official compendium or other FDA-recognized standard reference (e.g., *AOAC International Book of Methods*) to provide certain recommended CMC information (e.g., general methods, monograph standard) for an investigational drug substance or drug product, when applicable. Reference to DMFs or other existing INDs or NDAs, with an authorization letter from the holder, sponsor or applicant, can also be used to provide CMC information in support of the IND submission (21 CFR 312.23(b)).

Before the phase 3 studies, the sponsor can have an *end-of-phase-2* meeting, or during the phase 3 studies, a CMC specific *end-of-phase-2* meeting, with the Agency. As part of the preparation for that meeting, a background document is often provided that can be a valuable information amendment to the IND. The document would include updates to describe the materials already used and/or to be used in phase 3 studies, as well as put the studies performed to date in context with the prospective strategy for the ultimate NDA.

---

[7] Throughout section IV, the guidance indicates that updates of or changes in CMC safety information from that provided for phase 1 or phase 2 should be submitted. This statement refers to updates that would be warranted because changes are being made for materials intended for phase 3 studies. If changes relate to materials used during phase 1 or phase 2 studies, the changes should have been reported, as warranted, during those studies.

A. Drug Substance

1. General Information

General descriptive information on the physical, chemical, and biological characteristics of the drug substance should be provided in an annual report. This information, if not partition coefficient, dissociation constant (pKa), and isoelectric point (pI); (3) hygroscopicity; (4) crystal properties and morphology determined by thermal analysis (e.g., DSC, TGA),[8] powder X-ray diffraction and microscopy; (5) particle size and surface area; (6) melting point and boiling point; (7) optical rotation; (8) stereochemistry; and (9) biological activities, if applicable.

2. Manufacturer

a. Manufacturers

A list of all firms associated with the manufacture of the drug substance should be provided in an information amendment, including contract facilities used for manufacturing and/or testing (e.g., stability studies, quality control release testing).

b. Description of Manufacturing Process and Process Controls

An updated flow diagram should be provided in an information amendment when changes occur. A general step-by-step description of the synthesis and manufacturing processes, including the final isolation of the drug substance, should be provided in an annual report. Examples of relevant information that should be included in the description are as follows: (1) batch size (range); (2) relative ratios of reagents, solvents, and auxiliary materials; (3) process controls (brief description of the analytical procedures) and general operating conditions (time, temperature); (4) controls of critical steps and intermediates; (5) control of crystalline forms; and (6) literature references for any novel reactions or complex mechanisms. Reprocessing procedures and pertinent controls should be described in an annual report, except when reprocessing steps for fermentation or natural source drug substances are likely to affect safety (e.g., virus or impurity clearance). In this case, new or updates of previously submitted reprocessing information should be provided in an information amendment.

For sterile drug substances, updates on information from that provided in phase 1 and phase 2 should be submitted in an information amendment. The information should include a description of changes in the drug substance sterilization process (e.g., terminal sterilization to aseptic processing). Information related to the validation of the sterilization process need not be submitted at this time but should be submitted at the time of an NDA filing (see FDA guidance *Submission Documentation for Sterilization Process Validation in Applications for Human and Veterinary Drug Products*).

c. Control of Materials

In addition to the information provided during phase 1 and phase 2, analytical procedures and acceptance criteria for assessing the quality of starting materials should be provided in an

---

[8] Differential scanning calorimetry (DSC) and thermogravimetric analysis (TGA).

information amendment. Furthermore, a list of any new reagents, solvents, auxiliary materials, or biological raw materials should be provided in an information amendment. For critical, complex materials (e.g., monoclonal antibodies configured in affinity matrices), changes to the description of the manufacturing process and acceptance criteria should also be provided in an information amendment.

In an annual report, a table listing all reagents, solvents, and catalysts should be submitted that includes (1) a reference to a quality standard for each material used and (2) the specific identity test performed upon receipt of the material. When warranted, a more comprehensive list of tests and acceptance criteria should be submitted in an annual report for special reagents (e.g., reagents for kinetic resolution, sera, enzymes, or proteins).

Information should be provided in an information amendment to (1) update the information submitted in phase 1 and phase 2 regarding the origin of fermentation products or natural substances extracted from plant, human, or animal sources and (2) provide any new information to support the safety of materials of human or animal origin.

d. Controls of Critical Steps and Intermediates

Controls at critical steps in the synthesis or manufacturing process that ensure reaction completion, identity, purity or proper cell growth, and changes in critical controls reported during phase 2, should be described in an information amendment. For fermentation and natural source drug substances, changes to process controls that ensure performance of safety-related manufacturing steps (e.g., viral or impurity clearance) should be clearly described. Changes in controls for monitoring adventitious agents should be provided for fermentation and natural source drug substances, as appropriate.

For isolated intermediates that are controlled, the analytical procedures and tentative acceptance criteria should be described in an annual report. Tentative acceptance criteria can be used to allow for flexibility in the development process but should fulfill the primary purpose of quality control. The description of the analytical procedures can be brief, and appropriate validation information should be submitted upon request.

3. Characterization

Updates on the information previously provided during phase 2 should be provided in information amendments. The information amendment should include evidence to support the elucidation and characterization of the structure, which augments the information provided in phase 2. This information can include elemental analysis, conformational analysis, molecular weight determination, spectra from IR, NMR (1H & 13C), UV, MS, optical activity, and if available, single crystal X-ray diffraction data, if not previously provided.[9]

Analytical procedures used to characterize the primary reference material should also be provided in an information amendment. (See section IV.A.5)

---

[9] Infrared spectrometry (IR), nuclear magnetic resonance spectrometry (NMR), ultraviolet spectrometry (UV), and mass spectrometry (MS).

4. Control of Drug Substance

A detailed listing of all the tests performed on the drug substance (e.g., description, identity, assay, impurities, residual solvents) and the tentative acceptance criteria should be provided in an information amendment. A list should be provided for the testing performed by the sponsor and, if different, the drug product manufacturer. Test results and analytical data (e.g., infrared spectra, chromatograms) from batch release of representative clinical trial materials should also be provided in an information amendment initially and when any changes are made in the specification.

Information on the analytical procedures should be provided in an annual report. A general description of the analytical procedures should be provided that includes a citation to an official compendium, other FDA-recognized standard reference, or the sponsor's standard test procedure number, as appropriate. A description of analytical procedures with appropriate validation information should be provided for the analytical procedures that are not from an FDA-recognized standard reference (e.g., official compendium, *AOAC International Book of Methods*).

New impurities (e.g., from a change in synthetic pathway) should be identified, qualified, quantified, and reported, as appropriate. Procedures to evaluate impurities to support an NDA (e.g., recommended identification levels) may not be practical at this point in drug development. Suitable limits should be established based on manufacturing experience, available stability data, and safety considerations.

Suitable microbial limits should be established for nonsterile products that have potential to support microbial growth, if not previously submitted. These limits or changes in previously reported limits should be reported in an information amendment.

5. Reference Standards or Materials

If a national or international standard is not yet available, the sponsor should establish its own primary reference material during phase 3 studies. The manufacturer can continue to use the working standard from phase 2 or can establish a new working standard for lot release. The synthesis and purification of the primary reference material and/or working standard should be described in an information amendment if it differs from that of the investigational drug substance. The analytical procedures for and results from qualifying the working standard against the primary reference material should be provided in an annual report.

Where a recognized national or international standard is available and appropriate, the manufacturer's reference material and/or working standard should be qualified against this standard, and the results provided in an annual report.

6. Container Closure System

Any changes in the container closure system used to transport and/or store the bulk drug substance should be described in an annual report.

7. Stability

Changes in the drug substance stability program from that described for phase 2 (see section III.A.7) should be provided in an information amendment. Furthermore, the stability program should be updated to include descriptions of the stress and accelerated studies, if not previously described in phase 2. A container closure system that simulates the container closure system used to transport and/or store the bulk material can be used for the drug substance stability studies. Tests unique to the drug substance stability program (i.e., tests not included in section IV.A.4) should be defined and described.

Any stability data for the clinical material used in the phase 2 studies that were not reported during phase 2 should be provided in an information amendment. Stability data for representative clinical trial materials used in phase 3 should be provided in annual reports in tabular format as the data become available. The submitted stability information should include the lot number, manufacturer, manufacturing site, and the date of manufacture of the drug substance.

If not performed earlier, stress studies should be conducted during phase 3 to demonstrate the inherent stability of the drug substance, potential degradation pathways, and the capability and suitability of the proposed analytical procedures. The stress studies should assess the stability of the drug substance in different pH solutions, in the presence of oxygen and light, and at elevated temperatures and humidity levels. These one-time stress studies on a single batch are not considered part of the formal stability program. The results should be summarized and submitted in an annual report.

To ensure appropriate stability data are generated for filing at the NDA stage, a stability protocol that will be used for the formal stability studies should be developed.[10] The analytical procedures should be referenced to the drug substance specification section of the IND or an official compendium, if possible. Tests unique to the stability protocol should be defined and described. It is helpful if the stability protocol is submitted in an information amendment before or during phase 3 studies and is discussed at the end-ofphase- 2 meeting.

B. Drug Product

1. Description and Composition of the Drug Product

The sponsor should provide updated information regarding the components and composition in an information amendment if different from that reported in phase 1 and/or phase 2. The formulation for certain drug products delivered by devices (e.g., MDIs, dry powder inhalation (DPIs), and nasal sprays) should be similar to that intended for the marketed drug product.

---

[10] Applicants should refer to the forthcoming guidance Stability Testing of Drug Substances and Drug Products, when finalized, for information on stability protocols for formal stability studies. In June 1998 (63 FR 31224), the Agency made available a draft version of this guidance.

2. Manufacture

a. Manufacturers

A listing of all firms associated with the manufacture of the drug product should be provided in an information amendment, including any contract facilities used for manufacturing and/or testing (e.g., stability studies, packaging, labeling, quality control release testing).

b. Batch Formula

The sponsor should provide updated representative batch formula in an information amendment if different from that used in phase 1 and/or phase 2.

c. Description of Manufacturing Process and Process Controls

Changes in the manufacturing method for the drug product should be provided. An updated flow diagram and description of the manufacturing process (excluding packaging and labeling) should be provided in an information amendment. The description should indicate how the material is being processed and can be general enough to allow for flexibility in development. Reprocessing procedures and pertinent controls should be described in an information amendment, if applicable. A brief description of the packaging and labeling process for clinical supplies should be provided in an annual report.

For sterile products, updates on information from that provided for phase 1 and phase 2 should be submitted in an information amendment. The information should include a description of changes in the drug product sterilization process (e.g., terminal sterilization to aseptic processing). Information related to the validation of the sterilization process need not be submitted at this time but should be submitted at the time of an NDA filing (see FDA guidance for industry *Submission Documentation for Sterilization Process Validation in Applications for Human and Veterinary Drug Products*).

3. Control of Excipients

Updates on compendial excipient information previously provided should be submitted in information amendments. In certain cases, testing in addition to that specified in a compendium (e.g., functionality) can be useful and should be proposed.

For a noncompendial excipient, updates and a full description of the characterization, manufacture, control, analytical procedures, and acceptance criteria should be provided in an information amendment. Alternatively, a reference with authorization to a DMF can be provided. Information for excipients not used in previously approved drug products in the United States (e.g., novel excipients) should be equivalent to that submitted for new drug substances.

4. Control of Drug Product

A detailed listing of all the tests performed on the drug product and the tentative acceptance criteria should be provided in an information amendment. A summary table of test results and analytical data (e.g., chromatograms) from batch release of representative clinical trial materials should be provided in an information amendment initially and when any changes are made in the specification. Data on the particle size distribution and/or polymorphic form of the drug substance used in clinical trial materials should be included, when appropriate (e.g., inhalation, suspension, modified release solid dosage forms), so that relevant correlations can be established between data generated during early and late drug development and in vivo product performance.

A general description of the analytical procedures used should be provided in an annual report that includes a citation to an official compendium, other FDA-recognized standard reference, or the sponsor's standard test procedure number, as appropriate. A description of the analytical procedure with appropriate validation information should be provided for analytical procedures that are not from an FDA-recognized standard reference (e.g., official compendium, *AOAC International Book of Methods*).

Data updates on the degradation profile should be provided in an information amendment so safety assessments can be made. Degradation products should be identified, qualified, quantified, and reported, as appropriate. Evaluation procedures to support an NDA's degradants (e.g., recommended identification levels) may not be practical at this point in drug development. Suitable limits should be established based on manufacturing experience, stability data, and safety considerations.

For sterile-preserved products in multiple-dose containers or nonsterile-preserved products, a citation to the USP Antimicrobial Preservative-Effectiveness Test (APET) or a description of an equivalent procedure with the associated test validation information should be provided in an information amendment. This test should be performed at the lowest specified concentration of antimicrobial preservative specified for the drug product at release or at the end of the expiration dating period, whichever is less. The efficacy of preservative systems is evaluated based on their effect on inoculated microorganisms.

A dissolution testing program for oral immediate release dosage forms (e.g., tablets, capsules, suspensions) and a drug release program for modified release dosage forms (e.g., modified release tablets, capsules, suspensions, transdermal drug delivery systems) should be developed. Dissolution or drug release characteristics of a drug product, particularly the selection of the medium, are generally based on the pH solubility profile and pKa of the drug substance. Dissolution or drug release profiling should be performed in physiologically relevant media with reasonable speeds of rotation (e.g., basket at 50 or 100 rotations per minute (rpm), paddle at 50 rpm). The dissolution or drug release program at phase 3 should bring commonality to both the methodology and the proposed acceptance criteria by taking into consideration the results of dissolution or drug release testing of clinical, bioavailability, and bioequivalence batches (e.g., clinically studied formulations versus the *to-be-marketed* formulation) and relevant stability batches. The overall aim is to set in vitro dissolution or drug release acceptance criteria that ensure batch-to-batch and unit-to-unit consistency, post-NDA approval. The sponsor is encouraged to obtain concurrence on choice of apparatus, medium, rotation speed, and sampling time points from the Agency before the primary stability studies are initiated. Discussions with the Agency (e.g., at the end-of phase-2 meeting) can also include plans for establishing an in vivo-in vitro correlation (IVIVC) and characterizing the drug substance using the Biopharmaceutics Classification System (BCS).

5. Container Closure System

An update of the description of the container closure system should be provided in an information amendment if it differs from that reported during phase 2. When changes are made in the container closure system during phase 3 studies, information should be submitted in an information amendment if there can be an effect on product quality. Otherwise, the changes can be reported in an annual report.

For packaging components with compendial standards (e.g., glass, polyethylene containers), compliance with the appropriate compendial standards should be stated. If the sponsor refers to information in a Type III DMF, an authorization letter from the DMF holder should be provided. Additional information may be requested for atypical delivery systems (e.g., MDIs, disposable injection devices). The container closure system of certain drug products delivered by devices (e.g., MDIs, DPIs, nasal sprays) should be similar to that intended for the marketed drug product. A sponsor can consult with the appropriate CMC review team for additional guidance if it has any questions.

6. Stability

A stability program should be designed to monitor the chemical, physical, biological, or microbiological (if applicable) stability of the drug product throughout the clinical testing program. Changes in the drug product stability program from that described for phase 2 (see section III.B.7) should be provided in an information amendment. A brief description should be provided in an information amendment for each of the attributes being investigated in the stability program (i.e., long-term and accelerated), demonstrating that the appropriate controls and storage conditions are in place to ensure the quality of the product used in clinical trials. Furthermore, tests unique to the drug product stability program (i.e., tests not included in section IV.B.4) should be adequately defined and described.

Any stability data for the clinical material used in the phase 2 studies that were not reported during phase 2 should be provided in an information amendment. Stability data for representative clinical material used in phase 3 should be provided in annual reports in tabular format as the data become available. The submitted information should include the batch number, manufacturing site, date of manufacture of the drug product, and relevant information on the drug substance (e.g., lot number, manufacturer) used to manufacture the drug product. The analytical results for each test should be reported. Representative chromatograms should be provided in the annual report, if applicable.

For certain drug products, one-time stress testing can be warranted to assess the potential for changes in the physical (e.g., phase separation, precipitation, aggregation, changes in particular size distribution) and/or chemical (e.g., degradation and/or interaction of components) characteristics of the drug product. The studies could include testing to assess the effect of high temperature, humidity, oxidation, photolysis and/or thermal cycling. The relevant data should be provided in an annual report.

To ensure appropriate stability data are generated for filing at the NDA stage, a stability protocol that will be used for the formal stability studies should be developed.[11] The analytical procedures should be referenced to the control of drug product section of the IND or an official compendium, if possible. Tests unique to the stability protocol should be defined and described. It is helpful if the stability protocol is submitted in an information amendment before or during phase 3 studies and is discussed at the end-ofphase- 2 meeting, especially for those protocols including bracketing and matrixing approaches.

V. Placebo

A brief, general description of the composition, manufacture, and control of the placebo provided during phase 1 should be updated or provided for phase 2 and/or phase 3 if the placebo is being used for the first time. This information and any updates to this information should be provided in an information amendment. When placebos are used in clinical trials, the placebo clinical study materials should be tested to demonstrate the absence of the drug substance. The results from the placebo testing should be submitted in an annual report.

VI. Labeling

Updates of the information provided for phase 1 should be submitted in information amendments during phase 2 and phase 3.

VII. Environmental Assessments

Updates on information already submitted and on whether a claim for a previous categorical exclusion has changed should be provided in information amendments for phase 2 and phase 3 (see FDA guidance for industry on *Environmental Assessment of Human Drug and Biologics Applications*).

Resources

FDA continues to update existing and publish new guidance documents. An applicant should ensure that it is using current guidance when preparing a submission. CDER guidances are available on the Internet at http://www.fda.gov/cder/guidance/index.htm.

ICH Guidances

Although not intended to be applicable to IND applications, the International Conference on Harmonization (ICH) documents below can serve as valuable resources in guiding the course of product development.

ICH Q1A Stability Testing of New Drug Substances and Products

---

[11] Applicants should refer to the forthcoming guidance *Stability Testing of Drug Substances and Drug Products*, when finalized, for information on stability protocols for formal stability studies. In June 1998 (63 FR 31224), the Agency made available a draft version of this guidance.

ICH Q1B Photostability Testing of New Substances and Products

ICH Q1C Stability Testing for New Dosage Forms

ICH Q2A Validation of Analytical Procedures

ICH Q2B Validation of Analytical Procedures: Methodology

ICH Q3A Impurities in New Drug Substances

ICH Q3B Impurities in New Drug Products

ICH Q3C Impurities: Residual Solvents

ICH Q5A Viral Safety Evaluation of Biotechnology Products Derived From Cell Lines of Human or Animal Origin

ICH Q5B Quality of Biotechnological Products: Analysis of the Expression Construct in Cells Used for Production of r-DNA Derived Protein Products

ICH Q5D Quality of Biotechnological/Biological Products: Derivation and Characterization of Cell Substrates Used for Production of Biotechnological/Biological Products

ICH Q6A Specifications: Test Procedures and Acceptance Criteria for New Drug Substances and New Drug Products: Chemical Substances

ICH Q7A Good Manufacturing Practice Guide for Active Pharmaceutical Ingredients

FDA Guidances for Industry

Draft FDA guidances are cited for completeness of information and are not for implementation until finalized.

Analytical Procedures and Methods Validation — Chemistry, Manufacturing, and Controls Documentation, Draft. The Agency published this draft guidance in the Federal Register on August 30, 2000 (65 FR 52776).

Botanical Drug Products, Draft. The Agency published this draft guidance in the Federal Register on August 11, 2000 (65 FR 49247).

Container Closure Systems for Packaging Human Drugs and Biologics

Content and Format of Investigational New Drug Applications (INDs) for Phase 1 Studies of Drugs, Including Well-Characterized, Therapeutic, Biotechnology-derived Products

Drug Product: Chemistry, Manufacturing, and Controls Information, Draft. The Agency published this draft guidance in the Federal Register on January 28, 2003 (68 FR 4219).

Drug Substance: Chemistry, Manufacturing, and Controls Information (forthcoming guidance)

Environmental Assessment of Human Drug and Biologics Applications

Fast Track Drug Development Programs – Designation, Development, and Applications Review

IND Meetings for Human Drugs and Biologics — Chemistry, Manufacturing, and Controls Information

Monoclonal Antibody Used as Reagents in Drug Manufacturing

Stability Testing of Drug Substances and Drug Products, Draft. The Agency published this draft guidance in the Federal Register on June 8, 1998 (63 FR 31224).

Submission of Chemistry, Manufacturing, and Controls Information for Synthetic Peptide Substances

Submission Documentation for Sterilization Process Validation in Applications for Human and Veterinary Drug Products

Part VI

Selected FDA GCP/Clinical Trial Guidance Documents:
Electronic Data

Computerized Systems Used in Clinical Investigations

Computerized Systems Used in Clinical Investigations, Guidance for Industry [T, 1]

U.S. Department of Health and Human Services
Food and Drug Administration (FDA)
Office of the Commissioner (OC)

May 2007

Contains Nonbinding Recommendations

This guidance provides recommendations to sponsors, contract research organizations, data management centers, clinical investigators and institutional review boards regarding the use of computerized systems in clinical investigations.

Note: This guidance represents the Food and Drug Administration's (FDA's) current thinking on this topic. It does not create or confer any rights for or on any person and does not operate to bind FDA or the public. You can use an alternative approach if the approach satisfies the requirements of the applicable statutes and regulations. If you want to discuss an alternative approach, contact the appropriate FDA staff. If you cannot identify the appropriate FDA staff, call the appropriate number listed on the title page of this guidance.

I. Introduction

This document provides to sponsors, contract research organizations (CROs), data management centers, clinical investigators, and institutional review boards (IRBs), recommendations regarding the use of computerized systems in clinical investigations. The computerized system applies to records in electronic form that are used to create, modify, maintain, archive, retrieve, or transmit clinical data required to be maintained, or submitted to the FDA. Because the source data[2] are necessary for the reconstruction and evaluation of the study to determine the safety of food and color additives and safety and effectiveness of new

---

[T] Available on the FDA website at: http://www.fda.gov/downloads/Drugs/ GuidanceComplianceRegulatoryInformation/Guidances/UCM070266.pdf

[1] This guidance has been prepared by the Office of Critical Path Programs, the Good Clinical Practice Program, and the Office of Regulatory Affairs in cooperation with Bioresearch Monitoring Program Managers for each Center within the Food and Drug Administration.

[2] Under 21 CFR 312.62(b), reference is made to records that are part of case histories as "supporting data"; the ICH *E6 Good Clinical Practice* consolidated guidance uses the term "source documents." For the purpose of this guidance, these terms describe the same information and have been used interchangeably.

human and animal drugs,[3] and medical devices, this guidance is intended to assist in ensuring confidence in the reliability, quality, and integrity of electronic source data and source documentation (i.e., electronic records).

This guidance supersedes the guidance of the same name dated April 1999; and supplements the guidance for industry on *Part 11, Electronic Records; Electronic Signatures — Scope and Application* and the Agency's international harmonization efforts[4] when applying these guidances to source data generated at clinical study sites.

FDA's guidance documents, including this guidance, do not establish legally enforceable responsibilities. Instead, guidances describe the Agency's current thinking on a topic and should be viewed only as recommendations, unless specific regulatory or statutory requirements are cited. The use of the word should in Agency guidances means that something is suggested or recommended, but not required.

II. Background

There is an increasing use of computerized systems in clinical trials to generate and maintain source data and source documentation on each clinical trial subject. Such electronic source data and source documentation must meet the same fundamental elements of data quality (e.g., attributable, legible, contemporaneous, original,[5] and accurate) that are expected of paper records and must comply with all applicable statutory and regulatory requirements. FDA's acceptance of data from clinical trials for decision-making purposes depends on FDA's ability to verify the quality and integrity of the data during FDA on-site inspections and audits. (21 CFR 312, 511.1(b), and 812).

In March 1997, FDA issued 21 CFR part 11, which provides criteria for acceptance by FDA, under certain circumstances, of electronic records, electronic signatures, and handwritten signatures executed to electronic records as equivalent to paper records and handwritten signatures executed on paper. After the effective date of 21 CFR part 11, significant concerns regarding the interpretation and implementation of part 11 were raised by both FDA and industry. As a result, we decided to reexamine 21 CFR part 11 with the possibility of proposing additional rulemaking, and exercising enforcement discretion regarding enforcement of certain part 11 requirements in the interim.

This guidance finalizes the draft guidance for industry entitled *Computerized Systems Used in Clinical Trials*, dated September 2004 and supplements the guidance for industry entitled *Part*

---

[3] Human drugs include biological drugs.

[4] In August 2003, FDA issued the guidance for industry entitled *Part 11, Electronic Records; Electronic Signatures-Scope and Application* clarifying that the Agency intends to interpret the scope of part 11 narrowly and to exercise enforcement discretion with regard to part 11 requirements for validation, audit trails, record retention, and record copying. In 1996, the International Conference on Harmonisation of Technical Requirements for Registration of Pharmaceuticals for Human Use (ICH) issued *E6 Good Clinical Practice: Consolidated Guidance.*

[5] FDA is allowing original documents to be replaced by copies provided the copies are identical and have been verified as such (See, e.g., FDA Compliance Policy Guide # 7150.13). See Definitions section for a definition of original data.

11, Electronic Records; Electronic Signatures – Scope and Application (Scope and Application Guidance), dated August 2003. The Scope and Application Guidance clarified that the Agency intends to interpret the scope of part 11 narrowly and to exercise enforcement discretion with regard to part 11 requirements for validation, audit trails, record retention, and record copying. However, other Part 11 provisions remain in effect.

The approach outlined in the Scope and Application Guidance, which applies to electronic records generated as part of a clinical trial, should be followed until such time as Part 11 is amended.

III. Scope

The principles outlined in this guidance should be used for computerized systems that contain any data that are relied on by an applicant in support of a marketing application, including computerized laboratory information management systems that capture analytical results of tests conducted during a clinical trial. For example, the recommendations in this guidance would apply to computerized systems that create source documents (electronic records) that satisfy the requirements in 21 CFR 312.62(b) and 812.140(b), such as case histories. This guidance also applies to recorded source data transmitted from automated instruments directly to a computerized system (e.g., data from a chemistry autoanalyser or a Holter monitor to a laboratory information system). This guidance also applies when source documentation is created in hardcopy and later entered into a computerized system, recorded by direct entry into a computerized system, or automatically recorded by a computerized system (e.g., an ECG reading). The guidance does not apply to computerized medical devices that generate such data and that are otherwise regulated by FDA.

IV. Recommendations

This guidance provides the following recommendations regarding the use of computerized systems in clinical investigations.

A. Study Protocols

Each specific study protocol should identify each step at which a computerized system will be used to create, modify, maintain, archive, retrieve, or transmit source data. This information can be included in the protocol at the time the investigational new drug application (IND), Investigational Device Exemption (IDE), or Notice of Claimed Investigational Exemption for a New Animal Drug containing the protocols is submitted or at any time after the initial submission.

The computerized systems should be designed: (1) to satisfy the processes assigned to these systems for use in the specific study protocol (e.g., record data in metric units, blind the study), and (2) to prevent errors in data creation, modification, maintenance, archiving, retrieval, or transmission (e.g., inadvertently unblinding a study).

B. Standard Operating Procedures

There should be specific procedures and controls in place when using computerized systems to create, modify, maintain, or transmit electronic records, including when collecting source data at clinical trial sites. A list of recommended standard operating procedures (SOPs) is provided in Appendix A. Such SOPs should be maintained either on-site or be remotely accessible through electronic files as part of the specific study records, and the SOPs should be made available for use by personnel and for inspection by FDA.

C. Source Documentation and Retention

When original observations are entered directly into a computerized system, the electronic record is the source document. Under 21 CFR 312.62, 511.1(b)(7)(ii) and 812.140, the clinical investigator must retain records required to be maintained under part 312, § 511.1(b), and part 812, for a period of time specified in these regulations. This requirement applies to the retention of the original source document, or a copy of the source document.

When source data are transmitted from one system to another (e.g., from a personal data assistant to a sponsor's server), or entered directly into a remote computerized system (e.g., data are entered into a remote server via a computer terminal that is located at the clinical site), or an electrocardiogram at the clinical site is transmitted to the sponsor's computerized system, a copy of the data should be maintained at another location, typically at the clinical site but possibly at some other designated site. Copies should be made contemporaneously with data entry and should be preserved in an appropriate format, such as XML, PDF or paper formats.

D. Internal Security Safeguards

1. Limited Access

Access must be limited to authorized individuals (21 CFR 11.10(d). This requirement can be accomplished by the following recommendations. We recommend that each user of the system have an individual account. The user should log into that account at the beginning of a data entry session, input information (including changes) on the electronic record, and log out at the completion of data entry session. The system should be designed to limit the number of log-in attempts and to record unauthorized access log-in attempts.

Individuals should work only under their own password or other access key and not share these with others. The system should not allow an individual to log onto the system to provide another person access to the system. We also recommend that passwords or other access keys be changed at established intervals commensurate with a documented risk assessment.

When someone leaves a workstation, the person should log off the system. Alternatively, an automatic log off may be appropriate for long idle periods. For short periods of inactivity, we recommend that a type of automatic protection be installed against unauthorized data entry (e.g., an automatic screen saver can prevent data entry until a password is entered).

2. Audit Trails

It is important to keep track of all changes made to information in the electronic records that document activities related to the conduct of the trial (audit trails). The use of audit trails or other security measures helps to ensure that only authorized additions, deletions, or alterations of information in the electronic record have occurred and allows a means to reconstruct significant details about study conduct and source data collection necessary to verify the quality and integrity of data. Computer-generated, time-stamped audit trails or other security measures can also capture information related to the creation, modification, or deletion of electronic records and may be useful to ensure compliance with the appropriate regulation.

The need for audit trails should be determined based on a justified and documented risk assessment that takes into consideration circumstances surrounding system use, the likelihood that information might be compromised, and any system vulnerabilities. Should it be decided that audit trails or other appropriate security measures are needed to ensure electronic record integrity, personnel who create, modify, or delete electronic records should not be able to modify the documents or security measures used to track electronic record changes. Computergenerated, time-stamped electronic audits trails are the preferred method for tracking changes to electronic source documentation.

Audit trails or other security methods used to capture electronic record activities should describe when, by whom, and the reason changes were made to the electronic record. Original information should not be obscured though the use of audit trails or other security measures used to capture electronic record activities.

3. Date/Time Stamps

Controls should be established to ensure that the system's date and time are correct. The ability to change the date or time should be limited to authorized personnel, and such personnel should be notified if a system date or time discrepancy is detected. Any changes to date or time should always be documented. We do not expect documentation of time changes that systems make automatically to adjust to daylight savings time conventions.

We recommend that dates and times include the year, month, day, hour, and minute and encourage synchronization of systems to the date and time provided by international standardsetting agencies (e.g., U.S. National Institute of Standards and Technology provides information about universal time, coordinated (UTC)).

Computerized systems are likely to be used in multi-center clinical trials and may be located in different time zones. For systems that span different time zones, it is better to implement time stamps with a clear understanding of the time zone reference used. We recommend that system documentation explain time zone references as well as zone acronyms or other naming conventions.

E. External Security Safeguards

In addition to internal safeguards built into a computerized system, external safeguards should be put in place to ensure that access to the computerized system and to the data is restricted to authorized personnel. Staff should be kept thoroughly aware of system security measures and the importance of limiting access to authorized personnel.

Procedures and controls should be put in place to prevent the altering, browsing, querying, or reporting of data via external software applications that do not enter through the protective system software.

You should maintain a cumulative record that indicates, for any point in time, the names of authorized personnel, their titles, and a description of their access privileges. That record should be kept in the study documentation, accessible for use by appropriate study personnel and for inspection by FDA investigators.

We also recommend that controls be implemented to prevent, detect, and mitigate effects of computer viruses, worms, or other potentially harmful software code on study data and software.

F. Other System Features

1. Direct Entry of Data

We recommend that you incorporate prompts, flags, or other help features into your

computerized system to encourage consistent use of clinical terminology and to alert the user to data that are out of acceptable range. You should not use programming features that automatically enter data into a field when the field is bypassed (default entries). However, you can use programming features that permit repopulation of information specific to the subject. To avoid falsification of data, you should perform a careful analysis in deciding whether and when to use software programming instructions that permit data fields to be automatically populated.

2. Retrieving Data

The computerized system should be designed in such a way that retrieved data regarding each individual subject in a study is attributable to that subject. Reconstruction of the source documentation is essential to FDA's review of the clinical study submitted to the Agency. Therefore, the information provided to FDA should fully describe and explain how source data were obtained and managed, and how electronic records were used to capture data.

It is not necessary to reprocess data from a study that can be fully reconstructed from available documentation. Therefore, the actual application software, operating systems, and software development tools involved in the processing of data or records need not be retained.

3. Dependability System Documentation

For each study, documentation should identify what software and hardware will be used to create, modify, maintain, archive, retrieve, or transmit clinical data. Although it need not be submitted to FDA, this documentation should be retained as part of the study records and be available for inspection by FDA (either on-site or remotely accessible).

4. System Controls

When electronic formats are the only ones used to create and preserve electronic records, sufficient backup and recovery procedures should be designed to protect against data loss. Records should regularly be backed up in a procedure that would prevent a catastrophic loss and ensure the quality and integrity of the data. Records should be stored at a secure location specified in the SOP. Storage should typically be offsite or in a building separate from the original records.

We recommend that you maintain backup and recovery logs to facilitate an assessment of the nature and scope of data loss resulting from a system failure.

5. Change Controls

The integrity of the data and the integrity of the protocols should be maintained when making changes to the computerized system, such as software upgrades, including security and performance patches, equipment, or component replacement, or new instrumentation. The effects of any changes to the system should be evaluated and some should be validated depending on risk. Changes that exceed previously established operational limits or design specifications should be validated. Finally, all changes to the system should be documented.

G. Training of Personnel

Those who use computerized systems must determine that individuals (e.g., employees, contractors) who develop, maintain, or use computerized systems have the education, training and experience necessary to perform their assigned tasks (21 CFR 11.10(i)).

Training should be provided to individuals in the specific operations with regard to computerized systems that they are to perform. Training should be conducted by qualified individuals on a continuing basis, as needed, to ensure familiarity with the computerized system and with any changes to the system during the course of the study.

We recommend that computer education, training, and experience be documented.

Definitions

The following is a list of definitions for terms used in, and for the purposes of, this guidance document.

Audit Trail: For the purpose of this guidance, an audit trail is a process that captures details such as additions, deletions, or alterations of information in an electronic record without obliterating the original record. An audit trail facilitates the reconstruction of the course of such details relating to the electronic record.

Certified Copy: A certified copy is a copy of original information that has been verified, as indicated by a dated signature, as an exact copy having all of the same attributes and information as the original.

Computerized System: A computerized system includes computer hardware, software, and associated documents (e.g., user manual) that create, modify, maintain, archive, retrieve, or transmit in digital form information related to the conduct of a clinical trial.

Direct Entry: Direct entry is recording data where an electronic record is the original means of capturing the data. Examples are the keying by an individual of original observations into a system, or automatic recording by the system of the output of a balance that measures subject's body weight.

Electronic Record: An electronic record is any combination of text, graphics, data, audio, pictorial, or other information representation in digital form that is created, modified, maintained, archived, retrieved, or distributed by a computer system.

Original data: For the purpose of this guidance, original data are those values that represent the first recording of study data. FDA is allowing original documents and the original data recorded on those documents to be replaced by copies provided the copies are identical and have been verified as such (see FDA Compliance Policy Guide # 7150.13).

Source Documents: Original documents and records including, but not limited to, hospital records, clinical and office charts, laboratory notes, memoranda, subjects' diaries or evaluation checklists, pharmacy dispensing records, recorded data from automated instruments, copies or transcriptions certified after verification as being accurate and complete, microfiches, photographic negatives, microfilm or magnetic media, x-rays, subject files, and records kept at the pharmacy, at the laboratories, and at medico-technical departments involved in a clinical trial.

Transmit: Transmit is to transfer data within or among clinical study sites, contract research organizations, data management centers, sponsors, or to FDA.

References

FDA, 21 CFR Part 11, "Electronic Records; Electronic Signatures; Final Rule." *Federal Register* Vol. 62, No. 54, 13429, March 20, 1997.

FDA, *Compliance Program Guidance Manual,* "Compliance Program 7348.810 – Bioresearch Monitoring - Sponsors, Contract Research Organizations and Monitors," February 21, 2001.

FDA, *Compliance Program Guidance Manual,* "Compliance Program 7348.811 – Bioresearch Monitoring - Clinical Investigators," September 30, 2000.

FDA, *Good Clinical Practice VICH GL9.*

FDA, *Guideline for the Monitoring of Clinical Investigations.*

FDA, *Information Sheets for Institutional Review Boards and Clinical Investigators.* http://www.fda.gov/ic/ohrt/irbs/default.htm

FDA, *E6 Good Clinical Practice: Consolidated Guidance.* http://www.fda.gov/cder/guidance/959fnl.pdf.

FDA, *Part 11, Electronic Records; Electronic Signatures — Scope and Application*, 2003.

FDA, *General Principles of Software Validation; Guidance for Industry and FDA Staff.*

Part 11, Electronic Records; Electronic Signatures--Scope and Application

Guidance for Industry on Part 11, Electronic Records; Electronic Signatures--Scope and Application [Ŧ, 1]

U.S. Department of Health and Human Services
Food and Drug Administration
Center for Drug Evaluation and Research (CDER)
Center for Biologics Evaluation and Research (CBER)
Center for Devices and Radiological Health (CDRH)
Center for Food Safety and Applied Nutrition (CFSAN)
Center for Veterinary Medicine (CVM)
Office of Regulatory Affairs (ORA)

August 2003

Pharmaceutical CGMPs

Contains Nonbinding Recommendations

This guidance is intended to describe the FDA's current thinking regarding the scope and application of part 11 of Title 21 of the Code of Federal Regulations; Electronic Records; Electronic Signatures (21 CFR Part 11).

Note: This guidance represents the Food and Drug Administration's (FDA's) current thinking on this topic. It does not create or confer any rights for or on any person and does not operate to bind FDA or the public. You can use an alternative approach if the approach satisfies the requirements of the applicable statutes and regulations. If you want to discuss an alternative approach, contact the appropriate FDA staff. If you cannot identify the appropriate FDA staff, call the appropriate number listed on the title page of this guidance.

I. Introduction

This guidance is intended to describe the Food and Drug Administration's (FDA's) current thinking regarding the scope and application of part 11 of Title 21 of the Code of Federal Regulations; Electronic Records; Electronic Signatures (21 CFR Part 11).[2]

[Ŧ] Available on the FDA website at: http://www.fda.gov/downloads/RegulatoryInformation/ Guidances/UCM126953.pdf

[1] This guidance has been prepared by the Office of Compliance in the Center for Drug Evaluation and Research (CDER) in consultation with the other Agency centers and the Office of Regulatory Affairs at the Food and Drug Administration.

[2] 62 FR 13430

This document provides guidance to persons who, in fulfillment of a requirement in a statute or another part of FDA's regulations to maintain records or submit information to FDA,[3] have chosen to maintain the records or submit designated information electronically and, as a result, have become subject to part 11. Part 11 applies to records in electronic form that are created, modified, maintained, archived, retrieved, or transmitted under any records requirements set forth in Agency regulations. Part 11 also applies to electronic records submitted to the Agency under the Federal Food, Drug, and Cosmetic Act (the Act) and the Public Health Service Act (the PHS Act), even if such records are not specifically identified in Agency regulations (§ 11.1).

The underlying requirements set forth in the Act, PHS Act, and FDA regulations (other than part 11) are referred to in this guidance document as *predicate rules*.

As an outgrowth of its current good manufacturing practice (CGMP) initiative for human and animal drugs and biologics,[4] FDA is re-examining part 11 as it applies to all FDA regulated products. We anticipate initiating rulemaking to change part 11 as a result of that re-examination. This guidance explains that we will narrowly interpret the scope of part 11. While the re-examination of part 11 is under way, we intend to exercise enforcement discretion with respect to certain part 11 requirements. That is, we do not intend to take enforcement action to enforce compliance with the validation, audit trail, record retention, and record copying requirements of part 11 as explained in this guidance. However, records must still be maintained or submitted in accordance with the underlying predicate rules, and the Agency can take regulatory action for noncompliance with such predicate rules.

In addition, we intend to exercise enforcement discretion and do not intend to take (or recommend) action to enforce any part 11 requirements with regard to systems that were operational before August 20, 1997, the effective date of part 11 (commonly known as legacy systems) under the circumstances described in section III.C.3 of this guidance.

Note that part 11 remains in effect and that this exercise of enforcement discretion applies only as identified in this guidance.

FDA's guidance documents, including this guidance, do not establish legally enforceable responsibilities. Instead, guidances describe the Agency's current thinking on a topic and should be viewed only as recommendations, unless specific regulatory or statutory requirements are cited. The use of the word *should* in Agency guidances means that something is suggested or recommended, but not required.

---

[3] These requirements include, for example, certain provisions of the Current Good Manufacturing Practice regulations (21 CFR Part 211), the Quality System regulation (21 CFR Part 820), and the Good Laboratory Practice for Nonclinical Laboratory Studies regulations (21 CFR Part 58).

[4] See *Pharmaceutical CGMPs for the 21st Century: A Risk-Based Approach; A Science and Risk-Based Approach to Product Quality Regulation Incorporating an Integrated Quality Systems Approach* at www.fda.gov/oc/guidance/gmp.html.

II. Background

In March of 1997, FDA issued final part 11 regulations that provide criteria for acceptance by FDA, under certain circumstances, of electronic records, electronic signatures, and handwritten signatures executed to electronic records as equivalent to paper records and handwritten signatures executed on paper. These regulations, which apply to all FDA program areas, were intended to permit the widest possible use of electronic technology, compatible with FDA's responsibility to protect the public health.

After part 11 became effective in August 1997, significant discussions ensued among industry, contractors, and the Agency concerning the interpretation and implementation of the regulations. FDA has (1) spoken about part 11 at many conferences and met numerous times with an industry coalition and other interested parties in an effort to hear more about potential part 11 issues; (2) published a compliance policy guide, CPG 7153.17: Enforcement Policy: 21 CFR Part 11; Electronic Records; Electronic Signatures; and (3) published numerous draft guidance documents including the following:

- *21 CFR Part 11; Electronic Records; Electronic Signatures, Validation*

- *21 CFR Part 11; Electronic Records; Electronic Signatures, Glossary of Terms*

- *21 CFR Part 11; Electronic Records; Electronic Signatures, Time Stamps*

- *21 CFR Part 11; Electronic Records; Electronic Signatures, Maintenance of Electronic Records*

- *21 CFR Part 11; Electronic Records; Electronic Signatures, Electronic Copies of Electronic Records*

Throughout all of these communications, concerns have been raised that some interpretations of the part 11 requirements would (1) unnecessarily restrict the use of electronic technology in a manner that is inconsistent with FDA's stated intent in issuing the rule, (2) significantly increase the costs of compliance to an extent that was not contemplated at the time the rule was drafted, and (3) discourage innovation and technological advances without providing a significant public health benefit. These concerns have been raised particularly in the areas of part 11 requirements for validation, audit trails, record retention, record copying, and legacy systems.

As a result of these concerns, we decided to review the part 11 documents and related issues, particularly in light of the Agency's CGMP initiative. In the Federal Register of February 4, 2003 (68 FR 5645), we announced the withdrawal of the draft guidance for industry, 21 CFR Part 11; Electronic Records; Electronic Signatures, Electronic Copies of Electronic Records. We had decided we wanted to minimize industry time spent reviewing and commenting on the draft guidance when that draft guidance may no longer represent our approach under the CGMP initiative. Then, in the Federal Register of February 25, 2003 (68 FR 8775), we announced the withdrawal of the part 11 draft guidance documents on validation, glossary of terms, time stamps,[5] maintenance of electronic records, and CPG 7153.17. We received

---

[5] Although we withdrew the draft guidance on time stamps, our current thinking has not changed in that when using time stamps for systems that span different time zones, we do not expect you to record the signer's local time. When using time stamps, they should be implemented with a clear understanding of the

valuable public comments on these draft guidances, and we plan to use that information to help with future decision-making with respect to part 11. We do not intend to re-issue these draft guidance documents or the CPG.

We are now re-examining part 11, and we anticipate initiating rulemaking to revise provisions of that regulation. To avoid unnecessary resource expenditures to comply with part 11 requirements, we are issuing this guidance to describe how we intend to exercise enforcement discretion with regard to certain part 11 requirements during the re-examination of part 11. As mentioned previously, part 11 remains in effect during this re-examination period.

III. Discussion

A. Overall Approach to Part 11 Requirements

As described in more detail below, the approach outlined in this guidance is based on three main elements:

- Part 11 will be interpreted narrowly; we are now clarifying that fewer records will be considered subject to part 11.

- For those records that remain subject to part 11, we intend to exercise enforcement discretion with regard to part 11 requirements for validation, audit trails, record retention, and record copying in the manner described in this guidance and with regard to all part 11 requirements for systems that were operational before the effective date of part 11 (also known as legacy systems).

- We will enforce all predicate rule requirements, including predicate rule record and recordkeeping requirements.

It is important to note that FDA's exercise of enforcement discretion as described in this guidance is limited to specified part 11 requirements (setting aside legacy systems, as to which the extent of enforcement discretion, under certain circumstances, will be more broad). We intend to enforce all other provisions of part 11 including, but not limited to, certain controls for closed systems in § 11.10. For example, we intend to enforce provisions related to the following controls and requirements:

- limiting system access to authorized individuals

- use of operational system checks

- use of authority checks

- use of device checks

- determination that persons who develop, maintain, or use electronic systems have the education, training, and experience to perform their assigned tasks

---

time zone reference used. In such instances, system documentation should explain time zone references as well as zone acronyms or other naming conventions.

- establishment of and adherence to written policies that hold individuals accountable for actions initiated under their electronic signatures

- appropriate controls over systems documentation

- controls for open systems corresponding to controls for closed systems bulleted above (§ 11.30)

- requirements related to electronic signatures (e.g., §§ 11.50, 11.70, 11.100, 11.200, and 11.300)

We expect continued compliance with these provisions, and we will continue to enforce them. Furthermore, persons must comply with applicable predicate rules, and records that are required to be maintained or submitted must remain secure and reliable in accordance with the predicate rules.

B. Details of Approach – Scope of Part 11

1. Narrow Interpretation of Scope

We understand that there is some confusion about the scope of part 11. Some have understood the scope of part 11 to be very broad. We believe that some of those broad interpretations could lead to unnecessary controls and costs and could discourage innovation and technological advances without providing added benefit to the public health. As a result, we want to clarify that the Agency intends to interpret the scope of part 11 narrowly.

Under the narrow interpretation of the scope of part 11, with respect to records required to be maintained under predicate rules or submitted to FDA, when persons choose to use records in electronic format in place of paper format, part 11 would apply. On the other hand, when persons use computers to generate paper printouts of electronic records, and those paper records meet all the requirements of the applicable predicate rules and persons rely on the paper records to perform their regulated activities, FDA would generally not consider persons to be "using electronic records in lieu of paper records" under §§ 11.2(a) and 11.2(b). In these instances, the use of computer systems in the generation of paper records would not trigger part 11.

2. Definition of Part 11 Records

Under this narrow interpretation, FDA considers part 11 to be applicable to the following records or signatures in electronic format (part 11 records or signatures):

- Records that are required to be maintained under predicate rule requirements and that are maintained in electronic format *in place of paper format*. On the other hand, records (and any associated signatures) that are not required to be retained under predicate rules, but that are nonetheless maintained in electronic format, are not part 11 records. We recommend that you determine, based on the predicate rules, whether specific records are part 11 records. We recommend that you document such decisions.

- Records that are required to be maintained under predicate rules, that are maintained in electronic format *in addition to paper format*, and that *are relied on to perform regulated activities*.

In some cases, actual business practices may dictate whether you are using electronic records instead of paper records under § 11.2(a). For example, if a record is required to be maintained under a predicate rule and you use a computer to generate a paper printout of the electronic records, but you nonetheless rely on the electronic record to perform regulated activities, the Agency may consider you to be using the electronic record instead of the paper record. That is, the Agency may take your business practices into account in determining whether part 11 applies.

Accordingly, we recommend that, for each record required to be maintained under predicate rules, you determine in advance whether you plan to rely on the electronic record or paper record to perform regulated activities. We recommend that you document this decision (e.g., in a Standard Operating Procedure (SOP), or specification document).

- Records submitted to FDA, under predicate rules (even if such records are not specifically identified in Agency regulations) in electronic format (assuming the records have been identified in docket number 92S-0251 as the types of submissions the Agency accepts in electronic format). However, a record that is not itself submitted, but is used in generating a submission, is not a part 11 record unless it is otherwise required to be maintained under a predicate rule and it is maintained in electronic format.

- Electronic signatures that are intended to be the equivalent of handwritten signatures, initials, and other general signings required by predicate rules. Part 11 signatures include electronic signatures that are used, for example, to document the fact that certain events or actions occurred in accordance with the predicate rule (e.g. *approved*, *reviewed*, and *verified*).

C. Approach to Specific Part 11 Requirements

1. Validation

The Agency intends to exercise enforcement discretion regarding specific part 11 requirements for validation of computerized systems (§ 11.10(a) and corresponding requirements in § 11.30). Although persons must still comply with all applicable predicate rule requirements for validation (e.g., 21 CFR 820.70(i)), this guidance should not be read to impose any additional requirements for validation.

We suggest that your decision to validate computerized systems, and the extent of the validation, take into account the impact the systems have on your ability to meet predicate rule requirements. You should also consider the impact those systems might have on the, reliability, integrity, availability, and authenticity of required records and signatures. Even if there is no predicate rule requirement to validate a system, in some instances it may still be important to validate the system.

We recommend that you base your approach on a justified and documented risk assessment and a determination of the potential of the system to affect product quality and safety, and record integrity. For instance, validation would not be important for a word processor used only to generate SOPs.

For further guidance on validation of computerized systems, see FDA's guidance for industry and FDA staff *General Principles of Software Validation* and also industry guidance such as the *GAMP 4 Guide* (See References).

2. Audit Trail

The Agency intends to exercise enforcement discretion regarding specific part 11 requirements related to computer-generated, time-stamped audit trails (§ 11.10 (e), (k)(2) and any corresponding requirement in §11.30). Persons must still comply with all applicable predicate rule requirements related to documentation of, for example, date (e.g., § 58.130(e)), time, or sequencing of events, as well as any requirements for ensuring that changes to records do not obscure previous entries.

Even if there are no predicate rule requirements to document, for example, date, time, or sequence of events in a particular instance, it may nonetheless be important to have audit trails or other physical, logical, or procedural security measures in place to ensure the trustworthiness and reliability of the records.[6] We recommend that you base your decision on whether to apply audit trails, or other appropriate measures, on the need to comply with predicate rule requirements, a justified and documented risk assessment, and a determination of the potential effect on product quality and safety and record integrity. We suggest that you apply appropriate controls based on such an assessment. Audit trails can be particularly appropriate when users are expected to create, modify, or delete regulated records during normal operation.

3. Legacy Systems[7]

The Agency intends to exercise enforcement discretion with respect to all part 11 requirements for systems that otherwise were operational prior to August 20, 1997, the effective date of part 11, under the circumstances specified below.

This means that the Agency does not intend to take enforcement action to enforce compliance with any part 11 requirements if all the following criteria are met for a specific system:

- The system was operational before the effective date.

- The system met all applicable predicate rule requirements before the effective date.

- The system currently meets all applicable predicate rule requirements.

- You have documented evidence and justification that the system is fit for its intended use (including having an acceptable level of record security and integrity, if applicable).

If a system has been changed since August 20, 1997, and if the changes would prevent the system from meeting predicate rule requirements, Part 11 controls should be applied to Part 11 records and signatures pursuant to the enforcement policy expressed in this guidance.

---

[6] Various guidance documents on information security are available (see References).

[7] In this guidance document, we use the term legacy system to describe systems already in operation before the effective date of part 11.

4. Copies of Records

The Agency intends to exercise enforcement discretion with regard to specific part 11 requirements for generating copies of records (§ 11.10 (b) and any corresponding requirement in §11.30). You should provide an investigator with reasonable and useful access to records during an inspection. All records held by you are subject to inspection in accordance with predicate rules (e.g., §§ 211.180(c), (d), and 108.35(c)(3)(ii)).

We recommend that you supply copies of electronic records by:

- Producing copies of records held in common portable formats when records are maintained in these formats

- Using established automated conversion or export methods, where available, to make copies in a more common format (examples of such formats include, but are not limited to, PDF, XML, or SGML)

In each case, we recommend that the copying process used produces copies that preserve the content and meaning of the record. If you have the ability to search, sort, or trend part 11 records, copies given to the Agency should provide the same capability if it is reasonable and technically feasible. You should allow inspection, review, and copying of records in a human readable form at your site using your hardware and following your established procedures and techniques for accessing records.

5. Record Retention

The Agency intends to exercise enforcement discretion with regard to the part 11 requirements for the protection of records to enable their accurate and ready retrieval throughout the records retention period (§ 11.10 (c) and any corresponding requirement in §11.30). Persons must still comply with all applicable predicate rule requirements for record retention and availability (e.g., §§ 211.180(c),(d), 108.25(g), and 108.35(h)).

We suggest that your decision on how to maintain records be based on predicate rule requirements and that you base your decision on a justified and documented risk assessment and a determination of the value of the records over time.

FDA does not intend to object if you decide to archive required records in electronic format to nonelectronic media such as microfilm, microfiche, and paper, or to a standard electronic file format (examples of such formats include, but are not limited to, PDF, XML, or SGML). Persons must still comply with all predicate rule requirements, and the records themselves and any copies of the required records should preserve their content and meaning. As long as predicate rule requirements are fully satisfied and the content and meaning of the records are preserved and archived, you can delete the electronic version of the records. In addition, paper and electronic record and signature components can co-exist (i.e., a hybrid[8] situation) as long

---

[8] Examples of hybrid situations include combinations of paper records (or other nonelectronic media) and electronic records, paper records and electronic signatures, or handwritten signatures executed to electronic records.

as predicate rule requirements are met and the content and meaning of those records are preserved.

IV. References

Food and Drug Administration References

1. Glossary of Computerized System and Software Development Terminology (Division of Field Investigations, Office of Regional Operations, Office of Regulatory Affairs, FDA 1995) (http://www.fda.gov/ora/inspect_ref/igs/gloss.html)

2. General Principles of Software Validation; Final Guidance for Industry and FDA Staff (FDA, Center for Devices and Radiological Health, Center for Biologics Evaluation and Research, 2002) (http://www.fda.gov/cdrh/comp/guidance/938.html)

3. Guidance for Industry, FDA Reviewers, and Compliance on Off-The-Shelf Software Use in Medical Devices (FDA, Center for Devices and Radiological Health, 1999) (http://www.fda.gov/cdrh/ode/guidance/585.html)

4. Pharmaceutical CGMPs for the 21st Century: A Risk-Based Approach; A Science and Risk-Based Approach to Product Quality Regulation Incorporating an Integrated Quality Systems Approach (FDA 2002) (http://www.fda.gov/oc/guidance/gmp.html)

Industry References

1. The Good Automated Manufacturing Practice (GAMP) Guide for Validation of Automated Systems, GAMP 4 (ISPE/GAMP Forum, 2001) (http://www.ispe.org/gamp/)

2. ISO/IEC 17799:2000 (BS 7799:2000) Information technology – Code of practice for information security management (ISO/IEC, 2000)

3. ISO 14971:2002 Medical Devices- Application of risk management to medical devices (ISO, 2001)

Part VII

Combined Glossary and Index

Combined Glossary

505(b)(2) Application means an application submitted under section 505(b)(1) of the act for a drug for which the investigations described in section 505(b)(1)(A) of the act and relied upon by the applicant for approval of the application were not conducted by or for the applicant and for which the applicant has not obtained a right of reference or use from the person by or for whom the investigations were conducted. [21 CFR § 314]

510(k) – See Premarket Notification.

A

Abbreviated application means the application described under 314.94, including all amendments and supplements to the application. "Abbreviated application" applies to both an abbreviated new drug application and an abbreviated antibiotic application. [21 CFR § 314]

Ability to detect change — Evidence that a PRO instrument can identify differences in scores over time in individuals or groups who have changed with respect to the measurement concept. [Patient-Reported Outcome Measures Guidance]

Acceptance criteria means the product specifications and acceptance/rejection criteria, such as acceptable quality level and unacceptable quality level, with an associated sampling plan, that are necessary for making a decision to accept or reject a lot or batch (or any other convenient subgroups of manufactured units). [21 CFR § 210]

Act means the Federal Food, Drug, and Cosmetic Act (secs. 201-903 (21 U.S.C. 321-393)). [21 CFR § 11]

    Act means the Federal Food, Drug, and Cosmetic Act, as amended (secs. 201-902, 52 Stat. 1040et seq. as amended (21 U.S.C. 321-392)). [21 CFR § 50, § 56, § 58, § 312, § 812]

    Act means the Federal Food, Drug, and Cosmetic Act, as amended (21 U.S.C. 301et seq.). [21 CFR § 210]

    Act means the Federal Food, Drug, and Cosmetic Act (sections 201-901 (21 U.S.C. 301-392)). [21 CFR § 314]

Act means the Federal Food, Drug, and Cosmetic Act (sections 201-902, 52 Stat. 1040 et seq., as amended (21 U.S.C. 321-392)). [21 CFR § 814]

Active ingredient means any component that is intended to furnish pharmacological activity or other direct effect in the diagnosis, cure, mitigation, treatment, or prevention of disease, or to affect the structure or any function of the body of man or other animals. The term includes those components that may undergo chemical change in the manufacture of the drug product and be present in the drug product in a modified form intended to furnish the specified activity or effect. [21 CFR § 210]

Actual yield means the quantity that is actually produced at any appropriate phase of manufacture, processing, or packing of a particular drug product.

Adverse drug experience is any adverse event associated with the use of a new animal drug, whether or not considered to be drug related, and whether or not the new animal drug was used in accordance with the approved labeling (i.e., used according to label directions or used in an extralabel manner, including but not limited to different route of administration, different species, different indications, or other than labeled dosage). Adverse drug experience includes, but is not limited to: [21 CFR § 514]

(1) An adverse event occurring in animals in the course of the use of an animal drug product by a veterinarian or by a livestock producer or other animal owner or caretaker.

(2) Failure of a new animal drug to produce its expected pharmacological or clinical effect (lack of expected effectiveness).

(3) An adverse event occurring in humans from exposure during manufacture, testing, handling, or use of a new animal drug.

Agency means the Food and Drug Administration. [21 CFR § 11]

Agreement meeting – a meeting, under section 520(g)(7) of the Act (21 U.S.C. § 360j(g)(7)), that is available to anyone planning to investigate the safety or effectiveness of a class III device (see definition below) or any implant. The purpose of the meeting is to reach agreement on the key parameters of the investigational plan, including the study protocol. The meeting is to be held within 30 days of the receipt of a written request. FDA will document in writing any agreement reached and make it a part of the administrative record. The agreement is binding on FDA and can only be changed with the written agreement of the applicant or when there is a substantial scientific issue essential to determining the safety or effectiveness of the device. See 21 U.S.C. § 360j(g)(7). A guidance document regarding these meetings, "Early Collaboration Meetings Under the FDA Modernization Act (FDAMA); Final Guidance for Industry and for CDRH Staff," is available at http://www.fda.gov/MedicalDevices/DeviceRegulationandGuidance/GuidanceDocuments/ucm073604.htm.

ANADA is an abbreviated new animal drug application including all amendments and supplements. [21 CFR § 514]

Analyte specific reagent (ASR) - ASRs are defined as "antibodies, both polyclonal and monoclonal, specific receptor proteins, ligands, nucleic acid sequences, and similar reagents which, through specific binding or chemical reactions with substances in a specimen, are intended for use in a diagnostic application for identification and quantification of an individual chemical substance or ligand in biological specimens." 21 CFR 864.4020(a). ASRs are medical devices that are regulated by FDA. They are subject to general controls, including current Good Manufacturing Practices (cGMPs), 21 CFR Part 820, as well as the specific provisions of the ASR regulations (21 CFR 809.10(e), 809.30, 864.4020).

Applicant means the party who submits a marketing application to FDA for approval of a drug, device, or biologic product. The applicant is responsible for submitting the appropriate certification and disclosure statements required in this part. [21 CFR § 54]

> ***Applicant*** means any person who submits an application or abbreviated application or an amendment or supplement to them under this part to obtain FDA approval of a new drug or an antibiotic drug and any person who owns an approved application or abbreviated application. [21 CFR § 314]

> ***Applicant*** is a person or entity who owns or holds on behalf of the owner the approval for an NADA or an ANADA, and is responsible for compliance with applicable provisions of the act and regulations. [21 CFR § 514]

Application means the application described under 314.50, including all amendements and supplements to the application. [21 CFR § 314]

Application for research or marketing permit includes: [21 CFR § 50]

(1) A color additive petition, described in part 71.

(2) A food additive petition, described in parts 171 and 571.

(3) Data and information about a substance submitted as part of the procedures for establishing that the substance is generally recognized as safe for use that results or may reasonably be expected to result, directly or indirectly, in its becoming a component or otherwise affecting the characteristics of any food, described in 170.30 and 570.30.

(4) Data and information about a food additive submitted as part of the procedures for food additives permitted to be used on an interim basis pending additional study, described in 180.1.

(5) Data and information about a substance submitted as part of the procedures for establishing a tolerance for unavoidable contaminants in food and food-packaging materials, described in section 406 of the act.

(6) An investigational new drug application, described in part 312 of this chapter.

(7) A new drug application, described in part 314.

(8) Data and information about the bioavailability or bioequivalence of drugs for human use submitted as part of the procedures for issuing, amending, or repealing a bioequivalence requirement, described in part 320.

(9) Data and information about an over-the-counter drug for human use submitted as part of the procedures for classifying these drugs as generally recognized as safe and effective and not misbranded, described in part 330.

(10) Data and information about a prescription drug for human use submitted as part of the procedures for classifying these drugs as generally recognized as safe and effective and not misbranded, described in this chapter.

(11) [Reserved]

(12) An application for a biologics license, described in part 601 of this chapter.

(13) Data and information about a biological product submitted as part of the procedures for determining that licensed biological products are safe and effective and not misbranded, described in part 601.

(14) Data and information about an in vitro diagnostic product submitted as part of the procedures for establishing, amending, or repealing a standard for these products, described in part 809.

(15) An Application for an Investigational Device Exemption, described in part 812.

(16) Data and information about a medical device submitted as part of the procedures for classifying these devices, described in section 513.

(17) Data and information about a medical device submitted as part of the procedures for establishing, amending, or repealing a standard for these devices, described in section 514.

(18) An application for premarket approval of a medical device, described in section 515.

(19) A product development protocol for a medical device, described in section 515.

(20) Data and information about an electronic product submitted as part of the procedures for establishing, amending, or repealing a standard for these products, described in section 358 of the Public Health Service Act.

(21) Data and information about an electronic product submitted as part of the procedures for obtaining a variance from any electronic product performance standard, as described in 1010.4.

(22) Data and information about an electronic product submitted as part of the procedures for granting, amending, or extending an exemption from a radiation safety performance standard, as described in 1010.5.

(23) Data and information about a clinical study of an infant formula when submitted as part of an infant formula notification under section 412(c) of the Federal Food, Drug, and Cosmetic Act.

(24) Data and information submitted in a petition for a nutrient content claim, described in 101.69 of this chapter, or for a health claim, described in 101.70 of this chapter.

(25) Data and information from investigations involving children submitted in a new dietary ingredient notification, described in 190.6 of this chapter.

Application for research or marketing permit includes: [21 CFR § 56]

(1) A color additive petition, described in part 71.

(2) Data and information regarding a substance submitted as part of the procedures for establishing that a substance is generally recognized as safe for a use which results or may reasonably be expected to result, directly or indirectly, in its becoming a component or otherwise affecting the characteristics of any food, described in 170.35.

(3) A food additive petition, described in part 171.

(4) Data and information regarding a food additive submitted as part of the procedures regarding food additives permitted to be used on an interim basis pending additional study, described in 180.1.

(5) Data and information regarding a substance submitted as part of the procedures for establishing a tolerance for unavoidable contaminants in food and food-packaging materials, described in section 406 of the act.

(6) An investigational new drug application, described in part 312 of this chapter.

(7) A new drug application, described in part 314.

(8) Data and information regarding the bioavailability or bioequivalence of drugs for human use submitted as part of the procedures for issuing, amending, or repealing a bioequivalence requirement, described in part 320.

(9) Data and information regarding an over-the-counter drug for human use submitted as part of the procedures for classifying such drugs as generally recognized as safe and effective and not misbranded, described in part 330.

(10) An application for a biologics license, described in part 601 of this chapter.

(11) Data and information regarding a biological product submitted as part of the procedures for determining that licensed biological products are safe and effective and not misbranded, as described in part 601 of this chapter.

(12) An Application for an Investigational Device Exemption, described in part 812.

(13) Data and information regarding a medical device for human use submitted as part of the procedures for classifying such devices, described in part 860.

---

(14) Data and information regarding a medical device for human use submitted as part of the procedures for establishing, amending, or repealing a standard for such device, described in part 861.

(15) An application for premarket approval of a medical device for human use, described in section 515 of the act.

(16) A product development protocol for a medical device for human use, described in section 515 of the act.

(17) Data and information regarding an electronic product submitted as part of the procedures for establishing, amending, or repealing a standard for such products, described in section 358 of the Public Health Service Act.

(18) Data and information regarding an electronic product submitted as part of the procedures for obtaining a variance from any electronic product performance standard, as described in 1010.4.

(19) Data and information regarding an electronic product submitted as part of the procedures for granting, amending, or extending an exemption from a radiation safety performance standard, as described in 1010.5.

(20) Data and information regarding an electronic product submitted as part of the procedures for obtaining an exemption from notification of a radiation safety defect or failure of compliance with a radiation safety performance standard, described in subpart D of part 1003.

(21) Data and information about a clinical study of an infant formula when submitted as part of an infant formula notification under section 412(c) of the Federal Food, Drug, and Cosmetic Act.

(22) Data and information submitted in a petition for a nutrient content claim, described in 101.69 of this chapter, and for a health claim, described in 101.70 of this chapter.

(23) Data and information from investigations involving children submitted in a new dietary ingredient notification, described in 190.6 of this chapter.

Application for research or marketing permit includes: [21 CFR § 58]

(1) A color additive petition, described in part 71.

(2) A food additive petition, described in parts 171 and 571.

(3) Data and information regarding a substance submitted as part of the procedures for establishing that a substance is generally recognized as safe for use, which use results or may reasonably be expected to result, directly or indirectly, in its becoming a component or otherwise affecting the characteristics of any food, described in 170.35 and 570.35.

(4) Data and information regarding a food additive submitted as part of the procedures regarding food additives permitted to be used on an interim basis pending additional study, described in 180.1.

(5) Aninvestigational new drug application, described in part 312 of this chapter.

(6) Anew drug application, described in part 314.

(7) Data and information regarding an over-the-counter drug for human use, submitted as part of the procedures for classifying such drugs as generally recognized as safe and effective and not misbranded, described in part 330.

(8) Data and information about a substance submitted as part of the procedures for establishing a tolerance for unavoidable contaminants in food and food-packaging materials, described in parts 109 and 509.

(9) [Reserved]

(10) A Notice of Claimed Investigational Exemption for a New Animal Drug, described in part 511.

(11) A new animal drug application, described in part 514.

(12) [Reserved]

(13) An application for a biologics license, described in part 601 of this chapter.

(14) An application for an investigational device exemption, described in part 812.

(15) An Application for Premarket Approval of a Medical Device, described in section 515 of the act.

(16) A Product Development Protocol for a Medical Device, described in section 515 of the act.

(17) Data and information regarding a medical device submitted as part of the procedures for classifying such devices, described in part 860.

(18) Data and information regarding a medical device submitted as part of the procedures for establishing, amending, or repealing a performance standard for such devices, described in part 861.

(19) Data and information regarding an electronic product submitted as part of the procedures for obtaining an exemption from notification of a radiation safety defect or failure of compliance with a radiation safety performance standard, described in subpart D of part 1003.

(20) Data and information regarding an electronic product submitted as part of the procedures for establishing, amending, or repealing a standard for such product, described in section 358 of the Public Health Service Act.

(21) Data and information regarding an electronic product submitted as part of the procedures for obtaining a variance from any electronic product performance standard as described in 1010.4.

(22) Data and information regarding an electronic product submitted as part of the procedures for granting, amending, or extending an exemption from any electronic product performance standard, as described in 1010.5.

(23) A premarket notification for a food contact substance, described in part 170, subpart D, of this chapter.

Approval letter means a written communication to an applicant from FDA approving an application or an abbreviated application. [21 CFR § 314]

Assent means a child's affirmative agreement to participate in a clinical investigation. Mere failure to object may not, absent affirmative agreement, be construed as assent. [21 CFR § 50]

Assess the effects of the change means to evaluate the effects of a manufacturing change on the identity, strength, quality, purity, and potency of a drug product as these factors may relate to the safety or effectiveness of the drug product. [21 CFR § 314]

Audit Trail: For the purpose of this guidance, an audit trail is a process that captures details such as additions, deletions, or alterations of information in an electronic record without obliterating the original record. An audit trail facilitates the reconstruction of the course of such details relating to the electronic record. [Computerized Systems Used in Clinical Investigations Guidance]

Authorized generic drug means a listed drug, as defined in this section, that has been approved under section 505(c) of the act and is marketed, sold, or distributed directly or indirectly to retail class of trade with labeling, packaging (other than repackaging as the listed drug in blister packs, unit doses, or similar packaging for use in institutions), product code, labeler code, trade name, or trademark that differs from that of the listed drug. [21 CFR § 314]

B

Batch means a specific quantity or lot of a test or control article that has been characterized according to 58.105(a). [21 CFR § 58]

Batch means a specific quantity of a drug or other material that is intended to have uniform character and quality, within specified limits, and is produced according to a single manufacturing order during the same cycle of manufacture. [21 CFR § 210]

Bioavailability means the rate and extent to which the active ingredient or active moiety is absorbed from a drug product and becomes available at the site of action. For drug products that are not intended to be absorbed into the bloodstream, bioavailability may be assessed by measurements intended to reflect the rate and extent to which the active ingredient or active moiety becomes available at the site of action. [21 CFR § 320]

Bioequivalence means the absence of a significant difference in the rate and extent to which the active ingredient or active moiety in pharmaceutical equivalents or pharmaceutical alternatives becomes available at the site of drug action when administered at the same molar dose under similar conditions in an appropriately designed study. Where there is an intentional difference in rate (e.g., in certain extended release dosage forms), certain pharmaceutical equivalents or alternatives may be considered bioequivalent if there is no significant difference in the extent to which the active ingredient or moiety from each product becomes available at the site of drug action. This applies only if the difference in the rate at which the active ingredient or moiety becomes available at the site of drug action is intentional and is reflected in the proposed labeling, is not essential to the attainment of effective body drug concentrations on chronic use, and is considered medically insignificant for the drug. [21 CFR § 320]

Bioequivalence requirement means a requirement imposed by the Food and Drug Administration for in vitro and/or in vivo testing of specified drug products which must be satisfied as a condition of marketing. [21 CFR § 320]

Biological marker (biomarker): A characteristic that is objectively measured and evaluated as an indicator of normal biologic processes, pathogenic processes, or pharmacologic responses to a therapeutic intervention. [Pharmacogenomic Data Submissions Guidance]

Biometrics means a method of verifying an individual's identity based on measurement of the individual's physical feature(s) or repeatable action(s) where those features and/or actions are both unique to that individual and measurable. [21 CFR § 11]

C

Certified Copy: A certified copy is a copy of original information that has been verified, as indicated by a dated signature, as an exact copy having all of the same attributes and information as the original. [Computerized Systems Used in Clinical Investigations Guidance]

Children means persons who have not attained the legal age for consent to treatments or procedures involved in clinical investigations, under the applicable law of the jurisdiction in which the clinical investigation will be conducted. [21 CFR § 50]

Claim — A statement of treatment benefit. A claim can appear in any section of a medical product's FDA-approved labeling or in advertising and promotional labeling of prescription drugs and devices. [Patient-Reported Outcome Measures Guidance]

Class 1 resubmission means the resubmission of an application or efficacy supplement, following receipt of a complete response letter, that contains one or more of the following: Final printed labeling, draft labeling, certain safety updates, stability updates to support provisional or final dating periods, commitments to perform postmarketing studies (including proposals for such studies), assay validation data, final release testing on the last lots used to support approval, minor reanalyses of previously submitted data, and other comparatively minor information. [21 CFR § 314]

Class 2 resubmission means the resubmission of an application or efficacy supplement, following receipt of a complete response letter, that includes any item not specified in the definition of "Class 1 resubmission," including any item that would require presentation to an advisory committee. [21 CFR § 314]

Class I devices – devices for which the general controls of the Act are sufficient to provide reasonable assurance of their safety and effectiveness. They typically present minimal potential for harm to the user and the person being tested. They are subject to general controls, which include registration and listing, labeling, and adverse event reporting requirements (section 513(a)(1)(A) of the Act). Most Class I devices are exempt from premarket notification (see definition below), subject to certain limitations found in section 510(l) of the Act and in 21 CFR 862.9, 864.9, and 866.9. Some are also exempt from the "Quality Systems Regulation" found in 21 CFR Part 820. IVD examples of Class I devices include complement reagent, phosphorus (inorganic) test systems (21 CFR 862.1580), and E. coli serological reagents (21 CFR 866.3255).

Class II devices – devices for which general controls alone are insufficient to provide reasonable assurance of their safety and effectiveness and for which establishment of special controls can provide such assurances. Special controls may include special labeling, mandatory performance standards, risk mitigation measures identified in guidance, and postmarket surveillance (section 513(a)(1)(B) of the Act). Some Class II devices are exempt from premarket notification (see definition below), subject to limitation in 21 CFR 862.9, 864.9, and 866.9. IVD examples of Class II devices include glucose test systems (21 CFR 862.1345), antinuclear antibody immunological test systems (21 CFR 866.5100), and coagulation instruments (21 CFR 864.5400).

Class III devices – devices for which insufficient information exists to provide reasonable assurance of safety and effectiveness through general or special controls. Class III devices are usually those that support or sustain human life, are of substantial importance in preventing impairment of human health, or which present a potential, unreasonable risk of illness or injury (section 513(a)(1)(C) of the Act). Most Class III devices require premarket approval (PMA, see definition below). IVD examples of these include automated PAP smear readers, nucleic acid amplification devices for tuberculosis, and total prostate specific antigen (PSA) for the detection of cancer. A limited number of Class III devices that are equivalent to devices legally marketed before enactment of the Medical Device Amendments of 1976 may be marketed through the premarket notification (510(k)) process (see definition below), until FDA has published a requirement for manufacturers of that generic type of device to submit PMA data.

Closed system means an environment in which system access is controlled by persons who are responsible for the content of electronic records that are on the system. [21 CFR § 11]

Clinical investigation means any experiment that involves a test article and one or more human subjects and that either is subject to requirements for prior submission to the Food and Drug Administration under section 505(i) or 520(g) of the act, or is not subject to requirements for prior submission to the Food and Drug Administration under these sections of the act, but the results of which are intended to be submitted later to, or held for inspection by, the Food and Drug Administration as part of an application for a research or marketing permit. The term does not include experiments that are subject to the provisions of part 58 of this chapter, regarding nonclinical laboratory studies. [21 CFR § 50]

> *Clinical investigation* means any experiment that involves a test article and one or more human subjects, and that either must meet the requirements for prior submission to the Food and Drug Administration under section 505(i) or 520(g) of the act, or need not meet the requirements for prior submission to the Food and Drug Administration under these sections of the act, but the results of which are intended to be later submitted to, or held for inspection by, the Food and Drug Administration as part of an application for a research or marketing permit. The term does not include experiments that must meet the provisions of part 58, regarding nonclinical laboratory studies. The termsresearch, clinical research, clinical study, study, andclinical investigation are deemed to be synonymous for purposes of this part. [21 CFR § 56]

> *Clinical investigation* means any experiment in which a drug is administered or dispensed to, or used involving, one or more human subjects. For the purposes of this part, an experiment is any use of a drug except for the use of a marketed drug in the course of medical practice. [21 CFR § 312]

Clinical investigator means only a listed or identified investigator or subinvestigator who is directly involved in the treatment or evaluation of research subjects. The term also includes the spouse and each dependent child of the investigator. [21 CFR § 54]

> *Clinical Investigator* – An individual who actually conducts a clinical investigation (i.e., under whose immediate direction the drug is administered or dispensed to a subject) (21 CFR 312.3(b)). In this guidance, when a clinical investigation involves BA or BE studies, the clinical investigator has the responsibility of retaining the reserve samples at the testing facility or through an independent third party. [Bioavailability and Bioequivalence Testing Samples]

Cognitive interviewing — A qualitative research tool used to determine whether concepts and items are understood by patients in the same way that instrument developers intend. Cognitive interviews involve incorporating follow-up questions in a field test interview to gain a better understanding of how patients interpret questions asked of them. In this method, respondents are often asked to think aloud and describe their thought processes as they answer the instrument questions. [Patient-Reported Outcome Measures Guidance]

Compassionate use – The compassionate use provision allows access for patients with a serious disease or condition who do not meet the requirements for inclusion in the clinical investigation but for whom the treating physician believes the device may provide a benefit in treating and/or diagnosing their disease or condition. There must be no feasible alternative therapies/diagnostics available. Compassionate use is typically available only for individual patients but also may be used to treat a small group. Prior FDA approval is needed before compassionate use occurs.

Further information can be found at http://www.fda.gov/MedicalDevices/ DeviceRegulationandGuidance/HowtoMarketYourDevice/InvestigationalDeviceExempti onIDE/ucm051345.htm.

Compensation affected by the outcome of clinical studies means compensation that could be higher for a favorable outcome than for an unfavorable outcome, such as compensation that is explicitly greater for a favorable result or compensation to the investigator in the form of an equity interest in the sponsor of a covered study or in the form of compensation tied to sales of the product, such as a royalty interest. [21 CFR § 54]

Complete response letter means a written communication to an applicant from FDA usually describing all of the deficiencies that the agency has identified in an application or abbreviated application that must be satisfactorily addressed before it can be approved. [21 CFR § 314]

Component means any ingredient intended for use in the manufacture of a drug product, including those that may not appear in such drug product. [21 CFR § 210]

Computerized System: A computerized system includes computer hardware, software, and associated documents (e.g., user manual) that create, modify, maintain, archive, retrieve, or transmit in digital form information related to the conduct of a clinical trial. [Computerized Systems Used in Clinical Investigations Guidance]

Concept — The specific measurement goal (i.e., the thing that is to be measured by a PRO instrument). In clinical trials, a PRO instrument can be used to measure the effect of a medical intervention on one or more concepts. PRO concepts represent aspects of how patients function or feel related to a health condition or its treatment. [Patient-Reported Outcome Measures Guidance]

Conceptual framework of a PRO instrument — An explicit description or diagram of the relationships between the questionnaire or items in a PRO instrument and the concepts measured. The conceptual framework of a PRO instrument evolves over the course of instrument development as empiric evidence is gathered to support item grouping and scores. We review the alignment of the final conceptual framework with the clinical trial's objectives, design, and analysis plan. [Patient-Reported Outcome Measures Guidance]

Confidence interval – the range of plausible values for a statistical parameter, consistent with the observed data, which is computed with a sample estimate parameter (e.g., mean) and its standard deviation. For example, a "95% Confidence Interval" is computed such that, if the parameter was determined for repeated experiments, the resulting values would include the true parameter value 95% of the time. [IVD Device Studies Guidance]

Construct validity — Evidence that relationships among items, domains, and concepts conform to a priori hypotheses concerning logical relationships that should exist with other measures or characteristics of patients and patient groups. [Patient-Reported Outcome Measures Guidance]

Content validity — Evidence from qualitative research demonstrating that the instrument measures the concept of interest including evidence that the items and domains of an instrument are appropriate and comprehensive relative to its intended measurement concept, population, and use. Testing other measurement properties will not replace or rectify problems with content validity. [Patient-Reported Outcome Measures Guidance]

Continued access to investigational devices – FDA may allow sponsors to continue to enroll subjects under an IDE, after the trial has been completed, while a marketing application is prepared by the sponsor and/or reviewed by FDA. To continue enrolling subjects, a sponsor should show that there is a public health need for the device, that preliminary evidence indicates that the device is likely to be effective for the indications proposed, and that no significant safety concerns have been identified for the proposed indication. A guidance document, entitled "Continued Access to Investigational Devices During PMA Preparation and Review," can be found at http://www.fda.gov/MedicalDevices/DeviceRegulationandGuidance/ GuidanceDocuments/ucm080260.htm . [IVD Device Studies Guidance]

Contract research organization means a person that assumes, as an independent contractor with the sponsor, one or more of the obligations of a sponsor, e.g., design of a protocol, selection or monitoring of investigations, evaluation of reports, and preparation of materials to be submitted to the Food and Drug Administration. [21 CFR § 312]

Contract Research Organization (CRO) – An independent contractor of the sponsor or manufacturer that assumes one or more of the obligations of a sponsor (e.g., design of a protocol, selection or monitoring of investigations, evaluation of reports, and preparation of materials to be submitted to the FDA) (21 CFR 312.3(b)). This guidance addresses BA and BE studies submitted to support approvals of new and generic drugs. These studies are usually conducted by CROs under contract to study sponsors and/or drug manufacturers. Many CROs have their own testing facilities, with physicians (to serve as clinical investigators) and clinical support staff (e.g., nurses, medical technologists) to conduct the BA and BE studies. [Bioavailability and Bioequivalence Testing Samples]

Control article means any food additive, color additive, drug, biological product, electronic product, medical device for human use, or any article other than a test article, feed, or water that is administered to the test system in the course of a nonclinical laboratory study for the purpose of establishing a basis for comparison with the test article. [21 CFR § 58]

Co-variables – data elements relating to a subject that might affect how well a diagnostic test works, such as demographic status (age, gender, etc.), morbidity, or concurrent therapy. [IVD Device Studies Guidance]

Covered clinical study means any study of a drug or device in humans submitted in a marketing application or reclassification petition subject to this part that the applicant or FDA relies on to establish that the product is effective (including studies that show equivalence to an effective product) or any study in which a single investigator makes a significant contribution to the demonstration of safety. This would, in general, not include phase l tolerance studies or pharmacokinetic studies, most clinical pharmacology studies (unless they are critical to an efficacy determination), large open safety studies conducted at multiple sites, treatment protocols, and parallel track protocols. An applicant may consult with FDA as to which clinical studies constitute "covered clinical studies" for purposes of complying with financial disclosure requirements. [21 CFR § 54]

Criterion validity — The extent to which the scores of a PRO instrument are related to a known gold standard measure of the same concept. For most PROs, criterion validity cannot be measured because there is no gold standard. [Patient-Reported Outcome Measures Guidance]

Custom device means a device that: [21 CFR § 812]

(1) Necessarily deviates from devices generally available or from an applicable performance standard or premarket approval requirement in order to comply with the order of an individual physician or dentist;

(2) Is not generally available to, or generally used by, other physicians or dentists;

(3) Is not generally available in finished form for purchase or for dispensing upon prescription;

(4) Is not offered for commercial distribution through labeling or advertising; and

(5) Is intended for use by an individual patient named in the order of a physician or dentist, and is to be made in a specific form for that patient, or is intended to meet the special needs of the physician or dentist in the course of professional practice.

D

Design history file (DHF) means a compilation of records which describes the design history of a finished device. [21 CFR § 820.3(e)]

Design input means the physical and performance requirements of a device that are used as a basis for device design. [21 CFR § 820.3(f)]

Design output means the results of a design effort at each design phase and at the end of the total design effort. The finished design output is the basis for the device master record.

The total finished design output consists of the device, its packaging and labeling, and the device master record. [21 CFR § 820.3(g)]

Design review means a documented, comprehensive, systematic examination of a design to evaluate the adequacy of the design requirements, to evaluate the capability of the design to meet these requirements, and to identify problems. [21 CFR § 820.3(h)]

Determination Meeting – A Determination meeting under section 513(a)(3)(D) of the Act is available to anyone anticipating submitting a PMA or PDP and is intended to provide the applicant with the FDA's determination of the type of valid scientific evidence that will be necessary to demonstrate that the device is effective for its intended use. As a result of this meeting, FDA will determine whether clinical studies are needed to establish effectiveness and, in consultation with the applicant, determine the least burdensome way of evaluating device effectiveness that has a reasonable likelihood of success. The applicant can expect that FDA will determine if concurrent randomized controls, concurrent non-randomized controls, historical controls, or other types of evidence will be acceptable. FDA's determination is to be written, shared with the applicant within 30 days following the meeting, and is binding upon the FDA, unless it could be contrary to public health. 21 U.S.C. § 360c(a)(3)(D). A guidance document regarding these meetings, "Early Collaboration Meetings Under the FDA Modernization Act (FDAMA); Final Guidance for Industry and for CDRH Staff," is available at http://www.fda.gov/MedicalDevices/DeviceRegulationandGuidance/GuidanceDocuments/ucm073604.htm. [IVD Device Studies Guidance]

Device – as defined in the Act, section 201(h): an instrument, apparatus, implement, machine, contrivance, implant, in vitro reagent, or other similar or related article, including any component, part, or accessory, which is a) recognized in the official National Formulary, or the United States Pharmacopeia, or any supplement to them; b) intended for use in the diagnosis of disease or other conditions, or in the cure, mitigation, treatment, or prevention of disease, in man or other animals; or c) intended to affect the structure or any function of the body of man or other animals, and which does not achieve its primary intended purposes through chemical action within or on the body of man or other animals and which is not dependent upon being metabolized for the achievement of its primary intended purposes. [IVD Device Studies Guidance]

Digital signature means an electronic signature based upon cryptographic methods of originator authentication, computed by using a set of rules and a set of parameters such that the identity of the signer and the integrity of the data can be verified. [21 CFR § 11]

Direct Entry: Direct entry is recording data where an electronic record is the original means of capturing the data. Examples are the keying by an individual of original observations into a system, or automatic recording by the system of the output of a balance that measures subject's body weight. [Computerized Systems Used in Clinical Investigations Guidance]

Domain — A subconcept represented by a score of an instrument that measures a larger concept comprised of multiple domains. For example, psychological function is the larger concept containing the domains subdivided into items describing emotional function and cognitive function. [Patient-Reported Outcome Measures Guidance]

Drug product means a finished dosage form, for example, tablet, capsule, solution, etc., that contains an active drug ingredient generally, but not necessarily, in association with inactive ingredients. The term also includes a finished dosage form that does not contain an active ingredient but is intended to be used as a placebo. [21 CFR § 210]

> *Drug product* means a finished dosage form, for example, tablet, capsule, or solution, that contains a drug substance, generally, but not necessarily, in association with one or more other ingredients. [21 CFR § 314]

> *Drug product* means a finished dosage form, e.g., tablet, capsule, or solution, that contains the active drug ingredient, generally, but not necessarily, in association with inactive ingredients. [21 CFR § 320]

Drug substance means an active ingredient that is intended to furnish pharmacological activity or other direct effect in the diagnosis, cure, mitigation, treatment, or prevention of disease or to affect the structure or any function of the human body, but does not include intermediates use in the synthesis of such ingredient. [21 CFR § 314]

E, F

Efficacy supplement means a supplement to an approved application proposing to make one or more related changes from among the following changes to product labeling: [21 CFR § 314]

(1) Add or modify an indication or claim;

(2) Revise the dose or dose regimen;

(3) Provide for a new route of administration;

(4) Make a comparative efficacy claim naming another drug product;

(5) Significantly alter the intended patient population;

(6) Change the marketing status from prescription to over-the-counter use;

(7) Provide for, or provide evidence of effectiveness necessary for, the traditional approval of a product originally approved under subpart H of part 314; or

(8) Incorporate other information based on at least one adequate and well-controlled clinical study.

Electronic record means any combination of text, graphics, data, audio, pictorial, or other information representation in digital form that is created, modified, maintained, archived, retrieved, or distributed by a computer system. [21 CFR § 11]

Electronic signature means a computer data compilation of any symbol or series of symbols executed, adopted, or authorized by an individual to be the legally binding equivalent of the individual's handwritten signature. [21 CFR § 11]

Emergency use means the use of a test article on a human subject in a life-threatening situation in which no standard acceptable treatment is available, and in which there is not sufficient time to obtain IRB approval. [21 CFR § 56]

Endpoint — The measurement that will be statistically compared among treatment groups to assess the effect of treatment and that corresponds with the clinical trial's objectives, design, and data analysis. For example, a treatment may be tested to decrease the intensity of symptom Z. In this case, the endpoint is the change from baseline to time T in a score that represents the concept of symptom Z intensity. [Patient-Reported Outcome Measures Guidance]

Endpoint model — A diagram of the hierarchy of relationships among all endpoints, both PRO and non-PRO, that corresponds to the clinical trial's objectives, design, and data analysis plan.

Excess samples – remnants of human specimens collected for routine clinical care or analysis that would otherwise have been discarded, as well as specimens leftover from specimens previously collected for other unrelated research or investigations. Excess samples are also referred to as "surplus samples," "residual", "reserved samples," "library samples," and "leftover specimens." [IVD Device Studies Guidance]

Expedited review – review by an institutional review board (IRB) that does not require full board review or a convened meeting. Such a review may be carried out by the IRB chairperson or one or more experienced reviewers assigned by the IRB chairperson from among the members of the IRB. Reviewers may exercise all of the authorities of the IRB except they may not disapprove the study. Disapproval may only result through the IRB's non-expedited review process. Expedited review is reserved for minimal risk studies. (See 21 CFR 56.110.) [IVD Device Studies Guidance]

Family member means any one of the following legally competent persons: Spouse; parents; children (including adopted children); brothers, sisters, and spouses of brothers and sisters; and any individual related by blood or affinity whose close association with the subject is the equivalent of a family relationship. [21 CFR § 50]

FDA means the Food and Drug Administration. [21 CFR § 312, § 314, § 812, § 814]

Fiber means any particulate contaminant with a length at least three times greater than its width. [21 CFR § 210]

G, H

Gang-printed labeling means labeling derived from a sheet of material on which more than one item of labeling is printed. [21 CFR § 210]

General purpose reagents – chemical reagents that have general laboratory application and that are not labeled or otherwise intended for a specific diagnostic application. They are used to collect, prepare, and/or examine specimens from the human body for diagnostic purposes. (Example: reagents used for general staining in microscopic procedures.) General purpose reagents do not include laboratory machinery, automated or powered systems (21 CFR 864.4010(a)). [IVD Device Studies Guidance]

Gold standard – see truth standard. [IVD Device Studies Guidance]

Guardian means an individual who is authorized under applicable State or local law to consent on behalf of a child to general medical care when general medical care includes participation in research. For purposes of subpart D of this part, a guardian also means an individual who is authorized to consent on behalf of a child to participate in research. [21 CFR § 50]

Handwritten signature means the scripted name or legal mark of an individual handwritten by that individual and executed or adopted with the present intention to authenticate a writing in a permanent form. The act of signing with a writing or marking instrument such as a pen or stylus is preserved. The scripted name or legal mark, while conventionally applied to paper, may also be applied to other devices that capture the name or mark. [21 CFR § 11]

HDE means a premarket approval application submitted pursuant to this subpart seeking a humanitarian device exemption from the effectiveness requirements of sections 514 and 515 of the act as authorized by section 520(m)(2) of the act. [21 CFR § 814]

Health-related quality of life (HRQL) — HRQL is a multidomain concept that represents the patient's general perception of the effect of illness and treatment on physical, psychological, and social aspects of life. Claiming a statistical and meaningful improvement in HRQL implies: (1) that all HRQL domains that are important to interpreting change in how the clinical trial's population feels or functions as a result of the targeted disease and its treatment were measured; (2) that a general improvement was demonstrated; and (3) that no decrement was demonstrated in any domain. [Patient-Reported Outcome Measures Guidance]

HUD (humanitarian use device) means a medical device intended to benefit patients in the treatment or diagnosis of a disease or condition that affects or is manifested in fewer than 4,000 individuals in the United States per year. [21 CFR § 814]

Human subject means an individual who is or becomes a participant in research, either as a recipient of the test article or as a control. A subject may be either a healthy human or a patient. [21 CFR § 50]

> *Human subject* means an individual who is or becomes a participant in research, either as a recipient of the test article or as a control. A subject may be either a healthy individual or a patient. [21 CFR § 56]

Humanitarian use devices (HUDs) –HUDs are devices intended to diagnose a disease or condition in fewer than 4,000 patients in the U.S. per year. Such devices are regulated under 21 CFR Part 814, Subpart H. If a device receives a designation as an HUD, a Humanitarian Device Exemption request (HDE) can be submitted. HUDs that are approved for marketing under an HDE have specific labeling requirements. IRB approval is required for use of a HUD (21 CFR 814.124). [IVD Device Studies Guidance]

I

IDE means an approved or considered approved investigational device exemption under section 520(g) of the act and parts 812 and 813. [21 CFR § 814]

Implant means a device that is placed into a surgically or naturally formed cavity of the human body if it is intended to remain there for a period of 30 days or more. FDA may, in order to protect public health, determine that devices placed in subjects for shorter periods are also "implants" for purposes of this part. [21 CFR § 812]

Inactive ingredient means any component other than anactive ingredient. [21 CFR § 210]

Increased frequency of adverse drug experience is an increased rate of occurrence of a particular serious adverse drug event, expected or unexpected, after appropriate adjustment for drug exposure. [21 CFR § 514]

IND means an investigational new drug application. For purposes of this part, "IND" is synonymous with "Notice of Claimed Investigational Exemption for a New Drug." [21 CFR § 312]

Independent ethics committee (IEC) means a review panel that is responsible for ensuring the protection of the rights, safety, and well-being of human subjects involved in a clinical investigation and is adequately constituted to provide assurance of that protection. An institutional review board (IRB), as defined in 56.102(g) of this chapter and subject to the requirements of part 56 of this chapter, is one type of IEC. [21 CFR § 312]

Independent Third Party – In this guidance, independent third party indicates a person that has no affiliation with the study sponsor and/or drug manufacturer. [Bioavailability and Bioequivalence Testing Samples]

In-process material means any material fabricated, compounded, blended, or derived by chemical reaction that is produced for, and used in, the preparation of the drug product.

Institution means any public or private entity or agency (including Federal, State, and other agencies). The word facility as used in section 520(g) of the act is deemed to be synonymous with the term institution for purposes of this part. [21 CFR § 50, § 56]

Institution means a person, other than an individual, who engages in the conduct of research on subjects or in the delivery of medical services to individuals as a primary activity or as an adjunct to providing residential or custodial care to humans. The term includes, for example, a hospital, retirement home, confinement facility, academic establishment, and device manufacturer. The term has the same meaning as "facility" in section 520(g) of the act. [21 CFR § 812]

Institutional review board (IRB) means any board, committee, or other group formally designated by an institution to review biomedical research involving humans as subjects, to approve the initiation of and conduct periodic review of such research. The term has the same meaning as the phrase institutional review committee as used in section 520(g) of the act. [21 CFR § 50]

> *Institutional Review Board (IRB)* means any board, committee, or other group formally designated by an institution to review, to approve the initiation of, and to conduct periodic review of, biomedical research involving human subjects. The primary purpose of such review is to assure the protection of the rights and welfare of the human subjects. The term has the same meaning as the phrase institutional review committee as used in section 520(g) of the act. [21 CFR § 56]

> *Institutional review board (IRB)* means any board, committee, or other group formally designated by an institution to review biomedical research involving subjects and established, operated, and functioning in conformance with part 56. The term has the same meaning as "institutional review committee" in section 520(g) of the act. [21 CFR § 812]

Instrument — A means to capture data (i.e., a questionnaire) plus all the information and documentation that supports its use. Generally, that includes clearly defined methods and instructions for administration or responding, a standard format for data collection, and well-documented methods for scoring, analysis, and interpretation of results in the target patient population. [Patient-Reported Outcome Measures Guidance]

Investigation means a clinical investigation or research involving one or more subjects to determine the safety or effectiveness of a device. [21 CFR § 812]

Investigation – a clinical investigation or research involving one or more subjects to determine the safety or effectiveness of a device (21 CFR 812.3(h)). It is often referred to as a clinical trial and is sometimes referred to as a field trial. [IVD Device Studies Guidance]

Investigational device means a device, including a transitional device, that is the object of an investigation. [21 CFR § 812]

Investigational Device Exemption (IDE) – application which, when approved, allows the device to be used lawfully for the purpose of conducting studies regarding the safety and effectiveness of the device, without complying with certain requirements of the Act. (See 21 CFR 812.1 for specific exemptions.) For significant risk (SR) device studies (see definition below), a sponsor must apply to FDA to obtain approval for an IDE. (See 21 CFR 812.20.) For non-significant risk (NSR) device studies (see definition below), an IDE is considered approved when a sponsor meets the abbreviated requirements found in 21 CFR 812.2(b), which include approval from the reviewing Institutional Review Board(s) (IRB(s)). [IVD Device Studies Guidance]

Investigational new drug means a new drug or biological drug that is used in a clinical investigation. The term also includes a biological product that is used in vitro for diagnostic purposes. The terms "investigational drug" and "investigational new drug" are deemed to be synonymous for purposes of this part. [21 CFR § 312]

Investigational plan – sponsor's overall plan regarding the conduct of an investigational study. It includes, but is not limited to, the purpose of the study, a written protocol, a risk analysis, device description, labeling, written monitoring procedures, informed consent materials, and Institutional Review Board (IRB) information. (21 CFR 812.25.) For IVD studies, protocols should describe the study objectives, design, methodology, subject populations, types of specimens, data to be collected and planned data analysis. [IVD Device Studies Guidance]

Investigational Use in vitro diagnostic (IVD) product – an IVD product being used for product testing prior to full commercial marketing (e.g., for use on specimens derived from humans to compare the usefulness of the product with other products or procedures in current use or recognized as useful). These products must be labeled according to 21 CFR 812.5 for non-significant risk or significant risk devices and according to 21 CFR 809.10(c)(2)(ii) for devices that are exempt under 21 CFR 812.2(c). [IVD Device Studies Guidance]

Investigator means an individual who actually conducts a clinical investigation, i.e., under whose immediate direction the test article is administered or dispensed to, or used involving, a subject, or, in the event of an investigation conducted by a team of individuals, is the responsible leader of that team. [21 CFR § 50, § 56]

> *Investigator* means an individual who actually conducts a clinical investigation (i.e. , under whose immediate direction the drug is administered or dispensed to a subject). In the event an investigation is conducted by a team of individuals, the investigator is the responsible leader of the team. "Subinvestigator" includes any other individual member of that team. [21 CFR § 312]

Investigator means an individual who actually conducts a clinical investigation, i.e., under whose immediate direction the test article is administered or dispensed to, or used involving, a subject, or, in the event of an investigation conducted by a team of individuals, is the responsible leader of that team. [21 CFR § 812]

Investigator – an individual who actually conducts a clinical investigation, i.e., a person under whose immediate direction the investigational product is administered, dispensed, or used, provided that the investigation involves a subject. In the event of an investigation conducted by a team of individuals, the investigator is the responsible leader of that team (21 CFR 812.3(i)). [IVD Device Studies Guidance]

In vitro diagnostic (IVD) products – those reagents, instruments, and systems intended for use in the diagnosis of disease or other conditions, including a determination of the state of health, in order to cure, mitigate, treat, or prevent disease or its sequelae. Such products are intended for use in the collection, preparation, and examination of specimens taken from the human body. IVD products are devices as defined in section 201(h) of the Act and may also be biological products subject to section 351 of the Public Health Service Act. The regulatory definition of in vitro diagnostic products is found in 21 CFR 809.3(a). [IVD Device Studies Guidance]

IRB approval means the determination of the IRB that the clinical investigation has been reviewed and may be conducted at an institution within the constraints set forth by the IRB and by other institutional and Federal requirements. [21 CFR § 56]

Item — An individual question, statement, or task (and its standardized response options) that is evaluated by the patient to address a particular concept. [Patient-Reported Outcome Measures Guidance]

Item tracking matrix — A record of the development (e.g., additions, deletions, modifications, and the reasons for the changes) of items used in an instrument. [Patient-Reported Outcome Measures Guidance]

J, K, L

Leftover specimens -- remnants of specimens collected for routine clinical care or analysis that would otherwise have been discarded, or remnants of specimens previously collected for other unrelated research. These specimens may be obtained from a specimen repository -- a common site for storage of collections of human biological specimens available for study. See also Excess samples. [IVD Device Studies Guidance]

Legally authorized representative means an individual or judicial or other body authorized under applicable law to consent on behalf of a prospective subject to the subject's particpation in the procedure(s) involved in the research. [21 CFR § 50]

Listed drug means a new drug product that has an effective approval under section 505(c) of the act for safety and effectiveness or under section 505(j) of the act, which has not been withdrawn or suspended under section 505(e)(1) through (e)(5) or (j)(5) of the act, and which has not been withdrawn from sale for what FDA has determined are reasons of safety or effectiveness. Listed drug status is evidenced by the drug product's identification as a drug with an effective approval in the current edition of FDA's "Approved Drug Products with Therapeutic Equivalence Evaluations" (the list) or any current supplement thereto, as a drug with an effective approval. A drug product is deemed to be a listed drug on the date of effective approval of the application or abbreviated application for that drug product. [21 CFR § 314]

Lot means a batch, or a specific identified portion of a batch, having uniform character and quality within specified limits; or, in the case of a drug product produced by continuous process, it is a specific identified amount produced in a unit of time or quantity in a manner that assures its having uniform character and quality within specified limits. [21 CFR § 210]

Lot number, control number, or batch number means any distinctive combination of letters, numbers, or symbols, or any combination of them, from which the complete history of the manufacture, processing, packing, holding, and distribution of a batch or lot of drug product or other material can be determined. [21 CFR § 210]

M

Management with executive responsibility means those senior employees of a manufacturer who have the authority to establish or make changes to the manufacturer's quality policy and quality system. [21 CFR § 820.3 (n)]

Manufacture, processing, packing, or holding of a drug product includes packaging and labeling operations, testing, and quality control of drug products. [21 CFR § 210]

Marketing application means an application for a new drug submitted under section 505(b) of the act or a biologics license application for a biological product submitted under the Public Health Service Act. [21 CFR § 312]

Master file means a reference source that a person submits to FDA. A master file may contain detailed information on a specific manufacturing facility, process, methodology, or component used in the manufacture, processing, or packaging of a medical device. [21 CFR § 814]

Measurement properties — All the attributes relevant to the application of a PRO instrument including the content validity, construct validity, reliability, and ability to detect change. These attributes are specific to the measurement application and cannot be assumed to be relevant to all measurement situations, purposes, populations, or settings in which the instrument is used. [Patient-Reported Outcome Measures Guidance]

Medicated feed means any Type B or Type C medicated feed as defined in 558.3 of this chapter. The feed contains one or more drugs as defined in section 201(g) of the act. The manufacture of medicated feeds is subject to the requirements of part 225 of this chapter. [21 CFR § 210]

Medicated premix means a Type A medicated article as defined in 558.3 of this chapter. The article contains one or more drugs as defined in section 201(g) of the act. The manufacture of medicated premixes is subject to the requirements of part 226 of this chapter. [21 CFR § 210]

Minimal risk means that the probability and magnitude of harm or discomfort anticipated in the research are not greater in and of themselves than those ordinarily encountered in daily life or during the performance of routine physical or psychological examinations or tests. [21 CFR § 50, § 56]

Monitor, when used as a noun, means an individual designated by a sponsor or contract research organization to oversee the progress of an investigation. The monitor may be an employee of a sponsor or a consultant to the sponsor, or an employee of or consultant to a contract research organization.Monitor, when used as a verb, means to oversee an investigation. [21 CFR § 812]

N

NADA is a new animal drug application including all amendments and supplements. [21 CFR § 514]

Newly acquired information means data, analyses, or other information not previously submitted to the agency, which may include (but are not limited to) data derived from new clinical studies, reports of adverse events, or new analyses of previously submitted data (e.g., meta-analyses) if the studies, events or analyses reveal risks of a different type or greater severity or frequency than previously included in submissions to FDA. [21 CFR § 314, § 814]

Nonapplicant is any person other than the applicant whose name appears on the label and who is engaged in manufacturing, packing, distribution, or labeling of the product. [21 CFR § 514]

Nonclinical laboratory study means in vivo or in vitro experiments in which test articles are studied prospectively in test systems under laboratory conditions to determine their safety. The term does not include studies utilizing human subjects or clinical studies or field trials in animals. The term does not include basic exploratory studies carried out to determine whether a test article has any potential utility or to determine physical or chemical characteristics of a test article. [21 CFR § 58]

Nonfiber releasing filter means any filter, which after appropriate pretreatment such as washing or flushing, will not release fibers into the component or drug product that is being filtered. [21 CFR § 210]

Noninvasive, when applied to a diagnostic device or procedure, means one that does not by design or intention: (1) Penetrate or pierce the skin or mucous membranes of the body, the ocular cavity, or the urethra, or (2) enter the ear beyond the external auditory canal, the nose beyond the nares, the mouth beyond the pharynx, the anal canal beyond the rectum, or the vagina beyond the cervical os. For purposes of this part, blood sampling that involves simple venipuncture is considered noninvasive, and the use of surplus samples of body fluids or tissues that are left over from samples taken for noninvestigational purposes is also considered noninvasive. [21 CFR § 812]

Non-significant risk (NSR) device – a device that does not meet the definition of significant risk (SR) device (see definition below). An IDE is considered approved for a NSR investigational device study once sponsors meet the abbreviated requirements found in the "Investigational Device Exemptions" regulation at 21 CFR 812.2(b). The risk determination for an investigational device study should be based on the proposed use of the device in the investigation in addition to the characteristics of the device. [IVD Device Studies Guidance]

O, P

Open system means an environment in which system access is not controlled by persons who are responsible for the content of electronic records that are on the system. [21 CFR § 11]

Original application means a pending application for which FDA has never issued a complete response letter or approval letter, or an application that was submitted again after FDA had refused to file it or after it was withdrawn without being approved. [21 CFR § 314]

Original data: For the purpose of this guidance, original data are those values that represent the first recording of study data. FDA is allowing original documents and the original data recorded on those documents to be replaced by copies provided the copies are identical and have been verified as such (see FDA Compliance Policy Guide # 7150.13). [Computerized Systems Used in Clinical Investigations Guidance]

Outlier – a data observation whose value appears to be out of line with the main body of data that has been collected. [IVD Device Studies Guidance]

Parent means a child's biological or adoptive parent. [21 CFR § 50]

Patient-reported outcome (PRO) — A measurement based on a report that comes directly from the patient (i.e., study subject) about the status of a patient's health condition without amendment or interpretation of the patient's response by a clinician or anyone else. A

PRO can be measured by self-report or by interview provided that the interviewer records only the patient's response. [Patient-Reported Outcome Measures Guidance]

Percentage of theoretical yield means the ratio of the actual yield (at any appropriate phase of manufacture, processing, or packing of a particular drug product) to the theoretical yield (at the same phase), stated as a percentage. [21 CFR § 210]

Permission means the agreement of parent(s) or guardian to the participation of their child or ward in a clinical investigation. Permission must be obtained in compliance with subpart B of this part and must include the elements of informed consent described in 50.25. [21 CFR § 50]

Person includes an individual, partnership, corporation, association, scientific or academic establishment, government agency, or organizational unit thereof, and any other legal entity. [21 CFR § 58]

> ***Person*** includes any individual, partnership, corporation, association, scientific or academic establishment, Government agency or organizational unit of a Government agency, and any other legal entity. [21 CFR § 812]

> ***Person*** includes any individual, partnership, corporation, association, scientific or academic establishment, Government agency, or organizational unit thereof, or any other legal entity. [21 CFR § 814]

Pharmaceutical alternatives means drug products that contain the identical therapeutic moiety, or its precursor, but not necessarily in the same amount or dosage form or as the same salt or ester. Each such drug product individually meets either the identical or its own respective compendial or other applicable standard of identity, strength, quality, and purity, including potency and, where applicable, content uniformity, disintegration times and/or dissolution rates. [21 CFR § 320]

Pharmaceutical equivalents means drug products in identical dosage forms that contain identical amounts of the identical active drug ingredient,i.e. , the same salt or ester of the same therapeutic moiety, or, in the case of modified release dosage forms that require a reservoir or overage or such forms as prefilled syringes where residual volume may vary, that deliver identical amounts of the active drug ingredient over the identical dosing period; do not necessarily contain the same inactive ingredients; and meet the identical compendial or other applicable standard of identity, strength, quality, and purity, including potency and, where applicable, content uniformity, disintegration times, and/or dissolution rates. [21 CFR § 320]

Pharmacogenetic test: An assay intended to study interindividual variations in DNA sequence related to drug absorption and disposition (pharmacokinetics) or drug action pharmacodynamics), including polymorphic variation in the genes that encode the functions of transporters, metabolizing enzymes, receptors, and other proteins. [Pharmacogenomic Data Submissions Guidance]

Pharmacogenomic test: An assay intended to study interindividual variations in whole-genome or candidate gene, single-nucleotide polymorphism (SNP) maps, haplotype markers, or alterations in gene expression or inactivation that may be correlated with pharmacological function and therapeutic response. In some cases, the pattern or profile of change is the relevant biomarker, rather than changes in individual markers. [Pharmacogenomic Data Submissions Guidance]

PMA means any premarket approval application for a class III medical device, including all information submitted with or incorporated by reference therein. "PMA" includes a new drug application for a device under section 520(1) of the act. [21 CFR § 814]

PMA amendment means information an applicant submits to FDA to modify a pending PMA or a pending PMA supplement. [21 CFR § 814]

PMA supplement means a supplemental application to an approved PMA for approval of a change or modification in a class III medical device, including all information submitted with or incorporated by reference therein. [21 CFR § 814]

> *30-day PMA supplement* means a supplemental application to an approved PMA in accordance with 814.39(e). [21 CFR § 814]

Potential applicant means any person: [21 CFR § 514]

(1) Intending to investigate a new animal drug under section 512(j) of the Federal Food, Drug, and Cosmetic Act (the act),

(2) Investigating a new animal drug under section 512(j) of the act,

(3) Intending to file a new animal drug application (NADA) or supplemental NADA under section 512(b)(1) of the act, or

(4) Intending to file an abbreviated new animal drug application (ANADA) under section 512(b)(2) of the act.

Pre-amendment in vitro diagnostic (IVD) tests – IVD tests that were in commercial distribution before May 28, 1976. [IVD Device Studies Guidance]

Precision – the closeness of agreement between independent diagnostic test results obtained under stipulated conditions. For additional information refer to the Harmonized Technology Database, Clinical and Laboratory Standards Institute, available at http://www.clsi.org. [IVD Device Studies Guidance]

Premarket Approval Application (PMA) – the application for approval required prior to the marketing of most Class III medical devices (section 515 of the Act, 21 U.S.C. 360e). (See definitions of Class I, II, and III devices above.) PMA approval is based on a determination by FDA that the applicant's submission provides sufficient valid scientific evidence to provide reasonable assurance that the device is safe and effective for its intended use(s). The PMA regulation is 21 CFR Part 814. PMA information is available at

http://www.fda.gov/MedicalDevices/DeviceRegulationandGuidance/HowtoMarketYour
Device/PremarketSubmissions/PremarketApprovalPMA/default.htm. [IVD Device
Studies Guidance]

Premarket Notification – also referred to as a 510(k), is a submission to FDA that contains
information to demonstrate that a device is substantially equivalent (SE) to a legally
marketed (predicate) device. Governing regulations regarding premarket notification
procedures are found in Subpart E of the "Establishment Registration and Device Listing
for Manufacturers and Initial Importers of Devices" regulation (21 CFR Part 807). The
510(k) device advice page is available at
http://www.fda.gov/MedicalDevices/DeviceRegulationandGuidance/HowtoMarketYour
Device/PremarketSubmissions/PremarketNotification510k/default.htm. [IVD Device
Studies Guidance]

Presubmission conference means one or more conferences between a potential applicant
and FDA to reach a binding agreement establishing a submission or investigational
requirement. [21 CFR § 514]

Presubmission conference agreement means that section of the memorandum of
conference headed "Presubmission Conference Agreement" that records any agreement
on the submission or investigational requirement reached by a potential applicant and
FDA during the presubmission conference. [21 CFR § 514]

Product defect/manufacturing defect is the deviation of a distributed product from the
standards specified in the approved application, or any significant chemical, physical, or
other change, or deterioration in the distributed drug product, including any microbial or
chemical contamination. A manufacturing defect is a product defect caused or aggravated
by a manufacturing or related process. A manufacturing defect may occur from a single
event or from deficiencies inherent to the manufacturing process. These defects are
generally associated with product contamination, product deterioration, manufacturing
error, defective packaging, damage from disaster, or labeling error. For example, a labeling
error may include any incident that causes a distributed product to be mistaken for, or its
labeling applied to, another product. [21 CFR § 514]

Product Development Protocol (PDP) – FDA process of approval for marketing of medical
devices, usually reserved for Class III devices (see definitions of device classes above), by
which the sponsor and FDA agree on the product design and testing early in the concept
and planning stages of a product (section 515(f) of the Act). [IVD Device Studies
Guidance]

Proprietary interest in the tested product means property or other financial interest in the
product including, but not limited to, a patent, trademark, copyright or licensing
agreement. [21 CFR § 54]

Protocol – a document that contains a description of the objectives and design of an investigational study, methodology(s) to be used, and data to be collected. It may also contain information regarding the planned data analysis and study monitoring. For most studies in the development of an IVD product, it also contains information regarding types of specimens and subject populations. [IVD Device Studies Guidance]

Proxy-reported outcome — A measurement based on a report by someone other than the patient reporting as if he or she is the patient. A proxy-reported outcome is not a PRO. A proxy report also is different from an observer report where the observer (e.g., clinician or caregiver), in addition to reporting his or her observation, may interpret or give an opinion based on the observation. We discourage use of proxy-reported outcome measures particularly for symptoms that can be known only by the patient. [Patient-Reported Outcome Measures Guidance]

Q, R

Quality means the totality of features and characteristics that bear on the ability of a device to satisfy fitness-for-use, including safety and performance. [21 CFR § 820.3 (s)]

Quality assurance unit means any person or organizational element, except the study director, designated by testing facility management to perform the duties relating to quality assurance of nonclinical laboratory studies. [21 CFR § 58]

Quality control unit means any person or organizational element designated by the firm to be responsible for the duties relating to quality control. [21 CFR § 210]

Quality of life — A general concept that implies an evaluation of the effect of all aspects of life on general well-being. Because this term implies the evaluation of nonhealth-related aspects of life, and because the term generally is accepted to mean what the patient thinks it is, it is too general and undefined to be considered appropriate for a medical product claim.

Quality system means the organizational structure, responsibilities, procedures, processes, and resources for implementing quality management. [21 CFR § 820.3 (v)]

Questionnaire — A set of questions or items shown to a respondent to get answers for research purposes. Types of questionnaires include diaries and event logs. [Patient-Reported Outcome Measures Guidance]

Raw data means any laboratory worksheets, records, memoranda, notes, or exact copies thereof, that are the result of original observations and activities of a nonclinical laboratory study and are necessary for the reconstruction and evaluation of the report of that study. In the event that exact transcripts of raw data have been prepared (e.g., tapes which have been transcribed verbatim, dated, and verified accurate by signature), the exact copy or exact transcript may be substituted for the original source as raw data.Raw data may include photographs, microfilm or microfiche copies, computer printouts, magnetic media, including dictated observations, and recorded data from automated instruments. [21 CFR § 58]

Reasonable probability means that it is more likely than not that an event will occur. [21 CFR § 814]

Recall period — The period of time patients are asked to consider in responding to a PRO item or question. Recall can be momentary (real time) or retrospective of varying lengths. [Patient-Reported Outcome Measures Guidance]

Reference listed drug means the listed drug identified by FDA as the drug product upon which an applicant relies in seeking approval of its abbreviated application. [21 CFR § 314]

Reference Standard – In this guidance, reference standard refers to the reference product used in a BE study. It is usually the innovator's product or a marketed product of the drug under investigation. For BA studies, the reference standard can be an oral solution of the drug under investigation. [Bioavailability and Bioequivalence Testing Samples]

Reliability — The ability of a PRO instrument to yield consistent, reproducible estimates of true treatment effect. [Patient-Reported Outcome Measures Guidance]

Representative sample means a sample that consists of a number of units that are drawn based on rational criteria such as random sampling and intended to assure that the sample accurately portrays the material being sampled. [21 CFR § 210]

Reserve Samples – In this guidance, reserve samples and retention samples are used interchangeably. [Bioavailability and Bioequivalence Testing Samples]

Reserved samples – see excess samples. [IVD Device Studies Guidance]

Responder definition — A score change in a measure, experienced by an individual patient over a predetermined time period that has been demonstrated in the target population to have a significant treatment benefit. [Patient-Reported Outcome Measures Guidance]

Resubmission means submission by the applicant of all materials needed to fully address all deficiencies identified in the complete response letter. An application or abbreviated application for which FDA issued a complete response letter, but which was withdrawn before approval and later submitted again, is not a resubmission. [21 CFR § 314]

Right of reference or use means the authority to rely upon, and otherwise use, an investigation for the purpose of obtaining approval of an application, including the ability to make available the underlying raw data from the investigation for FDA audit, if necessary. [21 CFR § 314]

S

Same drug product formulation means the formulation of the drug product submitted for approval and any formulations that have minor differences in composition or method of manufacture from the formulation submitted for approval, but are similar enough to be relevant to the agency's determination of bioequivalence. [21 CFR § 320]

Saturation — When interviewing patients, the point when no new relevant or important information emerges and collecting additional data will not add to the understanding of how patients perceive the concept of interest and the items in a questionnaire. [Patient-Reported Outcome Measures Guidance]

Scale — The system of numbers or verbal anchors by which a value or score is derived for an item. Examples include VAS, Likert scales, and rating scales. [Patient-Reported Outcome Measures Guidance]

Score — A number derived from a patient's response to items in a questionnaire. A score is computed based on a prespecified, validated scoring algorithm and is subsequently used in statistical analyses of clinical trial results. Scores can be computed for individual items, domains, or concepts, or as a summary of items, domains, or concepts. [Patient-Reported Outcome Measures Guidance]

Sensitivity – the probability that a diagnostic test will yield a positive result when the disease or the target analyte is present. [IVD Device Studies Guidance]

Serious adverse drug experience is an adverse event that is fatal, or life-threatening, or requires professional intervention, or causes an abortion, or stillbirth, or infertility, or congenital anomaly, or prolonged or permanent disability, or disfigurement. [21 CFR § 514]

Serious, adverse health consequences means any significant adverse experience, including those which may be either life-threatening or involve permanent or long term injuries, but excluding injuries that are nonlife-threatening and that are temporary and reasonably reversible. [21 CFR § 814]

Sign — Any objective evidence of a disease, health condition, or treatment-related effect. Signs are usually observed and interpreted by the clinician but may be noticed and reported by the patient. [Patient-Reported Outcome Measures Guidance]

Significant equity interest in the sponsor of a covered study means any ownership interest, stock options, or other financial interest whose value cannot be readily determined through reference to public prices (generally, interests in a nonpublicly traded corporation), or any equity interest in a publicly traded corporation that exceeds $50,000 during the time the clinical investigator is carrying out the study and for 1 year following completion of the study. [21 CFR § 54]

Significant payments of other sorts means payments made by the sponsor of a covered study to the investigator or the institution to support activities of the investigator that have a monetary value of more than $25,000, exclusive of the costs of conducting the clinical study or other clinical studies, (e.g., a grant to fund ongoing research, compensation in the form of equipment or retainers for ongoing consultation or honoraria) during the time the clinical investigator is carrying out the study and for 1 year following the completion of the study. [21 CFR § 54]

Significant risk device means an investigational device that: [21 CFR § 812]

(1) Is intended as an implant and presents a potential for serious risk to the health, safety, or welfare of a subject;

(2) Is purported or represented to be for a use in supporting or sustaining human life and presents a potential for serious risk to the health, safety, or welfare of a subject;

(3) Is for a use of substantial importance in diagnosing, curing, mitigating, or treating disease, or otherwise preventing impairment of human health and presents a potential for serious risk to the health, safety, or welfare of a subject; or

(4) Otherwise presents a potential for serious risk to the health, safety, or welfare of a subject.

Significant risk (SR) device – an investigational device that presents a potential for serious risk to the health, safety, or welfare of a subject and:

1. is intended as an implant;

2. is purported or represented to be for use in supporting or sustaining human life;

3. is for a use of substantial importance in diagnosing, curing, mitigating, or treating disease, or otherwise preventing impairment of human health; or

4. otherwise presents a potential for serious risk to the health, safety, or welfare of a subject.

The risk determination for an investigational device study should be based on the proposed use of the device in the investigation in addition to the device characteristics. Sponsors of significant risk device studies must apply to FDA for an Investigational Device Exemption (IDE) (see definition above). (21 CFR 812.3(a), 812.3(m); 812.20.) [IVD Device Studies Guidance]

Simulated specimens – specimens made in the laboratory by adding the analyte of interest in known concentrations to a medium that simulates the natural matrix. [IVD Device Studies Guidance]

Site Management Organization (SMO) – In this guidance, site management organization refers to an organization that manages clinical study sites on behalf of the sponsor and/or drug manufacturer. [Bioavailability and Bioequivalence Testing Samples]

Source Documents: Original documents and records including, but not limited to, hospital records, clinical and office charts, laboratory notes, memoranda, subjects' diaries or evaluation checklists, pharmacy dispensing records, recorded data from automated instruments, copies or transcriptions certified after verification as being accurate and complete, microfiches, photographic negatives, microfilm or magnetic media, x-rays, subject files, and records kept at the pharmacy, at the laboratories, and at medico-technical departments involved in a clinical trial. [Computerized Systems Used in Clinical Investigations Guidance]

Specification means the quality standard (i.e. , tests, analytical procedures, and acceptance criteria) provided in an approved application to confirm the quality of drug substances, drug products, intermediates, raw materials, reagents, components, in-process materials, container closure systems, and other materials used in the production of a drug substance or drug product. For the purpose of this definition,acceptance criteria means numerical limits, ranges, or other criteria for the tests described. [21 CFR § 314]

> ***Specification*** means any requirement with which a product, process, service, or other activity must conform. [21 CFR §820.3(y)]

Specificity – the probability that a diagnostic test will yield a negative result when the disease or target analyte is absent. [IVD Device Studies Guidance]

Specimen means any material derived from a test system for examination or analysis. [21 CFR § 58]

Sponsor means a person who initiates a clinical investigation, but who does not actually conduct the investigation, i.e., the test article is administered or dispensed to or used involving, a subject under the immediate direction of another individual. A person other than an individual (e.g., corporation or agency) that uses one or more of its own employees to conduct a clinical investigation it has initiated is considered to be a sponsor (not a sponsor-investigator), and the employees are considered to be investigators. [21 CFR § 50]

> ***Sponsor*** means a person or other entity that initiates a clinical investigation, but that does not actually conduct the investigation, i.e., the test article is administered or dispensed to, or used involving, a subject under the immediate direction of another individual. A person other than an individual (e.g., a corporation or agency) that uses one or more of its own employees to conduct an investigation that it has initiated is considered to be a sponsor (not a sponsor-investigator), and the employees are considered to be investigators. [21 CFR § 56]

Sponsor means: [21 CFR § 58]

> (1) A person who initiates and supports, by provision of financial or other resources, a nonclinical laboratory study;

(2) A person who submits a nonclinical study to the Food and Drug Administration in support of an application for a research or marketing permit; or

(3) A testing facility, if it both initiates and actually conducts the study.

Sponsor means a person who takes responsibility for and initiates a clinical investigation. The sponsor may be an individual or pharmaceutical company, governmental agency, academic institution, private organization, or other organization. The sponsor does not actually conduct the investigation unless the sponsor is a sponsor-investigator. A person other than an individual that uses one or more of its own employees to conduct an investigation that it has initiated is a sponsor, not a sponsor-investigator, and the employees are investigators. [21 CFR § 312]

Sponsor means a person who initiates, but who does not actually conduct, the investigation, that is, the investigational device is administered, dispensed, or used under the immediate direction of another individual. A person other than an individual that uses one or more of its own employees to conduct an investigation that it has initiated is a sponsor, not a sponsor-investigator, and the employees are investigators. [21 CFR § 812]

Sponsor – a person who initiates, but who does not actually conduct, the investigation, i.e., the investigational device is administered, dispensed, or used under the immediate direction of another individual. A person other than an individual that uses one or more of its own employees to conduct an investigation that it has initiated is a sponsor, not a sponsor-investigator (see next definition), and the employees are investigators (see definition above) (21 CFR 812.3(n)). [IVD Device Studies Guidance]

Sponsor of the covered clinical study means the party supporting a particular study at the time it was carried out. [21 CFR § 54]

Sponsor-investigator means an individual who both initiates and actually conducts, alone or with others, a clinical investigation, i.e., under whose immediate direction the test article is administered or dispensed to, or used involving, a subject. The term does not include any person other than an individual, e.g., corporation or agency. [21 CFR § 50]

Sponsor-investigator means an individual who both initiates and actually conducts, alone or with others, a clinical investigation, i.e., under whose immediate direction the test article is administered or dispensed to, or used involving, a subject. The term does not include any person other than an individual, e.g., it does not include a corporation or agency. The obligations of a sponsor-investigator under this part include both those of a sponsor and those of an investigator. [21 CFR § 56]

Sponsor-Investigator means an individual who both initiates and conducts an investigation, and under whose immediate direction the investigational drug is administered or dispensed. The term does not include any person other than an individual. The requirements applicable to a sponsor-investigator under this part include both those applicable to an investigator and a sponsor. [21 CFR § 312]

Sponsor-Investigator – An individual who both initiates and conducts an investigation, and under whose immediate direction the investigational drug is administered or dispensed. The term does not include any person other than an individual (21 CFR 312.3(b)). [Bioavailability and Bioequivalence Testing Samples]

Sponsor-investigator means an individual who both initiates and actually conducts, alone or with others, an investigation, that is, under whose immediate direction the investigational device is administered, dispensed, or used. The term does not include any person other than an individual. The obligations of a sponsor-investigator under this part include those of an investigator and those of a sponsor. [21 CFR § 812]

Sponsor-investigator – an individual who both initiates and actually conducts, alone or with others, an investigation, i.e., under whose immediate direction the investigational device is administered, dispensed, or used. The term does not include any person other than an individual. The obligations of a sponsor-investigator include both those of an investigator and those of a sponsor (21 CFR 812.3(o)). [IVD Device Studies Guidance]

Statement of material fact means a representation that tends to show that the safety or effectiveness of a device is more probable than it would be in the absence of such a representation. A false affirmation or silence or an omission that would lead a reasonable person to draw a particular conclusion as to the safety or effectiveness of a device also may be a false statement of material fact, even if the statement was not intended by the person making it to be misleading or to have any probative effect. [21 CFR § 814]

Statistical hypothesis – a statement about some state of nature that a proposed study or set of studies will either accept or reject on the basis of the experimental data. The hypothesis is usually broken down into a null hypothesis (a statement of what the testing results will hopefully reject) and an alternative hypothesis (a statement of what the testing results will hopefully accept). [IVD Device Studies Guidance]

Strength means: [21 CFR § 210]

(i)   The concentration of the drug substance (for example, weight/weight, weight/volume, or unit dose/volume basis), and/or

(ii) The potency, that is, the therapeutic activity of the drug product as indicated by appropriate laboratory tests or by adequately developed and controlled clinical data (expressed, for example, in terms of units by reference to a standard).

Study – as used in this document, covers the systematic evaluations conducted in the development of an IVD product, including the feasibility, analytical assessments, method comparison, and evaluations to determine clinical utility of a product. [IVD Device Studies Guidance]

Study completion date means the date the final report is signed by the study director. [21 CFR § 58]

Study director means the individual responsible for the overall conduct of a nonclinical laboratory study. [21 CFR § 58]

Study initiation date means the date the protocol is signed by the study director. [21 CFR § 58]

Study Sponsor – A person who takes responsibility for and initiates a clinical investigation. The sponsor may be an individual or pharmaceutical company, governmental agency, academic institution, private organization, or other organization. The sponsor does not actually conduct the investigation unless the sponsor is a sponsor-investigator (21 CFR 312.3 (b)).

In this guidance, the term *study sponsor and/or drug manufacturer* is used in recognition of the fact that most study sponsors are pharmaceutical companies that manufacture the drugs under investigation. [Bioavailability and Bioequivalence Testing Samples]

Subject means a human who participates in an investigation, either as a recipient of the investigational new drug or as a control. A subject may be a healthy human or a patient with a disease. [21 CFR § 312]

    Subject means a human who participates in an investigation, either as an individual on whom or on whose specimen an investigational device is used or as a control. A subject may be in normal health or may have a medical condition or disease. [21 CFR § 812]

    Subject – a human who participates in an investigation, either as an individual on whom or on whose specimen an investigational device is used or as a control. A subject may be in normal health or may have a medical condition or disease (21 CFR 812.3(p)). [IVD Device Studies Guidance]

Surplus samples – see excess samples. [IVD Device Studies Guidance]

Symptom — Any subjective evidence of a disease, health condition, or treatment-related effect that can be noticed and known only by the patient. [Patient-Reported Outcome Measures Guidance]

T

Target product profile (TPP) — A clinical development program summary in the context of labeling goals where specific types of evidence (e.g., clinical trials or other sources of data) are linked to the targeted labeling claims or concepts. [Patient-Reported Outcome Measures Guidance]

Termination means a discontinuance, by sponsor or by withdrawal of IRB or FDA approval, of an investigation before completion. [21 CFR § 812]

Test article means any drug (including a biological product for human use), medical device for human use, human food additive, color additive, electronic product, or any other article subject to regulation under the act or under sections 351 and 354-360F of the Public Health Service Act (42 U.S.C. 262 and 263b-263n). [21 CFR § 50]

> ***Test article*** means any drug for human use, biological product for human use, medical device for human use, human food additive, color additive, electronic product, or any other article subject to regulation under the act or under sections 351 or 354-360F of the Public Health Service Act. [21 CFR § 56]

> ***Test article*** means any food additive, color additive, drug, biological product, electronic product, medical device for human use, or any other article subject to regulation under the act or under sections 351 and 354-360F of the Public Health Service Act. [21 CFR § 58]

Test system means any animal, plant, microorganism, or subparts thereof to which the test or control article is administered or added for study. Test system also includes appropriate groups or components of the system not treated with the test or control articles. [21 CFR § 58]

Testing facility means a person who actually conducts a nonclinical laboratory study, i.e., actually uses the test article in a test system. Testing facility includes any establishment required to register under section 510 of the act that conducts nonclinical laboratory studies and any consulting laboratory described in section 704 of the act that conducts such studies. Testing facility encompasses only those operational units that are being or have been used to conduct nonclinical laboratory studies. [21 CFR § 58]

> ***Testing Facility*** – The entity performing the BA or BE (in vivo or in vitro) study. The testing facility can be a CRO, university, hospital, clinic of a clinical investigator, or in-house clinical study unit of a study sponsor and/or drug manufacturer, where dosing and sampling (i.e., blood, urine, or clinical endpoints) are performed. In issuing the final rule, the Agency intended that reserve samples should be kept at the testing facility. [Bioavailability and Bioequivalence Testing Samples]

The list means the list of drug products with effective approvals published in the current edition of FDA's publication "Approved Drug Products with Therapeutic Equivalence Evaluations" and any current supplement to the publication. [21 CFR § 314]

Theoretical yield means the quantity that would be produced at any appropriate phase of manufacture, processing, or packing of a particular drug product, based upon the quantity of components to be used, in the absence of any loss or error in actual production. [21 CFR § 210]

Transitional device means a device subject to section 520(l) of the act, that is, a device that FDA considered to be a new drug or an antibiotic drug before May 28, 1976. [21 CFR § 812]

> *Transitional device* – a product defined as a device as of May 28, 1976, but previously considered by FDA to be a new drug or an antibiotic drug (21 CFR 812.3(r)). [IVD Device Studies Guidance]

Transmit: Transmit is to transfer data within or among clinical study sites, contract research organizations, data management centers, sponsors, or to FDA. [Computerized Systems Used in Clinical Investigations Guidance]

Treatment benefit — The effect of treatment on how a patient survives, feels, or functions. Treatment benefit can be demonstrated by either an effectiveness or safety advantage. For example, the treatment effect may be measured as an improvement or delay in the development of symptoms or as a reduction or delay in treatment-related toxicity. Measures that do not directly capture the treatment effect on how a patient survives, feels, or functions are surrogate measures of treatment benefit. [Patient-Reported Outcome Measures Guidance]

Treatment IDE – use of an unapproved investigational device for the treatment or diagnosis of patients during the clinical trial or prior to final FDA action on the marketing application, if during the course of the clinical trial the data suggest that the device is effective. A treatment IDE may cover a large number of patients that exceeds the number of clinical sites and patients stipulated in the original IDE. The device must be for treatment or diagnosis of a serious or immediately life-threatening disease or condition; there must be no comparable or satisfactory alternative device or therapy available; the device must be under investigation in a controlled clinical study for the same use under an approved IDE, or such clinical studies have been completed; and the sponsor must be actively pursuing marketing approval or clearance of the device. Requirements for an application for a treatment IDE are found in the Investigational Device Exemptions regulation at 21 CFR 812.36. [IVD Device Studies Guidance]

Truth Standard ("Gold" Standard) – any medical procedure or laboratory method or combination of procedures and methods that the clinical community relies upon for diagnosis, that is accepted by FDA, and that is regarded as having negligible risk of either a false positive or a false negative result. The truth standard result should be definitive (positive/negative, present/absent, or diseased/non-diseased), and should not give an indeterminate result. As science and technology improve, newer, more reliable standards may replace previous standards, particularly in the case of new disease markers. [IVD Device Studies Guidance]

U, V

Unanticipated adverse device effect means any serious adverse effect on health or safety or any life-threatening problem or death caused by, or associated with, a device, if that effect, problem, or death was not previously identified in nature, severity, or degree of incidence in the investigational plan or application (including a supplementary plan or application), or any other unanticipated serious problem associated with a device that relates to the rights, safety, or welfare of subjects. [21 CFR § 812]

Unexpected adverse drug experience is an adverse event that is not listed in the current labeling for the new animal drug and includes any event that may be symptomatically and pathophysiologically related to an event listed on the labeling, but differs from the event because of greater severity or specificity. For example, under this definition hepatic necrosis would be unexpected if the labeling referred only to elevated hepatic enzymes or hepatitis. [21 CFR § 514]

Usability testing — A formal evaluation with documentation of respondents' abilities to use the instrument, as well as comprehend, retain, and accurately follow instructions. [Patient-Reported Outcome Measures Guidance]

Validation means confirmation by examination and provision of objective evidence that the particular requirements for a specific intended use can be consistently fulfilled. [21 CFR § 820.3(z)]

(1) ***Process Validation*** means establishing by objective evidence that a process consistently produces a result or product meeting its predetermined specifications.

(2) ***Design Validation*** means establishing by objective evidence that device specifications conform with user needs and intended use(s).

Valid biomarker: A biomarker that is measured in an analytical test system with wellestablished performance characteristics and for which there is an established scientific framework or body of evidence that elucidates the physiologic, toxicologic, pharmacologic, or clinical significance of the test results. The classification of biomarkers is context specific. Likewise, validation of a biomarker is context-specific and the criteria for validation will varywith the I ntended use of the biomarker. The clinical utility (e.g., predict toxicity, effectiveness or dosing) and use of epidemiology/population data (e.g., strength of genotype-phenotype associations) are examples of approaches that can be used to determine the specific context and the necessary criteria for validation. [Pharmacogenomic Data Submissions Guidance]

• ***Known valid biomarker:*** A biomarker that is measured in an analytical test system with well-established performance characteristics and for which there is widespread agreement in the medical or scientific community about the physiologic, toxicologic, pharmacologic, or clinical significance of the results

- *Probable valid biomarker:* A biomarker that is measured in an analytical test system with well-established performance characteristics and for which there is a scientific framework or body of evidence that appears to elucidate the physiologic, toxicologic, pharmacologic, or clinical significance of the test results. A probable valid biomarker may not have reached the status of a known valid marker because, for example, of any one of the following reasons:

  – The data elucidating its significance may have been generated within a single company and may not be available for public scientific scrutiny.

  – The data elucidating its significance, although highly suggestive, may not be conclusive.

  – Independent verification of the results may not have occurred.

Verification means confirmation by examination and provision of objective evidence that specified requirements have been fulfilled. [21 CFR §820.3(aa)]

Voluntary genomic data submission (VGDS): The designation for pharmacogenomic data submitted voluntarily to FDA. [Pharmacogenomic Data Submissions Guidance]

W, X, Y, Z

Ward means a child who is placed in the legal custody of the State or other agency, institution, or entity, consistent with applicable Federal, State, or local law. [21 CFR § 50]

Index

About the author

Mindy J. Allport-Settle was born in Beckley and raised in Oak Hill, West Virginia. She moved to North Carolina to attend the N.C. School of the Arts for high school and now lives near Raleigh. Following in the footsteps of Gordon Allport, all of her books are built on a foundation of psychology and sociology with a focus on improving some aspect of industry through research and education.

Her career in healthcare began when she was a teenager working as an emergency medical technician. Since then, she has joined the U.S. Navy's advanced hospital corps, worked in organ and human tissue procurement, specialized in ophthalmology, and moved on to serve as a key executive, board member, and consultant for some of the best companies in the pharmaceutical, medical device, and biotechnology industry. She has provided guidance in regulatory compliance, corporate structuring, restructuring and turnarounds, new drug submissions, research and development and product commercialization strategies, and new business development. Her experience and dedication have resulted in international recognition as the developer of the only FDA-recognized and benchmarked quality systems training and development business methodology.

Her education includes a Bachelor's degree from the University of North Carolina, an MBA in Global Management from the University of Phoenix, and completion of the corporate governance course series in audit committees, compensation committees, and board effectiveness at Harvard Business School.

About PharmaLogika

Since 2002, PharmaLogika, Inc has established itself as one of the world's premier consulting firms for Pharmaceutical, Biotech, and Medical Device companies across the globe. In so doing, it has earned the trust of executives in Life Sciences for its integrity, accuracy, and unwavering commitment to independent thought with regard to its products and services as well as those of its customers. Through www.PharmaLogika.com, its involvement in sponsored events, and personal references it has reached millions in print and online. Its mission, to provide flawlessly designed and executed products and services to startups as well as established industry leaders to facilitate their growth from discovery and clinical trial navigation to the commercialization and marketing of their products.

PharmaLogika consults with pharmaceutical, biotech, and medical device quality units to provide third party audits, training, pre approval inspections (PAIs), compliance remediation, and a portfolio of related quality and regulatory affairs products and services. Those products include but are not limited to Quality Assurance Forms, SOP and clinical templates, and the highly successful Integrated Development Training System.

Regulatory action guidance as well as quality systems guidance are delivered as part of its standard products and services. Through the use of highly skilled resources throughout the process, each offering is designed to enact a comprehensive quality systems approach in addressing Quality Assurance (QA) issues. The results insure a close adherence to current Good Manufacturing Practice (cGMP) standards.

PharmaLogika also has a Research and Development OTC line for human consumption that targets alpha 1-antitrypsin deficiency, Fibromyalgia, Restless Legs Syndrome, and Attention Deficit Disorder. A veterinary OTC is currently available that provides canine and feline oral debriding and cleansing agents as well as a stain remover and topical antiseptic. These products combine to provide a strong pipeline of both current and future deliverables.

Other books available

Current Good Manufacturing Practices: Pharmaceutical, Biologics, and Medical Device Regulations and Guidance Documents Concise Reference

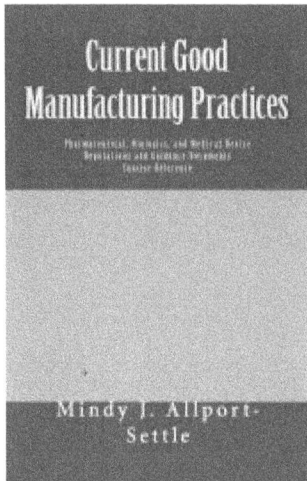

Good Manufacturing Practice (GMP) Guidelines: The Rules Governing Medicinal Products in the Eurpean Union, EudraLex Volume 4 Concise Reference

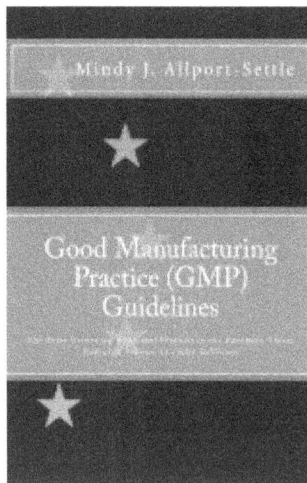

International Conference on Harmonisation (ICH) Quality Guidelines: Pharmaceutical, Biologics, and Medical Device Guidance Documents Concise Reference

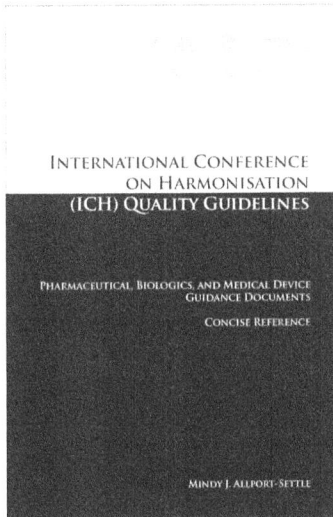

FDA Acronyms, Abbreviations and Terminology: Human and Veterinary Regualtory Reference

Course Development 101: Developing Training Programs for Regulated Industries

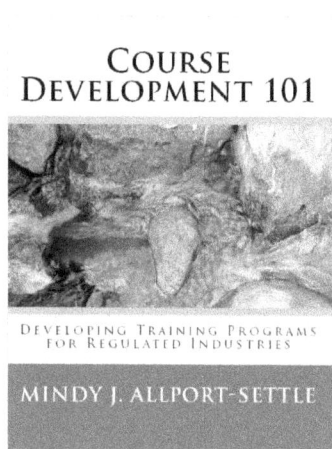

Compliance Remediation for Pharmaceutical Manufacturing: A Project Management Guide for Re-establishing FDA Compliance

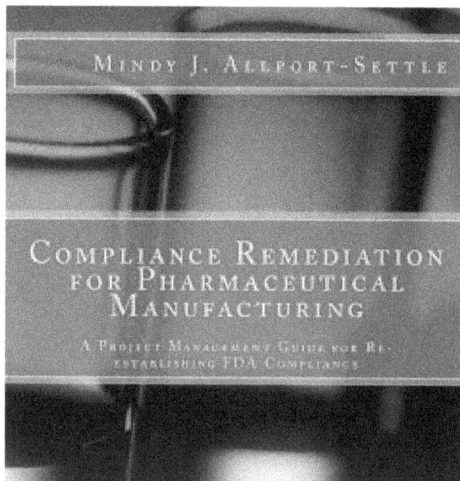

Investigations Operations Manual: FDA Field Inspection and Investigation Policy and Procedure Concise Reference

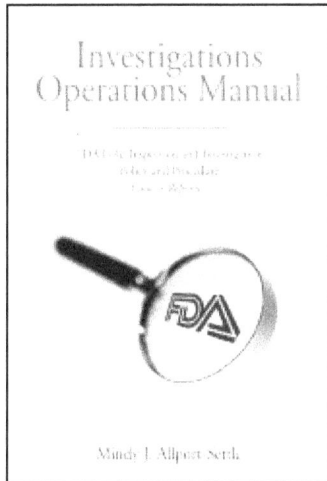

Canadian Good Manufacturing Practices: Pharmaceutical, Biotechnology, and Medical Device
Regulations and Guidance Concise Reference

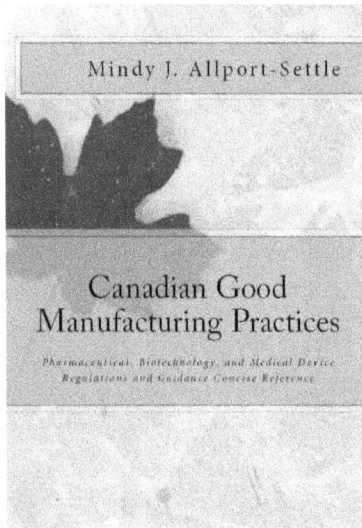

**Please visit www.PharmaLogika.com for additional titles
and for a list of resellers**

or visit your favorite local or internet book seller.

* Companion products and bulk discounts are only available at

www.PharmaLogika.com

www.ingramcontent.com/pod-product-compliance
Lightning Source LLC
Chambersburg PA
CBHW082101220326

41598CB00066BA/4547